ADVANCES IN MACROFUNGI
Pharmaceuticals and Cosmeceuticals

Series: Progress in Mycological Research

ADVANCES IN MACROFUNGI
Pharmaceuticals and Cosmeceuticals

Editors

Kandikere R. Sridhar

Department of Biosciences
Mangalore University, Mangalore, India

Sunil K. Deshmukh

Nano Biotechnology Centre
The Energy and Resources Institute, New Delhi, India

CRC Press
Taylor & Francis Group
Boca Raton London New York

CRC Press is an imprint of the
Taylor & Francis Group, an **informa** business

A SCIENCE PUBLISHERS BOOK

First edition published 2021
by CRC Press
6000 Broken Sound Parkway NW, Suite 300, Boca Raton, FL 33487-2742

and by CRC Press
2 Park Square, Milton Park, Abingdon, Oxon, OX14 4RN

© 2021 Taylor & Francis Group, LLC

CRC Press is an imprint of Taylor & Francis Group, LLC

ISBN: 978-1-032-04277-0 (hbk)
ISBN: 978-1-032-04281-7 (pbk)
ISBN: 978-1-003-19127-8 (ebk)

Typeset in Times New Roman
by Radiant Productions

Preface

Mycology is a fascinating branch of biology expanding its territory in almost all environmental and human affairs. Diversity of fungi is an imminent matter of debate for the last three decades and a revised estimate reveals global occurrence between 2.2–3.8 million fungal species (Hawksworth and Lucking, 2017). Global statistics reveal that only about 7–8 per cent of fungi are known, which denotes fungal vastness as well as our limited knowledge on their functional significance. A recent estimate disclosed global occurrence macrofungi up to 0.14–1.25 million (Azeem et al., 2020). Filamentous fungi are vital components of the ecosystem and precious research tool in the field of mycology and biotechnology. However, a handful of them are in cultivation, used for human consumption and adapted in production of value-added components or metabolites as health protectants. There are several challenges to explore the unknown macrofungi as well as utilisation of their potential (e.g., cataloguing, barcoding, collections, conservation and guidance).

Macrofungi as non-conventional diets, serve as prospective sources of nutritional components (high protein, high fibre, low fat and low calorie) including amino acids and vitamins. Although many mushrooms are inedible or poisonous, they are resourceful as ectomycorrhizas, endowed with bioactive compounds and produce functional metabolites. Medicinal macrofungi possess many value-added bioactive components (glucans, polysaccharides, complexes of polysaccharide-protein, enzymes and pigments) and produce metabolites of therapeutic interest (pharmaceuticals and immunoceuticals). Recent studies also demonstrated that macrofungi can produce specific cosmeceuticals which are valuable in dermal care and useful to synthesise nanoparticles with multifaceted applications. To date, about 30–700 species of mushrooms have been recognised as promising medicinal and nutraceutical significance, respectively. Besides, macrofungal mycelia serve as appropriate source for production of eco-friendly biomaterials of industrial value (Meyer et al., 2020). In the recent past, fungal biotechnology focused towards the production of economically versatile and ecofriendly biodegradable industrial products through bio-based sustainable merchandise (Meyer et al., 2020). Many macrofungi are prospective candidates in solid waste management and desired breakthrough in bioremediation.

In the 21st century, macrofungi have become an indispensable component of basic as well as applied research to connect different disciplines of applied science. There is tremendous expansion in acquaintance as well as application of macrofungi in agriculture, health industry and environmental bioremediation. Owing to advances in macrofungal research in the fields of pharmaceuticals and cosmeceuticals, the present contribution collated different sections, such as bioactive potential, therapeutics, nutraceuticals and immunoceuticals. To echo the recent progress in macrofungal research, several experienced scientists have contributed versatile chapters of therapeutic interest.

This book deals with (1) bioactive metabolites of various wild and cultivated mushrooms; (2) pharmaceutical potential of epigeous and hypogeous mushrooms; (3) healing properties of edible mushrooms; (4) medicinal attributes of *Cordyceps* spp.; (5) neuroprotective capabilities of medicinal mushrooms; (6) pharmacokinetic interactions of mushroom-derived clinical drugs; (7) food and pharmaceutical features of mushrooms; (8) cosmeceutical characteristics of selected mushrooms; (9) mushrooms as source of flavours and scents; (10) immunoceuticals of mushrooms in cancer therapy. This book also deals with nutraceuticals, medicinal properties and specific methods of evaluation of nutritional traits of mushrooms.

The various articles tend to draw one's attention towards understanding the potential role of macrofungi while serving as a valuable resource to students at graduate, post-graduate and research levels in various disciplines (mycology, microbiology, biotechnology, pharmacology, entomology, immunology, food science and medicine). We are thankful to all the contributors for their meticulous correspondence, timely submission and revision of their chapters. Our appreciation is due to the amicable gesture of the reviewers for assessment of the chapters in a short time-frame. Dr Ulrike Lindequist has drafted a thoughtful overview on the contents of the book; we profusely express our gratitude to her. The CRC Press has extended all possible co-operation during the pandemic COVID-19 for readjustment of deadlines and ease of other official formalities to present this book.

Mangalore

New Delhi

Kandikere R. Sridhar

Sunil K. Deshmukh

References

Azeem, U., Hakkem, K.R. and Ali, M. (2020). Fungi for Human Health—Current Knowledge and Further Perspectives. Springer Nature, Switzerland.

Hawksworth, D.L. and Lücking, R. (2017). Microbiology Spectrum. 5: FUNK-0052-2016.

Meyer, V., Basenko, E.Y., Benz, J.P., Braus, G.H., Caddick, M.X. et al. (2020). Fungal Biol. Biotechnol., 7: 5. 10.1186/s40694-020-00095-z.

Contents

List of Contributors

Kodye L Abbott

Department of Anatomy, Physiology and Pharmacology, Auburn University, Auburn, AL-36849, USA.

Naif Abdullah Al-Dhabi

Department of Botany and Microbiology, College of Science, King Saud University, P.O. Box 2455, Riyadh-11451, Saudi Arabia.

Mariadhas Valan Arasu

Department of Botany and Microbiology, College of Science, King Saud University, P.O. Box 2455, Riyadh-11451, Saudi Arabia.
Xavier Research Foundation, St. Xavier's College, Palayamkottai, Thirunelveli, Tamil Nadu, India.

Nur Izyan Wan Azelee

School of Chemical and Energy Engineering, Faculty of Engineering, Universiti Teknologi Malaysia (UTM), 81310 Skudai, Johor, Malaysia.
Institute of Bioproduct Development (IBD), Universiti Teknologi Malaysia (UTM), 81310 Skudai, Johor, Malaysia.

PB Benil

Department of Agadatantra, Vaidyaratnam PS Varier Ayurveda College, P.O. Edarikode, Kottakkal, Kerala, India.

Ekta Chaudhary

Department of Microbiology, Panjab University, Chandigarh-160014, India.

Sujata Chaudhuri

Department of Botany, University of Kalyani, B-16/212, Kalyani-741235, West Bengal, India.

Ghoson Daba

Chemistry of Natural and Microbial Products Department, Pharmaceutical Industries Researches Division, National Research Centre, El Buhouth St., Dokki, 12311, Giza, Egypt.

Daniel Joe Dailin

School of Chemical and Energy Engineering, Faculty of Engineering, Universiti Teknologi Malaysia (UTM), 81310 Skudai, Johor, Malaysia.
Institute of Bioproduct Development (IBD), Universiti Teknologi Malaysia (UTM), 81310 Skudai, Johor, Malaysia.

Hemanta Kumar Datta

School of Chemical Sciences, Indian Association for the Cultivation of Science, 2A & 2B Raja S.C. Mullick Road, Kolkata-700032, India.

Sunil K Deshmukh

TERI-Deakin Nano Biotechnology Centre, The Energy and Resources Institute, Darbari Seth Block, IHC Complex, Lodhi Road, New Delhi-110 003, India.

Muralikrishnan Dhanasekaran

Department of Drug Discovery and Development, Harrison School of Pharmacy, Auburn University, Auburn, AL-36849, USA.

Julia Dickenson

Polish-American Fulbright Commission, K.I. Gałczyńskiego St. 4, 00-362 Warsaw, Poland.

Waill Elkhateeb

Chemistry of Natural and Microbial Products Department, Pharmaceutical Industries Researches Division, National Research Centre, El Buhouth St., Dokki, 12311, Giza, Egypt.

Marwa Elnahas

Chemistry of Natural and Microbial Products Department, Pharmaceutical Industries Researches Division, National Research Centre, El Buhouth St., Dokki, 12311, Giza, Egypt.

Hesham Ali El Enshasy

School of Chemical and Energy Engineering, Faculty of Engineering, Universiti Teknologi Malaysia (UTM), 81310 Skudai, Johor, Malaysia.
Institute of Bioproduct Development (IBD), Universiti Teknologi Malaysia (UTM), 81310 Skudai, Johor, Malaysia.
City of Scientific Research and Technology Applications (SRTA), New Burg Al Arab, Alexandria, Egypt.

Mary Fabbrini

Department of Drug Discovery and Development, Harrison School of Pharmacy, Auburn University, Auburn, AL-36849, USA.

Patrick C Flannery

Department of Anatomy, Physiology and Pharmacology, Auburn University, Auburn, AL-36849, USA.

Shin-Yee Fung

Medicinal Mushroom Research Group (MMRG), Department of Molecular Medicine, Faculty of Medicine, University of Malaya, 50603 Kuala Lumpur, Malaysia.

Joe Gallagher

Institute of Biological, Environmental and Rural Sciences, Aberystwyth University, Aberystwyth, UK.

Kristina S Gill

Department of Anatomy, Physiology and Pharmacology, Auburn University, Auburn, AL-36849, USA.

Manoj Govindarajulu

Department of Drug Discovery and Development, Harrison School of Pharmacy, Auburn University, Auburn, AL-36849, USA.

Manish K Gupta

SGT College of Pharmacy, SGT University, Gurugram-122505, Haryana, India.

Dorota Hilszczańska

Forest Ecology Department, Forest Research Institute in Sękocin Stary, BraciLeśnej 3, 05-090 Raszyn, Poland.

Savarimuthu Ignacimuthu

Xavier Research Foundation, St. Xavier's College, Palayamkottai, Thirunelveli, Tamil Nadu, India.

Ameer Khusro

Research Department of Plant Biology and Biotechnology, Loyola College, Nungambakkam, Chennai, India.

Hak-Jae Kim

Department of Clinical Pharmacology, College of Medicine, Soonchunhyang University, Cheonan, Republic of Korea.

Young Ock Kim

Department of Clinical Pharmacology, College of Medicine, Soonchunhyang University, Cheonan, Republic of Korea.

R Lekshmi

Department of Botany and Microbiology, MSM College, Kayamkulam, Kerala, India.

Ong Mei Leng

Harita Go Green Sdn. Bhd., Johor Bahru, Johor, Malayisa.

Ulrike Lindequist

Institute of Pharmacy, Pharmaceutical Biology, University of Greifswald, Institute of Pharmacy, University of Greifswald, Friedrich-Ludwig-Jahn Str. 17, D-17489 Greifswald, Germany.

Roslinda Malek

School of Chemical and Energy Engineering, Faculty of Engineering, Universiti Teknologi Malaysia (UTM), 81310 Skudai, Johor, Malaysia.

Nor Hasmaliana Abdul Manas

School of Chemical and Energy Engineering, Faculty of Engineering, Universiti Teknologi Malaysia (UTM), 81310 Skudai, Johor, Malaysia.
Institute of Bioproduct Development (IBD), Universiti Teknologi Malaysia (UTM), 81310 Skudai, Johor, Malaysia.

Grace McKerley

Department of Drug Discovery and Development, Harrison School of Pharmacy, Auburn University, Auburn, AL-36849, USA.

Ewa Moliszewska

University of Opole, B. Kominka St. 6A, 45-035 Opole, Poland.

Neo Moloi

Sawubone Mycelium Co., Centurion, Gauteng, South Africa.

Timothy Moore

Department of Drug Discovery and Development, Harrison School of Pharmacy, Auburn University, Auburn, AL-36849, USA.

Małgorzata Nabrdalik

University of Opole, B. Kominka St. 6A, 45-035 Opole, Poland.

Rishi M Nadar

Department of Drug Discovery and Development, Harrison School of Pharmacy, Auburn University, Auburn, AL-36849, USA.

Satyanarayana Pondugula

Department of Anatomy, Physiology and Pharmacology, Auburn University, Auburn, AL-36849, USA.

Deepak K Rahi

Department of Microbiology, Panjab University, Chandigarh-160014, India.

Sonu Rahi

Department of Botany, Government Girls College, A.P.S. University, Rewa-486003, India.

Madaiah Rajashekhar

Department of Biosciences, Mangalore University, Mangalagangotri, Mangalore, Karnataka, India.

R Rajakrishnan

Department of Botany and Microbiology, College of Science, King Saud University, P.O. Box 2455, Riyadh-11451, Saudi Arabia.

Sindhu Ramesh

Department of Drug Discovery and Development, Harrison School of Pharmacy, Auburn University, Auburn, AL-36849, USA.

Venugopalan Ravikrishnan

Department of Biosciences, Mangalore University, Mangalagangotri, Mangalore, Karnataka, India.

Muhammad Fazril Razif

Medicinal Mushroom Research Group (MMRG), Department of Molecular Medicine, Faculty of Medicine, University of Malaya, 50603 Kuala Lumpur, Malaysia

Julia M Salamat

Department of Anatomy, Physiology and Pharmacology, Auburn University, Auburn, AL-36849, USA.

S Shishupala

Department of Microbiology, Davangere University, Shivagangothri, Davangere-577007, Karnataka, India.

Anna Solomonik

Department of Drug Discovery and Development, Harrison School of Pharmacy, Auburn University, Auburn, AL-36849, USA.

Kandikere R Sridhar

Department of Biosciences, Mangalore University, Mangalagangotri, Mangalore, Karnataka, India.
Centre for Environmental Studies, Yenepoya (deemed to be) University, Mangalore, Karnataka, India.

Paul Thomas

Mycorrhizal Systems Ltd, Lancashire, PR25 2SD, UK.
University of Stirling, Stirling, FK9 4LA, UK.

Shilpa A Verekar

Parle Agro Pvt. Ltd., Off Western Express Highway, Sahar-Chakala Road, Parsiwada, Andheri (East) Mumbai-400099, India.

János Vetter

Department of Botany, University of Veterinary Science, Budapest, Hungary.

Ana Winters

Institute of Biological, Environmental and Rural Sciences, Aberystwyth University, Aberystwyth, UK.

About the Editors

Kandikere R Sridhar

Dr. Kandikere R Sridhar is an adjunct professor in the Department of Biosciences, Mangalore University and Yenepoya (deemed to be) University. His main areas of research are 'Diversity and Ecology of Fungi of the Western Ghats, Mangroves and Marine Habitats'. He has been NSERC postdoctoral fellow/visiting professor in Mount Allison University, Canada; Helmholtz Centre for Environmental Research-UFZ and Martin Luther University, Germany; Centre of Biology, University of Minho, Portugal. He is Fellow of the Indian Mycological Society, Kolkata (2014), Distinguished Asian Mycologist (2015), and is considered one among the world's top scientists in the field of mycology.

Sunil K Deshmukh

Dr. Sunil Kumar Deshmukh is a veteran industrial mycologist who spent a substantial part of his career at Hoechst Marion Roussel Limited and Piramal Enterprises Limited. He is a fellow of Mycological Society of India (MSI) and served as Adjunct Associate Professor at Deakin University, Australia and worked towards the development of natural food colors, antioxidants and biostimulants through nanotechnology intervention.

Macrofungi in Pharmacy, Medicine, Cosmetics and Nutrition

An Appraisal

Ulrike Lindequist

1. INTRODUCTION

Macrofungi or higher fungi can be defined as fungi that form epigeal/hypogeal, large spore-bearing structures, called sporocarps. The more common term 'mushroom' means a macrofungus with a distinctive fruit body large enough to be seen with the naked eye and to be picked by hand (Chang and Wasser, 2012). Medicinal mushrooms are macroscopic fungi, which are used in the form of extracts or powder for prevention, alleviation, or healing of diseases and/or to fulfil nutrition (Lindequist, 2013).

The number of species of fungi is commonly cited 1.5 but actual estimated at 2.2–3.8 million. The number of currently accepted species is 120,000, suggesting that at best just 8 per cent, and in the worst case scenario just 3 per cent are named so far (Hawksworth and Lücking, 2016). An estimated number of macrofungal species in the world have risen from 140,000 to 1,250,000 (Azeem et al., 2020). From a taxonomic point of view, mainly basidiomycetes but also some species of ascomycetes, e.g., *Gyromitra esculenta* (PERS.: FR.) FR., belong to macrofungi or mushrooms.

To survive in their natural environment, macrofungi are constrained to produce bioactive secondary metabolites, e.g., such with antimicrobial or cytostatic activities. This ecological need together with the experience in ethnomedicinal use

Institute of Pharmacy, Pharmaceutical Biology, University of Greifswald, Friedrich-Ludwig-Jahn Str. 17, D-17489 Greifswald, Germany.
Email: lindequi@uni-greifswald.de

of mushrooms and the knowledge about the great potential of microscopic fungi, e.g., *Penicillium* and *Aspergillus* species, for production of potential drugs justify the great potential of macrofungi in pharmacy, medicine, cosmetics and nutrition. Improved possibilities for genetic, pharmacological and chemical analyses and for cultivation allowed great progress in the exploration and application of useful properties of macrofungi. Otherwise, challenges like resistance development against available antibiotics and cytostatics, insufficient treatment options for diseases like cancer or dementia and increasing age of human beings require new and better drugs and healthy nutrition.

Fungi have played an important role as food, medicine, poison and for religious and other purposes in the life of man since prehistoric times. The medicinal use of mushrooms has a very long tradition in the East Asian countries. In other parts of the world, e.g., Middle Europe, the tradition was interrupted during the last few centuries. Meanwhile, the use of mushrooms because of their health-promoting properties is increasing worldwide. The global mushroom market had a value of 35 billion US$ in 2015. It is expected that the market will reach nearly 60 billion US$ in 2021 (Raman et al., 2018; Azeem et al., 2020). The total world production of cultivated mushrooms was nearly 10 million tons in 2013, whereas it was only 4.2 million tons in 2000 (Kalač, 2016; Gargano et al., 2017). The consumption per head of macrofungi worldwide increased from 1 kg in 1997 to 3.6 kg in 2018 (Lelley, 2018). The mushroom production is dominated largely by China.

2. Relevant Mushroom Species

More than 30 species of medicinal mushrooms are currently identified as sources for biologically active metabolites. Most important medicinal mushrooms with long tradition in Asian ethnomedicine are *Ganoderma lucidum* (CURTIS) P. KARST (Ling Zhi, Reishi), *Lentinula edodes* (BERK.) PEGLER (Shiitake), *Grifola frondosa* (DICKS.: FR.) GRAY (Maitake), *Hericium erinaceus* (BULL.) PERS. and *Ophiocordyceps sinensis* (BERK.) G.H. SUNG, J.M. SUNG, HYWEL-JONES and SPATAFORA (*Cordyceps sinensis*). Mushrooms from other continents, like *Agaricus subrufescens* PECK (*Agaricus brasiliensis* WASSER ET AL., *A. blazei* Murrill sensu HEINEM, South America), *Inonotus obliquus* (PERS.: FR.) PILÁT (Chaga, Eastern Europe) or *Fomitopsis betulina* (Bull.) B.K. CUI, M.L. HAN and Y.C. DAI (Europe) attract more and more interest (Grienke et al., 2014; Gründemann et al., 2020). Further relevant mushroom species are, e.g., *Flammulina velutipes* (CURTIS: FR.) SINGER, *Pleurotus ostreatus* (JACQ.: FR.) P. KUMM., *Wolfiporia extensa* (PECK) GINNS (*W. cocos*), *Cordyceps militaris* (L.) LINK, Phellinus spec. *Taiwanofungus camphoratus* (M.Z. HANG and C.H. SU) SHENG H. WU, Z.H. YU, Y.C. DAI and C.H. SU, *Auricularia auricula judae* (BULL.: FR.) QUÉL., *Trametes versicolor* (L.: FR.) LLOYD, *Polyporus umbellatus* (PERS.: FR.) FR. and other species from the genera *Ganoderma*, *Pleurotus* or *Grifola*. The most cultivated species is *Agaricus bisporus* (J.E. LANGE) IMBACH, dominating worldwide, followed by *Pleurotus ostreatus*, *Lentinula edodes*, *Auricularia auricula judae* and *Flammulina velutipes* (Kalač, 2016; Lelley, 2018). Not only fruit bodies but also mycelial biomass, growing in submerged culture, and spores are exploited.

3. Overview about Possible Applications and Responsible Compounds

It is estimated that more than 700 taxa out from 2,000 known safe mushroom species contain functional nutraceutical or medicinal properties (Wasser, 2010; Gargano et al., 2017). According to the present knowledge, medicinal mushrooms and fungi exhibit more than 130 medicinal functions (Wasser, 2017). Pharmacological investigations have been mostly done with extracts (hot water or organic solvents) from fungal fruit bodies. Strong evidences exist for their immunomodulatory (Mizuno and Nishitani, 2013; Guggenheim et al., 2014), antimicrobial (Alves et al., 2012) including antiviral (Linnakoski et al., 2018) and cytostatic/antitumor (Joseph et al., 2018; Stajic et al., 2019; Zmitrovich et al., 2019) activities. Hypoglycemic (Lindequist and Haertel, 2020), hepatoprotective (Soares et al., 2013), osteoprotective (Erjavec et al., 2016), neuroprotective (Phan et al., 2015, 2017), adaptogenic and antifatigue (Geng et al., 2017), anti-inflammatory (Du et al., 2018), hypotensive (Yahaya et al., 2014), antiallergic (Merdivan and Lindequist, 2017) and antioxidative (Islam et al., 2019) activities have also been demonstrated (Lindequist et al., 2005; Wasser et al., 2010, 2017; De Silva et al., 2013; Gargano et al., 2017; Glamočlia and Soković, 2017; Chaturvedi et al., 2018; Azeem et al., 2020). In opposite to a high number of results from *in vitro* and animal assays, the number of qualitatively good clinical studies published in English language is unfortunately low. Most of the products are recommended for supportive therapy, healthy nutrition and prophylaxis but not for direct cure.

From a chemical point of view, mainly polysaccharides (ß-glucans, heteropolysaccharides and polysaccharide-protein-complexes), proteins, terpenes (triterpenes, diterpenes, meroterpenoids), steroids, nucleoside derivatives, amino acids and some other compounds are responsible for the observed pharmacological effects (De Silva et al., 2013; Gargano et al., 2017; Sandargo et al., 2019).

In the last few years, many efforts have been carried out for elucidation of mode of action. Today we know that ß-glucans, for instance, exhibit the immunomodulating effects by reaction with dectin 1 and other receptors on cells of the unspecific immune system in the intestine (Wasser et al., 2017). Many low molecular compounds target pathways of cell proliferation and differentiation or influence the activity of enzymes. The influence of mushroom components on gut microbiome and metabolites produced by microflora seems to support positive fungal effects (Liu et al., 2020).

Macrofungi are also a source of pure bioactive compounds, which are licenced as drugs. Two examples are the diterpene antibiotic retapamulin from *Clitopilus passeckerianus* PILÁT and the immunosuppressive compound fingolimod, a chemical modification of myriocin from the ascomycete *Isaria sinclairii* BERK. Fingolimod was the first drug for the oral treatment of multiple sclerosis and retapamulin is applied as antibiotic on the skin (Lindequist, 2013). Anti-inflammatory, antioxidative, UV-protecting and wound-healing properties of mushroom extracts and components are useful for the application as cosmeceuticals and other skin products (Hyde et al., 2010).

Fresh and preserved edible macrofungi are components of human foods since thousands of years (Chang and Wasser, 2012; Kalač, 2016; Gargano et al., 2017; Chaturvedi et al., 2018; Azeem et al., 2020). They are rich in proteins (containing all essential amino acids), carbohydrates (high content in dietary fibre), vitamins and vitamin precursors (ergosterol as precursor of vitamin D_2) and minerals, poor in fats with a high proportion of polyunsaturated fatty acids and purins. Most edible mushrooms possess a pleasant taste and flavour. They serve not only as delicious foodstuff but also as a source of food-flavouring substances.

4. Conclusions and Outlook

The present book demonstrates that macrofungi have an immense potential in pharmacy, medicine, cosmetics and nutrition. Depending on the ever-increasing knowledge about chemistry, biotechnology and molecular biology of mushrooms with improvement of screening and cultivation methods, a further increase in the applications of mushrooms and mushroom-derived compounds can be expected. To explore this potential, it is necessary to strengthen the efforts especially in the following fields:

- Continuous production of mushrooms (fruit bodies or mycelia or spores) in high amounts and in a standardised quality.
- Development and formulation of innovative mushroom products, probably opening the way for patent protection.
- Establishment of suitable quality parameters and of analytical methods to control these parameters.
- Legal regulations for authorisation as drugs, dietary supplements or cosmetics/cosmeceuticals. As prerequisite for authorisation as drugs, we need good and meaningful clinical trials.

Besides, it is necessary to extend our knowledge about the pharmakokinetic behaviour of mushroom components, right dosages and time schedule, about possible interactions between mushrooms and conventional drugs and about the influence of mushrooms on the microbes in intestinum. Only a small fraction of the estimated fungal biodiversity has been investigated for bioactivities. Mushrooms from other geographic regions, 'forgotten' mushrooms, or mushrooms previously not investigated may also represent promising potentials for drug development.

It should be mentioned that macrofungi are of interest also for technical purposes, e.g., production of leather and other materials (Cerimi et al., 2019), as source of enzymes, for bioremediation, for conservation of forests, plant protection and in animal breeding. Last but not least, mushrooms can exhibit a great impact on agriculture, environment and economic development in society, especially in less developed countries (Gargano et al., 2017).

References

Alves, M.J., Ferreira, I.C.F.R., Dias, J., Teixeira, V., Martins, A. and Pintado, M. (2012). A review on antimicrobial activity of mushroom (*Basidiomycetes*) extracts and isolated compounds. Planta Med., 78: 1707–1718.

Azeem, U., Hakeem, K.R. and Ali, M. (2020). Fungi for Human Health, Current Knowledge and Further Perspectives. Springer Nature, Switzerland, Cham.

Chang, S.T. and Wasser, S.P. (2012). The role of culinary medicinal mushrooms on human welfare with a pyramid model for human health. Int. J. Med. Mushrooms, 14: 95–134.

Chaturvedi, V.K., Agarwal, S., Gupta, K.K., Ramteke, P.W. and Singh, M.P. (2018). Medicinal mushroom: boon for therapeutic applications. 3, Biotech., 8: 334. 10.1007/s13205-018-1358-0.

Cerimi, K., Akkaya, K.C., Pohl, C., Schmidt, P. and Neubauer, P. (2019). Fungi as source for new bio-based materials: A patent review. Fungal. Biol. Biotechnol., 6: 17. 10.1186/s40694-019-0080-y.

De Silva, D.G., Rapior, S., Sudarman, E., Stadler, M., Xu, J., Alias, S.A. and Hyde, K.D. (2013). Bioactive metabolites from macrofungi: Ethnopharmacology, biological activities and chemistry. Fungal Diversity, 62: 1–40.

Du, B., Zhu, F. and Xu, B. (2018). An insight into the anti-inflammatory properties of edible and medicinal mushrooms. J. Funct. Foods, 47: 334–342.

Erjavec, I., Brkljacic, J., Vukicevic, S., Jakopovic, B. and Jakopovich, I. (2016). Mushroom extracts decrease bone resorption and improve bone formation. Int. J. Med. Mushrooms, 18: 559–569.

Gargano, M.L., van Griensven, L.J.L.D., Omoanghe, S.I., Lindequist, U., Venturella, G. et al. (2017). Medicinal mushrooms: Valuable biological resources of high exploitation potential. Pl. Biosys., 151: 548–565.

Geng, P., Siu, K.C., Wang, Z. and Wu, J.Y. (2017). Antifatigue functions and mechanisms of edible and medicinal mushrooms. Biomed. Res. Int., 2017: 9648496. 10.1155/2017/9648496.

Glamočlia, J. and Soković, M. (2017). Fungi as source with huge potential for mushroom pharmaceuticals. Lekovite Sirovine, 37: 50–56.

Grienke, U., Zöll, M., Peintner, U. and Rollinger, J.M. (2014). European medicinal polypores—a modern view on traditional uses. J. Ethnopharmacol., 154: 564–583.

Gründemann, C., Reinhard, J.K. and Lindequist, U. (2020). European medicinal mushrooms. Do they have potential for modern medicine?—An update. Phytomedicine, 66: 153131. 10.1016/j.phymed.2019.153131.

Guggenheim, A., Wright, K.M. and Zwickey, H.L. (2014). Immune modulation from five major mushrooms: Application to integrative oncology. Integr. Med., 13: 32–44.

Hawksworth, D.L. and Lücking, R. (2017). Fungal diversity revisited: 2.2 to 3.8 million species. Microbiol. Spectrum, 5: FUNK-0052-2016.

Hyde, K.D., Bahkali, A.H. and Moslem, M.A. (2010). Fungi—An unusual source for cosmetics. Fungal Diversity, 43: 1–9.

Islam, T., Ganesan, K. and Xu, B. (2019). New insight into mycochemical profiles and antioxidant potential of edible and medicinal mushrooms: A review. Int. J. Med. Mushrooms, 21: 237–251.

Joseph, T.P., Chanda, W., Padhiar, A.A., Batool, S., Li Qun, S. et al. (2018). A pre clinical evaluation of the antitumor activities of edible and medicinal mushrooms: A molecular insight. Integr. Cancer Ther., 17: 200–209.

Kalač, P. (2016). Edible Mushrooms—Chemical Composition and Nutritional Value. First ed., Elsevier Amsterdam, p. 236.

Lelley, J.I. (2018). No fungi—No future. Wie Pilze die Welt retten können, Springer, Berlin (in German).

Lindequist, U., Niedermeyer, T.H.J. and Jülich, W.D. (2005). The pharmacological potential of mushrooms. eCAM, 2(3): 285–299.

Lindequist, U. (2013). The merit of medicinal mushrooms from a pharmaceutical point of view. Int. J. Med. Mushrooms, 15: 517–523.

Lindequist, U. and Haertel, B. (2020). Medicinal mushrooms for treatment of type-2 diabetes—An update of clinical trials. Int. J. Med. Mushrooms, 22: 845–854.

Linnakoski, R., Reshamwala, D., Veteli, P., Cortina-Escribano, M., Vanhanen, H. and Marjomäki, V. (2018). Antiviral agents from fungi: Diversity, mechanisms and potential applications. Front. Microbiol., 9: 2325. 10.3389/fmicb.2018.02325.

Liu, X., Yu, Z., Jia, W., Wu, Y., Wu, D. et al. (2020). A review on linking the medicinal functions of mushroom prebiotics with gut microbiota. Int. J. Med. Mushrooms, 22: 943–951.

Merdivan, S. and Lindequist, U. (2017). Medicinal mushrooms with antiallergic activities. pp. 93–110. *In*: Agrawal, D.C., Tsay, H.S., Shyur, L.F., Wu, Y.C. and Wang, S.Y. (eds.). Medicinal Plants and Fungi—Recent Advances in Research and Development. Springer Nature, Singapore.

Mizuno, M. and Nishitani, Y. (2013). Immunomodulating compounds in *Basidiomycetes*. J. Clin. Biochem. Nutr., 52: 202–207.

Phan, C.W., David, P. and Sabaratnam, V. (2017). Edible and medicinal mushrooms: Emerging brain food for the mitigation of neurodegenerative diseases. J. Med. Food, 20: 1–10.

Phan, C.W., David, P., Naidu, M., Wong, K.H. and Sabaratnam, V. (2015). Therapeutic potential of culinary-medicinal mushrooms for the management of neurodegenerative diseases: Diversity, metabolite, and mechanism. Crit. Rev. Biotechnol., 35: 355–368.

Raman, J., Lee, S.K., Im, J.H., Oh, M.J., Oh, Y.L. and Jang, K.Y. (2018). Current prospects of mushroom production and industrial growth in India. J. Mushrooms, 16: 239–249.

Sandargo, B., Chepkirui, C., Cheng, T., Chaverra-Munoz, L., Thongbai, B. et al. (2019). Biological and chemical diversity go hand in hand: Basidiomycota as source of new pharmaceuticals and agrochemicals. Biotechnol. Adv., 37: 107344. 10.1016/j.biotechadv.2019.01.011.

Soares, A.A., de Sa-Nakanishi, A.B., Bracht, A., Gomes da Costa, S.M., Kochulein, E.A. et al. (2013). Hepatoprotective effect of mushrooms. Molecules, 18: 7609–7630.

Stajić, M., Vukojević, J. and Ćilerdžić, J. (2019). Mushrooms as potential natural cytostatics. pp. 143–168. *In*: Agrawal, D.C. and Dhanasekaran, M. (eds.). Medicinal Mushrooms, Springer Nature, Singapore.

Wasser, S.P. (2010). Medicinal mushroom science: History, current status, future trends, and unsolved problems. Int. J. Med. Mushrooms, 12: 1–16.

Wasser, S.P. (2017). Medicinal mushrooms in human clinical studies. Part I, Anticancer, oncoimmunological, and immunomodulatory activities: A review. Int. J. Med. Mushrooms, 19: 279–317.

Yahaya, N.F.M., Rahmann, M.A. and Abdullah, N. (2014). Therapeutic potential of mushrooms in preventing and ameliorating hypertension. Tr. Food Sci. Technol., 39: 104–115.

Zmitrovich, I.V., Belova, N.V., Balandaykin, M.E., Bondartseva, M.A. and Wasser, S.P. (2019). Cancer without pharmacological illusions and a niche for mycotherapy. Int. J. Med. Mushrooms, 21: 105–119.

2

Hypogeous and Epigeous Mushrooms in Human Health

Waill Elkhateeb,[1,*] *Paul Thomas,*[2,3] *Marwa Elnahas*[1] and *Ghoson Daba*[1]

1. INTRODUCTION

Dikarya, the fungal subkingdom, consists of the majority of higher fungi. It gains its name from *dikaryons*, which is a stage of life that contains two genetically distinct nuclei. This subkingdom does not have any flagellated life stages. It is divided into two main phyla: Basidiomycota and Ascomycota (Wellehan and Divers, 2019). The phylum Basidiomycota consists of the most complex and evolutionarily advanced members of this subkingdom, which includes some of the most iconic fungal species, such as the gilled mushrooms, puffballs and bracket fungi. Members of the Basidiomycota have a typical fungal morphology with diploid zygotes (called basidiospores) and they are characterised with large fruit bodies (Wellehan and Divers, 2019). The basidiospores are sexual spores formed outside special reproductive cells and called basidia. Another unique character for this group is the clamp connections. These are structures that are formed during the nuclear division on the growing hyphal tip (Alexopoulos et al., 1996). The phylum Ascomycota or sac fungi are monophyletic, accounting for approximately 75 per cent of all described fungi. It includes most of the fungi which associate with algae to form lichens and the majority of fungi lack morphological evidence of sexual reproduction. Ascospores are formed within the ascus by an enveloping membrane system, which packages each nucleus with its adjacent cytoplasm and provides the site for ascospore wall

[1] Chemistry of Natural and Microbial Products Department, Pharmaceutical Industries Researches Division, National Research Centre, El Buhouth St., Dokki, 12311, Giza, Egypt.
[2] Mycorrhizal Systems Ltd., Lancashire, PR25 2SD, UK.
[3] University of Stirling, Stirling, FK9 4LA, UK.
* Corresponding author: Waillahmed@yahoo.com

formation. Like other fungi, Ascomycota are heterotrophs and obtain nutrients from dead or living organisms.

2. Basidiomycota

Basidiomycota inhabit a wide range of ecological niches, carrying out vital ecosystem roles, particularly in carbon cycling and serve as symbiotic partners with a range of tree species. The majority of Basidiomycota are usually associated with plants as pathogens or decomposers. Specifically in the context of human use, the Basidiomycetes are a highly valuable food source and are increasingly medicinally valued. The Basidiomycetes mushrooms are famous for their use as sources of therapeutic bioactive compounds, such as *Geastrum fimbriatum* and *Hydnellum peckii* which exhibit promising anticoagulant activity (Elkhateeb et al., 2019a). *Handkea utriformis, Hericium erinaceus, Sparassis crispa, Agaricus blazei* and *Ganoderma oregonense* have wound-healing capabilities (Elkhateeb et al., 2019b), *Trametes versicolor* and *Dictyophora indusiata* show promising antioxidant, antimicrobial, antihyperlipidemia, antitumor and immunity enhancement effects (Elkhateeb et al., 2020a), *Fomes fomentarius* and *Polyporus squamosus* have significant importance as antifungal, antibacterial, anti-inflammatory, antioxidant, antitumor and antiviral agents (Elkhateeb et al., 2019c; Elkhateeb et al., 2020b). Many mushroom genera are famous for their promising therapeutic capabilities and one of the mushrooms attracting attention is *Inonotus obliquus*, which has been used for maintaining long and healthy human life (Thomas et al., 2020). All these vital activities have been reported from extracts of fruit bodies of these mushrooms or their biologically active isolated compounds.

2.1 Fomes fomentarius

Fomes fomentarius is a white rot fungus that causes heartrot of wood; it is also known as tinder or iceman's fungus (Fig. 1a). This perennial woody Basidiomycetes has a large size and grows as a parasite or saprophyte on deciduous tree species in Europe; it also grows on birch, oak and beech (Větrovský et al., 2011). It mainly appears in the northern hemisphere and in areas, such as China, Japan, Europe and North America. The saprotrophic Basidiomycetes that inhibit wood are considered as the major decomposers of cellulose and lignin in dead wood (Ruel et al., 1994). The *F. fomentarius* fruit bodies have historically been used in traditional Chinese medicine for treating inflammations, oral ulcer, gastroenteric disorder, as well as various cancers (Chen et al., 2008). Moreover, the fungus was also used as a styptic by dentists and surgeons (Grienke et al., 2014).

Several studies have reported that *F. fomentarius* exhibits many vital biological activities, such as anti-inflammatory, hypoglycemic, anti-infective and antitumor activities (Park et al., 2004; Yang et al., 2008; Seniuk et al., 2011). These activities are related to the presence of various classes of bioactive metabolites, which could be primary metabolites, such as proteins, polysaccharides (β-glucans), polysaccharide-protein complexes or secondary metabolites like glycosides and triterpene, esters and lactones, ketones (protocatechualdehyde; (22E)-ergosta-7,22-dien-3-one),

Fig. 1. Examples of some important Basidiomycetes mushrooms: (a) *Fomes fomentarius* (photographs taken by Paul W. Thomas; Locality: Pitlochry, Scotland UK); (b) Chaga (*Inonotus obliquus*) (photographs taken by Paul W. Thomas, Near Pitlochry, Scotland).

benzofurans (paulownin), organic acids and coumarins (daphnetin) (Grienke et al., 2014). Interestingly, optimisation of the nutritional requirements and submerged culture conditions for mycelial biomass as well as exopolysaccharide production from *F. fomentarius* was studied. It was reported the extracellular polysaccharide from *F. fomentarius* has antiproliferative effect on SGC-7901 (human gastric cancer cell line) and this activity is dose-dependent, where the polysaccharide concentration of 0.25 mg/ml was able to sensitise doxorubicin (Dox) induced growth of SGC-7901 cells after 24 hours of incubation (Chen et al., 2008). Thus, it is crucial to optimise the bioreactor fermentation conditions to achieve higher scale production by *F. fomentarius*.

2.2 *Inonotus obliquus*

Inonotus obliquus (Chaga mushroom) is a wood-rot fungus in live trees (Fig. 1b). The fungus enters through wounds within the tree and develops by causing 10 to 80+ years of decay through mycelial mass (Lee et al., 2008). Chaga mushroom is parasitic fungus growing on birches and used in traditional medicine to treat various human health problems (Géry et al., 2018). Numerous independent studies document its valuable role in prevention and healing cancer; it beneficially activates the immune system, inhibits cellular degeneration due to oxidation, suppresses inflammation, kills and/or inhibits the growth of viruses and supports diabetes treatment by preventing hyperglycaemia. Remarkably, this fungus demonstrates virtually no side effects during its use in disease treatment (Wasser, 2002; Choi et al., 2010). Other polypore fungi have been used medicinally for many years due to the presence of a variety of biologically active compounds that occur in their fruiting bodies. Chaga mushroom is a host of pharmacologically active compounds which are beneficial to human health (Zjawiony, 2004; De Silva et al., 2013). In the last decade, several studies have reported biological activities of *Inonotus obliquus*, such as anticancer, antioxidation, antiinflammatory, antidiabetic and enhancement of immunity (Choi et al., 2010). Chaga mushroom should be considered as an important issue in medicinal mushroom science. The prevalence of polyphenolic composites in this mushroom indicates its clear antioxidant and anticancer, antimicrobial, antihyperglycemic activities and other protective functions. The glucan and triterpenoid profile of this mushroom allows its use directly as an antitumor agent (Elkhateeb, 2020; Thomas et al., 2020).

3. Ascomycota

Ascomycota mushrooms play the biggest role in recycling dead plant material. As biotrophs, they associate symbiotically with algae (lichens), plant roots (mycorrhizae) or the leaves and stems (endophytes). Ascomycetes have been important throughout history in food production and as sources of medicinal compounds. Ascomycetes may also form large fruit bodies and many of these serve as food sources (Kirk et al., 2008). There are many well known genera in Ascomycetes as medicine via their therapeutic or bioactive compounds (e.g., Cordyceps, *Morchella* and truffles and truffle-like fungi).

3.1 Cordyceps—The Golden Ascomycetous Mushroom

The genus *Cordyceps* is entomopathogenic consisting of about 450 species associated with diverse insect hosts worldwide. However, only *Cordyceps sinensis* has been reported to be used in traditional Chinese medicine. It is also known as Dongchongxiacao (Lin and Li, 2011). The name Cordyceps is derived from Latin words meaning 'club and head'. Natural *Cordyceps sinensis* yields have been declining and this has been driven by anthropogenic interference, such as overharvesting practices. In the previous 25 years, the yield has dropped by more than 90 per cent and hence *C. sinensis* prices have skyrocketed to the extent that they exceeded US$25,000/kg in 2007 (Feng et al., 2008). Due to the rarity of *C. sinensis*, other natural substitutes for *C. sinensis* have been found in markets and these include *C. liangshanensis, C. militaris* (Fig. 2a), and *C. cicadicola* (Yang et al., 2009). Moreover, cultured mycelia of *C. sinensis* have been widely used as a commercial product. Since 2002, the State Food and Drug Administration of China has approved about 50 medicines and two dietary supplements from cultured *Cordyceps* (Feng et al., 2008). *Cordyceps sinensis* showed several biological activities and it has gained importance for the treatment of asthenia after illness, renal failure, renal dysfunction, fatigue and cough (Feng et al., 2008).

Cordyceps have also been used to treat hepatic-related diseases (Zhao, 2000). It has been reported that *Cordyceps* enhances the immunological function of chronic hepatitis B patients (Gong et al., 2000) and from post-hepatic cirrhosis (Zhu and Liu, 1992). In addition, *Cordyceps* are capable of inhibiting and reversing liver fibrosis through collagen degradation in dimethylnitrosamine-induced liver cirrhosis in rats (Li et al., 2006). It shows an inhibition effect on the hepatic stellate cell proliferation *in vitro* (Chor et al., 2005); it regulates CD126 and intercellular adhesion molecule-I in human fibroblasts (Wachtel-Galor, 2004). Interestingly, *Cordyceps* reduces the level of hepatic tissues and serum lipid peroxide (Zeng et al., 2001). The *C. sinensis* also gained its biological significance due to its role in treating various renal diseases including chronic pyelonephritis, chronic renal dysfunction or failure, chronic nephritis and nephritic syndrome (Feng et al., 2008). The *C. sinensis* extract plays a vital role in regulating the apoptotic caspase-3 gene and decreasing other inflammatory genes, such as TNF-α, and MCP-1, suggesting that it has an important therapeutic role in renal transplantation (Shahed, 2001).

Many biologically active compounds have been isolated from different *Cordyceps* spp. including cordycepin, ergosterol and polysaccharides (Elkhateeb et al., 2019d; El-Hagrassi, 2020). Nucleosides are considered as one of the main components in *Cordyceps*. More than 10 nucleosides and their derivatives have been isolated from *Cordyceps* (adenine, adenosine, cytidine, cytosine, 2'-deoxyuridine, 2'-deoxyadenosine, guanine, guanosine, hypoxanthine, inosine, N^6-methyladenosine, 6-hydroxyethyl-adenosine, thymine, thymidine, uracil, uridine and cordycepin) (Feng et al., 2008). Adenosine receptors are found in the brain, heart, lung, kidney, liver and are very important in many central nervous system-mediated events, like sleep. They are also involved in immunological response, cardiovascular function, respiratory regulation, kidney and liver activities (Li and Yang, 2008). Interestingly, the pharmacological effects of *Cordyceps* match well with the distribution and physiological roles of adenosine receptors, including anticancer, antiaging, antithrombosis, antiarrhythmias, antihypertension, immunomodulatory activity and has protective effects on the kidney, liver and lungs (Li and Yang, 2008).

Ten free fatty acids (FFA) have been found in the extract of natural *C. sinensis*, *C. gunnii* and *C. liangshanensis* which has been found in cultured *C. sinensis* and *C. militaris*. These are lauric acid, stearic acid, myristic acid, linoleic acid, oleic acid, pentadecanoic acid, palmitoleic acid, palmitic acid, docosanoic acid and lignoceric acid. The stearic acid, palmitic acid, oleic acid and linoleic acid are the major free fatty acids components in both natural and cultured *Cordyceps* (Yang et al., 2009). The importance of these free fatty acids is not only restricted to nutrition, but they also serve as modulators of many vital cellular functions through their receptors (Rayasam et al., 2007; Hirasawa et al., 2008). *Cordyceps* contain polysaccharides in high amounts, ranging between 3–8 per cent of the total dry weight. Additionally, it contains a high amount of D-mannitol. In 1957, D-mannitol was isolated from *C. sinensis* called cordycepic acid. It contributes over 3.4 per cent and 2.4 per cent of dry weight of natural and cultured *Cordyceps*, respectively (Feng et al., 2008; Li and Yang, 2008). The D-mannitol shows an osmotic activity and hence it has been used for the treatment of cerebral edema, intracranial hypertension in brain injury and stroke (Rangel-Castillo et al., 2008). The mannitol has also showed its usefulness in treating patients with cystic fibrosis and bronchiectasis by inhaling its dry powdered form as it rehydrates the airway and increases the mucociliary clearance (Jaques et al., 2008). Due to D-mannitol pharmacological effects, *Cordyceps* gains its importance in treatment of some respiratory diseases, such as chronic bronchitis as well as asthma.

3.2 *Morchella esculenta*

Morchella esculenta is one of the most economically beneficial and edible wild species of mushrooms (Fig. 2b). It has several common names, like guchi, morel, yellow morel, sponge morel, common morel, true morel, morel mushroom and so on. It has a broad global distribution, occurring in both the northern and southern hemispheres, but in North America it is found early in the year (March to July), often in forests frequently at altitudinal range of 2500–3500 m asl. It occurs in close proximity to both coniferous and hardwood trees (Hamayun et al., 2006). The world production of *Morchella esculenta* is about 150 tons dry weight and this is equivalent to 1.5 million tons of fresh weight. The major morel-producing countries are Pakistan and India, each of them producing around 50 tons of dry weight (Ciesla, 2002). It is an expensive mushroom and so it is locally called 'growing gold of mountains'. The *M. esculenta* often grows in forests with loamy soil that is rich in humus. It has both nutritional and medicinal importance because of its unique bioactive compounds, such as dietary fibres, vitamins, polysaccharides, proteins and trace elements (Ali et al., 2011; Elkhateeb et al., 2020b).

The fruit bodies of *M. esculenta* exhibit antioxidant activity (Elmastas et al., 2006). Moreover, its mycelia also shows antioxidant activities because of its linoleic acid and beta-carotene contents. Other biological importance have been reported for *M. esculenta*, which shows anti-inflammatory and antitumor activities (Nitha et al., 2007; Nitha et al., 2013) owing to the presence of polysaccharides (Yang et al., 2015). The *M. esculenta* being a rare natural resource, its production via submerged fermentation has been introduced and the polysaccharides show promising antitumor activities (Li et al., 2013). Additionally, the mushroom extracts have shown antimicrobial activity against several bacteria, such as *Staphylococcus aureus*, *Escherichia coli*, *Enterobacter cloacae*, *Salmonella typhimurium* and *Listeria monocytogenes* (Heleno et al., 2013). Powder of *M. esculenta* shows antiseptic activity and is very useful in healing wounds. Besides, it has been used widely for treatment of stomach ache (Mahmood et al., 2011).

Fig. 2. Examples of some important Aascomycetous mushrooms: (a) *Cordyceps militaris* (photographs taken by Ting-Chi Wen, Guizhou University, Guiyang, Guizhou Province, China); (b) *Morchella esculenta* (photographs taken by Andreas Kunze, hosted by: https://commons.wikimedia.org).

4. Truffles and Truffle-like Fungi

Many fungi develop hypogeal fruit bodies showing dependence on mycophagy for spore dispersal. This group is dominated by Ascomycetes (Trappe and Claridge, 2010). As such, Ascomycetes have large hypogeous fruit bodies known as truffles or truffle-like mushrooms (Thomas et al., 2019). Dependence on mycophagy by animals may have a resultant impact on nutritional composition, with compounds being used as an attractant for spore dispersal or reveal the presence of compounds with therapeutic benefit.

As outlined, it is clear that mushrooms are a rich source of therapeutic compounds and historical/traditional knowledge by local populations can provide a good guidance to where to search for such mushrooms for their versatile compounds. However, according to Thomas et al. (2019), such knowledge is fragile and in many areas of continental Africa the traditional knowledge is often passed orally from generation to generation and much of it is documented in written media and may be getting lost. Despite this, most of what we know about the nutritive and medicinal values of truffles and truffle-like fungi arises from the research that has been carried out on species that are found within continental Africa and Middle East. In southern continental Africa, there is a common and valued species of truffle endemic to the Kalahari region and is widely known as Kalahari truffle or *Kalahari tuberpfeilii* (Ferdman et al., 2005). Kalahari truffle is widely respected and frequently used by local people, primarily as a food source with a well-documented history of use (Trappe et al., 2008). However, the Kalahari truffle is also supposedly has medicinal properties, for example, some San hunters may have carried dried pieces of its fruit bodies in the belief that they may be used as an oral antidote to the poison of their arrows (Trappe et al., 2010a). The Khoisan people have also claimed aphrodisiac properties (Trappe et al., 2010b) and in Botswana, dried and powdered fruit bodies are reportedly used medicinally to induce birth in humans and livestock (Khonga et al., 2007).

In northern Africa, the known diversity of truffle and truffle-like fungi is greater and many of these are claimed to possess medicinal properties (Thomas et al., 2019). For instance, *Picoa* is a genus often small but has appreciably edible species, known to be high in antioxidants and its identified compounds possess antimutagenic and anticarcinogenic properties (Murcia et al., 2002). Species within the genus *Picoa* may also be surprisingly high in protein with nutritional profile: 22.54 per cent protein, 19.94 per cent fat, 36.66 per cent carbohydrates, 13.04 per cent fibre and 8.21 per cent ash (Murcia et al., 2002).

Perhaps the most commonly cited truffle species with assumed medicinal qualities are those from the genus *Terfezia*, which is distributed over North Africa, Middle East and Southern Europe (Thomas et al., 2019). *Terfezia boudieri* is a widely consumed species and nutritionally contains all the essential amino acids along with antioxidants. The composition of *T. boudieri* shows 14 per cent protein, 8 per cent fat and 54 per cent carbohydrates (Dundar et al., 2012). Slama et al. (2009) observed that the fruit bodies are rich in Ca^{2+} (1,423), K^+ (1,346), P (346), Mg^{2+} (154) and Na^+ (77 mg/100 g dry weight). Further, *T. boudieri* displays antimicrobial properties, for example, in 2013, Doðan and Aydin (2013) tested extracts of *T. boudieri* against nine bacteria and one yeast. Extracts showed antimicrobial properties against all the tested bacteria and the most pronounced significant impact was on the yeast, *Candida albicans*. Extracts of *Terfezia claveryi* Chatin have particularly noteworthy impacts on *Chlamydia trachomatis* and also aiding healing open wounds and stomach ulcers (Dabbour and Takruri, 2002). Nutritionally, *T. claveryi* has a composition of 16 per cent protein, 7 per cent fat, 65 per cent carbohydrate, 8 per cent crude fibre and 4 per cent ash, whilst also being relatively rich in vitamins like thiamin, riboflavin, niacin, pantothenic acid and pyridoxine (Al-Marzooky, 1981). Additionally, *T. claveryi* seems to be a rich source of Fe and a reasonably rich source of Zn as well as Mg (Al-Marzooky, 1981).

Another highly appreciated, common, frequently abundant genus with a distribution across the Saharan and sub-Saharan regions of Africa is *Tirmania*. *Tirmania nivea* contains 40–60 per cent carbohydrate, 20–27 per cent protein, 2.4–7.5 per cent lipid, 7–13 per cent crude fibre and 7.4–9.6 per cent ash (Fig. 3a) (Hussain and Al-Ruqaie, 1999). Al-Laith (2010) also states that the antioxidant levels of this species are high. The *T. nivea* and *Tirmania pinoyi* contain all essential amino acids including those that are often limiting in foods of plant origin (methionine, cysteine, tryptophan and lysine). The *T. pinoyi* is another desert truffle that appears to have potent antimicrobial activities. For instance, extracts are powerful against Gram-positive bacteria (*B. subtilis* and *S. aureus*) and are known to cause eye infections (Dib-Bellahouel and Fortas, 2011). This is of particular relevance as Omer et al. (1994) reported that desert truffles (*Tirmania* and *Terfezia* spp.) have a long history of utilisation in folk medicine which has mostly been recorded for the treatment of ophthalmic diseases. It should also be noted that ergosteroids have been identified in desert truffle species. These widespread fungal sterols are known to be transformed into vitamin D in the human body (Doğan and Aydin, 2013).

Finally, perhaps the most widely known and revered species of truffle and truffle-like fungi are those belonging to the genus *Tuber*. One species, *Tuber melanosporum* is in such high demand for culinary purposes that its price regularly exceeds 1000 EUR/kg and the future of this species is uncertain due to climate change (Thomas and Büntgen, 2019). This reverence has resulted in a number of in-depth studies. For example, the vitamin D2 precursor ergosterol (ergosta-5,7,22-trienol) has been identified along with brassica sterol (ergosta-5,22-dienol) within fruit bodies (Harki et al., 1996). Further *T. melanosporum* is also a source of potentially therapeutic metabolic enzymes. Recently, it has been reported that *T. melanosporum* consists of prominent members of the endocannabinoid system, such as anandamide, whose concentration may depend on the maturity of the fruit body (Pacioni et al., 2015). Besides humans, many animals are known to consume truffles that are known to have endocannabinoid-binding receptors (Trappe and Claridge, 2010). Endocannabinoids are linked to food seeking, eating and the hedonic evaluation of food during eating with a potential role in treating dietary disorders (De Luca et al., 2012). Endocannabinoids may also have potential therapeutic uses in multiple sclerosis, modification of various pain states, Alzheimer's disease, Parkinson's disease, Huntington's disease and epilepsy (Maccarrone et al., 2017). Further, *T. melanosporum* (Fig. 3b) amongst other *Tuber* species is rich in l-tyrosine, which is suggested to be beneficial in the management of stress, depression and bipolar

Fig. 3. Examples of some important Ascomycetes mushrooms: (a) *Tirmania nivea* (photographs taken by Waill A. Elkhateeb, National Research Centre, Egypt); (b) *Tuber melanosporum*; (c) *Tuber aestivum* (photographs taken by Paul W. Thomas, University of Stirling, Stirling, UK).

disorder (Patel et al., 2017), but the degree of impact is currently under debate. The *Tuber* genus may also be a useful source of polysaccharide fractions which could promote antitumor activity. For instance, one study on *T. melanosporum, T. indicum, T. sinense, T. aestivum* (Fig. 3c) and *T. himalayense* isolated 52 polysaccharides with the significant potential to promote antitumor activity (Zhao et al., 2014).

5. Conclusion

Ascomycetous and Basidiomycetous mushrooms represented by *Cordyceps* spp., *Morchella esculenta,* Truffles and truffle-like fungi, *Fomes fomentarius* and *Inonotus obliquus* have a rich history of use as a food source and well-claimed medicinal properties. This chapter summarises a number of sources with details of nutritional content and beneficial compounds (antimicrobial properties to antitumor and health-promoting sterols). The reliance of many of these species on mycophagy for spore dispersal may be linked to the development of other potentially therapeutic metabolic enzymes, specially those of the endocannabinoid system. Such enzymes could be useful in treating a range of conditions ranging from dietary disorders to Alzheimer's disease and epilepsy. Despite these advances, there is much we have yet to understand and these hypogeal fruiting Ascomycetes and Basidiomycetes prove to be a fruitful source of novel medicinal compounds.

References

Alexopoulos, C.J., Mims, C.W. and Blackwell, M. (1996). Introductory Mycology. John Wiley and Sons.

Ali, H., Sannai, J., Sher, H. and Rashid, A. (2011). Ethnobotanical profile of some plant resources in Malam Jabba valley of Swat, Pakistan. J. Med. Pl. Res., 5: 4676–4687.

Al-Laith, A.A. (2010). Antioxidant components and antioxidant/antiradical activities of desert truffle (*Tirmania nivea*) from various Middle Eastern origins. J. Food Comp. Anal., 23(1): 15–22.

Al-Marzooky, M.A. (1981). Truffles in eye disease. Proc. Int. Islamic Med., 353–357.

Chen, W., Zhao, Z., Chen, S.-F. and Li, Y.-Q. (2008). Optimisation for the production of exopolysaccharide from *Fomes fomentarius* in submerged culture and its antitumor effect *in vitro*. Biores. Technol., 99: 3187–3194.

Choi, S.Y., Hur, S.J., An, C.S., Jeon, Y.H., Jeoung, Y.J., Bak, J.P. and Lim, B.O. (2010). Anti-inflammatory effects of *Inonotus obliquus* in colitis induced by dextran sodium sulphate. BioMed Research Int., Article ID 943516: 1–5.

Chor, S., Hui, A., To, K., Chan, K., Go, Y., Chan, H., Leung, W. and Sung, J. (2005). Anti-proliferative and pro-apoptotic effects of herbal medicine on hepatic stellate cell. J. Ethnopharmacol., 100: 180–186.

Ciesla, W.M. (2002). Non-wood Forest Products from Temperate Broad-leaved Trees. vol. 15, Food and Agricultural Organisation, Rome.

Dabbour, I.R. and Takruri, H.R. (2002). Protein digestibility using corrected amino acid score method (PDCAAS) of four types of mushrooms grown in Jordan. Pl. Foods Hum. Nutr., 57(1): 13–24.

De Luca, M.A., Solinas, M., Bimpisidis, Z., Goldberg, S.R. and Di Chiara, G. (2012). Cannabinoid facilitation of behavioral and biochemical hedonic taste responses. Neuropharmacol., 63(1): 161–8.

De Silva, D., Rapior, S., Sudarman, E., Stadler, M., Jianchu, X., Aisyah, A. and Kevin, D. (2013). Bioactive metabolites from macrofungi: Ethnopharmacology, biological activities and chemistry. Fungal Diversity, 62(1): 1–40.

Dib-Bellahouel, S. and Fortas, Z. (2011). Antibacterial activity of various fractions of ethyl acetate extract from the desert truffle, *Tirmania pinoyi*, preliminarily analysed by gas chromatography-mass spectrometry (GC-MS). Afr. J. Biotechnol., 10(47): 9694–9699.

Doğan, H.H. and Aydin, S. (2013). Determination of antimicrobial effect, antioxidant activity and phenolic contents of desert truffle in Turkey. Afr. J. Trad. Compl. Alt. Med., 10(4): 52–8.

Dundar, A., Yesil, O.F., Acay, H., Okumus, V., Ozdemir, S. and Yildiz, A. (2012). Antioxidant properties, chemical composition and nutritional value of *Terfezia boudieri* (Chatin) from Turkey. Food Sci. Technol. Int., 18(4): 317–28.

El-Hagrassi, A., Daba, G., Elkhateeb, W., Ahmed, E., El-Dein, A.N., Fayad, W., Shaheen, M., Shehata, R., El-Manawaty, M. and Wen, T.C. (2020). *In vitro* bioactive potential and chemical analysis of the n-hexane extract of the medicinal mushroom, *Cordyceps militaris* Malay. J. Microbiol., 16(1): 40–48.

Elkhateeb, W.A., Daba, G.M., Elnahas, M.O. and Thomas, P.W. (2019a). Anticoagulant capacities of some medicinal mushrooms. ARC J. Pharmaceut. Sci., 5(4): 1–9.

Elkhateeb, W.A., Elnahas, M.O., Thomas, P.W. and Daba, G.M. (2019b). To heal or not to heal? Medicinal mushrooms wound healing capacities. ARC J. Pharmaceut. Sci., 5(4): 28–35.

Elkhateeb, W.A., Daba, G.M., Elmahdy, E.M., Thomas, P.W., Wen, T.C. and Shaheen, M.N. (2019c). Antiviral potential of mushrooms in the light of their biological active compounds. ARC J. Pharmaceut. Sci., 5(2): 45–49.

Elkhateeb, W.A., Daba, G.M., Thomas, P.W. and Wen, T.C. (2019d). Medicinal mushrooms as a new source of natural therapeutic bioactive compounds. Egyp. Pharmaceut. J., 18(2): 88–101.

Elkhateeb, W.A. (2020). What medicinal mushroom can do? Chem. Res. J., 5(1): 106–118.

Elkhateeb, W.A., Elnahas, M.O., Thomas, P.W. and Daba, G.M. (2020a). *Trametes versicolor* and *Dictyophora indusiata* champions of medicinal mushrooms open access. J. Pharmaceut. Res., 4(1): 1–7.

Elkhateeb, W.A., Elnahas, M.O., Thomas, P.W. and Daba, G.M. (2020b). *Fomes fomentarius* and *Polyporus squamosus* models of marvel medicinal mushrooms. Biomed. Res. Rev., 3(1): 1–4.

Elmastas, M., Turkekul, I., Ozturk, L., Gulcin, I., Isildak, O. and Aboul-Enein, H.Y. (2006). Antioxidant activity of two wild edible mushrooms (*Morchella vulgaris* and *Morchella esculanta*) from North Turkey. Combinat. Chem. High Throughput Screen., 9: 443–448.

Feng, K., Yang, Y. and Li, S. (2008). Renggongchongcao: Pharmacological Activity-based Quality Control of Chinese Herbs, 155–178.

Ferdman, Y., Aviram, S., Nurit, R.B., Trappe, J.M. and Kagan-Zur, V. (2005). Phylogenetic studies of *Terfezia pfeilii* and *Choiromyces echinulatus* (Pezizales) support new genera for southern African truffles: Kalarituber and Eremiomyces. Mycol. Res., 109(2): 237–45.

Géry, A., Dubreule, C., André, V., Rioult, J.P., Bouchart, V., Heutte, N., Philippe Eldin de Pécoulas, Tetyana, K. and Garon, D. (2018). Chaga (*Inonotus obliquus*), a future potential medicinal fungus in oncology? A chemical study and a comparison of the cytotoxicity against human lung adenocarcinoma cells (A549) and human bronchial epithelial cells (BEAS-2B). Integr. Can. Ther., 17(3): 832–843.

Gong, H., Wang, K. and Tang, S. (2000). Effects of *Cordyceps sinensis* on T lymphocyte subsets and hepatofibrosis in patients with chronic hepatitis B. Hunan Yi Ke Da Xue Xue Bao, 25(3): 248–250.

Grienke, U., Zöll, M., Peintner, U. and Rollinger, J.M. (2014). European medicinal polypores—A modern view on traditional uses. J. Ethnopharmacol., 154: 564–583.

Hamayun, M., Khan, S.A., Ahmad, H., Shin, D.-H. and Lee, I.-J. (2006). Morel collection and marketing: A case study from the Hindu-Kush mountain region of Swat, Pakistan. Lyonia, 11: 7–13.

Harki, E., Klaebe, A., Talou, T. and Dargent, R. (1996). Identification and quantification of *Tuber melanosporum* Vitt. Sterols. Steroids, 61(10): 609–12.

Heleno, S.A., Stojković, D., Barros, L., Glamočlija, J., Soković, M., Martins, A., Queiroz, M.J.R. and Ferreira, I.C. (2013). A comparative study of chemical composition, antioxidant and antimicrobial properties of *Morchella esculenta* (L.) Pers. from Portugal and Serbia. Food Res. Int., 51: 236–243.

Hirasawa, A., Hara, T., Katsuma, S., Adachi, T. and Tsujimoto, G. (2008). Free fatty acid receptors and drug discovery. Biol. Pharmaceut. Bull., 31: 1847–1851.

Hussain, G. and Al-Ruqaie, I.M. (1999). Occurrence, chemical composition, and nutritional value of truffles: An overview. Pak. J. Biol. Sci., 2(2): 510–514.

Jaques, A., Daviskas, E., Turton, J.A., McKay, K., Cooper, P., Stirling, R.G., Robertson, C.F., Bye, P.T., LeSouëf, P.N. and Shadbolt, B. (2008). Inhaled mannitol improves lung function in cystic fibrosis. Chest, 133(6): 1388–1396.

Khonga, E.B. and Mogotsi, K.K. (2007). Indigeous knowledge on edible, poisonous, and medicinal macrofungi in Botswana. Int. J. Med. Mush., 9(3-4): 262.

Kirk, P.M., Cannon, P.F., Minter, D.W. and Stalpers, J.A. (2008). Dictionary of the Fungi. 10th ed., Wallingford, UK.

Lee, M.W., Hur, H., Chang, K.C., Lee, T.S., Ka, K.H. and Jankovsky, L. (2008). Introduction to distribution and ecology of sterile conks of *Inonotus obliquus*. Mycobiol., 36(4): 199–202.

Li, F.-H., Liu, P., Xiong, W. and Xu, G. (2006). Effects of *Cordyceps sinensis* on dimethylnitrosamine-induced liver fibrosis in rats. Zhong xi yi jie he xue bao= J. Chin. Integra. Med., 4: 514–517.

Li, S. and Yang, Y. (2008). Dongchongxiacao. Pharmacological activity-based quality control of Chinese herbs. Nova Science Publishers Inc., New York, 139–156.

Li, S., Sang, Y., Zhu, D., Yang, Y., Lei, Z. and Zhang, Z. (2013). Optimisation of fermentation conditions for crude polysaccharides by *Morchella esculenta* using soybean curd residue. Ind. Crops Prod., 50: 666–672.

Lin, B. and Li, S. (2011). *Cordyceps* as an herbal drug. *In*: Benzie, I.F.F. and Wachtel-Galor, S. (eds.). Herbal Medicine: Biomolecular and Clinical Aspects. 2nd ed., CRC Press. https://www.ncbi.nlm. nih.gov/books/NBK92758/.

Maccarone, M., Maldonado, R., Casas, M., Henze, T. and Centonze, D. (2017). Cannabinoids therapeutic use: What is our current understanding following the introduction of THC, THC: CBD or mucosal spray and others? Exp. Rev. Clin. Pharmacol., 10(4): 443–55.

Mahmood, A., Malik, R.N., Shinwari, Z.K. and Mahmood, A. (2011). Ethnobotanical survey of plants from Neelum, Azad Jammu and Kashmir, Pakistan. Pak. J. Bot., 43: 105–110.

Murcia, M.A., Martinez-Tome, M., Jimenez, A.M., Vera, A.M., Honrubia, M. and Parras, P. (2002). Antioxidant activity of edible fungi (truffles and mushrooms): Losses during industrial processing. J. Food Prot., 65(10): 1614–1622.

Nitha, B., Meera, C. and Janardhanan, K. (2007). Anti-inflammatory and antitumour activities of cultured mycelium of morel mushroom, *Morchella esculenta*. Curr. Sci., 235–239.

Nitha, B., Fijesh, P. and Janardhanan, K. (2013). Hepatoprotective activity of cultured mycelium of Morel mushroom, *Morchella esculenta*. Exp. Toxicol. Pathol., 65: 105–112.

Omer, E.A., Smith, D.L., Wood, K.V. and El-Menshawi, B.S. (1994). The volatiles of desert truffle: *Tirmania nivea*. Pl. Foods Hum. Nutr., 45(3): 247–9.

Pacioni, G., Rapino, C., Zarivi, O., Falconi, A., Leonardi, M. et al. (2015). Truffles contain endocannabinoid metabolic enzymes and anandamide. Phytochem., 110: 104–110.

Patel, S., Rauf, A., Khan, H., Khalid, S. and Mubarak, M.S. (2017). Potential health benefits of natural products derived from truffles: A review. Tr. Food Sci. Technol., 70: 1–8.

Park, Y.M., Kim, I.T., Park, H.J., Choi, J.W., Park, K.Y. et al. (2004). Anti-inflammatory and anti-nociceptive effects of the methanol extract of *Fomes fomentarius*. Biol. Pharmaceut. Bull., 27(10): 1588–1593.

Rangel-Castillo, L., Gopinath, S. and Robertson, C.S. (2008). Management of intracranial hypertension. Neurol. Clin., 26: 521–541.

Rayasam, G.V., Tulasi, V.K., Davis, J.A. and Bansal, V.S. (2007). Fatty acid receptors as new therapeutic targets for diabetes. Exp. Op. Ther. Targ., 11: 661–671.

Ruel, K., Ambert, K. and Joseleau, J.-P. (1994). Influence of the enzyme equipment of white-rot fungi on the patterns of wood degradation. FEMS Microbiol. Rev., 13: 241–254.

Seniuk, O.F., Gorovoj, L.F., Beketova, G.V., Savichuk, H.O., Rytik, P.G. et al. (2011). Anti-infective properties of the melanin-glucan complex obtained from medicinal tinder bracket mushroom, *Fomes fomentarius* (L.: Fr.) Fr. (Aphyllophoromycetideae). Int. J. Med. Mushr., 13(1): 7–18.

Shahed, A. (2001). Down regulation of apoptotic and inflammatory genes by *Cordyceps sinensis* extract in rat kidney following ischemia/reperfusion. Transpl. Proc., 33: 2986–2987.

Slama, A., Neffati, M. and Boudabous, A.M. (2009). Biochemical composition of desert truffle *Terfezia boudierii* Chatin. Acta Hortic., 853: 285–290.

Thomas, P. and Büntgen, U. (2019). A risk assessment of Europe's black truffle sector under predicted climate change. Sci. Tot. Environ., 655: 27–34.

Thomas, P.W., Elkhateeb, W.A. and Daba, G. (2019). Truffle and truffle-like fungi from continental Africa. Acta Mycol., 1, 54(2): 1–15.

Thomas, P.W., Elkhateeb, W.A. and Daba, G.M. (2020). Chaga (*Inonotus obliquus*): A medical marvel mushroom becomes a conservation dilemma. Sydowia, 72: 123–130.

Trappe, J.M., Claridge, A.W., Arora, D. and Smit, W.A. (2008). Desert truffles of the African Kalahari: Ecology, ethnomycology and taxonomy. Econ. Bot., 62: 521–529.

Trappe, J.M. and Claridge, A.W. (2010). The hidden life of truffles. Sci. Am., 302(4): 78–82.

Trappe, J.M., Kovacs, G.M. and Claridge, A.W. (2010a). Comparative taxonomy of desert truffles of the Australian Outback and the African Kalahari. Mycol. Prog., 9: 131–143.

Trappe, J.M., Kovacs, G.M. and Claridge, A.W. (2010b). Validation of the new combination *Mattirolomycesaustro africanus*. Mycol. Prog., 9: 145.

Větrovský, T., Voříšková, J., Šnajdr, J., Gabriel, J. and Baldrian, P. (2011). Ecology of coarse wood decomposition by the saprotrophic fungus *Fomes fomentarius*. Biodegradation, 22: 709–718.

Wachtel-Galor, S. (2004). The biological and pharmacological properties of *Cordyceps sinensis*, a traditional Chinese medicine that has broad clinical applications. pp. 657–682. *In*: Herbal and Traditional Medicine. Marcel Dekker, New York.

Wasser, S.P. (2002). Medicinal mushrooms as a source of antitumor and immunomodulating polysaccharides. Appl. Microbiol. Biotechnol., 60: 258–274.

Wellehan, J.F. and Divers, S.J. (2019). Bacteriology. pp. 235–246. *In*: Divers, S. and Stahl, S. (eds.). Mader's Reptile and Amphibian Medicine and Surgery. 3rd ed., Elsevier.

Yang, F., Feng, K., Zhao, J. and Li, S. (2009). Analysis of sterols and fatty acids in natural and cultured *Cordyceps* by one-step derivatisation followed with gas chromatography–mass spectrometry. J. Pharmaceut. Biomed. Anal., 49: 1172–1178.

Yang, H., Yin, T. and Zhang, S. (2015). Isolation, purification, and characterisation of polysaccharides from wide *Morchella esculenta* (L.) Pers. Int. J. Food Prop., 18: 1385–1390.

Zeng, X., Tang, Y. and Yuan, S. (2001). The protective effects of CS and CN80-2 against the immunological liver injury in mice. Zhongguo yao xue za zhi (Zhongguo yao xue hui: 1989), 36: 161–164.

Zhao, S. (2000). Advance of treatment for *Cordyceps* on chronic hepatic diseases. Shanxi Zhong Yi, 16: 59–60.

Zhao, W., Wang, X.H., Li, H.M., Wang, S.H., Chen, T., Yuan, Z.P. and Tang, Y.J. (2014). Isolation and characterisation of polysaccharides with the antitumor activity from Tuber fruiting bodies and fermentation system. Appl. Microbiol. Biotechnol., 98(5): 1991–2002.

Zhu, J. and Liu, C. (1992). Modulating effects of extractum semen Persicae and cultivated *Cordyceps* hyphae on immuno-dysfunction of inpatients with post hepatitic cirrhosis. Chin. J. Int. Trad. West. Med., 12: 207–209.

Zjawiony, J.K. (2004). Biologically active compounds from Aphyllophorales (polypore) fungi. J. Nat. Prod., 67: 300–310.

3

Bioactive Attributes of Edible Wild Mushrooms of the Western Ghats

Venugopalan Ravikrishnan,[1] *Kandikere R Sridhar*[1,2,]*
and *Madaiah Rajashekhar*[1]

1. INTRODUCTION

Cultivated as well as wild edible mushrooms have drawn major attention worldwide as potential sources of bioactive substances with health-promoting benefits (Chang and Wasser, 2012). Edible mushrooms possess low fat and low calorie diet with abundant carbohydrates, proteins, dietary fibre and vitamins (Cheung, 2010). Apart from their nutritional qualities, they have immense therapeutic importance for their antitumor, antiviral and antioxidant potential. Thus, there is increased interest to adapt them for the benefit of human nutrition, health promotion and disease prevention. The current consumer interest on mushrooms is to develop functional foodstuffs which serve as drugs as well as nutraceuticals (Hobbs, 2000). Several functional components of mushrooms (e.g., polysaccharides, phenolics, flavonoids, sterols, terpenes and ceramide) are of immense value in health-promotion towards antitumor and immune-modulation properties.

Soluble sugars and organic acids in mushrooms are known for their organoleptic qualities. Soluble sugars in mushrooms possess therapeutic and commercial significance. Organic acids in mushrooms, in addition to add flavour and taste, serve as antioxidants, neuroprotective compounds, anti-inflammatory and anti-microbial agents (Barros et al., 2013; Leal et al., 2013). Polyphenols of mushrooms serve as potential natural antioxidants in place of synthetic antioxidants, like butylated

---
[1] Centre for Environmental Studies, Yenepoya (deemed to be) University, Mangalore, Karnataka, India.
[2] Department of Biosciences, Mangalore University, Mangalagangotri, Mangalore, Karnataka, India.
* Corresponding author: kandikere@gmail.com

hydroxyanisole (BHA), butylated hydroxytoluene (HBT), propyl gallate (PG) and tertiary butylhydroxyquinone (TBHQ) (Abdalla et al., 1999; Lee et al., 2007). Mushrooms being a rich source of polysaccharides facilitate production of functional foods with health-protective attributes. Besides, polysaccharides, mushrooms are well known for a wide range of health-promoting functions, especially for their prebiotic, antioxidant, immune modulation and antitumor properties (Xu et al., 2009; Stachowiak and Reguła, 2012). For instance, lentinan (*Lentinus edodes*), pleuran (*Pleurotus* spp.), schizophyllan (*Schizophyllum commune*), grifolan (*Grifola frondosa*), ganoderan/ganopoly (*Ganoderma lucidum*) and krestin (*Trametes versicolor*) are interesting polysaccharides (Gonzaga et al., 2004; Sun, 2014; Wang et al., 2017). Although active polysaccharides could be derived from culture mycelium and broths, high quantity as well as different polysaccharides are available in fruit bodies of basidiomycetous mushrooms (Reshetnikov et al., 2001). Polysaccharide-protein complexes of several medicinal mushrooms serve as important bioactive components (Ooi and Liu, 2000; Zhang et al., 2007). Mushroom polysaccharides are known for anticancer, immune modulation, antioxidant, radical scavenging and cytotoxic properties (Zhang et al., 2007; Thetsrimuang et al., 2011; Siu et al., 2014). Polysaccharides of mushrooms are relatively non-toxic and will not result in significant adverse effects that facilitate development of natural antioxidants (Wasser and Weis, 1999).

Fair knowledge about sugars, organic acids, polyphenols and polysaccharides of edible and medicinal mushrooms is of immense value in developing value-added foodstuffs and medicines. To go forward, it is necessary to evaluate biochemical components (e.g., sugars, organic acids, free amino acids, phenolics and polysaccharides) and functional attributes (e.g., antioxidant, immune modulation and antimicrobial potential) of cultivated and wild mushrooms. Wild mushrooms are becoming more popular owing to their nutritional (proteins, amino acids and fatty acids), organoleptic (flavour and taste) and pharmaceutical (antioxidant, cholesterol lowering and immunomodulatory) properties. Hence, the current chapter documents some of the above components of six edible wild mushrooms occurring in the Western Ghats of India and which include ectomycorrhizal, wood-preferring, leaf litter-preferring and termite mound-preferring species. Composition of extracted polysaccharides and antioxidant properties were also evaluated in two *Lentinus* species.

2. Mushrooms and Processing

2.1 Sampling and Processing

Six wild edible mushrooms (ectomycorrhizal, *Amanita hemibapha*, *Boletus edulis*; leaf litter-preferring *Hygrocybe alwisii*; wood preferring *Lentinus polychrous*, *L. squarrosulus*; termite mound-inhabiting *Termitomyces schimperi*) were collected from two forest locations adjacent to River Sita during the southwest monsoon. Hebri forest (13°28'N, 74°59'E) and Someshwara Wildlife Sanctuary (13°29'N, 74°50'E) in the Agumbe Ghat (Western Ghats) were the potential locations of edible wild mushrooms. These forests, known for semi-evergreen and moist mixed deciduous tree species, encounter heavy rainfall during the southwest monsoon season.

Harvested fruit bodies of mushrooms from three different sites (as replicates) in cool packs were transferred to the laboratory for processing. After removing the debris, fruit bodies were rinsed and blotted to reduce the moisture content. They were weighed, spread on aluminum foils and dried at $58 \pm 2°C$ in hot-air oven until attaining a constant weight. Dried fruit bodies were pulverised and preserved in air-tight containers for analysis.

3. Assessment of Bioactive Components

3.1 Soluble Sugars

Mushroom flour (1 g) was sonicated with distilled water and methanol (5 ml) separately at 40 per cent amplitude followed by centrifugation (10,000 rpm; 15 min). The recovered supernatant was concentrated by vacuum concentrator and the final volume was made up to 1 ml by Millipore water and methanol, respectively. The samples were used for sugar analysis after filtering through 0.45 µm PTFE membrane filters. They were analysed in HPLC (Shimadzu LC10A) coupled with a refractive index detector. Analysis was performed on the amino column, using acetonitrile and water (3:1) as the mobile phase, at the flow rate of 1 ml/min (Reis et al., 2012).

3.2 Organic Acids

Mushroom flour (1 g) was sonicated with distilled water and methanol (5 ml) separately at 40 per cent amplitude followed by centrifugation (10,000 rpm; 15 min). The recovered supernatant was concentrated by vacuum concentrator and the final volume was made up to 1 ml by Millipore water and methanol, respectively. The samples were used for sugar analysis after filtering through 0.45 µm PTFE membrane filters. They were analysed in HPLC (Shimadzu LC10A) coupled with a refractive index detector. Analysis was performed on reverse phase (C18 column), using an injection volume of 5 µl and detection at 210 nm. The mobile phase employed was 0.008 M sulphuric acid at a flow rate of 0.7 ml/min in an isocratic mode of elution (Pereira et al., 2013).

3.3 Total Free Amino Acids

To determine the total free amino acids in mushroom flours, a method proposed by Sadashivam and Manickam (1992) was employed. Mushroom flour (500 mg) was homogenised with 5–10 ml of 80 per cent ethanol and centrifuged. The extraction was repeated twice and supernatants were pooled and the volume was reduced by evaporation using rotoevaporator. To an aliquot of 0.1 ml of extract, 1 ml of ninhydrin solution [0.8 g of $SnCl_2.2H_2O$ (stannous chloride) in 500 ml of 0.2 M citrate buffer (pH 5.0) was added to 20 g of ninhydrin in 500 ml of methyl cellosolve (2 methoxyethanol)] was added, followed by addition of distilled water to make up the volume to 2 ml. The contents were heated in a boiling waterbath for 20 min. Later 5 ml of diluent (equal volumes of water and n-propanol were mixed and used) was added and mixed. The intensity of colour developed was read at 570 nm. A standard graph was prepared by using leucine (20–100 µg/ml).

3.4 Polyphenols

Mushroom flour (1 g) was sonicated with 5 ml of methanol at 40 per cent amplitude followed by centrifugation (10,000 rpm; 15 min). The recovered supernatant was concentrated by vacuum concentrator and the final volume was made up to 1 ml by methanol. The samples were filtered with 0.45 µm PTFE membrane filters prior to being subjected to high pressure liquid chromatography (HPLC) analysis. The system was equipped with an auto-injector and photodiode array detector (Waters, Milford). The analysis was performed on RP C-18 column, using an injection volume of 20 µl, followed by detection at the range of 280 and 320 nm. The mobile phases employed were A (acetonitrile) and B (0.1 per cent orthophosphoric acid) at a flow rate of 1 ml/min (Dasgupta et al., 2015).

3.5 Polysaccharides

The water soluble and ethanol fractions of polysaccharides of two *Lentinus* spp. were prepared following the procedure by Cheng et al. (2008) with minor modifications (Fig. 1). The dried mushroom powder was defatted by immersion in petroleum ether at room temperature for 24 hours. The solid residue was collected by filtration and

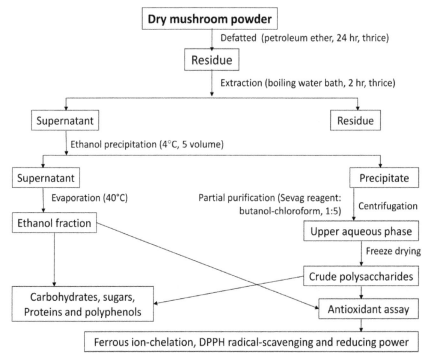

Fig. 1. Schematic processing of dry mushrooms powder (*Lentinus polychrous* and *L. squarrosulus*) for extraction of crude polysaccharides and assessment of its qualities.

the procedure was repeated twice to make the sample devoid of lipids. The filtered residue was air-dried at room temperature followed by hot water extraction (Stajić et al., 2013; Sun, 2014; Afshari et al., 2015). The extraction is carried out with boiling water (liquid/solid, 20:1 v/w; 2 hr) under agitation on a magnetic stirrer. Later, the filtered residue was extracted twice by applying the same method. Pooled extract was centrifuged to remove the insoluble materials followed by filtration through Whatman # 4 paper. The volume of aqueous extracts was concentrated by using the rotoevaporator. The concentrated extracts were precipitated with four-fold volume of ethanol overnight (4°C). The precipitate was collected by centrifugation (4000 rpm; 15 min; 4°C) (Hu et al., 2013). Later, the precipitate was solubilised in deionised water and deproteinised with Sevag method (chloroform/butyl alcohol; 4:1 v/v). The aqueous phase was collected and lyophilised to yield polysaccharides (Luo et al., 2011). The polysaccharides yield was calculated in percentage.

Polysaccharide (%) = $m_1/m_0 \times 100$

(where m_1 is the weight of dried polysaccharide; m_0 is the weight of dried mushroom powder)

Infrared spectra of the dried water-soluble and ethanol extracts of polysaccharide fractions were recorded on a Fourier Transform Infrared (FTIR) Spectrophotometer (Shimadzu IR Prestige 21 Japan) in transmittance mode was in the range of frequency 40004000 cm^{-1}.

3.6 Total Carbohydrates

The total carbohydrates were estimated by phenol-sulphuric acid method (Dubois et al., 1956) for water soluble polysaccharides (1 mg/ml) and ethanolic fractions (10 mg/ml). Each sample (0.1 ml) was made up to 1 ml and phenol solution (1 ml) was added to each tube followed by 5 ml of 96 per cent sulphuric acid and the mixture was incubated for 20 min at room temperature. The absorbance was read at 490 nm and total carbohydrate content was calculated by comparison with a glucose standard curve.

3.7 Reducing Sugars

Dinitrosalicylic acid (DNS) reagent method was employed to estimate the reducing sugars present in the polysaccharide extracts (Miller, 1959). One ml of the crude water-soluble polysaccharides (1 mg/ml) and 0.5 ml of ethanolic fraction (10 mg/ml) were made up to 3 ml with water and ethanol respectively and DNS reagent (3 ml) was added. The contents were heated in a boiling water bath for 5 min, followed by addition of 1 ml of 40 per cent Rochelle salt solution. The absorbance was read at 510 nm and the concentration of reducing sugar was estimated from a glucose standard graph.

3.8 Total Proteins

The protein content in the crude water-soluble polysaccharide and ethanolic fractions was determined by alkaline copper solution (2 per cent sodium carbonate in 0.1 N sodium hydroxide mixed with 0.5 per cent of copper sulphate in 1 per cent potassium sodium tartrate) and Folin-Ciocalteu's reagent. Sample (0.2 ml) was used from a stock concentration (1 mg/ml) for crude water-soluble polysaccharides and 0.1 ml of ethanolic fraction taken from a stock concentration (10 mg/ml). The samples were made to 1 ml and 5 ml by addition of alkaline copper solution, incubated (10 min) at room temperature. Folin-Ciocalteu's solution (0.5 ml) of was added. The absorbance was read at 660 nm after the incubation (30 min in dark at room temperature) and a standard graph was prepared using BSA to find the concentration of protein (Lowry et al., 1951).

3.9 Total Phenolics

The total phenolic content was determined according to Rosset et al. (1982). An aliquot of 0.5 ml of sample was used from a stock concentration (1 mg/ml) for water-soluble and 0.1 ml from stock concentration (10 mg/ml) for ethanolic fraction of polysaccharides. The samples were mixed with distilled water and treated with 5 ml Na_2CO_3 (in 0.1 N NaOH). After 10 min of incubation, 0.5 ml Folin-Ciocalteu's reagent (diluted; 1:2 v/v) was added and the absorbance was read at 725 nm. Gallic acid was used to prepare a standard curve.

4. Antioxidant Activities

Antioxidant potential of water and ethanol fractions of polysaccharides were determined based on the ferrous ion-chelating capacity (FCC), the DPPH radical-scavenging activity and the reducing power.

4.1 Ferrous Ion-chelating Capacity

The Fe^{2+} chelating capacity (FCC) was determined according to the method by Hsu et al. (2003). To 1 ml of the extract, 0.1 ml each of 2 mM $FeCl_2$ and 0.2 ml of 5 mM ferrozine were added. The final volume was made up to 5 ml using methanol. The mixture was incubated for 10 min at laboratory temperature followed by determination of absorbance of Fe^{2+}-ferrozine complex at 562 nm. The sample without the extract served as control to calculate ferrous ion-chelating capacity.

Ferrous ion chelating capacity (%) = $[1 - (A_{s562}/A_{c562})] \times 100$

(where, A_c is absorbance of the control; A_s is absorbance of sample)

4.2 DPPH Free Radical-scavenging Activity

Free radical scavenging activity of the extract was measured according to Singh et al. (2002). To 1 ml of the extracts, 4 ml of 0.01 mM 2,2-diphenyl-1-picrylhydrazyl (DPPH) was added and allowed to react at room temperature for 20 min. Reagents without addition of sample served as control and absorbance of the reaction mixture was measured at 517 nm to calculate the free radical-scavenging activity.

Free radical-scavenging activity (%) = $[(A_{c517} - A_{s517})/(A_{c517})] \times 100$

(where, A_c is absorbance of the control; A_s is absorbance of sample)

4.3 Reducing Power

Reducing power (RP) of the extracts was measured by the method by Oyaizu (1986) with a slight modification. Different concentrations (0.2–1 ml: 200–1,000 µg) of mushroom polysaccharides extracted in water as well as its ethanolic fraction were taken and 0.2 M phosphate buffer (pH, 6.6; 2.5 ml) was added followed by addition of potassium ferricyanide (2.5 ml; 1%). The contents were mixed and incubated at 50°C for 20 min. After incubation, Trichloroacetic acid (2.5 ml; 10% w/v) was added and the mixture was centrifuged (3000 rpm; 10 min). The supernatant (2.5 ml) was mixed with equal volume of distilled water. To the above mixture, ferric chloride (0.1%; 0.5 ml) was added and the absorbance was measured at 700 nm. Increase in the absorbance of the reaction mixture indicated increased reducing power.

5. Results and Discussion

5.1 Bioactive Components

5.1.1 Soluble Sugars

Among the six wild mushrooms evaluated, except for *L. polychrous*, as many as seven reducing sugars were determined (Table 1). Some sugars were extracted in water as well as methanol (e.g., arabinose in *A. hemibapha* and *H. alwisii*; galactose in *H. alwisii*; glucose, maltose and trehalose in *B. edulis*; galactose, glucose and trehalose in *T. schimperi*). Major sugars include arabinose and maltose (in *A. hemibapha*), maltose and glucose (in *B. edulis*), arabinose (*H. alwisii*) and glucose (*T. schimperi*). Overall, the glucose content is highest in aqueous extract of *T. schimperi* (10.7 mg/g) followed by arabinose in methanol extract of *A. hemibapha* (6.2 mg/g) and maltose in methanol extract of *B. edulis* (4.9 mg/g). The total sugar content was highest in *T. schimperi* followed by *A. hemibapha* in aqueous extract, while it was highest in *B. edulis* followed by *A. hemibapha* in methanol extract. Sucrose was not detectable in the mushrooms assessed. Except for *L. polychrous, L. squarrosulus* and *T. schimperi*, in the rest of the mushrooms, methanol possessed a significantly higher quantity of sugars.

Being non-volatile components, soluble sugars of mushrooms contribute to the sweetness (Litchfield, 1967). Accumulation of sugars in the fruit bodies of mushrooms has been reported by several researchers (e.g., Harada et al., 2004; Barros et al., 2007; Jedidi et al., 2016). In the present study, the quantity of different sugars differs among the six mushrooms studied, thus exhibiting different tastes. It is likely *T. schimperi* is more sweetish as it has the highest quantity of sugars. Maltose was not detected in *T. schimperi* and it is quite likely broken down to glucose during the drying process as predicted by Barros et al. (2008). Edible wild mushrooms in Europe are known for trehalose as a principal carbohydrate (Kalač, 2009), so also are the commercially cultivated four mushrooms in China (Li et al., 2014). However, trehalose content ranged between 0 and 1.53 mg/g. Unlike in this study, wild *B. edulis* of Tunisia was composed of high quantity of trehalose (0.41 vs. 22.6 mg/g) (Jedidi et al., 2016) and similarly was wild *B. edulis* of Turkey also composed of high quantity of glucose (0.12 vs. 19.5 mg/g) (Turfan et al., 2018). According to Turfan et al. (2018), sugar composition of wild mushrooms is dependent on several

Table 1. Soluble sugar content (mg/g dry mass) in water and methanol extracts of six wild edible mushrooms of the Western Ghats (n = 3 ± SD; figures between water and methanol extracts significantly differed, t-test: *p < 0.01, **p < 0.001) (BDL, below detectable level).

	Extract	*Amanita hemibapha*	*Boletus edulis*	*Hygrocybe alwisii*	*Lentinus polychrous*	*Lentinus squarrosulus*	*Termitomyces schimperi*
Rhamnose	Water	BDL	BDL	BDL	BDL	BDL	0.25 ± 0.04**
	Methanol	0.79 ± 0.00**	BDL	0.38 ± 0.00**	BDL	BDL	BDL
Arabinose	Water	0.61 ± 0.11	BDL	0.25 ± 0.00	BDL	BDL	0.39 ± 0.10**
	Methanol	6.23 ± 0.12**	1.23 ± 0.05**	3.71 ± 0.20**	BDL	BDL	BDL
Fructose	Water	BDL	BDL	BDL	BDL	BDL	BDL
	Methanol	1.56 ± 0.00**	BDL	BDL	BDL	BDL	BDL
Galactose	Water	0.44 ± 0.06**	BDL	0.69 ± 0.00**	BDL	0.98 ± 0.0**	1.01 ± 0.05
	Methanol	BDL	BDL	0.18 ± 0.00	BDL	BDL	1.13 ± 0.11*
Glucose	Water	0.41 ± 0.00**	0.12 ± 0.00	0.52 ± 0.01**	BDL	BDL	10.68 ± 0.45**
	Methanol	BDL	3.42 ± 0.26**	BDL	BDL	0.96 ± 0.01**	4.35 ± 0.12
Maltose	Water	3.35 ± 0.20**	0.77 ± 0.04	0.31 ± 0.05**	BDL	BDL	BDL
	Methanol	BDL	4.85 ± 0.1**	BDL	BDL	BDL	BDL
Trehalose	Water	0.48 ± 0.01**	0.41 ± 0.00	0.13 ± 0.00**	BDL	BDL	1.53 ± 0.00**
	Methanol	BDL	0.49 ± 0.08	BDL	BDL	BDL	0.87 ± 0.05
Total	Water	5.30 ± 0.38	1.30 ± 0.04	1.90 ± 0.10	-	1.00 ± 0.00	18.13 ± 0.64**
	Methanol	8.60 ± 0.12**	10.0 ± 0.49**	4.30 ± 0.20**	-	1.00 ± 0.01	6.40 ± 0.28

factors (e.g., genetic, stage of development and pre- and post-harvest conditions). Surprisingly, in *L. polychrous* none of the sugars were found at detectable quantity as per our study.

5.1.2 *Organic Acids*

Among the six wild mushrooms, as many as eight organic acids were determined (Table 2). Dominance of organic acids was seen in aqueous as well methanol extracts. In water extract, acetic acid (*L. squarrosulus*), citric acid (*B. edulis*), lactic acid (*A. hemibapha* and *T. schimperi*) and succinic acid (*B. edulis, H. alwisii* and *L. squarrosulus*) were dominant. In methanol extract, acetic acid (*H. alwisii*), citric acid (*L. squarrosulus*), lactic acid (*B. edulis*) and succinic acid (*A. hemibapha, H. alwisii* and *L. polychrous*) were dominant. Succinic acid was dominant in aqueous as well as methanol extracts of *H. alwisii*. Overall, succinic acid was highest in

Table 2. Organic acid content (mg/g dry mass) in water and methanol extracts of six wild edible mushrooms of the Western Ghats (n = 3 ± SD; figures between water and methanol extracts significantly differed, t-test: *p < 0.01, **p < 0.001) (BDL, below detectable level).

	Extract	*Amanita hemibapha*	*Boletus edulis*	*Hygrocybe alwisii*	*Lentinus polychrous*	*Lentinus squarrosulus*	*Termitomyces schimperi*
Acetic acid	Water	BDL	BDL	BDL	2.38 ± 0.15**	3.73 ± 0.00**	BDL
	Methanol	BDL	BDL	3.86 ± 0.24**	BDL	0.39 ± 0.01**	0. 95 ± 0.34**
Ascorbic acid	Water	0.36 ± 0.11**	BDL	BDL	BDL	BDL	1.27 ± 0.42**
	Methanol	BDL	BDL	0.79 ± 0.00**	0.16 ± 0.011**	0.08 ± 0.00**	BDL
Citric acid	Water	BDL	6.79 ± 0.00**	BDL	BDL	BDL	BDL
	Methanol	BDL	BDL	0.64 ± 0.00**	BDL	3.24 ± 0.09**	BDL
Lactic acid	Water	14.24 ± 0.74**	1.94 ± 0.03	BDL	BDL	BDL	22.36 ± 0.95**
	Methanol	BDL	3.28 ± 0.65**	BDL	BDL	BDL	2.0 ± 0.08
Malic acid	Water	BDL	1.26 ± 0.03	BDL	BDL	BDL	BDL
	Methanol	BDL	1.61 ± 0.07*	BDL	BDL	BDL	BDL
Pyruvic acid	Water	BDL	BDL	1.30 ± 0.10**	BDL	2.32 ± 0.22**	BDL
	Methanol	BDL	BDL	BDL	0.14 ± 0.04**	BDL	BDL
Succinic acid	Water	BDL	8.48 ± 0.18**	24.42 ± 0.80**	BDL	3.61 ± 0.04**	BDL
	Methanol	11.67 ± 0.25**	2.36 ± 0.10	11.66 ± 0.65	10.57 ± 0.41**	0.62 ± 0.03**	BDL
Tartaric acid	Water	BDL	BDL	BDL	0.89 ± 0.02**	BDL	0.73 ± 0.05
	Methanol	BDL	0.97 ± 0.00**	0.73 ± 0.01**	0.34 ± 0.00	0.79 ± 0.13**	1.81 ± 0.00
Total	Water	14.60 ± 0.85*	18.50 ± 0.23**	25.72 ± 0.90**	3.27 ± 0.17	9.70 ± 0.26**	24.40 ± 1.42**
	Methanol	11.70 ± 0.25	8.22 ± 0.82	17.70 ± 0.91	11.21 ± 0.46**	1.80 ± 0.26	4.80 ± 0.42

H. alwisii in aqueous extract (24.4 mg/g) followed by lactic acid in aqueous extract of *T. schimperi* (22.4 mg/g) and *A. hemibapha* (14.2 mg/g). Total organic acids content in aqueous extract was dominant in *H. alwisii* followed by *T. schimperi*, while in methanol extract, it was dominant in *H. alwisii* followed by *A. hemibapha*. Unlike soluble sugars, except for *L. polychrous*, in the rest of the mushrooms the total organic acid content was higher in aqueous than methanol extract.

Similar to fruits and vegetables, organic acids in mushrooms play a vital role in offering flavour as well as antioxidant capacity (Vaughan and Giessler, 1997; Silva et al., 2004). Unlike pigments and flavour components, organic acids are not vulnerable to changes owing to processing as well as storage (Cámara et al., 1994). Organic acids are also known for their ability to improve the shelf-life of food by preventing spoilage by various bacteria (Ouattara et al., 1997; Singla et al., 2012). According to Ribeiro et al. (2007), oxalic, citric, malic, quinic and fumaric acids are common in many mushroom species. Compared to earlier studies, although *Amanita caesarea* of Portugal possesses citric and malic acids, *A. hemibapha* was devoid of these acids in our study (Valentão et al., 2005). Unlike *A. hemibapha* of Portugal, succinic and citric acids were the major organic acids in *A. hemibapha* in this study.

5.1.3 Total Free Amino Acids

Assessment of total free amino acids of wild mushrooms revealed that they were highest in quantity in *T. schimperi* followed by *B. edulis* (> 25 mg/g), *H. alwisii* and *A. hemibapha* (> 10 mg/g) (Fig. 2). They were low in *L. squarrosulus* as well as *L. polychrous* (< 5 mg/g). The total free amino acid content of *B. edulis* is higher than the same species in Turkey (26 vs. 3.6 mg/g) (Turfan et al., 2018). Further, except for *B. edulis* and *T. schimperi*, total free amino acid content is comparable to 11 species of wild mushrooms in Turkey.

Siberian boletes (*Boletus edulis, B. sipellis, B. scaber* and *B. variegatus*) is composed of 23 classes of free amino acids with dominance of arginine, asparagine, glutamine, leucine, lysine, threonine, tryptophan and valine (Dembitsky et al., 2010). Among them, *B. edulis* and *B. sipellis* were rich in free amino acids. Free amino acids are known to contribute the characteristic flavour of edible mushrooms (Yang et al., 2001). Glutamic acid in adequate quantity in mushrooms is accountable for the '*umami*' taste (meaning, palatable), which serves as a natural potentiator of flavour similar to the additive like monosodium glutamate (Yamaguchi, 1971). In this study, *B. edulis* after *T. schimperi* was dominant in total free amino acids and is thus naturally a more preferred wild mushroom by tribals.

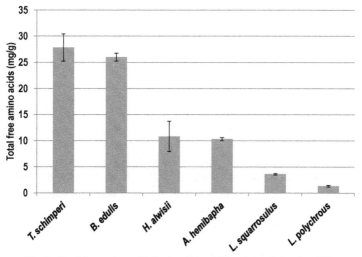

Fig. 2. Total free amino acids in six edible wild mushrooms ($n = 3 \pm$ SD).

5.1.4 Polyphenols

Among the six wild mushrooms evaluated, as many as six polyphenols were determined (Table 3). Ethyl catechol was dominant in *B. edulis* followed by *T. schimperi*, while epicatechin was dominant in *L. squarrosulus*. Overall, the ethyl catechol was highest in *B. edulis* (3.3 mg/g) followed by epicatechin in *L. squarrosulus* (3.2 mg/g) and ethyl catechol in *T. schimperi* (3.18 mg/g). Polyphenols were highest in *L. squarrosulus* followed by *B. edulis* and *T. schimperi*.

In addition to vegetables and fruits, mushrooms are a valuable source of polyphenols (Barros et al., 2009). Among the bioactive components of mushrooms, there is a growing interest in polyphenol content owing to their therapeutic potential. Several studies have correlated antioxidant potential of wild mushrooms with that of phenolic contents (Ren et al., 2014; Lin et al., 2015; Smolskaité et al., 2015).

Table 3. Polyphenols in methanol extract (mg/g dry mass) of six wild edible mushrooms of the Western Ghats (n = 3 ± SD) (BDL-below detectable level).

	Amanita hemibapha	*Boletus edulis*	*Hygrocybe alwisii*	*Lentinus polychrous*	*Lentinus squarrosulus*	*Termitomyces schimperi*
Vanillin	BDL	BDL	BDL	0.14 ± 0.01	BDL	BDL
Epicatechin	BDL	BDL	BDL	BDL	3.20 ± 0.00	BDL
Methyl catechol	1.61 ± 0.04	1.55 ± 0.03	2.32 ± 0.00	0.30 ± 0.01	1.51 ± 0.11	1.46 ± 0.02
Ethyl catechol	BDL	3.28 ± 0.17	BDL	0.77 ± 0.12	BDL	3.18 ± 0.05
Syringic acid	BDL	BDL	BDL	0.42 ± 0.00	1.54 ± 0.08	BDL
Ferulic acid	BDL	BDL	BDL	BDL	0.84 ± 0.00	BDL
Total	1.61 ± 0.04	4.83 ± 0.20	2.32 ± 0.00	1.63 ± 0.14	7.10 ± 0.19	4.64 ± 0.07

5.2 Polysaccharides

Extraction of polysaccharides in two *Lentinus* spp. varied and it was higher in *L. squarrosulus* than *L. polychrous* in aqueous as well as ethanol extraction (Fig. 3). In both mushrooms, aqueous extract was significantly higher than ethanolic extract ($p < 0.01$). Aqueous extract of cultivated *L. polychrous* of Thailand possesses crude polysaccharides in increasing order: fresh fruit bodies (5.2 per cent) < dried fruit bodies (5.7 per cent) < mycelia (5.9 per cent) (Thetsrimuang et al., 2011). In our study, the aqueous extract of whole fresh fruit bodies of *L. polychrous* yielded higher than the Thailand strain (11.2 vs. 5.2–5.9 per cent).

Crude polysaccharides are composed of mainly carbohydrates (51–62 per cent), proteins (15–28 per cent) and phenolic compounds (< 5 per cent) (Siu et al., 2014). Those polysaccharides possess high protein content and are known as PS-protein (PSP) complexes (e.g., proteoglycans and glycoproteins). In spite of precipitation of high molecular weight components in ethanol, low molecular weight phenolic compounds (3–4 per cent) persist (Siu et al., 2014). It has been predicted that owing to the powerful bonding of phenolics with polysaccharides, proteins will remain along with high molecular compounds (McManus et al., 1985).

Aqueous and ethanol extracted polysaccharides were subjected to the FTIR to follow its gross chemical composition followed by chemical analysis and assessment of antioxidant potential. To follow the gross chemical composition, crude polysaccharides isolated from two wild mushrooms (*L. squarrosulus* and

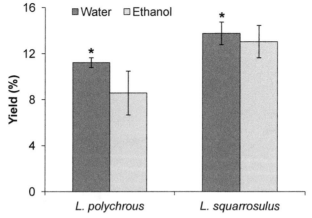

Fig. 3. Yield of polysaccharides (%) in water and ethanol of two edible wild mushrooms ($n = 3 \pm$ SD) (*t*-test: *, p < 0.01).

L. polychrous) were subjected to the FTIR analysis. The FTIR spectra of the water-soluble polysaccharide and ethanolic fractions (Figs. 4, 5) of *Lentinus polychrous* and *L. squarrosulus* showed the characteristic strong broad absorption in the range of 3,200–3,400 cm^{-1} arising from O-H stretching vibrations of hydroxyls of polysaccharides and water and asymmetric and symmetric stretching of the N-H bonds in amino groups (Salvador et al., 2015). The bands in the region of 2850–2900 cm^{-1} are due to the CH$_3$, CH$_2$ stretching vibrations of proteins and lipids. Likewise, protein patterns were also observed with characteristic absorption in the range of 1800–1500 cm^{-1}, thus indicating C=O stretching (1658–1625 cm^{-1} corresponds to amide I and also represents chitin) and N-H bending (1510–1580 cm^{-1} corresponds to amide II and chitosan) (D'Souza and Kamat, 2017). The bands at 1410–1310 cm^{-1} were ascribed to OH groups of phenolic compounds (Kozarski et al., 2012). Strong IR bands obtained in the region of 900–1200 cm^{-1} are characteristic of polysaccharides and can be attributed to coupled CO and C=C stretching and COH bending vibrations

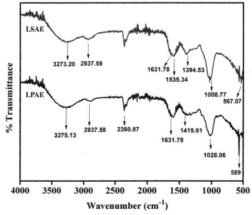

Fig. 4. The FTIR spectrum of water soluble fractions of edible wild mushroom *Lentinus squarrosulus* (upper) and *L. polychrous* (lower).

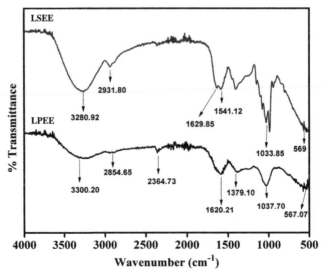

Fig. 5. The FTIR spectrum of ethanol soluble fractions of edible wild mushroom *Lentinus squarrosulus* (upper) and *L. polychrous* (lower).

(Synytsya et al., 2009; Cheung et al., 2012). All the samples had characteristic peaks, which indicate the presence of polysaccharides, proteins and phenolics.

The data obtained from FTIR spectra showed that crude polysaccharides isolated by sequential extraction by hot water and ethanol precipitation of mushrooms were mainly composed of polysaccharides, proteins and a small amount of phenolic compounds. Hence, the antioxidant activities of polysaccharides can be attributed to the presence of proteins and phenolic compounds (Siu et al., 2014). In the FTIR spectra, the presence of bands specific for proteins are results of proteins which are covalently bound to polysaccharides but also may result from protein impurities occurring in crude polysaccharide fractions (Gonzaga et al., 2004; Synytsya et al., 2009). It has been suggested that this extraction technique might not be able to remove the polysaccharide-bound proteins (Synytsya et al., 2009).

5.2.1 *Components of Polysaccharides*

Crude polysaccharides have been subjected to chemical analysis to detect total carbohydrates, total sugars, total proteins and polyphenols.

The total carbohydrate content of polysaccharides was higher in *L. polychrous* than *L. squarrosulus* (Fig. 6a). In both mushrooms, aqueous extract showed a significantly higher quantity of carbohydrates ($p < 0.05$). The total carbohydrates of *L. polychrous* fresh fruit bodies of our study are comparable to those of Thailand (47 vs. 46 mg/g) (Thetsrimuang et al., 2011). Reducing sugars of polysaccharides were higher in *L. polychrous* than in *L. squarrosulus* (Fig. 6b). The aqueous extract showed a significantly higher quantity of reducing sugars in both mushrooms ($p < 0.05$). The reducing sugar content of *L. polychrous* was higher as compared to the Thailand strain (8.5 vs. 3.9 mg/g) (Thetsrimuang et al., 2011).

The protein content in polysaccharides of *Lentinus* spp. gave varied results (Fig. 6c). It was higher in *L. squarrosulus* than *L. polychrous*. In both mushrooms,

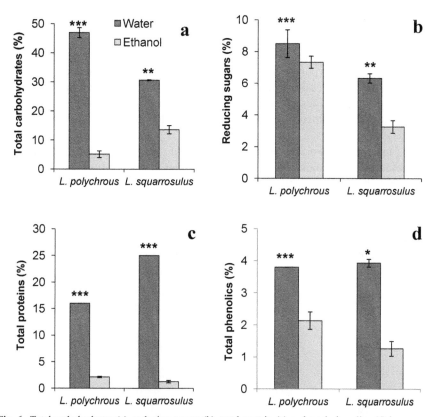

Fig. 6. Total carbohydrates (a), reducing sugars (b), total protein (c) and total phenolics (d) in aqueous and ethanol extracts of polysaccharides of two edible wild mushrooms (n = 3 ± SD) (t-test: *, p < 0.05; **, p < 0.01; ***, p < 0.001).

aqueous extract showed a significantly higher protein content than ethanol extract (p < 0.001). Total protein content of *L. polychrous* was about four-fold higher as compared to the Thailand strain (16 vs. 3.9 mg/g) (Thetsrimuang et al., 2011).

The total phenolics content of polysaccharides in *L. polychrous* and *L. squarrosulus* was almost similar (Fig. 6d). Aqueous extracts of both mushrooms possess higher quantity of total phenolics (p < 0.05). The total phenolics content of *L. polychrous* is lower than the Thailand strain (3.8 vs. 5.8 mg/g) (Thetsrimuang et al., 2011). Interestingly, total phenolic content of water-soluble carbohydrates of *Lentinus edode* grown in different media as well as time differs (Wu and Hansen, 2008).

Composition of crude polysaccharides showed a major difference between aqueous and ethanol extracts. The crude polysaccharides of aqueous extract of *L. squarrosulus* and *L. polychrous* are composed of the highest quantity of carbohydrates (31–47 per cent) followed by proteins (16–25 per cent) and reducing sugars (6.3–8.5 per cent). Ethanol extract of crude polysaccharides showed the highest quantity of carbohydrates (5.1–13.6 per cent) and reducing sugars (3.3–7.3 per cent) while almost the same quantities of proteins and total phenolics (1.3–2.1 per cent).

5.2.2 *Antioxidant Potential of Polysaccharides*

As the crude polysaccharides of *Lentinus* spp. are known for their antioxidant activity, their ferrous ion-chelating capacity (FCC), DPPH radical-scavenging activity (DPPH) and reducing power (RP) were evaluated. Polysaccharides of mushrooms are known for several bioactivities like, anticancer, anticoagulant, antimicrobial and immunological properties. Thus, polysaccharides of mushroom origin have been an attraction for commercial pharmaceutics (e.g., lentinan from *Ganoderma* and *Lentinus*).

In aqueous as well as ethanol extracts of polysaccharides, the FCC of polysaccharides was higher in *L. squarrosulus* than *L. polychrous* (Fig. 7a). It was significantly higher in aqueous than ethanol extract in *L. squarrosulus*. Unlike FCC, the DPPH of polysaccharides was higher in *L. polychrous* than *L. squarrosulus*

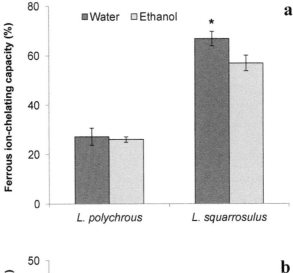

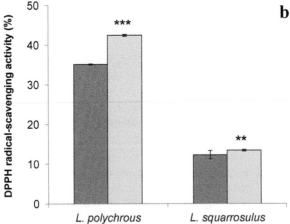

Fig. 7. Ferrous ion-chelating capacity (a) and DPPH radical-scavenging activity (b) in aqueous and ethanol extracts of polysaccharides of two edible wild mushrooms ($n = 3 \pm$ SD) (*t*-test: *, p < 0.05; **, p < 0.01; ***, p < 0.001).

(Fig. 7b). In both mushrooms, ethanol extract was significantly higher than the aqueous extract (p < 0.05).

The RP of polysaccharides of *L. polychrous* and *L. squarrosulus* was higher in aqueous extract than in the ethanol extract (p < 0.001) (Fig. 8a, b). The overall reducing power is higher in *L. polychrous* than *L. squarrosulus* and reducing power showed dose-dependence in both mushrooms.

The radical-scavenging and reducing power of crude polysaccharides of Thailand *L. polychrous* in increasing order is as follows: fresh fruit bodies < dried fruit bodies < mycelia (Thetsrimuang et al., 2011). The radical-scavenging activities of aqueous as well as methanol extracts of Thailand strain of *Lentinus edode* showed variability depending on the time of harvest (Wu and Hensen, 2008). The antioxidant activity of crude polysaccharides of three mushrooms of China showed significant correlation with total phenolic content but a moderate correlation with protein content (Siu et al., 2014). The present study corroborates with Siu et al. (2014), showing the relationship with antioxidant activity of *L. squarrosulus* and *L. polychrous* with phenolics as well as protein content of crude polysaccharides.

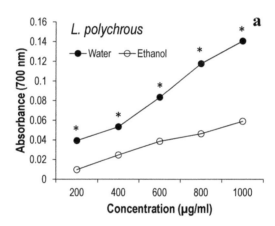

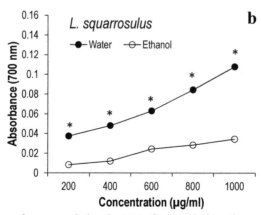

Fig. 8. Reducing power of aqueous and ethanol extracts of polysaccharides of two edible wild mushrooms (*n* = 3 ± SD) (*t*-test: *, p < 0.001).

6. Conclusion

Mushrooms constitute the potential alternative source of food, medicine and industrial resource similar to plants and animals. Besides cultivated mushrooms, wild mushrooms are a centre of attraction for production of functional foods and health products owing to their versatile nutritional and pharmaceutical properties. The present study offers bioactive attributes of six wild mushrooms of the Western Ghats composed of ectomycorrhizal, wood-preferring, leaf litter-preferring and termite mound-preferring species. They are the potential source of total free amino acids, soluble sugars, organic acids and polyphenols. Mushroom polysaccharides have a potential role in anticancer, immune modulation, antioxidant, radical scavenging and cytotoxic effects. The FTIR spectra of crude polysaccharides of two *Lentinus* spp. extracted by hot water and ethanol revealed they were composed of a high quantity of carbohydrates followed by proteins and a small amount of phenolics. Extracted polysaccharides also showed potential antioxidant properties. As several valuable wild mushrooms are available in bulk quantity, their methods of harvest, processing, preservation and trade need to be improved. Systematic survey of wild mushrooms in the foothills of the Western Ghats with the help of ethnic knowledge of tribals provides further insight into the new taxa of nutritional, medicinal and industrial significance of mushrooms.

References

Abdalla, A.E., Tirzite, D., Tirzitis, G. and Roozen, J.P. (1999). Antioxidant activity of 1,4-dihydropyridine derivatives in-carotenemethyl linoleate, sunflower oil and emulsions. Food Chem., 66: 189–195.

Afshari, K., Samavati, V. and Shahidi, S.A. (2015). Ultrasonic-assisted extraction and *in vitro* antioxidant activity of polysaccharide from Hibiscus leaf. Int. J. Biol. Macromol., 74: 558–567.

Barros, L., Baptista, P., Correia, D.M. et al. (2007). Fatty acid and sugar compositions and nutritional value of five wild edible mushrooms from northeast Portugal. Food Chem., 105: 140–145.

Barros, L., Cruz, T., Baptista, P. et al. (2008). Wild and commercial mushrooms as source of nutrients and nutraceuticals. Food Chem. Toxicol., 46: 2742–2747.

Barros, L., Dueñas, M. and Ferreira, I.C.F.R. (2009). Phenolic acids determination by HPLC-DAD-ESI/MS in sixteen different Portuguese wild mushrooms species. Food Chem. Toxicol., 47: 1076–1079.

Barros, L., Pereira, C. and Ferreira, I.C.F.R. (2013). Optimised analysis of organic acids in edible mushrooms from Portugal by ultra fast liquid chromatography and photodiode array detection. Food Anal. Meth., 6: 309–316.

Cámara, M., Díez, C., Torija, M.E. and Cano, M.P. (1994). HPLC determination of organic acids in pineapple juices and nectars. Z. Lebensm.-Unters.-Forsch., 198: 52–56.

Chang, S.T. and Wasser, S.P. (2012). The role of culinary-medicinal mushrooms on human welfare with a pyramid model for human health. Int. J. Med. Mushrooms, 14: 95–134.

Cheng, J.J., Lin, C.Y., Lur, H.S. et al. (2008). Properties and biological functions of polysaccharides and ethanolic extracts isolated from medicinal fungus, *Fomitopsis pinicola*. Proc. Biochem., 43: 829–834.

Cheung, P.C.K. (2010). The nutritional and health benefits of mushrooms. Nutr. Bull., 35: 292–299.

Cheung, Y.C., Siu, K.C., Liu, Y.S. and Wu, J.Y. (2012). Molecular properties and antioxidant activities of polysaccharide-protein complexes from selected mushrooms by ultrasound-assisted extraction. Proc. Biochem., 47: 892–895.

Dasgupta, A., Dutta, A.K., Halder, A. and Acharya, K. (2015). Mycochemicals, phenolic profile and antioxidative activity of a wild edible mushroom from Eastern Himalaya. J. Biol. Active Prod. Nat., 5: 373–382.

Dembitsky, V.M., Alexander, Terent'ev, O. and Levitsky, D.O. (2010). Amino and fatty acids of wild edible mushrooms of the genus *Boletus*. Rec. Nat. Prod., 4: 218–223.

D'Souza, R.A. and Kamat, N.M. (2017). Potential of FTIR spectroscopy in chemical characterisation of termitomyces pellets. J. Appl. Biol. Biotechnol., 5: 80–84.

Dubois, M., Gilles, K.A., Hamilton, J.K. et al. (1956). Colorimetric method for determination of sugars and related substances. Anal. Chem., 28: 350–356.

Gonzaga, M., Ricardo, N., Heatley, F. and Soares. S. (2004). Isolation and characterisation of polysaccharides from *Agaricus blazei* Murill. Carbohyd. Polym., 60: 43–49.

Harada, A., Gisusi, S., Yoneyama, S. and Aoyama, M. (2004). Effects of strain and cultivation medium on the chemical composition of the taste components in fruit-body of *Hypsizygus marmoreus*. Food Chem., 84: 265–270.

Hobbs, C. (2000). Medicinal values of *Lentinus edodes* (Berk.) Sing. (Agaricomycetideae). A literature review. Int. J. Med. Mushrooms, 2: 287–297.

Hsu, C.L., Chen, W., Weng, Y.M. and Tseng, C.Y. (2003). Chemical composition, physical properties, and antioxidant activities of yam flours as affected by different drying methods. Food Chem., 83: 85–92.

Hu, J., Pang, W., Chen, J. et al. (2013). Hypoglycemic effect of polysaccharides with different molecular weight of *Pseudostellaria heterophylla*. BMC Complement, Altern. Med., 13: 267–276.

Jedidi, I.K., Ayoub, I.K., Philipe, T. and Bouzouita, N. (2016). Chemical composition and non-volatile components of three wild edible mushrooms collected from northwest Tunisia. Med. J. Chem., 5: 434–441.

Kalac, P. (2009). Chemical composition and nutritional value of European species of wild growing mushrooms: A review. Food Chem., 113: 9–16.

Kozarski, M., Klaus, A., Nikšić, M. et al. (2012). Antioxidative activities and chemical characterisation of polysaccharide extracts from the widely used mushrooms *Ganoderma applanatum*, *Ganoderma lucidum, Lentinus edodes* and *Trametes versicolor*. J. Food Comp. Anal., 26: 144–153.

Leal, A.R., Barros, L., Barreira, J.C.M. et al. (2013). Portuguese wild mushrooms at the 'pharma-nutrition' interface: Nutritional characterisation and antioxidant properties. Food Res. Int., 50: 1–9.

Lee, I.-K., Kim, Y.-S., Jang, Y.-W. et al. (2007). New antioxidant polyphenols from the medicinal mushrooms *Inonotus obliquus*. Bioorg. Med. Chem. Lett., 17: 6678–6681.

Li, W., Gu, Z., Yang, Y. et al. (2014). Non-volatile taste components of several cultivated mushrooms. Food Chem., 143: 427–431.

Lin, S., Ching, L.T., Chen, J. and Cheung, P.C.K. (2015). Antioxidant and anti-angiogenic effects of mushroom phenolics-rich fractions. J. Func. Foods, 17: 802–815.

Litchfield, J.H. (1967). Morel mushroom mycelium as a food-flavouring material. Biotech. Bioeng., 9: 289–304.

Lowry, O.H., Rosebrough, N.J., Farr, A.L. and Randall, R.J. (1951). Protein measurement with the Folin phenol reagent. J. Biol. Chem., 193: 265–275.

Luo, Q., Zhang, J., Yan, L. et al. (2011). Composition and antioxidant activity of water-soluble polysaccharides from *Tuber indicum*. J. Med. Food., 14: 1609–1616.

McManus, J.P., Davis, K.G., Beart, J.E. et al. (1985). Polyphenol interactions. Part 1. Introduction; Some observations on the reversible complexation of polyphenols with proteins and polysaccharides. J. Chem. Soc. Perkin Trans., 2: 1429–1438.

Miller, G.L. (1959). Use of dinitrosalicylic acid reagent for determination of reducing sugar. Anal. Chem., 31: 426–428.

Ooi, V.E.C. and Liu, F. (2000). Immunomodulation and anti-cancer activity of polysaccharide-protein complexes. Curr. Med. Chem., 7: 715–729.

Ouattara, B.O., Simard, R.E., Holley, R.A. et al. (1997). Inhibitory effect of organic acids upon meat spoilage bacteria. Food Prot., 60: 246–253.

Oyaizu, M. (1986). Studies on products of browning reaction. Jap. J. Nutr. Diet., 44: 307–315.

Pereira, C., Barros, L., Carvalho, A.M. and Ferreira, I.C. (2013). Use of UFLC-PDA for the analysis of organic acids in thirty-five species of food and medicinal plants. Food Anal. Meth., 6: 1337–1344.

Reis, F.S., Barros, L., Martins, A. and Ferreira, I.C. (2012). Chemical composition and nutritional value of the most widely appreciated cultivated mushrooms: An inter-species comparative study. Food Chem. Toxicol., 50: 191–197.

Ren, L., Hemar, Y., Perera, C.O. et al. (2014). Antibacterial and antioxidant activities of aqueous extracts of eight mushrooms. Bioact. Carbohy. Diet. Fibre, 3: 41–51.

Reshetnikov, S.V., Wasser, S.P. and Tan, K.K. (2001). Higher Basidiomycetes as a source of antitumour and immunostimulating polysaccharides. Int. J. Med. Mushrooms, 3: 361–394.

Ribeiro, B., Valentao, P., Baptista, P. et al. (2007). Phenolic compounds, organic acids profiles and antioxidative properties of beefsteak fungus (*Fistulina hepatica*). Food Chem. Toxicol., 45: 1805–1813.

Rosset, J., Bärlocher, F. and Oertli, J.J. (1982). Decomposition of conifer needles and deciduous leaves in two Black Forest and two Swiss Jura streams. Int. Rev. Hydrobiol., 67: 695–711.

Sadasivam, S. and Manickam, A. (1992). Biochemical Methods for Agricultural Sciences. Wiley Eastern Ltd. and Tamil Nadu Agricultural University, Coimbatore, India.

Salvador, C., Martins, M.R. and Caldeira, A.T. (2015). Microanalysis characterisation of bioactive protein-bound polysaccharides produced by *Amanita ponderosa* cultures. Microsc. Microanal., 21: 84–90.

Silva, B.M., Andrade, P.B., Valentão, P. et al. (2004). Quince (*Cydonia oblonga* Miller) fruit (pulp, peel, and seed) and jam: Antioxidant activity. J. Agric. Food Chem., 52: 4705–4712.

Singh, R.P., Murthy, C.K.N. and Jayaprakasha, G.K. (2002). Studies on antioxidant activity of pomegranate (*Punica granatum*) peel and seed extracts using *in vitro* methods. J. Agric. Food Chem., 50: 81–86.

Singla, R., Ganguli, A. and Ghose, M. (2012). Physicochemical and nutritional characteristics of organic acid-treated button mushrooms (*Agaricus bisporous*). Food Bioproc. Technol., 5: 808–815.

Siu, K.-C., Chen, X. and Wu, J.-Y. (2014). Constituents actually responsible for the antioxidant activities of crude polysaccharides isolated from mushrooms. J. Func. Foods, 11: 448–556.

Smolskaité, L., Venskutonis, P.R. and Talou, T. (2015). Comprehensive evaluation of antioxidant and antimicrobial properties of different mushroom species. LWT Food Sci. Technol., 60: 462–471.

Stachowiak, B. and Reguła, J. (2012). Health-promoting potential of edible macromycetes under special consideration of polysaccharides: A review. Eur. Food Res. Technol., 234: 369–380.

Stajić, M., Vukojević, J., Knežević, A. et al. (2013). Antioxidant protective effects of mushroom metabolites. Curr. Top. Med. Chem., 13: 2660–2676.

Sun, Y. (2014). Biological activities and potential health benefits of polysaccharides from *Poria cocos* and their derivatives. Int. J. Biol. Macromol., 68: 131–134.

Synytsya, A., Mičková, K., Synytsya, A. et al. (2009). Glucans from fruit bodies of cultivated mushrooms *Pleurotus ostreatus* and *Pleurotus eryngii*: Structure and potential prebiotic activity. Carbohyd. Polym., 76: 548–556.

Thetsrimuang, C., Khammuang, S., Chiablaem, K. et al. (2011). Antioxidant properties and cytotoxicity of crude polysaccharides from *Lentinus polychrous* Lév. Food Chem., 128: 634–639.

Turfan, N., Pekşen, A., Kibar, B. and Ünal, S. (2018). Determination of nutritional and bioactive properties in some selected wild growing and cultivated mushrooms from Turkey. Acta Sci. Pol. Hor. Cul., 17: 27–72.

Valentão, P., Lopes, G., Valente, M. et al. (2005). Quantitation of nine organic acids in wild mushrooms. J. Agric. Food Chem., 53: 3626–3630.

Vaughan, J.G. and Geissler, C.A. (1997). The New Oxford Book of Food Plants. Oxford University Press, New York.

Wang, Q., Wang, F., Xu, Z. and Ding, Z. (2017). Bioactive mushroom polysaccharides: A review on monosaccharide composition, biosynthesis and regulation. Molecules, 22: 222. 10.3390/molecules22060955.

Wasser, S.P. and Weis, A.L. (1999). Medicinal properties of substances occurring in higher basidiomycetes mushroom: current perspective. Int. J. Med. Mushrooms, 1: 31–51.

Wu, X.J. and Hansen, C. (2008). Antioxidant capacity, phenolic content, and polysaccharide content of *Lentinus edodes* grown in whey permeate-based submerged culture. Food Microbiol. Safe., 73: M1–M8.

Xu, W., Zhang, F., Luo, Y. et al. (2009). Antioxidant activity of a water-soluble polysaccharide purified from *Pteridium aquilinum*. Carbohyd. Res., 344: 217–222.

Yamaguchi, S., Yoshikawa, T., Ikeda, S. and Ninomiya, T. (1971). Measurement of the relative taste intensity of some a-amino acids and 50-nucleotides. J. Food Sci., 36: 846–849.

Yang, J.H., Lin, H.C. and Mau, J.L. (2001). Non-volatile taste components of several speciality mushrooms. Food Chem., 72: 465–471.

Zhang, M., Cui, S.W., Cheung, P.C.K. and Wang, Q. (2007). Antitumor polysaccharides from mushrooms: A review on their isolation process, structural characteristics and antitumor activity. Tr. Food Sci. Technol., 18: 4–19.

Healing Properties of Edible Mushrooms

Dorota Hilszczańska

1. INTRODUCTION

Mushrooms are cultivated and consumed by humans since ages because of their sensory characteristics and their attractive culinary values. Taxonomically they belong to classes Basidiomycetes and Ascomycetes, but the majority belong to the former. The knowledge about the real number of macrofungi is still in debate. Hawksworth (2001) reports that the number ranges between 14,000–22,000. It is estimated that 2,000 species could be eaten without harm to health, while 700 species have scientifically proven medical properties (Wasser, 2002). The most consumed mushrooms worldwide are *Agaricus bisporus*, *Pleurotus* spp. and *Lentinus edodes* (Roncero-Ramos and Delgado-Andrade, 2017). Mushrooms contain bioactive compounds of high medicinal value, such as lectins, polysaccharides, phenolics/polyphenolics, terpenoids, ergosterols and volatile organic compounds. These compounds are considered responsible agents for health protection activities, such as antitumor, immunomodulating, antioxidant, radical scavenging, antihypercholesterolemia, antiviral, antibacterial, hepatoprotective and antidiabetic effects (Wasser, 2014). According to numerous studies, different mushroom species are beneficial for the prevention and treatment of several chronic diseases, such as cancer, cardiovascular diseases, diabetes mellitus and neurodegenerative diseases (Guillamon et al., 2010; Meng et al., 2016; Kundakovic and Kolundzic, 2013; Sabaratnam et al., 2013). Based on nutritional values, mushrooms are valuable health foods. They have a significant amount of dietary fibre and are poor in calories and fat. Also a good protein content (20–30 per cent of dry matter) provides a nutritionally

Forest Ecology Department, Forest Research Institute in Sękocin Stary, Braci Leśnej 3, 05-090 Raszyn, Poland.
Email: d.hilszczanska@ibles.waw.pl

significant amount of vitamins and trace minerals (Thatoi and Singdevsachan, 2014). History of using many different types of fungal extracts demonstrating immunostimulatory, anti-inflammatory and anticancer activities in folk medicine dates back to ancient Japan, China and other countries from the Far East (Hobbs, 1995; Wasser and Weis, 1999). In some countries (USA and Israel), scientific achievements from these regions are used to implement complementary therapeutic methods (Wasser, 2002).

In the last few decades, mushrooms have become increasingly attractive as functional foods for their potential beneficial effects on human health (Roupas et al., 2012). The potential uses of the mushrooms have appeared as nutraceutical, nutritional therapy, phytonutrients, phytotherapy, and pharmaceutical owing to the accumulated number of secondary metabolites (Chaturvedi et al., 2018). Biologically active compounds, such as polysaccharides (beta 1-3, 1-4, 1-6 glucans, hetero-beta glucans, proteo-glucans), krestin, lentinan, coriolan, schyzophillan, sesquiterpenes, quinones, hydrophobins, galectins, sterols, ergothionin, tri-teripenes, sterols, germanium, nucleotide, drosophilin, armillasin, amphalone, eloporoside and volatile (skatole) were found in medicinal mushrooms (Lindequist et al., 2005). Moreover, mushrooms produce extracellular proteolytic enzymes which possess fibrinolytic and thrombolytic activities (El-Batal et al., 2015).

In highly valuated fungi of *Tuber* genus (truffles) due to their unique flavour and high nutritional value, some bioactive components, such as phenolics, terpenoids, polysaccharides, anandamide, fatty acids and ergosterols have been identified. A significant amount of unsaturated fatty acid (UFA), viz. oleic acid and linoleic acid (Yan et al., 2017; Angelini et al., 2015) exhibited potential reducing cholesterol levels, preventing cardiovascular and inhibiting cancer progression (Pacheco et al., 2008; Puiggrós et al., 2002). In terms of their high UFA content, as well as lipids, proteins, phenols, saccharides, total sterols, ergosterol and flavonoids (Hilszczańska et al., 2016), truffles might become a potential medicinal food (Lee et al., 2020). In this context, methods which enhance their shelf-life whilst maintaining their bioactive compounds and causing no harmful effects are promising (Tejedor-Calvo et al., 2019).

This review contains the present state of knowledge about the valuable compounds associated with selected edible mushrooms towards protecting human health.

2. Anticancer Compounds

Compounds which are able to interact with the immune system in order to upregulate or downregulate specific aspects of the host response can be classified as immunomodulators or biologic response modifiers. Many immonumodulators isolated from macrofungi are related to polysaccharides and their peptide or protein derivatives. So far, more than 50 species have shown immunomodulatory and antitumor effects *in vitro* and *in vivo* and also in human cancers. The biologically-active substances include lectins, polysaccharides, polysaccharides-peptides, polysaccharide-protein complexes, like lentinan, schizophyllan, polysaccharide-K, polysaccharide-P, active hexose correlated compounds (AHCC) and maitake D fraction. For example, species such as *Sparassis crispa, Pleurotus tuberregium, P. rhinoceros, Lentinus edodes, Grifola frondosa* and *Flammulina velutipes* are rich in compounds associated with the treatment of various cancers including breast, colorectal, cervical, skin, liver, ovarian, bladder, prostate, gastric, skin, lung, leukemia and stomach cancers (Wasser, 2002; Hilszczańska, 2012). Polysaccharides are carbohydrate polymers found abundantly in fruit bodies, mycelial mass and pure cultures of macrofungi. They have different chemical compositions including β-glucans, hetero-β-glucans, heteroglycans, α-manno-β-glucan complexes and they have been considered as major sources of therapeutic agents for immunomodulatory and antitumor properties (Gao and Zhou, 2002; Mizuno, 2002). Some immunomodulating and antitumor polysaccharides isolated from macrofungi are shown in Table 1.

Table 1. Some immunomodulating and antitumor polysaccharides from fungi.

Polysaccharides	Fungi	Reference
(1→6)-ß-glucan	*Agaricus blazei, A. brasiliensis* and *Lyophyllum decastes*	Kobayashi et al., 2005; Camelini et al., 2005 Angeli et al., 2006, 2009; Ukawa et al., 2000
(1→3)-ß-glucan	*Collybia dryophila, Cordyceps sinensis, Flammulina velutipes, Grifola frondosa, Heiricium erinaceus, Lentinus edodes, Pleurotus ostreatus* and *Sparassis crispa*	Pacheco-Sanchez et al., 2006 Yalin et al., 2005 Smiderle et al., 2006 Kodama et al., 2002; Poucheret et al., 2006 Dong et al., 2006 Surenjav et al., 2005; Yahaya et al., 2014 Carbonero et al., 2006; Jesenak et al., 2014 Ohno et al., 2000
ß-glucan	*Agaricus blazei* and *Volvariellaa volvacea*	Hetland et al., 2020 Kishida et al., 1989; Das et al., 2010
Mannogalactoglucan	*Agaricus blazei, Pleurotus cornucopiae* and *P. pulmonarius*	Cho et al., 1999; Biedroń et al., 2012 Gutierrez et al., 1996
Xsylo-glucan	*Pleurotus pulmonarius*	Gutierrez et al., 1996; Hereher et al., 2018
Galacto-xyloglucan	*Hericium erinaceus*	Mizuno et al., 1992; Khan et al., 2013
Heterogalactan	*Agaricus bisporus, A. blazei, Flammulina velutipes, Pleurotus eryngii* and *P. ostreatus*	Shida and Sakai, 2004; Smiderle et al., 2008
Fucogalactan	*Hericium erinaceus*	Shida and Sakai, 2004
Glucogalactan	*Tricholoma lobayense*	Chen et al., 2017; Mehmood et al., 2019
Galactomannan	*Collybia maculata*	Lim et al., 2005
Mannogalactofucan	*Grifola frondosa*	Zhuang et al., 1994
(1→6)-ß-D-glucan–protein	*Agaricus blazei*	Hong and Choi, 2007
α-glucan–protein	*Tricholoma matsutake*	Hoshi et al., 2005
Polysaccharide–protein	*Tricholoma lobayense*	Liu et al., 1996
Heteroglican–protein	*Grifora frondosa* and *Pleurotus sajor-caju*	Zhuang et al., 1993, 1994; Friedman, 2015

Mushroom polysaccharides can inhibit tumour growth by stimulating the immune system via effects on the natural killer cells, macrophages via T cells and their cytokine production (Meng et al., 2016) . The study by Li et al. (2014a) has shown that after consuming *Pleurotus eryngii* extract for eight weeks, the immune system was enhanced through Th1 phenotype potentiation as the macrophage-IL-12-IFN-g pathway leading to the activation of the cell-mediated immune system as exemplified by upregulation of natural killer cell activity. Dai et al. (2015) have also found that a regular intake of *Lentinus edodes* resulted in improved human immunity function by increased cell proliferation and activation and the production of higher levels of secretory immunoglobulin A.

Among the polysaccharides, β-glucans are the most important polysaccharides with immunomodulating and antitumor activity. For example, promising effects using grifolan, a b-glucan extracted from *Grifola frondosa*, have been reported in gastrointestinal, lung, liver and breast cancers (Poucheret et al., 2006). Grifolan is a macrophage activator, which augments cytokine production without dependence on endotoxins (Adachi et al., 1994) and also enhances mRNA level of IL-6, IL-1 and TNF-α of macrophages (Ooi and Liu, 1999).

Zhao et al. (2014) revealed that 52 polysaccharides isolated from fruit bodies of *Tuber* species (*Tuber aestivum, T. indicum, T. melanosporum*, and *T. sinense*) showed antitumor activities against A549, HCT-116, HepG2, HL-60, and SK-BR-3 cell lines. So also the oleic acid content in truffles contributed to anticancer activities. Oleic acid could suppress the over-expression of HER2, an oncogene correlated to the invasion and metastasis of cancer cells. Moreover, they could increase the intracellular ROS production and caspase 3 activity and consequently induce the cancer cell apoptosis (Carrillo et al., 2012).

In the fruit bodies of *T. melanosporum*, metabolic enzymes, such as *N*-acylphosphatidylethanolamine-specific phospholipase D (NAPE-PLD), fatty acidamide hydrolase (FAAH), diacylglycerol lipase (DAGL), monoacylglycerol lipase (MAGL) and anandamide were found. All of them showed anticancer activity (Pacioni et al., 2015).

3. Anti-metabolic Syndrome Compounds

Metabolic syndrome is a medical term for combination of diabetes, high blood pressure (hypertension) and obesity. Some compounds isolated from edible mushrooms have hypoglycemic, cholesterol- and triglyceride-lowering abilities and hypotensive effects. The most active compounds are ß-glucans, lectins and eritadenine, triterpenes, sterols and phenolic compounds (Kundakovic and Kolundzic, 2013).

3.1 Diabetes

A great number of studies, using control and diabetic animals have shown hypoglycemic effects of mushrooms and mushroom components. The effects occur to be mediated by polysaccharides by a direct interaction with insulin receptors on target tissues, although this mechanism remains still unknown (Roupas et al., 2012). The strong hypoglycemic activity has been associated with the bioactive components isolated from the fruit bodies of *Pleurotus* spp. (Roncero-Ramos and Delgado-Andrade, 2017). Li et al. (2014b) found that oral administration of *P. eryngii* extracts reduced the blood-glycated hemoglobin and serum glucose levels in alloxan-induced hyperglycemic mice. According to the authors, extracts from *P. eryngii* may become a new potential hypoglycemic food for people suffering from hyperglycemia. Studies with diabetic rats have also revealed a potent antidiabetic effect in aqueous extract of *P. pulmonarius* (Badole et al., 2008), *P. sajor-caju* (Ng et al., 2015) and polysaccharides from *P. citrinopileatus* (Hu et al., 2006). These authors suggest that these mushrooms have protective effects on cells of the islets of Langerhans, which is responsible for insulin secretion. Clinical investigation in diabetic patients has shown that *Pleurotus* consumption significantly reduced systolic and diastolic blood pressure, lowered total cholesterol, triglycerides and plasma glucose without any significant change in body weight as well as no deleterious effects on the functioning of liver or kidney (Khatun et al., 2007).

Anti-hyperglycemic properties have also been found in *Agaricus* spp. (Jeong et al., 2010; Mascaro et al., 2014; Niwa et al., 2011). The decrease in plasma glucose, cholesterol and triglyceride concentrations was observed after *A. bisporus* consumption by rats (Jeong et al., 2010). Results obtained by Niwa et al. (2011) suggest that *A. Blazei* has potential beneficial effect in the control of type 1 diabetes by reducing blood glucose, cholesterol and triglyceride levels, and increasing HDL cholesterol. Significant decrease in the serum glucose level as well as insulin level has also been demonstrated after the administration of an aqueous extract of *Hericium erinaceus* (100 and 200 mg/kg body weight) to diabetic rats for 28 days (Liang et al., 2013).

Aqueous extracts of various mushrooms possess hypoglycemic activity and antihyperglycemic activity against diabetes-inducing compounds in obese and diabetic animal models. For instance, an aqueous extract of *G. Lucidum* (0.03 and 0.3 g/kg) decreased the level of serum glucose in diabetic mice through suppression of hepatic PEPCK gene expression (Seto et al., 2009). In addition, aqueous extracts of *Pleurotus pulmonarius* have been found to be hypoglycemic together with having synergistic anti-hyperglycemic effects with acarbose (Badole et al., 2008).

An alpha-glucan isolated from *G. frondosa* affected a series of diabetes markers in KK-Ay mice, which may be linked with an effect on insulin receptors via increasing insulin sensitivity in peripheral target tissues (Lei et al., 2007). Beta-glucans from *A. blazei* possess anti-hyperglycemic and anti-arteriosclerotic activities showing generic anti-diabetic activity in diabetic rats (Kim et al., 2005). Extracellular polysaccharides (EPS) obtained from *Laetiporus sulphureus* stimulated insulin secretion (Hwang et al., 2008) as well as insulin sensitivity possibly by regulation of lipid metabolism (Cho et al., 2007) in the diabetic mouse. In turn, a polysaccharide isolated from *P. linteus* inhibited the development of autoimmune diabetes in non-obese diabetic mice (Kim et al., 2010).

3.2 Hypertension

For treating hypertension, edible mushrooms, such as *A. bisporus, G. frondosa, Hypsizygus marmoreus, Lentinula edodes* and genus *Pleurotus* have been considered as an excellent food alternative (Yahaya et al., 2014; Lau et al., 2014). In the last few decades, researchers have become interested in replacing synthetic antihypertensive drugs by natural sources of these compounds obtained from mushrooms. Aqueous extracts of these mushrooms contain active antihypertensive constituents, for example, peptides, D-mannitol, D-glucose, D-galactose, D-mannose, triterpenes and potassium. Hence, these different components could prevent and ameliorate hypertension based on various mechanisms, predominantly through inhibition of renin-angiotensin-aldosterone system (RAS) by interaction at the active site of the angiotension-converting enzyme (ACE) (Yahaya et al., 2014). Lau et al. (2014) identified two bioactive peptides from *A. bisporus* having the amino acid sequences Ala-His-Glu-Pro-Val-Lys and Arg-Ile-Gly-Leu-Phe. Both peptides exhibited potentially high ACE inhibitory activity even after *in vitro* gastrointestinal digestion. A study conducted by Khatun et al. (2007) showed that consumption of *Pleourotus* spp. by 89 diabetic patients reduced the systolic and diastolic blood pressure, total cholesterol and triglycerides. Another study carried out in this field demonstrated that *P. pulmonarius* mycelium contains proteins with potential antihypertensive effect via the angiotensin-converting enzyme inhibitory activity (Ibadallah et al., 2015). Kang et al. (2013) showed that the water-extract containing ACE inhibitor from *H. marmoreus* possesses clear antihypertensive effect on a spontaneously hypertensive rat. Reduction in blood pressure was also observed in Zucker fatty rats following oral administration of *Grifola frondosa* mushroom fractions (Talpur et al., 2002).

The earlier study with rats conducted by Kabir and Kumura (1989) revealed that mushrooms of *Lentinus* genus can lower the blood pressure and the free cholesterol level in plasma and can accelerate the accumulation of lipids in the liver by removing them from circulation.

3.3 Obesity

The positive effects of edible mushrooms and their polysaccharides on the intestinal microbiota (Delzenne et al., 2015; Chang et al., 2015) is highly associated with obesity, which is at present a dynamic area of research (Murphy et al., 2015). A study on mice showed that the extracts of *G. lucidum* help regulating dysbiosis and increase antiobesity effects (Chang et al., 2015). Huang et al. (2014) found out that the polysaccharides obtained from *Pleurotus tuber-regium* showed antihyperglycemic and antihyperlipidemic potential, and reduced the oxidative stress in obese diabetic rats. The study carried out with *L. edodes* showed that its compounds possess antihyperlipidemic activity and prevent gain in body weight. Rats fed with a high fat diet enriched with *L. edodes* significantly lowered the plasma triacylglycerol (TAG) and fat deposition compared to rats fed with high fat diet without the *L. edodes* (Handayani et al., 2011). The genus *Pleurotus* has also been pointed out for its effects on preventing weight gain and hyperlipidemia. Beta glucans from *Pleurotus sajor-caju* prevent the development of obesity and oxidative stress in mice fed on a high fat diet owing to their capacity to induce lipolysis and inhibit adipocyte differentiation (Kanagasabapathy et al., 2013). Chen et al. (2013) have indicated the hypolipidemic effects of polysaccharides from *Pleurotus eryngii* and which were able to decrease the lipid level in a high-fat-loaded mouse model. Rats fed with different concentrations of *P. eryngii* cellulose during six weeks had an enhanced antioxidant capacity of hepatic tissue, improved hepatic lipase activity and reduced hepatic fat deposition, thereby playing a role in hepato-protection and lowering the lipid levels (Huang et al., 2016).

In humans, long-term (one year) and short-term (four days) clinical studies with obese or diabetic participants showed that consumers of *Agaricus bisporus* in a diet had lesser BMI, decreased belly circumference and increased satiety without diminishing palatability (Cheskin et al., 2008; Poddar et al., 2013). The authors of this study concluded that the consumption of *A. bisporus* has potential as an antidiabetic and antiobesity. Similarly, this research has been further extended to include other extremely health-promoting mushroom species, such as *Hericium erinaceus* and *Lentinus edodes* (Kim et al., 2013; Friedman et al., 2015).

3.4 *Other Health Activities*

Mushrooms and their components possess antioxidant, antiinflammatory, hepatoprotective, antiallergic, antimicrobial and antiviral properties (Roupas et al., 2012; Zhang et al., 2016). In edible mushrooms, a whole range of antioxidant compounds, such as phenolics, polysaccharides, tocopherols, ergothioneine, carotenoids and ascorbic acid are abundantly found. Their antioxidant activity is mainly associated with the ability to lipid peroxidation inhibition as well as increase of antioxidant enzymes activity (Kozarski et al., 2015). Several aqueous extracts of mushrooms have been characterised as hepatoprotectors and probably their potency against liver damage is due to their antioxidant activity (Zhang et al., 2012). In addition, endopolysaccharides extracted from *H. erinaceus* seem to exert a considerable effect in the prevention of hepatic diseases. Soares et al. (2013) demonstrated that an aqueous extract of *A. blazei* protects against injury in hepatic tissue induced by paracetamol. Some mushrooms and their isolated compounds could be of potential interest in the treatment of allergic diseases. Ethanol extracts of *H. marmoreus*, *Flammulina velutipes*, *Pholiota nameko* and *Pleurotus eryngii* showed significant antiallergic effects in mice (oxazolone-induced type IV allergy) (Sano et al., 2002). Study by Jesenak et al. (2014) proved the anti-allergic effect of pleuran (beta-glucan from *P. ostreatus*) in children with recurrent respiratory tract infections by reducing the peripheral blood eosinophilia and stabilising the levels of total IgE in serum.

Various species of mushrooms have demonstrated the potential for antibacterial, antifungal and antiviral activities (Hassan et al., 2016). Although the antiviral effect of mushrooms does not seem to be related to viral adsorption (i.e., they do not kill the virus), numerous studies have reported the inhibitory effects at the initial stage of virus replication (Faccin et al., 2007). Hence, bioactive compounds from mushrooms may be the future appropriate tool as adjuvants to antiretroviral treatment. Bioactive compounds extracted from *G. frondosa*, in combination with human interferon alpha-2b might provide a potentially effective therapy against chronic hepatitis B virus infections (Gu et al., 2006). Moreover, some studies also indicate a promising activity of mushroom on the treatment of DNA damage, leukemia, rheumatoid arthritis, eye health and wound cure (Roupas et al., 2012).

Hetland et al. (2020), based on literature, suggest the possibility that species like *Agaricus blazei*, *Hericium erinaceum* and *Grifola frondosa* have the merit of prophylactic or therapeutic add-on remedies in COVID-19 infection, especially as a countermeasure against a pneumococcal super infection.

4. Conclusion

Many mysteries of fungi have not yet been fully discovered and the number of studies on acquisition from new valuable medicinal compounds is still growing as evidenced in many publications dedicated to mushroom studies. In the last few decades, much attention has been paid to the use of biological substances present in mushrooms for treatment, as compulsory therapy or as health diet aids. Biologically-active polysaccharides in fungi can be mostly found in carpophores, mycelium, sclerota and filtrates. One of the most promising features of polysaccharide-protein compounds present in fungi is the activity of immunostimulation and anticancer. This review summarises the use of mushrooms and their isolated compounds in the prevention and treatment of several chronic diseases, such as cancer, diabetes, hypertension and obesity. It may be concluded that the consumption of mushrooms as part of the daily diet could help in the treatment and prevention of several chronic diseases. Moreover, their bioactive compounds could also be potential nutraceuticals for different disorders, although the detailed mechanism of the various health benefits of mushrooms on humans still requires further investigation.

References

Adachi, Y., Okazaki, M., Ohno, N. and Yadomae, T. (1994). Enhancement of cytokine production by macrophages stimulated with (1→3)-Beta-D-glucan, Grifolan (GRN), isolated from *Grifola frondosa*. Biol. Pharm. Bull., 17: 1554–1560.

Angeli, J.P., Ribeiro, L.R., Gonzaga, M.L.C., Soares, S.D., Ricardo, M.P.S.N. et al. (2006). Protective effects of β-glucanextracted from *Agaricus brasiliensis* against chemically induced DNA damage in human lymphocytes. Cell Biol. Toxicol., 22: 285–291.

Angeli, J.P., Ribeiro, L.R., Bellini, M.F. and Mantovani, M.S. (2009). Beta-glucan extracted from the medicinal mushroom *Agaricus blazei* prevents the genotoxic effects of benzo[a]pyrene in the human hepatoma cell line HepG2. Arch. Toxicol., 83: 81–86.

Angelini, P., Tirillini, B., Properzi, A., Rol. C. and Venanzoni, R. (2015). Identification and bioactivity of the growth inhibitors in *Tuber* spp. methanolic extracts. Plant Biosyst., 149: 1000–1009.

Badole, S.L., Patel, N.M., Thakurdesai, P.A. and Bodhankar, S.L. (2008). Interaction of aqueous extract of *Pleurotus pulmonarius* (Fr.) Quel-Champ with glyburide in alloxan-induced diabetic mice. Evid. Based Complement. Altern. Med., 5: 159–164.

Biedroń, R., Tangen, J.M., Maresz, K. and Hetland, G. (2012). *Agaricus blazei* Murill–immunomodulatory properties and health benefits. Funct. Food Health Dis., 2(11): 428–447.

Camelini, C.M., Maraschin, M., de Mendonca, M.M., Zucco C., Ferreira, A.G. and Tavares, L.A. (2005). Structural characterisation of β-glucans of *Agaricus brasiliensis* in different stages of fruiting body maturity and their use in nutraceutical products. Biotechnol. Lett., 27: 1295–1299.

Carbonero, E.R., Gracher, A.H.P., Smiderle, F.R., Rosado, F.R., Sassaki, G.L. et al. (2006). A α-glucan from the fruit bodies of edible mushrooms *Pleurotus eryngii* and *Pleurotus ostreatoroseus*. Carbohydr. Polym., 66: 252–257.

Carrillo, C., del Cavia, M. and Alonso-Torre, S.R. (2012). Antitumor effect of oleic acid; mechanisms of action. A review. Nutr. Hosp., 27: 1860–1865.

Chang, C.J., Lin, C.S., Lu, C.C., Martel, J., Ko, Y.F. et al. (2015). *Ganoderma lucidum* reduces obesity in mice by modulating the composition of the gut microbiota. Nat. Comm., 6: 7489.

Chaturvedi, V.K., Agarwal, S., Gupta, K.K., Ramteke, P.W. and Singh, M.P. (2018). Medicinal mushroom: Boon for therapeutic applications. 3, Biotech., 8(8): 334.10.1007/s13205-018-1358-0.

Chen, J., Yong, Y., Xing, M., Gu, Y., Zhang, Z. et al. (2013). Characterisation of polysaccharides with marked inhibitory effect on lipid accumulation in *Pleurotus eryngii*. Carbohydr. Polym., 97: 604–613.

Chen, Y., Li, X.H., Zhou, L.Y., Li, W., Liu, L. et al. (2017). Structural elucidation of three antioxidative polysaccharides from *Tricholomalobayense*. Carbohydr. Polym., 157: 484–492.

Cheskin, L.J., Davis, L.M., Lipsky, L.M., Mitola, A.H., Lycan, T. et al. (2008). Lack of energy compensation over four days when white button mushrooms are substituted for beef. Appetite, 51: 50–57.

Cho, S.M., Park, J.S., Kim, K.P., Cha, D.Y., Kim, H.M. and Yoo, I.D. (1999). Chemical features and purification of immunostimulating polysaccharides from the fruit bodies of *Agaricus blazei*. Taehan Uijinkyun Hakhoe Chi, 27(2): 170–174.

Cho, J.H., Min, B.J., Chen, Y.J., Yoo, J.S., Wang, Q. et al. (2007). Evaluation of FSP (fermented soy protein) to replace soybean meal in weaned pigs: Growth performance, blood urea nitrogen and total protein concentrations in serum and nutrient digestibility. Asian Austral. J. Anim., 20(12): 1874–1879.

Dai, X., Stanilka, J.M., Rowe, C.A., Esteves, E.A., Nieves, C. Jr. et al. (2015). Consuming *Lentinula edodes* (shiitake) mushrooms daily improves human immunity: A randomized dietary intervention in healthy young adults. J. Am. Coll. Nutr., 34: 478–487.

Das, D., Mondal, S., Roy, S.K., Maiti, D., Bhunia, B., Maiti, T.K., Sikdar, S.R. and Islam, S.S. (2010). A (1→6)-β-glucan from a somatic hybrid of *Pleurotus florida* and *Volvariella volvacea*: isolation, characterisation, and study of immunoenhancing properties. Carbohydr. Res., 345: 974–978.

Delzenne, N.M. and Bindels, L.B. (2015). Gut microbiota: *Ganoderma lucidum*, a new prebiotic agent to treat obesity? Nat. Rev. Gastroenterol. Hepatol., 12: 553–554.

Dong, Q., Jia, L.M. and Fan, J.N. (2006). A β-D-glucan isolated from the fruiting bodies of *Hericium erinaceus* and its aqueous conformation. Carbohydr. Res., 34: 1791–795.

El-Batal, A.I., El-Kenawy, N.M., Yassin, A.S. and Amin, M.A. (2015). Laccase production by *Pleurotus ostreatus* and its application in synthesis of gold nanoparticles. Biotechnol. Rep., 5: 31–39.

Faccin, L.C., Benati, F., Rincao, V.P., Mantovani, M.S., Soares, S.A. et al. (2007). Antiviral activity of aqueous and ethanol extracts and of an isolated polysaccharide from *Agaricus brasiliensis* against poliovirus type 1. Lett. Appl. Microbiol., 45: 24–28.

Friedman, M. (2015). Chemistry, nutrition and health-promoting properties of *Hericium erinaceus* (Lion's Mane) mushroom fruiting bodies and mycelia and their bioactive compounds. J. Agric. Food Chem., 63: 7108–7123.

Gao, Y. and Zhou S. (2002). The immunomodulating effects of *Ganoderma lucidum* (Curt.: Fr.) P. Karst. (Ling Zhi, Reishi mushroom) (Aphyllophoromycetideae). Int. J. Med. Mushrooms, 4(1): 11.

Gu, C.Q., Li, J. and Chao, F.H. (2006). Inhibition of hepatitis B virus by D-fraction from *Grifola frondosa*: synergistic effect of combination with interferon-alpha in HepG2 2.2.15. Antivir. Res., 72: 162–165.

Guillamon, E., Garcia-Lafuente, A., Lozano, M., DArrigo, M., Rostagno, M.A. et al. (2010). Edible mushrooms: Role in the prevention of cardiovascular diseases. Fitoterapia, 81(7): 715–723.

Gutierrez, A., Prieto, A. and Martinez, A.T. (1996). Structural characterisation of extra-cellular polysaccharides produced by fungi from the genus *Pleurotus*. Carbohydr. Res., 281: 143–154.

Hassan, M.A.A., Rouf, R., Tiralongo, E., May, T.W. and Tiralongo, J. (2016). Mushroom lectins: Specificity, structure and bioactivity relevant to human disease. Int. J. Mol. Sci., 16: 7802–7838.

Handayani, D., Chen, J., Meyer, B.J. and Huang, X.F. (2011). Dietary Shiitake mushroom (*Lentinus edodes*) prevents fat deposition and lowers triglyceride in rats fed a high-fat diet. J. Obes., 258051: 1–8.

Hawksworth, D.L. (2001). Mushrooms: The extent of the unexplored potential. Int. J. Med. Mushrooms, 3: 333–340.

Hereher, F., El Fallal, A., Toson, E., Abou-Dobara, M. and Abdelaziz, M. (2018). Pilot study: Tumor suppressive effect of crude polysaccharide substances extracted from some selected mushroom. Beni-Suef Univ. J. Basic Appl. Sci., 7: 767–775.

Hetland, G., Tangen, J.M., Mahmood, F., Mirlashari, M.R., Nissen-Meyer, L.S.H., Nentwich, I., Therkelsen, S.P., Tjønnfjord, G.E. and Johnson, E. (2020). Antitumor, anti-inflammatory and antiallergic effects of *Agaricus blazei* mushroom extract and the related medicinal Basidiomycetes mushrooms, *Hericium erinaceus* and *Grifola frondosa*: A review of preclinical and clinical studies. Nutrients, 12: 1339.

Hetland, G., Johnson, E., Bernardshaw. S.V. and Grinde, B. (2020). Can medicinal mushrooms have prophylactic or therapeutic effect against COVID-19 and its pneumonic superinfection and complicating inflammation? Scand. J. Immunol., e12937. doi:10.1111/sji.12937.

Hilszczańska, D. (2012). Medicinal properties of macrofungi. For. Res. Pap., 73(4): 347–353.

Hilszczańska, D., Siebyła, M., Horak, J., Król, M., Podsadni, P., Steckiewicz, P., Bamburowicz-Klimkowska, M., Szutowski, M. and Turło. J. (2016). Comparison of chemical composition in *Tuber aestivum* Vittad of different geographical origin. Chem. Biodiversity, 13(12): 1617–1629, doi: 10.1002/cbdv.201600041.

Hobbs, Ch. (1995). Medicinal Mushrooms: An Exploration of Tradition, Healing and Culture. Botanica Press, Santa Cruz, California.

Hong, J.H. and Choi, Y.H. (2007). Physico-chemical properties of protein-bound polysaccharide from *Agaricus blazei* Murill prepared by ultra filtration and spray drying process. Carbohydr. Polym., 42: 1–8.

Hoshi, H., Yagi, Y., Iijima, H., Matsunaga, K., Ishihara, Y. and Yasuhara, T. (2005). Isolation and characterisation of a novel immunomodulatory α-glucan-protein complex from the mycelium of *Tricholoma matsutake* in *Basidiomycetes*. J. Agric Food Chem., 53: 8948–8956.

Hu, S.H., Wang, J.C., Lien, J.L., Liaw, E.T. and Lee, M.Y. (2006). Antihyperglycemic effect of polysaccharide from fermented broth of *Pleurotus citrinopileatus*. Appl. Microbiol. Biotechnol., 70: 107–113.

Huang, H.Y., Korivi, M., Yang, H.T., Huang, C.C., Chaing, Y.Y. and Tsai, Y.C. (2014). Effect of *Pleurotus tuber-regium* polysaccharides supplementation on the progression of diabetes complications in obese-diabetic rats. Chin. J. Physiol., 57(4): 198–208.

Huang, J.F., Zhan, T., Yu, X.L., He, Q.A., Huang, W.J. et al. (2016). Therapeutic effect of *Pleurotus eryngii* cellulose on experimental fatty liver in rats. Genet. Mol. Res., 15: 15017805.

Hwang, H.S., Lee, S.H., Baek, Y.M., Kim, S.W., Jeong, Y.K. and Yun, J.W. (2008). Production of extracellular polysaccharides by submerged mycelial culture of *Laetiporus sulphureus* var. *miniatus* and their insulinotropic properties. Appl. Microbiol. Biotechnol., 78: 419–429.

Ibadallah, B.X., Abdullah, N. and Shuib, A.S. (2015). Identification of angiotensin-converting enzyme inhibitory proteins from mycelium of *Pleurotuspulmonarius* (Oyster mushroom). Planta Med., 81: 123–129.

Jeong, S.C., Yang, B.K., Islam, R., Koyyalamudi, S.R., Pang, G. et al. (2010). White button mushroom (*Agaricus bisporus*) lowers blood glucose and cholesterol levels in diabetic and hypercholesterolemic rats. Nutr. Res., 30: 49–56.

Jesenak, M., Hrubisko, M., Majtan, J., Rennerova, Z. and Banovcin, P. (2014). Anti-allergic effect of Pleuran (b-glucan from *Pleurotus ostreatus*) in children with recurrent respiratory tract infections. Phytother. Res., 28: 471–474.

Kabir, Y. and Kimura, S. (1989). Dietary mushrooms reduce blood pressure in spontaneously hypertensive rats. J. Nutr. Sci. Vitaminol., 35: 91–94.

Kang, M.G., Kim, Y.H., Bolormaa, Z., Kim, M.K., Seo, G.S. and Lee, J.S. (2013). Characterisation of an antihypertensive angiotensin I-converting enzyme inhibitory peptide from the edible mushroom *Hypsizygus marmoreus*. Bio Med Res. Int., 2013: 283964.

Kanagasabapathy, G., Malek, S.N., Mahmood, A.A., Chua, K.H., Vikineswary, S. and Kuppusamy, U.R. (2013). Beta-glucan-rich extract from *Pleurotus sajor-caju* (Fr.) singer prevents obesity and oxidative stress in C57BL/6J mice fed on a high-fat diet. Evid. Based Compl. Altern. Med., 185259: 1–10.

Khan, M.A., Tania, M., Liu, R. and Rahman, M.M. (2013). *Hericium erinaceus*: An edible mushroom with medicinal values. J. Complement. Integr. Med., 10: 253–258.

Khatun, K., Mahtab, H., Khanam, P.A., Sayeed, M.A. and Khan, K.A. (2007). Oyster mushroom reduced blood glucose and cholesterol in diabetic subjects. Mymensingh Med. J., 16: 94–99.

Kim, Y.W., Kim, K.H., Choi, H.J. and Lee, D.S. (2005). Anti-diabetic activity of β-glucans and their enzymatically hydrolysed oligosaccharides from *Agaricus blazei*. Biotechnol. Lett., 27(7): 483–487.

Kim, H.M., Kang, J.S., Kim, J.Y., Park, S.K., Kim, H.S. et al. (2010). Evaluation of antidiabetic activity of polysaccharide isolated from *Phellinus linteus* in non-obese diabetic mouse. Int. Immunopharmacol., 10(1): 72–78.

Kim, S.P., Park, S.O., Lee, S.J., Nam, S.H. and Friedman, M. (2013). Apolysaccharide isolated from the liquid culture of *Lentinus edodes* (Shiitake) mushroom mycelia containing black rice bran protects mice against a Salmonella lipopolysaccharide-induced endotoxemia. J. Agric. Food Chem., 61(46): 10987–10994.

Kishida, E., Sone, Y. and Misaki, A. (1989). Purification of an antitumor-active, branched (1→3)-β-D-glucan from *Volvariella volvacea* and elucidation of its fine structure. Carbohydr. Res., 193: 227–239.

Kobayashi, H., Yoshida, R., Kanada, Y., Fukuda, Y., Yagyu, T. et al. (2005). Suppressing effects of daily oral supplementation of beta-glucan extracted from *Agaricus blazei* Murill on spontaneous and peritoneal disseminated metastasis in mouse model. J. Cancer Res. Clin., 131: 527–538.

Kodama, N., Komuta, K., Sakai, N. and Nanba, H. (2002). Effects of D-fraction, a polysaccharide from *Grifola frondosa* on tumor growth involve activation of NK cells. Biol. Pharm. Bull., 25: 1647–1650.

Kozarski, M., Klaus, A., Jakovljevic, D., Todorovic, N., Vunduk, J. et al. (2015). Antioxidants of edible mushrooms. Molecules, 20: 19489–19525.

Kundakovic, T. and Kolundzic, M. (2013). Therapeutic properties of mushrooms in managing adverse effects in the metabolic syndrome. Curr. Top. Med. Chem., 13: 2734–2744.

Lau, C.C., Abdullah, N., Shuib, A.S. and Aminudin, N. (2014). Novel angiotensin I-converting enzyme inhibitory peptides derived from edible mushroom *Agaricus bisporus* (J.E. Lange) imbach identified by LC–MS/MS. Food Chem., 148: 396–401.

Lee, H., Kyungmin Nam, K., Zahra, Z. and Farooqi, M.Q.U. (2020). Potentials of truffles in nutritional and medicinal applications: A review. Fungal Biol. Biotechnol., 7: 9.

Lei, H., Ma, X. and Wu, W.T. (2007). Anti-diabetic effect of an alphaglucan from fruit body of Maitake (*Grifola frondosa*) on Kk-Ay mice. J. Pharm. Pharmacol., 59: 575–582.

Li, J.P., Lei, Y.L. and Zhan, H. (2014a). The effects of the king oyster mushroom *Pleurotus eryngii* (higher Basidiomycetes) on glycemic control in alloxan-induced diabetic mice. Int. J. Med. Mushrooms, 16: 219–225.

Li, J., Zou, L., Chen, W., Zhu, B., Shen, N. et al. (2014b). Dietary mushroom intake may reduce the risk of breast cancer: evidence from a meta-analysis of observational studies. PLoS One, 9: e93437.

Liang, B., Guo, Z., Xie, F. and Zhao, A. (2013). Antihyperglycemic and antihyperlipidemic activities of aqueous extract of *Hericium erinaceus* in experimental diabetic rats. BMC Complment. Altern. Med., 13: 253.

Lim, J.M., Joo, J.H., Kim, H.O., Kim, H.M., Kim, S.W. et al. (2005). Structural analysis and molecular characterisation of exopolysaccharides produced by submerged mycelial culture of *Collybia maculate* TG1. Carbohydr. Polym., 61: 296–303.

Lindequist, U., Timo, H.J., Niedermeyer, Y. and Julich, W.D. (2005). The pharmacological potential of mushroom. Evid-based Compl. Alt., 3: 285–299.

Liu, F., Ooi, V.E.C., Liu, W.K. and Chang, S.T. (1996). Immunomodulation and antitumor activity of polysaccharide-protein complex from the culture filtrates of a local edible mushroom *Tricholoma lobayense*. Gen. Pharmacol., 27: 621–624.

Mehmood, S., Zhou, L.Y., Wang, X.F., Cheng, X.D., Meng, F.J., Ya Wang, Y., Lu, Y.M. and Chen, Y. (2019). Structural elucidation and antioxidant activity of a novel heteroglycan from *Tricholoma lobayense*. J. Carbohydr. Chem., 38(3): 192–211.

Mascaro, M.B., Franca, C.M., Esquerdo, K.F., Lara, M.A., Wadt, N.S. and Bach, E.E. (2014). Effects of dietary supplementation with *Agaricus sylvaticus* Schaeffer on glycemia and cholesterol after streptozotocin-induced diabetes in rats. Evid. Based Complement. Altern. Med., 107629: 1–10.

Meng, X., Liang, H. and Luo, L. (2016). Antitumor polysaccharides from mushrooms: A review on the structural characteristics, antitumor mechanisms and immunomodulating activities. Carbohydr. Res., 424: 30–41.

Mizuno, T., Wasa, T., Ito, H., Suzuki, C. and Ukai, N. (1992). Antitumor active polysaccharides isolated from the fruiting body of *Hericium erinaceum*, an edible and medicinal mushroom called yamabushitake or houtou. Biosci. Biotech. Bioch., 56: 347–348.

Mizuno, T. (2002). Medicinal properties and clinical effects of culinary medicinal mushroom *Agaricus blazei* Murrill (Agaricomycetideae) (Review). Int. J. Med. Mushrooms, 4(4): 1–14.

Murphy, E.A., Velazquez, K.T. and Herbert, K.M. (2015). Influence of high-fat diet on gut microbiota: A driving force for chronic disease risk. Curr. Opin. Clin. Nutr. Metab. Care, 18: 515–520.

Ng, S.H., Zain, M.S., Zakaria, F., Ishak, W.R.W. and Ahmad, W.A.N.W. (2015). Hypoglycemic and antidiabetic effect of *Pleurotus sajor-caju* aqueous extract in normal and streptozotocin-induced diabetic rats. Bio. Med. Res. Int., 2015: 214918.

Niwa, A., Tajiri, T. and Higashino, H. (2011). *Ipomoea batatas* and *Agaricus blazei* ameliorate diabetic disorders with therapeutic antioxidant potential in streptozotocin-induced diabetic rats. J. Clin. Biochem. Nutr., 48: 194–202.

Ohno, N., Miura, N.N., Nakajima, M. and Yadomae, T. (2000). Antitumor 1,3-α-glucan from cultured fruit body of *Sparassis crispa*. Biol. Pharm. Bull., 23: 866–872.

Ooi, V.E.C. and Liu, F. (1999). A review of pharmacological activities of mushroom polysaccharides. Int. J. Med. Mushroom, 1: 195–206.

Pacheco, Y.M., López, S., Bermúdez, B., Abia, R., Villar, J. and Muriana, F.J.G. (2008). A meal rich in oleic acid beneficially modulates postprandial sICAM-1 and sVCAM-1 in normotensive and hypertensive hypertriglyceridemic subjects. J. Nutr. Biochem., 19: 200–205.

Pacheco-Sanchez, M., Boutin, Y., Angers, P., Gosselin, A. and Tweddell, R.J. (2006). Abioactive (1→3)-, (1→4)-β-D-glucan from *Collybia dryophila* and other mushrooms. Mycologia, 98: 180–185.

Pacioni, G., Rapino, C., Zarivi, O., Falconi, A., Leonardi, M. et al. (2015). Truffles contain endocannabinoid metabolic enzymes and anandamide. Phytochemistry, 110: 104–110.

Poddar, K.H., Ames, M., Hsin-Jen, C., Feeney, M.J., Wang, Y. and Cheskin, L.J. (2013). Positive effect of mushrooms substituted for meat on body weight, body composition, and health parameters: A 1-year randomised clinical trial. Appetite, 71: 379–387.

Poucheret, P., Fons, F. and Rapior, S. (2006). Biological and pharmacological activity of higher fungi: 20-year retrospective analysis. Cryptogam. Mycol., 27: 311–333.

Puiggrós, C., Chacón, P., Armadans, L.I., Clapés, J. and Planas, M. (2002). Effects of oleic-rich and omega-3-rich diets on serum lipid pattern and lipid oxidation in mildly hypercholesterolemic patients. Clin Nutr., 21: 79–87. https ://doi.org/10.1054/clnu.2001.0511.

Roncero-Ramos, I. and Delgado-Andrade, C. (2017). The beneficial role of edible mushrooms in human health. Curr. Opin. Food Sci., 14: 122–128.

Roupas, P., Keogh, J., Noakes, M., Margetts, C. and Taylor, P. (2012). The role of edible mushrooms in health: Evaluation of the evidence. J. Funct. Foods, 4: 687–709.

Sabaratnam, V., Kah-Hui, W., Naidu, M. and David, P.R. (2013). Neuronal health—Can culinary and medicinal mushrooms help? J. Tradit. Compl. Med., 3: 62–68.

Sano, M., Yoshino, K., Matsuzawa, T. and Ikekawa, T. (2002). Inhibitory effects of edible higher Basidiomycetes mushroom extracts on mouse type IV allergy. Int. J. Med. Mushrooms, 4: 37–41.

Seto, S.W., Lam, T.Y., Tam, H.L., Au, A.L.S., Chan, S.W. et al. (2009). Novel hypoglycemic effects of *Ganoderma lucidum* water-extract in obese/diabetic (+ db/+ db) mice. Phytomedicine, 16(5): 426–436.

Shida, M. and Sakai, N. (2004). Heterogalactans obtained from some typical edible mushrooms. J. JPN. Soc. Food Sci., 51: 559–562.

Smiderle, F.., Carbonero, E.R., Mellinger, C.G., Sassaki, G.L., Gorin, P.A.J. and Iacomini, M. (2006). Structural characterisation of a polysaccharide and a α-glucan isolated from the edible mushroom *Flammulina velutipes*. Phytochemistry, 67: 2189–2196.

Smiderle, F.R., Carbonero, E.R., Sassaki, G.L., Gorin, P.A.J. and Iacomini, M. (2008). Characterisation of a heterogalactan: Some nutritional values of the edible mushroom *Flammulina velutipes*. Food Chem., 108: 329–333.

Soares, A.A., de Oliveira, A.L., SaNakanishi, A.B., Comar, J.F., Rampazzo, A.P.S. et al. (2013). Effects of an *Agaricus blazei* aqueous extract pretreatment on paracetamol-induced brain and liver injury in rats. Bio Med. Res. Int., 469180: 1–12.

Surenjav, U., Zhang, L.N., Xu, X.J., Zhang, M., Cheung, P.C.K. and Zeng, F.B. (2005). Structure, molecular weight and bioactivities of (1→3)-β-D-glucans and its sulfate derivatives from four kinds of *Lentinus edodes*. Chinese J. Polym. Sci., 23: 327–336.

Talpur, N.A., Echard, B.W., Fan, A.Y., Jaffari, O., Bagchi, D. and Preuss, H.G. (2002). Antihypertensive and metabolic effects of whole Maitake mushroom powder and its fractions in two rat strains. Mol. Cell. Biochem., 237(1-2): 129–136.

Tejedor-Calvo, E., Morales, D., Maro, P., Venturini, M.E., Blanco, D. and Soler-Rivas, C. (2019). Effects of combining electron-beam or gamma irradiation treatments with further storage under modified

atmospheres on the bioactive compounds of *Tuber melanosporum* truffles. Postharvest Biol,
Technol., 155: 149–55. https ://doi.org/10.1016/j.posth arvbio.2019.05.022.

Thatoi, H.N. and Singdevsachan, S.K. (2014). Diversity, nutritional composition and medicinal potential
of Indian mushrooms: A review. Afr. J. Biotechnol., 13: 523–545.

Ukawa, Y., Ito, H. and Hisamatsu, M. (2000). Antitumor effects of (1→3)-β-D-glucan and (1→6)-β-D-
glucan purified from newly cultivated mushroom, Hatakeshimeji (*Lyophyllum decastes* Sing.).
J. Biosci. Bioeng., 90: 98–104.

Wasser, S.P. (2002). Medicinal mushrooms as a source of antitumor and immunomodulating
polysaccharides. Appl. Microbiol. Biotechnol., 60(3): 258–274.

Wasser, S. (2014). Medicinal mushroom science: Current perspectives, advances, evidences, and
challenges. Biomed. J., 37(6): 345–356.

Wasser, S.P. and Weis, A.L. (1999). Medicinal properties of substances occurring in higher Basidiomycetes
mushrooms: Current perspectives. Int. J. Med. Mushrooms, 1: 47–50.

Yahaya, N.F.M., Rahman, M.A. and Abdullah, N. (2014). Therapeutic potential of mushrooms in
preventing and ameliorating hypertension. Trends Food Sci. Technol., 39: 104–115.

Yalin, W., Ishurd, O., Cuirong, S. and Yuanjiang, P. (2005). Structure analysis and anti-tumor activity
of ((1→3)-α-D-glucans (Cordyglucans) from the mycelia of *Cordyceps sinensis*. Planta Med.,
71: 381–384.

Yan, X., Wang, Y., Sang, X. and Fan, L. (2017). Nutritional value, chemical composition and antioxidant
activity of three *Tuber* species from China. AMB Express, 7: 136.

Zhang, Z., Lv, G., Pan, H., Pandey, A., He, W. and Fan, L. (2012). Antioxidant and hepatoprotective
potential of endo-polysaccharides from *Hericium erinaceus* grown on tofu whey. Int. J. Biol.
Macromol., 51: 1140–1146.

Zhang, J.J., Li, Y., Zhou, T., Xu, D.P., Zhang, P. et al. (2016). Bioactivities and health benefits of
mushrooms mainly from China. Molecules, 21: 938.

Zhao, W., Wang, X.H., Li, H.M., Wang, S.H., Chen, T. et al. (2014). Isolation and characterisation of
polysaccharides with the antitumor activity from Tuber fruiting bodies and fermentation system.
Appl. Microbiol. Biotechnol., 98: 1991–2002.

Zhuang, C., Mizuno, T., Shimada, A., Ito, H., Suzuki, C. et al. (1993). Antitumor protein-containing
polysaccharides from a Chinese mushroom Fengweigu or Houbitake *Pleurotus sajor-caju* (Fr)
Sings. Biosci. Biotechnol. Biochem., 57: 901–906.

Zhuang, C., Mizuno, T., Ito, H., Shimura, K., Sumiya, T. and Kawade, M. (1994). Fractionation and
antitumor activity of polysaccharides from *Grifola frondosa* mycelium. Biosci. Biotechnol.
Biochem., 58: 185–188.

5

Fomitopsis betulina
A Rich Source of Diverse Bioactive Metabolites

Shilpa A Verekar,[1] *Manish K Gupta*[2] *and Sunil K Deshmukh*[3,*]

1. INTRODUCTION

Carl Linnaeus in the year 1753 described this fungus as *Boletus suberosus* and later specific epithet was changed to *betulinus* as it was reported from the birch trees (*Betula* spp.). In the year 1821, this was transferred to the genus *Polyporus* by the French mycologist Jean Baptiste Francois (Pierre) Bulliard. In the year 1881, birch polypore was moved to a new genus, *Piptoporus* with two other species, by the Finnish mycologist Petter Adolf Karsten (1834–1917), and *Piptoporus betulinus* as the type species. Based on molecular phylogenetic studies, *P. betulinus* was found more closely related to *Fomitopsis* than to *Piptoporus* and hence it was reclassified as *Fomitopsis* with the current name *Fomitopsis betulina* (Bull.) B.K. Cui, M.L. Han and Y.C. Dai (Kim et al., 2005; Ortiz-Santana et al., 2013; Han et al., 2016; Pleszczyńska et al., 2017). The common names for the fungus include birch polypore, razor strop, or birch bracket mushroom. It is a hardwood-specific weak parasite that occurs exclusively throughout the range of birches (black, grey, paper white and yellow) in Asia, North America and Europe. It commonly grows on dead trees and

[1] Parle Agro Pvt. Ltd., Off Western Express Highway, Sahar-Chakala Road, Parsiwada, Andheri (East) Mumbai 400099, India.
[2] SGT College of Pharmacy, SGT University, Gurugram 122505, Haryana, India.
[3] TERI-Deakin Nano Biotechnology Centre, The Energy and Resources Institute, Darbari Seth Block, IHC Complex, Lodhi Road, New Delhi 110003, India.
* Corresponding author: sunil.deshmukh1958@gmail.com

rarely on living ones. The mushroom parasitises on the birch trees rendering tree trunks brittle as a result of which they can be destroyed by even a mild wind.

Piptoporus betulinus (birch polypore) is one of the few edible polypores. Its young fruit bodies have a strong and pleasant odour and an astringent, bitter taste (Wasson, 1969). However, on the aging of fruit body, it becomes tough, bitter and hence, inedible. The bracket-like fruit body of the mushroom can last for a year and is used for topical applications (as band-aid) for the treatment of inflammation. This review focuses on giving an insight into the traditional uses of birch polypore as well as the different strains of *P. betulinus* identified for their roles in eliciting bioactive compounds. Major emphasis is given to the pharmacological activities of various metabolites of this fungus. Literature also revealed that polysaccharides present in the fruit body of this fungus account for diverse biological properties (de Jesus et al., 2018; Sari et al., 2020) but are not covered in this review.

2. Traditional Uses of Birch Polypore

In the year 1991, a 5,300-year old mummified body was discovered in the Val Senales glacier in Italy and the man was named Ötzian Iceman. Two fragments of the two different-shaped fruit body's pieces of the polypores, *Fomitopsis betulina* and *Fomes fomentarius*, were found with the Iceman. The reason behind the Iceman carrying the *Fomitopsis* is not clear—whether it was for spiritual or medicinal concerns? An autopsy of Ötzi revealed that he suffered from parasitic intestinal whipworms (*Trichuris trichiura*) that cause diarrhoea and stomach pain; this could be the cause for the use of this fungus (Peintner et al., 1998). *Fomitopsis betulina* is traditionally used as an antimicrobial and antiparasitic agent in the treatment of wounds to arrest bleeding (Stamets, 1993).

The use of *F. betulina* in folk medicine is known since ages (Reshetnikov et al., 2001; Wasser, 2010). The infusion from fruit bodies of *F. betulina* was popular, especially in Russia, Baltic countries, Hungary and Romania for its nutritional and calming properties. Fungal tea was used against various cancer types, as an immuno-enhancing, antiparasitic agent and as a remedy for gastrointestinal disorders (Lucas, 1960; Semerdžieva and Veselský, 1986; Peintner and Pöder, 2000; Shamtsyan et al., 2004; Grienke et al., 2014). Fresh *F. betulina* fruit body was used to make antiseptic and anti-bleeding dressings to be applied to wounds and the powder obtained from dried ones was used as an analgesic (Rutalek, 2002; Grienke et al., 2014; Papp et al., 2015). It is reported that the extract obtained from fruit bodies of *F. betulina* can cure vaginal cancer after weeks of therapy in dogs in Poland (Grienke et al., 2014).

In Northern America and Siberia, snuff prepared from the ash of *P. betulinus* is used as a pain reliever, whereas the powder obtained from fruit bodies of *P. betulinus* is used as snuff in Austria (Rutalek, 2002). Strips of *P. betulinus* fruit bodies are used externally to stop the flow of blood from cuts (as styptic) and charcoal of this polypore is used as an antiseptic (Thoen, 1982; Hobbs, 2002).

Apart from its nutritional and medicinal use, *P. betulinus* was also used in several other applications. The velvety surface of the fruit body was used as a strop for smoothing razor edges (Thoen, 1982; Pegler, 2000). People from the Scottish Highlands were also using fruit bodies of *P. betulinus* as packing material for the back of their circular shields or targets (Marsh, 1973). In Styria (Austria), carved *P. betulinus* fruit bodies were used to protect farm animals from bad luck (Lohwag, 1965). Artistic and medicinal-spiritual applications of *P. betulinus* might be the reason for carrying the carved fruit bodies of *P. betulinus* by the Iceman on his clothes over the Alps journey (Peintner et al., 1998; Pöder and Peintner, 1999).

The methanolic extract of *P. betulinus* was found active against *Staphylococcus aureus* and *Bacillus subtilis* (Gram-positive) and *Escherichia coli* (Gram-negative) bacteria, whereas its *n*-hexane extract was found active against the only Gram-positive bacteria. None of the extract is found active against the yeast *Candida maltose* (Alresly, 2019).

3. Biological Investigations of Extracts of *Fomitopsis betulina*

Fomitopsis betulina has been tested *in vitro* for its activity against cancer cells. Hot water extract of fruit bodies exhibited moderate cytotoxic activity with an $IC_{50} = 0.1$ mg/ml on HeLa cells (Vunduk et al., 2015). Ethyl extract fraction prepared from dried fruit bodies was tested against A549, HT-29 human lung carcinoma, colon adenocarcinoma and C6 rat glioma cell lines. The extract significantly decreases viability, proliferation and migration of tumor cells, affects the stimulatory effect of IGF-I and is non-toxic to normal cells (Lemieszek et al., 2009).

Mycelia of *F. betulina* grown *in vitro* were extracted with ether and ethanol and tested for anti-proliferative activity against cancer cell lines. The extracts were found active against cell lines tested and could reduce the viability of cancer cells, slightly inhibit proliferation and tumor cell adhesion in a time- and dose-dependent manner (Cyranka et al., 2011). The fruit bodies of *F. betulina* were produced artificially and the water and ethanol extracts of these cultivated fruit bodies were found active against A549, HT-29 and T47D cancer cell lines (Pleszczyńska et al., 2016). The ethanolic extracts of *F. betulina* also significantly increased phagocytosis in granulocytes by 158 per cent (Doskocil et al., 2016).

Ether extract obtained from *P. betulinus* possesses interference inducer activity (Kandefer-Szerszeń and Kawecki, 1974). Methanol and chloroform extracts of fruit bodies of *F. betulina* possess antibacterial ability against *B. subtilis* and *E. coli* (Keller et al., 2002; Dresch et al., 2015). The hot alkali extracts of *F. betulina* significantly inhibit the angiotensin converting enzyme (ACE) (Vunduk et al., 2015). The alkali extracts also exhibited antioxidant activity in FRAP assay (Vunduk et al., 2015).

The ethyl acetate mycelium extract of *P. betulinus* increases the cell viability of human Keratinocyte cell line (HaCaT) in a dose-dependent and reduced a serum-deprivation-induced arrest of G_0/G_1 cell cycle. The HaCaT cells stressed by UVB broad-band irradiation (20 mJ/cm^2) showed a strong cell cycle arrest in G_2/M phase. The incubation of irradiated HaCaT cells with *P. betulinus* extract diminished the effect by UV-induced DNA damage by 20 per cent. Confirmation of these results by gel-free high-resolution mass spectrometry-based proteome analysis is currently under way. The first results indicated an increase in cellular oxido-reductase levels during treatment with *P. betulinus* extract (Harms et al., 2013).

3.1 Bioactive Metabolites from Fomitopsis betulina

Polyporenic acids A (**1**) and C (**2**) (Fig. 1) along with polyporenic acids B, were amongst the first few compounds isolated from birch polypore in the year 1940 (Cross et al., 1940). Later on, Bryce et al. (1967) reported methyl polyporenate C (**3**), 3α,12α-dihydroxy-24-methylene-Lanost-8-en-26-oic acid methyl ester (**4**), 3α,12α-dihydroxy-24-methylene-Lanost-8-en-26-oic acid methyl ester 3-acetate (**5**) (Fig. 1) along with polyporenic acids A (**1**) and C (**2**) from birch polypore. Polyporenic acid A (**1**) is reported in *F. betulina* in the form of conjugates, in which the 3-α-hydroxyl group is esterified with biologically important molecules as malonic and hydroxymethyl glutaric acids (Bryce et al., 1967) and later this was confirmed by Kamo et al. (2003) and Wangun et al. (2004). Polyporenic acid A (**1**), exhibited potent anti-inflammatory activity by inhibiting 3α-hydroxysteroid dehydrogenase, inhibiting bacterial hyaluronate lyase and suppressing the TPA-induced edema (Kamo

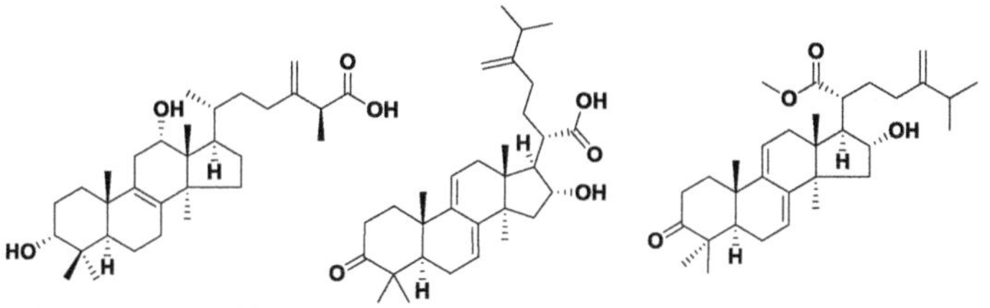

Polyporenic acid A (1)   Polyporenic acid C (2)   Methyl polyporenate C (3),

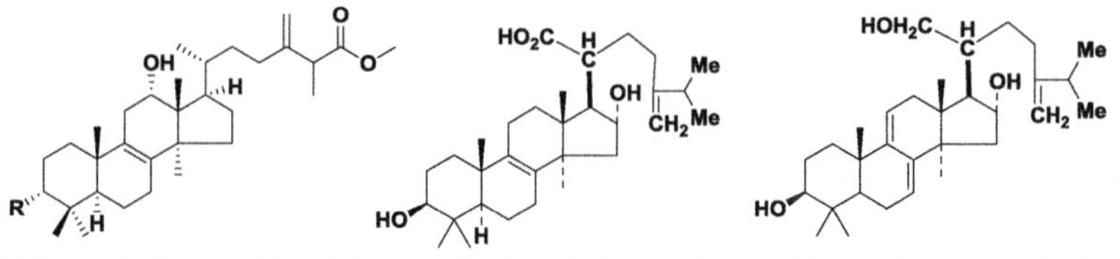

R=OH ; 3α,12α-dihydroxy-24-methylene
-Lanost-8-en-26-oic acid methyl ester (4),
R= OCOCH3 ; 3α,12α-dihydroxy-24-methylene
-Lanost-8-en-26-oic acid methyl ester 3-acetate (5)

3β : 16α -dihydroxyeburico-8
: 24(28)-dien-21-oic acid (6)

3β : 16α- dihydroxyeburico-7 : 9
(11) : 24(28)-trien-21-oic acid (7)

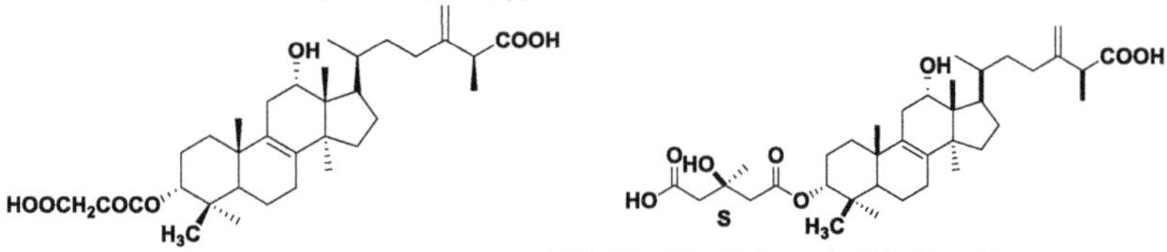

(25S)-(+)-12α-Hydroxy-3α-malonyloxy-24
-methyllanosta-8,24(31)-dien-26-oic acid (8)

(25S,3'S)-(+)-12α-Hydroxy-3α-(3'-hydroxy-3'-
methylglutaryloxy)
-24-methyllanosta-8,24(31)-dien-26-oic acid (9)

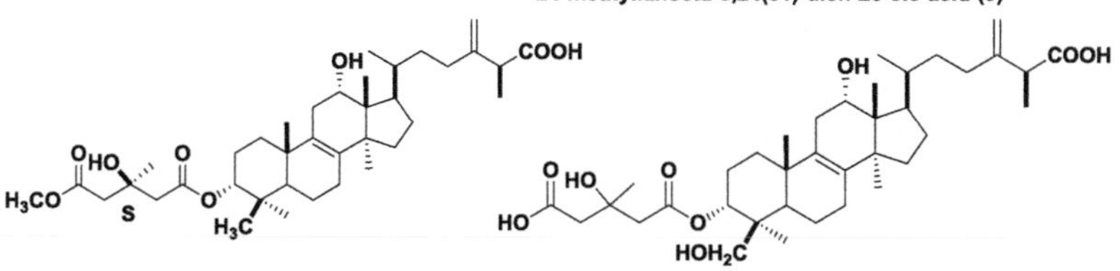

(25S,3 S)-(+)-12α-Hydroxy-3α-(3'-hydroxy
-4'-methoxycarbonyl-3 -methylbutyryloxy) -24-
methyllanosta-8,24-(31)-dien-26-oic acid (10)

(+)-12α,28-Dihydroxy-3α-(3'-hydroxy-3' methylglutaryloxy)
-24-methyllanosta-8,24(31)-dien-26-oic acid (11)

Fig. 1. Structure of compounds 1–11.

et al., 2003; Wangun et al., 2004). Guider et al. (1954) reported that polyporenic acid B is a mixture of 3β: 16 α-dihydroxyeburico-8: 24(28)-dien-21-oic acid (6) and 3β: 16α-dihydroxyeburico-7: 9(11): 24(28)-trien-21-oic acid (7) (Fig. 1).

It is reported that polyporenic acid C (2) possesses antibacterial, antivirus, anti-inflammatory, cytotoxic and antioxidant activities (Borenfreund and Puerner, 1985; Stamets, 1993; Chandramu et al., 2003). Polyporenic acid C (2) suppresses the proliferation of Panc-1, MiaPaca-2, AsPc-1 and BxPc-3 human pancreatic cancer cell lines (Cheng et al., 2013) and suppresses A549 cell proliferation by induction of caspase-8-mediated apoptosis. The inhibition of Akt activation and enhancement of p53 function may also contribute to apoptosis induced by polyporenic acids C (2) (Ling et al., 2009). It also inhibits DNA topoisomerase I and II (Li et al., 2004), which are important molecular targets for anticancer drugs. The polyporenic acid C (2) exhibited superior antibacterial activity against *Mycobacterium phlei* and poor activity against *S. aureus* and *Bacillus coli* after 24 hours and after 72 hours with a dilution of 1:1,280,000 of the antibiotic, respectively. In this case, the activity disappeared on further incubation up to seven days (Marcus, 1952). Polyporenic acid C (2) also inhibits the growth of *Bacterium racemosum* (Ying et al., 1987).

Later six lanostane-type triterpene acids, namely polyporenic acids A (1) and C (2), three derivatives of polyporenic acid A (1), (25S)-(+)-12 α-Hydroxy-3α-malonyloxy-24-methyllanosta-8,24(31)-dien-26-oic acid (8), (25S,3¢S)-(+)-12 α-Hydroxy-3 α-(3'-hydroxy-3'-methylglutaryloxy)-24-methyllanosta-8,24(31)-dien-26-oicacid (9), (25S,3¢S)-(+)-12 α-Hydroxy-3 α-(3'-hydroxy-4'-methoxycarbonyl-3'-methylbutyryloxy)-24-methyllanosta-8,24-(31)-dien-26-oic acid (10) and a novel compound, (+)-12 α,28-Dihydroxy-3 α-(3'-hydroxy-3'-methylglutaryloxy)-24-methyllanosta-8,24(31)-dien-26-oic acid (11) (Fig. 1) were isolated from the methanolic extract of fruit bodies of *P. betulinus* collected from Nagano, Japan. Compounds (1,2, 8–11) suppressed the 12-O-tetradecanoylphorbol-13-acetate (TPA) induced edema on mouse ears by 49–86 per cent with a 400 nmol/ear application (Kamo et al., 2003). Zwolińska (2004) reported polyporenic acid A (1) from birch polypore *Piptoporus betulinus* with anticancer activity.

Further, derivatives of polyporenic acids, 3α-acetylpolyporenic acid A (12) and (25S)-(+)-12α-hydroxy3α-methylcarboxyacetate-24-methyllanosta-8,24(31)-diene26-oic acid (13) together with the known lanostanoid, (25S,3'S)-(+)-12α-hydroxy-3α-(3'-hydroxy4'methoxycarbonyl-3'-methylbutyryloxy)-24-methyllanosta8, 24(31)-dien-26-oic acid (14) (Fig. 2) and polyporenic acid C (2) were isolated from fruit bodies of *P. betulinus* collected in a forest district near Jena (Thuringia, Germany). Compounds (12–14, 2) showed weak cyclooxygenase I inhibition potential. These compounds exhibited strong inhibition against 3α-hydroxysteroid dehydrogenase activity with IC_{50} value of 8.5, 4.0, 5.5 and 17.5 μg/ml, while positive control indomethacin exhibited activity with IC_{50} value of 6.5 μg/ml. Compounds also inhibited hyaluronatelyase from *Streptococcus agalactiae* with IC_{50} of 40, 3.5, 51.0 and 12.5 μM, respectively (Wangun et al., 2004).

The novel hydroquinone, (E)-2-(4-hydroxy-3-methyl-2-butenyl)-hydroquinone (15) (Fig. 2) and known compound, polyporenic acid C (2), were isolated as matrix metallo-proteinase inhibitors from the mushroom *P. betulinus*. Compound (15) showed inhibitory activity against MMP-1 with IC_{50} value of 28 μM. Compound (15)

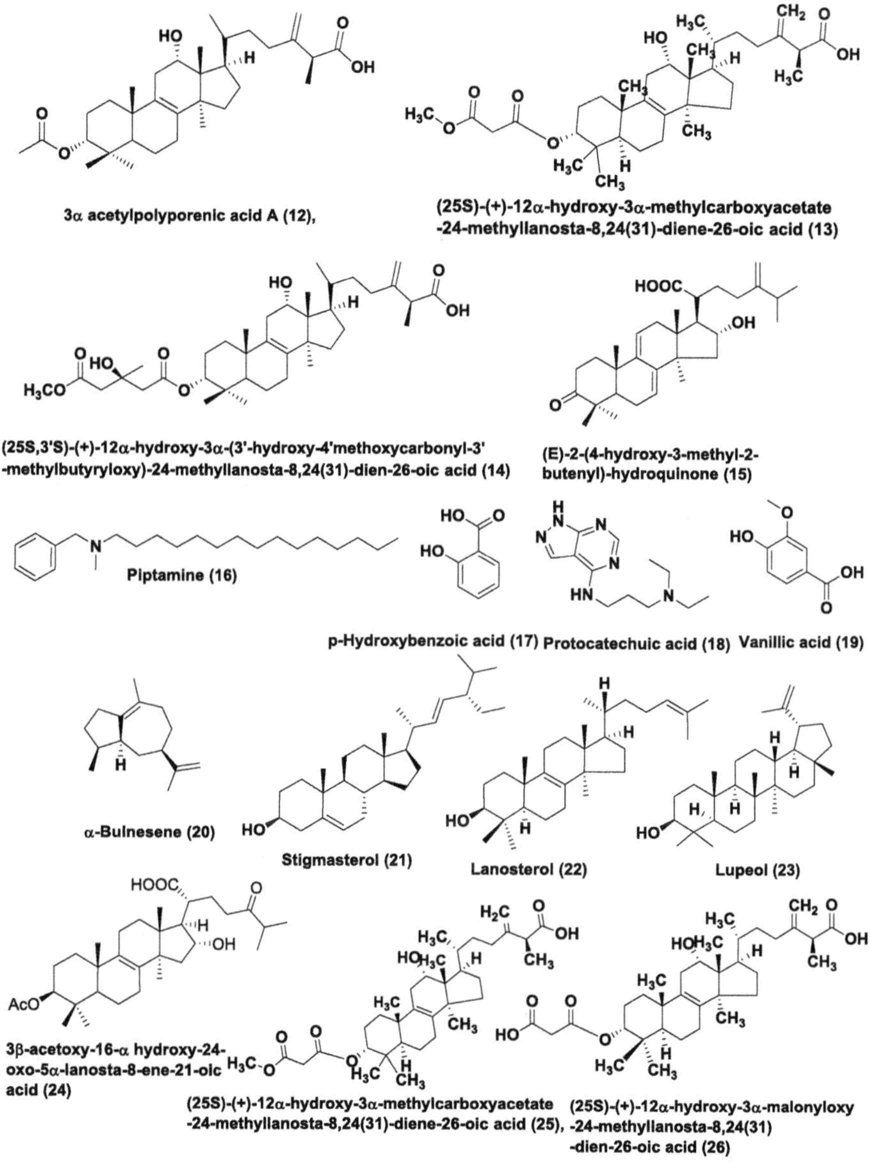

3α acetylpolyporenic acid A (12),

(25S)-(+)-12α-hydroxy-3α-methylcarboxyacetate
-24-methyllanosta-8,24(31)-diene-26-oic acid (13)

(25S,3'S)-(+)-12α-hydroxy-3α-(3'-hydroxy-4'methoxycarbonyl-3'
-methylbutyryloxy)-24-methyllanosta-8,24(31)-dien-26-oic acid (14)

(E)-2-(4-hydroxy-3-methyl-2-
butenyl)-hydroquinone (15)

Piptamine (16)

p-Hydroxybenzoic acid (17) Protocatechuic acid (18) Vanillic acid (19)

α-Bulnesene (20)

Stigmasterol (21)

Lanosterol (22)

Lupeol (23)

3β-acetoxy-16-α hydroxy-24-
oxo-5α-lanosta-8-ene-21-oic
acid (24)

(25S)-(+)-12α-hydroxy-3α-methylcarboxyacetate
-24-methyllanosta-8,24(31)-diene-26-oic acid (25),

(25S)-(+)-12α-hydroxy-3α-malonyloxy
-24-methyllanosta-8,24(31)
-dien-26-oic acid (26)

Fig. 2. Structure of compounds 12–26.

also inhibited MMP-3 (IC$_{50}$ of 23 μM) and MMP-9 (IC$_{50}$ of 37 μM). Polyporenic acid C (**2**) showed inhibitory activity against MMP-1 inhibitor with IC$_{50}$ of 126 μM (Kawagishi et al., 2002). Piptamine (**16**) (Fig. 2), an N-containing compound, was isolated from a submerged culture of the mushroom, with antimicrobial activity against a series of Gram-positive bacteria, yeasts and other fungi (Schlegel et al., 2000).

Sułkowska-Ziaja et al. (2012) studied total phenolic compounds in extracts of the fruit bodies of *P. betulinus* and isolated p-hydroxybenzoic acid (**17**), protocatechuic acid (**18**), vanillic acid (**19**) (Fig. 2). The total content of phenolic compounds was 13.84 mg/g dry mass. These compounds exhibited significant antioxidant activity in FRAP assay in comparison to trolox.

Compounds polyporenic acid C (**2**), piptamine (**16**), α-bulnesene (**20**), stigmasterol (**21**), lanosterol (**22**) and lupeol (**23**) (Fig. 2) were isolated from fruit bodies of *F. betulina* and collected in taiga zones of Kadinskii Natural Area, Irkutsk Oblast, Russia by using supercritical CO_2 extraction (Grishin et al., 2016).

The ethyl acetate extract obtained from fruit bodies of *F. betulina* were the source of a new lanostane triterpene, 3β-acetoxy-16-hydroxy-24-oxo-5α-lanosta-8-ene-21-oic acid (**24**) along with known triterpenes, polyporenic acid A (**1**), polyporenic acid C (**2**) and three derivatives of polyporenic acid A, namely (25S)-(+)-12α-hydroxy-3α-methylcarboxyacetate-24-methyllanosta-8,24(31)-diene-26-oic acid (**25**), (25S)-(+)-12α-hydroxy-3α-malonyloxy-24-methyllanosta-8,24(31)-dien-26-oic acid (**26**) (Fig. 2) and (25S,3'S)-(+)-12α-hydroxy-3α-(3'-hydroxy-3'-methyl-glutaryloxy)-24-methyllanosta-8,24(31)-dien-26-oic acid (**27**), betulin (**28**), betulinic acid (**29**), ergosterol peroxide (**30**), 9,11-dehydroergosterol peroxide (**31**) and fomefficinic acid (**32**) (Fig. 3). The fruit body was collected from *Betula pendula* from the Greifswald in the northeast of Germany. Compound 3β-acetoxy-16-hydroxy-24-oxo-5α-lanosta-8-ene-21-oic acid (**24**) was found active against *S. aureus* and *B. subtilis* with MIC value of 98 and 200 μg/ml, respectively. Compounds polyporenic acid C (**2**), ergosterol peroxide (**30**), 9,11-dehydroergosterol peroxide (**31**), (25S)-(+)-12α-hydroxy-3α-methylcarboxyacetate-24-methyllanosta-8,24(31)-diene-26-oic acid (**25**) and (25S,3'S)-(+)-12α-hydroxy-3α-(3'-hydroxy-3'-methyl-glutaryloxy)-24-methyllanosta-8,24(31)-dien-26-oic acid (**27**) exhibit poor to very poor activity against *S. aureus* and *B. subtilis* and no activity against *E. coli* and *Pseudomonas aeruginosa* (Alresly et al., 2016).

Betulinic acid (**29**), showed activity against neuroectodermal tumor cells including neuro-blastoma, medulloblastoma, glioblastoma and Ewing sarcoma cells. The mechanism of action of betulinic acid was different from conventional anti-cancer drugs. Betulinic acid (**29**) did not induce activation of the CD95 system or accumulation of wild type p53 protein. Betulinic acid (**29**) acts by directly targeting mitochondria. Mitochondrial perturbations on treatment with betulinic acid resulted in the release of cytochrome C (second mitochondria-derived activator of caspase) Smac apoptosis-inducing factor (AIF) from mitochondria into cytosol, where they induced activation of caspases and nuclear fragmentation leading to cell death (Fulda and Debatin, 2000). It also inhibits the growth of cancer cells, without affecting normal cells (Eiznhamer and Xu, 2004). Betulinic acid (**29**) shows antibacterial, antifungal, anti-HIV-1, antiplasmodial and anti-inflammatory activities (Chandramu et al., 2003; Yogeeswari and Sriram, 2005).

In addition, betulinic acid (**29**) showed strong anthelmintic activities, which was comparable to piperazine (Enwerem et al., 2001) besides showing activity against herpes simplex virus (Pavlova et al., 2003). It also protects mice against cerebral ischemia-reperfusion injury (Lu et al., 2011). Liu et al. (2004), reported betulinic acid (**29**) as an apoptosis inducer in skin cancer cells and causes differentiation in normal human keratinocytes, which supports its application not only for drugs but also for cosmetics. Betulinic acid (**29**) exhibited poor antibacterial activity against

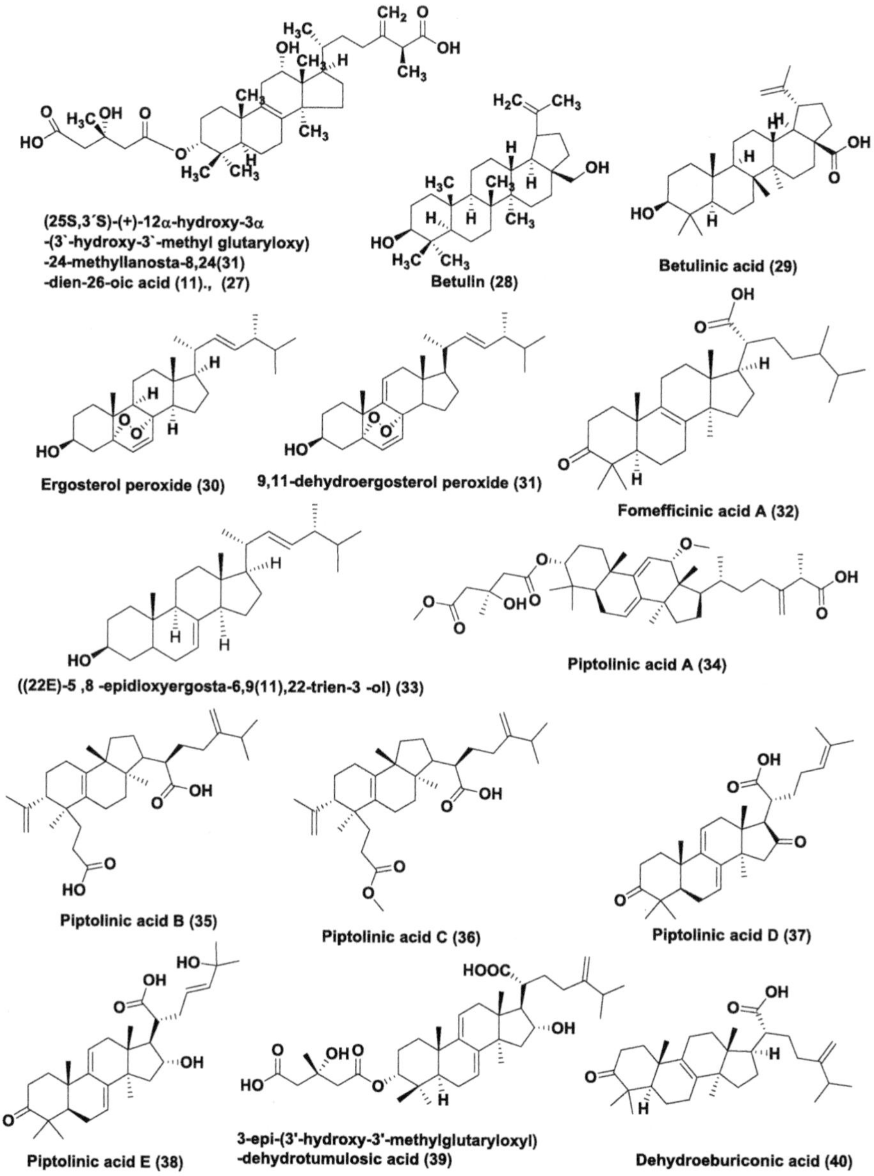

Fig. 3. Structure of compounds 27–40.

S. aureus (Alresly, 2019) and mild antibacterial activity against *B. subtilis* (Chandramu et al., 2003).

Betulin (**28**) exhibited potent anti-inflammatory effect in the carrageenan-induced paw edema in rats, comparable to that of the positive control phenylbutazone and dexamethasone (Patočka, 2003). Moreover, Betulin (**28**) also exhibited selective inhibitory effect on DNA topoisomerase I and II and also antiviral activities against a number of viruses (Shun-ichi et al., 2001; Gong et al., 2003).

Betulin (**28**) possesses antimicrobial and antifungal pathogenic with low-irritation properties, hence can be used in cosmetics. For the first time in the year 2016, betulin gel was approved by the European Commission with clinical benefit of accelerated wound healing as the topical therapeutic agent. The active ingredient of the gel consists of > 80 per cent betulin (**28**) (Fig. 2) and < 20 per cent of other triterpenes, mainly betulinic acid (**29**), lupeol (**23**), erythrodiol and oleanolic acid (Kindler et al., 2016). Cosmetic developers believe that betulinic acid (**29**) (at 50–500 mg per gram of cosmetic) may prevent and help treating UV-induced skin cancer (Krasutsky, 2006).

Ergosterol peroxide (**30**), 9,11-dehydroergosterol peroxide (**31**), (22E)-5α,8α-epidioxyergosta-6,9(11),22-trien-3β-ol) (**33**) (Fig. 2) and polyporenic acid C (**2**) were isolated from *P. betulinus* which were collected in Krc's Forest in Prague, Czech Republic (Hybelbauerova et al., 2008). Ergosterol peroxide (**30**), is found in several mushrooms with anticancer and immunosuppressive properties (Merdivan et al., 2017). It was active against various tumor cell lines of human and murine origin (Jong and Donovick, 1989). The cytotoxicity is assumed to be relevant to its conversion to ergosterol accompanied by the formation of highly toxic peroxide products (Nam et al., 2001). Ergosterol peroxide (**30**) also showed cytotoxicity against HL60 cells by induction of apoptosis (Takei, 2005). Its cytotoxicity was with IC_{50} values of 19.4 µg/ml in Hep 3B cell lines and less in normal immortalised cells than in cancer cell lines (Chen et al., 2009). Ergosterol peroxide (**30**) exhibited some anti-inflammatory property and could be used for prevention and treatment of rheumatoid arthritis (Gao et al., 2007) as it showed immunosuppressive effect (Kuo et al., 2003). The 9,11-dehydroergosterol peroxide (**31**) has immunosuppressive effects (Fujimoto et al., 1994). It also selectively suppresses the growth of human HT29 cell lines, but not the WI38 cell lines which are normal human fibroblast cells (Kobori et al., 2006). Recent studies show that ergosterol peroxide (**30**) could be a promising new source of an anticancer compound that can help in overcoming the drug-resistance of a tumor (Wu et al., 2012).

Fruit bodies of *P. betulinus* extracted in methanol yielded five new lanostane triterpenoids, piptolinic acids A–E (**34–38**), along with known lanostane triterpenoids 3-epi-(3'-hydroxy-3'-methylglutaryloxyl)-dehydrotumulosic acid (**39**), dehydroeburiconic acid (**40**) (Fig. 3), polyporenic acid C (**2**), 6α-hydroxypolyporenic acid C (**41**) and 3-epidehydropachymic acid (**42**) (Fig. 4). Piptolinic acid A (**34**) showed anticancer activity with IC_{50} values of 1.77 and 8.21 µM, respectively against HL-60 and THP-1 cell lines, while positive control fluorouracil exhibited IC_{50} values of 6.38 and 4.41 µM, respectively. The remaining compounds displayed poor or no activity ($IC_{50} > 10$ µM) against tested cell lines (Tohtahon et al., 2017).

Five new lanostanetriterpenoids, namely piptolinic acids F–J (**43–47**), along with its seven known analogues—(25S'3,S)-(þ)-12a-hydroxy-3a-(3'-hydroxy-40-methoxycarbonyl-3-methylbutyryloxy)-24-methyllanosta-8,24(31)-dien-26-oic acid (**48**), dehydrotumulosic acid (**49**), 3-epi-dehydrotumulosic acid (**50**), 16α-hydroxyeburiconic acid (**51**) polyporenic acid A (**1**), 3α,16α-dihydroxy-7-oxo-24-methyllanosta-8,24(31)-dien-21-oic acid (**52**), 16α-hydroxy-3-oxo-lanosta-7,9(11), 24-trien-21-oic acid (**53**) (Fig. 4) were isolated from methanolic extract of the fruit bodies of *P. betulinus*. Compounds (**43–46**) were 24-methyl-

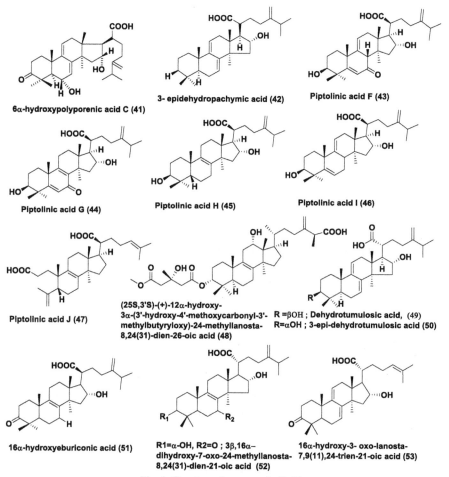

Fig. 4. Structure of compounds 41–53.

lanostane triterpenoids, while compound (**47**) was a 3,4-seco-lanostane derivative. In cytotoxicity screening, the only compound (**48**) showed moderate cytotoxic activity with IC_{50} values of 42.8 and 56.5 µM against A-375 and 786-O cell lines respectively (Khalilov et al., 2019).

4. Concluding Remarks

The therapeutic use of *Fomitopsis betulina* was known for several thousands of years for treating various human ailments and diseases. The prehistoric man referred as Iceman used this fungus to get rid of intestinal parasites. The fungus is also used historically in traditional medicine as an antimicrobial, anticancer and anti-inflammatory agent. A wide range of bioactive metabolites from *Fomitopsis betulina* have been isolated and characterised through modern instrumentation techniques. These include polyporenic acid A, polyporenic acid C, piptamine, betulin, betulinic acid, ergosterol peroxide, p-hydroxybenzoic acid, protocatechuic acid, stigmasterol,

3α-acetylpolyporenic acid A, fomefficinic acid, 9,11-dehydroergosterol peroxide, piptolinic acids A–J, dehydroeburiconic acid, dehydrotumulosic acid and 3-epi-dehydrotumulosic acid. These metabolites possess diverse pharmacological properties, especially antitumor, anti-inflammatory, antioxidant, antibacterial, antiviral, immune stimulating, antifungal and inhibitory activity against Matrix metalloproteinases, namely MMP-1, MMP-3 and MMP-9.

Polyporenic acids A (**1**) and C (**2**) are the most active metabolites of *F. betulina* and polyporenic acid C (**2**) possesses antibacterial, antiviral, anti-inflammatory, anticancer and antioxidant properties. The polyporenic acid C (**2**) shows antibacterial activity against Gram-positive bacteria. It also exhibits inhibitory action against DNA topoisomerase I and II, molecular target for anticancer screening. Polyporenic acid C (**2**) has been found active against human pancreatic cancer cell lines and suppresses A549 cell proliferation by the induction of caspase-8-mediated apoptosis. The inhibition of Akt activation and enhancement of p53 function may also contribute to apoptosis induced by polyporenic acids C. Polyporenic acid A (**1**) exhibited potent anti-inflammatory activity by inhibition of 3α-hydroxysteroid dehydrogenase, inhibiting bacterial hyaluronate lyase and suppressing the TPA-induced edema. The antimicrobial properties of betulin could be used in low-irritation cosmetics and anti-pathogenic fungal cosmetics. Betulinic acid (**29**) induces apoptosis in skin cancer cells and causes differentiation in normal human keratinocytes, facilitating its application in pharmaceutical and cosmeceutical applications. The diverse scaffolds (metabolites) from *F. betulina* associated with various biological activities have opened the door for researchers to develop new compounds with drug-like properties. The novel compounds can be assessed for their pharmacological activities in the treatment of various diseases, including cancer. Apart from this, a combination of existing drugs with metabolites of *F. betulina* may be investigated for their synergistic effects in the treatment of specific diseases.

References

Alresly, Z., Lindequist, U., Lalk, M., Porzel, A., Arnold, N. and Wessjohann, L. (2016). Bioactive triterpenes from the fungus *Piptoporus betulinus*. Rec. Nat. Prod., 10(1): 103.

Alresly, Z. (2019). Chemical and pharmacological investigations of *Fomitopsis betulina* (formerly: *Piptoporus betulinus*) and *Calvatia gigantean*. Mathematisch-Naturwissenschaftliche Fakultät der Universität Greifswald, pp. 1–241.

Borenfreund, E. and Puerner, J.A. (1985). Toxicity determined *in vitro* by morphological alterations and neutral red absorption. Toxicol. Lett., 24(2-3): 119–124.

Bryce, T.A., Campbell, I.M. and McCorkindale, N.J. (1967). Metabolites of the polyporaceae-I: Novel conjugates of polyporenic acid A from *Piptoporus betulinus*. Tetrahedron, 23: 3427–3434.

Chandramu, C., Manohar, R.D., Krupadanam, D.G. and Dashavantha, R.V. (2003). Isolation, characterisation and biological activity of betulinic acid and ursolic acid from *Vitex negundo* L. Phytother. Res., 17(2): 129–134.

Chen, Y.K., Kuo, Y.H., Chiang, B.H., Lo, J.M. and Sheen, L.Y. (2009). Cytotoxic activities of 9, 11-dehydroergosterol peroxide and ergosterol peroxide from the fermentation mycelia of *Ganoderma lucidum* cultivated in the medium containing leguminous plants on Hep 3B cells. J. Agric. Food Chem., 57(13): 5713–5719.

Cheng, S., Eliaz, I., Lin, J., Thyagarajan-Sahu, A. and Sliva, D. (2013). Triterpenes from *Poriacocos* suppress growth and invasiveness of pancreatic cancer cells through the downregulation of MMP-7. Int. J. Oncol., 42: 1869–1874.

Cross, L.C., Eliot, C.G., Heilborn, I.M. and Jones, E.R.H. (1940). Constituents of the higher fungi. Part I. The triterpene acids of *Polyporus betulinus*. Fr. J. Chem. Soc., 1940: 632–636.

Cyranka, M., Graz, M., Kaczor, J., Kandefer-Szerszen, M., Walczak, K., Kapka-Skrzypczak, L. and Rzeski, W. (2011). Investigation of antiproliferative effect of ether and ethanol extracts of Birch polypore medicinal mushroom, *Piptoporus betulinus* (Bull.: Fr.) P. Karst. (Higher Basidiomycetes) *in vitro* grown mycelium. Int. J. Med. Mush., 13(6): 525–533.

de Jesus, L.I., Smiderle, F.R., Ruthes, A.C., Vilaplana, F., Dal'Lin, F.T., Maria-Ferreira, D., Werner, M.F., Van Griensven, L.J. and Iacomini, M. (2018). Chemical characterisation and wound healing property of a β-D-glucan from edible mushroom *Piptoporus betulinus*. Int. J. Biol. Macromol., 117: 1361–1366.

Doskocil, I., Havlik, J., Verlotta, R., Tauchen, J., Vesela, L., Macakova, K., Opletal, L., Kokoska, L. and Rada, V. (2016). *In vitro* immunomodulatory activity, cytotoxicity and chemistry of some central European polypores. Pharmaceut. Biol., 54(11): 2369–2376.

Dresch, P., Rosam, K., Grienke, U., Rollinger, J.M. and Peintner, U. (2015). Fungal strain matters: Colony growth and bioactivity of the European medicinal polypores *Fomes fomentarius*, *Fomitopsis pinicola* and *Piptoporus betulinus*. AMB Express, 5(1): 1–14.

Eiznhamer, D.A. and Xu, Z.Q. (2004). Betulinic acid: A promising anticancer candidate. I Drugs, 7: 359–373.

Enwerem, N.M., Okogun, J.I., Wambebe, C.O., Okorie, D.A. and Akah, P.A. (2001). Anthelmintic activity of the stem bark extracts of *Berlina grandiflora* and one of its active principles, Betulinic acid. Phytomedicine, b(2): 112–114.

Fujimoto, H., Nakayama, M., Nakayama, Y. and Yamazaki, M. (1994). Isolation and characterisation of immunosuppressive components of three mushrooms, *Pisolithus tinctorius*, *Microporus flabelliformis* and *Lenzites betulina*. Chem. Pharmaceut. Bull., 42(3): 694–697.

Fulda, S. and Debatin, K.M. (2000). Betulinic acid induces apoptosis through a direct effect on mitochondria in neuroectodermal tumors. Med. Ped. Oncol., 35(6): 616–618.

Gao, J.M., Wang, M., Liu, L.P., Wei, G.H., Zhang, A.L., Draghici, C. and Konishi, Y. (2007). Ergosterol peroxides as phospholipase A2 inhibitors from the fungus *Lactarius hatsudake*. Phytomed., 14(12): 821–824.

Gong, Y., Raj, K.M., Luscombe, C.A., Gadawski, I., Tam, T., Chu, J., Gibson, D., Carlson, R. and Sacks, S.L. (2004). The synergistic effects of betulin with acyclovir against herpes simplex viruses. Antiviral Res., 64(2): 127–130.

Grienke, U., Zöll, M., Peintner, U. and Rollinger, J.M. (2014). European medicinal polypores—A modern view on traditional uses. J. Ethnopharmacol., 154(3): 564–583.

Grishin, A.A., Lutskii, V.I., Penzina, T.A., Dudareva, L.V., Zorina, N.V., Polyakova, M.S. and Osipenko, S.N. (2016). Composition of the supercritical CO_2 extract of the fungus *Piptoporus betulinus*. Chem. Nat. Comp., 52(3): 436–440.

Guider, M., Halsall, T.G., Hodges, R. and Jones, E.R.H. (1954). The chemistry of the triterpenes and related compounds. Part XXVI. The nature of polyporenic acid B. J. Chem. Soc., 3234–3238.

Han, M.L., Chen, Y.Y., Shen, L.L., Song, J., Vlasák, J., Dai, Y.C. and Cui, B.K. (2016). Taxonomy and phylogeny of the brown-rot fungi: *Fomitopsis* and its related genera. Fungal Diversity, 80(1): 343–373.

Harms, M., Lindequist, U., Al-Resly, Z. and Wende, K. (2013). Influence of the mushroom *Piptoporus betulinus* on human keratinocytes. Planta Medica, 79(13): PC4. 10.1055/s-0033-1351998.

Hobbs, C. (2002). Medicinal Mushrooms: An Exploration of Tradition, Healing and Culture (Herbs and Health Series). Botanica Press, USA.

Hybelbauerová, S., Sejbal, J., Dračínský, M., Hahnová, A. and Koutek, B. (2008). Chemical constituents of stereum subtomentosum and two other birch-associated Basidiomycetes: An interspecies comparative study. Chem. Biodivers., 5(5): 743–750.

Jong, S.C. and Donovick, R. (1989). Antitumor and antiviral substances from fungi. Adv. Appl. Microbiol., 34: 183–262.

Kamo, T., Asanoma, M., Shibata, H. and Hirota, M. (2003). Anti-inflammatory lanostane-type triterpene acids from *Piptoporus betulinus*. J. Nat. Prod., 66(8): 1104–1106.

Kandefer-Szerszeń, M. and Kawecki, Z. (1974). Ether extracts from the fruiting body of *Piptoporus betulinus* as interference inducers. Acta Microbiolo. Polon. Series A, Microbiol. Gen., 6(2): 197–200.

Karsten, P.A. (1881). Enumeratio Hydnearum Fr. Fennicarum, systemate novo dispositarum. Revue Mycologique, Toulouse (in Latin), 3 (9): 17.

Kawagishi, H., Hamajima, K. and Inoue, Y. (2002). Novel hydroquinone as a matrix metallo-proteinase inhibitor from the mushroom, *Piptoporus betulinus*. Biosci. Biotechnol. Biochem., 66(12): 2748–2750.

Keller, C., Maillard, M., Keller, J. and Hostettmann, K. (2002). Screening of European fungi for antibacterial, antifungal, larvicidal, molluscicidal, antioxidant and free-radical scavenging activities and subsequent isolation of bioactive compounds. Pharmaceut. Biol., 40(7): 518–525.

Khalilov, Q., Li, L., Liu, Y., Tohtahon, Z., Chen, X., Aisa, H.A. and Yuan, T. (2019). Piptolinic acids F–J, five new lanostane-type triterpenoids from *Piptoporus betulinus*. Nat. Prod. Res., 33(21): 3044–3051.

Kim, K.M., Yoon, Y.G. and Jung, H.S. (2005). Evaluation of the monophyly of *Fomitopsis* using parsimony and MCMC methods. Mycologia, 97(4): 812–822.

Kindler, S., Schuster, M., Seebauer, C., Rutkowski, R., Hauschild, A., Podmelle, F., Metelmann, C., Metelmann, B., Müller-Debus, C., Metelmann, H.R. and Metelmann, I. (2016). Triterpenes for well-balanced scar formation in superficial wounds. Molecules, 21(9): 1129.10.3390/molecules21091129.

Kobori, M., Yoshida, M., Ohnishi-Kameyama, M., Takei, T. and Shinmoto, H. (2006). 5α, 8α-Epidioxy-22E-ergosta-6, 9 (11), 22-trien-3β-ol from an edible mushroom suppresses growth of HL60 leukemia and HT29 colon adenocarcinoma cells. Biol. Pharmaceut. Bull., 29(4): 755–759.

Krasutsky, P.A. (2006). Birch bark research and development. Nat. Prod. Rep., 23(6): 919–942.

Kuo, Y.C., Weng, S.C., Chou, C.J., Chang, T.T. and Tsai, W.J. (2003). Activation and proliferation signals in primary human T lymphocytes inhibited by ergosterol peroxide isolated from *Cordyceps cicadae*. Br. J. Pharmacol., 140: 895–906

Lemieszek, M.K., Langner, E., Kaczor, J., Kandefer-Szerszen, M., Sanecka, B., Mazurkiewicz, W. and Rzeski, W. (2009). Anticancer effect of fraction isolated from medicinal Birch polypore mushroom, *Piptoporus betulinus* (Bull.: Fr.) P. Karst. (Aphyllophoromycetideae): *In vitro* studies. Int. J. Med. Mush., 11(4): 351–364.

Li, G., Xu, M.L., Lee, C.S., Woo, M.H., Chang, H.W. and Son, J.K. (2004). Cytotoxicity and dna topoisomerases inhibitory activity of constituents from the sclerotium of *Poria cocos*. Arch. Pharm. Res., 27(8): 829–833.

Ling, H., Zhou, L., Jia, X., Gapter, L.A., Agarwal, R. and Ng, K.Y. (2009). Polyporenic acid C induces caspase-8-mediated apoptosis in human lung cancer A549 cells. Mol. Carcinog., 48(6): 498–507.

Liu, W., Ho, J., Cheung, F. and Liu, B. (2004). Apoptotic activity of betulinic acid derivatives on murine melanoma B16 cell line. Eur. J. Pharmacol., 498: 71–78.

Lohwag, K. (1965). Birkenschwammschnitzereien aus der Steiermark. Mitteilungsblatt des Naturwissenschaftlichen Vereins für Steiermark 95: 136–139.

Lu, Q., Xia, N., Xu, H., Guo, L., Wenzel, P., Daiber, A., Münzel, T., Förstermann, U. and Li, H. (2011). Betulinic acid protects against cerebral ischemia–reperfusion injury in mice by reducing oxidative and nitrosative stress. Nitric Oxide, 24(3): 132–138.

Lucas, E.H. (1960). Folklore and Plant Drugs. Papers of the Michigan Academy of Sience, Arts, and Letters XLV, 127–136.

Marcus, S. (1952). Antibacterial activity of the triterpenoid acid (polyporenic acid C) and of ungulinic acid, metabolic products of *Polyporus benzoinus* (Wahl.) Fr. Biochem. J., 50(4): 516–517.

Marsh, R.W. (1973). Micological millinery. Bull. Br. Mycol. Soc., 7: 35.

Merdivan, S. and Lindequist, U. (2017). Ergosterol peroxide: a mushroom-derived compound with promising biological activities—A review. Int. J. Med. Mush., 19: 93–105.

Nam, K.S., Jo, Y.S., Kim, Y.H., Hyun, J.W. and Kim, H.W. (2001). Cytotoxic activities of acetoxyscirpenediol and ergosterol peroxide from *Paecilomyces tenuipes*. Life Sci., 9(2): 229–237.

Ortiz-Santana, B., Lindner, D.L., Miettinen, O., Justo, A. and Hibbett, D.S. (2013). A phylogenetic overview of the antrodia clade (Basidiomycota, Polyporales). Mycologia, 105(6): 1391–1411.

Papp, N., Rudolf, K., Bencsik, T. and Czégényi, D. (2015). Ethnomycological use of *Fomes fomentarius* (L.) Fr. and *Piptoporus betulinus* (Bull.) P. Karst. in Transylvania, Romania. Genet. Resour. Crop Evol., 64: 101–111.

Patocka, J. (2003). Biologically active pentacyclic triterpenes and their current medicine signification. J. Appl. Biomed., 1(1): 7–12.

Pavlova, N.I., Savinova, O.V., Nikolaeva, S.N., Boreko, E.I. and Flekhter, O.B. (2003). Antiviral activity of betulin, betulinic and betulonic acids against some enveloped and non-enveloped viruses. Fitoterapia, 74(5): 489–492.

Pegler, D.N. (2000). Useful fungi of the world: some use of bracket fungi. Mycologist, 14: 6–7.

Peintner, U., Pöder, R. and Pümpel, T. (1998). The iceman's fungi. Mycological Research, 102(10): 1153–1162.

Pleszczyńska, M., Wiater, A., Siwulski, M., Lemieszek, M.K., Kunaszewska, J., Kaczor, J., Rzeski, W., Janusz, G. and Szczodrak, J. (2016). Cultivation and utility of *Piptoporus betulinus* fruiting bodies as a source of anticancer agents. World J. Microbiol. Biotechnol., 32(9): 151.10.1007/s11274-016-2114-4.

Pleszczyńska, M., Lemieszek, M.K., Siwulski, M., Wiater, A., Rzeski, W. and Szczodrak, J. (2017). *Fomitopsis betulina* (formerly *Piptoporus betulinus*): The Iceman's polypore fungus with modern biotechnological potential. World J. Microbiol. Biotechnol., 33(5): 83.10.1007/s11274-017-2247-0.

Reshetnikov, S.V., Wasser, S.P. and Tan, K.K. (2001). Higher basidiomycota as source of antitumor and immunostimulating polysaccharides. Int. J. Med. Mushr., 3: 361–394.

Rutalek, R. (2002). Ethnomykologie–Eine Übersicht. Österr Z Pilzkd, 11: 79–94.

Sari, M., Toepler, K., Roth, C., Teusch, N. and Hambitzer, R. (2020). The birch bracket medicinal mushroom, *Fomitopsis betulina* (Agaricomycetes)–Bioactive source for beta-glucan fraction with tumor cell migration blocking ability. Int. J. Med. Mush., 22(1): 1–13.

Schlegel, B., Luhmann, U., Haertl, A. and Graefe, U. (2000). Piptamine, a new antibiotic produced by *Piptoporus betulinus* Lu 9-1. J. Antibiot., 53(9): 973–974.

Semerdžieva, M. and Veselský, J. (1986). Léčivé houby dříve a nyní. Academia Praha, Praha, pp. 177.

Shamtsyan, M., Konusova, V., Maksimova, Y., Goloshchev, A., Panchenko, A. et al. (2004). Immunostimulating and anti-tumor action of extracts of several mushrooms. J. Biotechnol., 13: 77–83.

Shun-ichi, W., Iida, A. and Tanaka, R. (2001). Screening of triterpenoids isolated from *Phyllanthus flexuosus* for DNA topoisomerase inhibitory activity. J. Nat. Prod., 64: 1545–1547.

Stamets, P. (2011). Growing Gourmet and Medicinal Mushrooms. Ten Speed Press, Barkeley.

Takei, T., Yoshida, M., Ohnishi-Kameyama, M. and Kobori, M. (2005). Ergosterol peroxide, an apoptosis-inducing component isolated from *Sarcodon aspratus* (Berk.). S. Ito. Biosci. Biotechnol. Biochem., 69(1): 212–215.

Thoen, D. (1982). Usage et le!gendes lie!s aux polypores. Note d'ethnomycology. Bulletin Trimestriel de la SocieUteU Mycologique de France, 98: 289–318.

Tohtahon, Z., Xue, J., Han, J., Liu, Y., Hua, H. and Yuan, T. (2017). Cytotoxic lanostane triterpenoids from the fruiting bodies of *Piptoporus betulinus*. Phytochem., 143: 98–103.

Vunduk, J., Klaus, A., Kozarski, M., Petrovic, P., Zizak, Z., Niksic, M. and Van Griensven, L.J.L.D. (2015). Did the Iceman know better? Screening of the medicinal properties of the birch polypore medicinal mushroom, *Piptoporus betulinus* (Higher Basidiomycetes). Int. J. Med. Mush., 17(12): 1113–25.

Wandokanty, F., Utzig, J. and Kotz, J. (1955). The effect of *Poria obliqua* and *Polyporus betulinus* on spontaneous cancer of the dog with respect to breast cancer in dogs. Med. Weter, 3: 148–151.

Wangun, H.V.K., Berg, A., Hertel, W., Nkengfack, A.E. and Hertweck, C. (2004). Anti-inflammatory and anti-hyaluronate lyase activities of lanostanoids from *Piptoporus betulinus*. J. Antibiot., 57(11): 755–758.

Wasser, S.P. (2010). Medicinal mushroom science: history, current status, future trends, and unsolved problems. Int. J. Med. Mushr., 12: 1–16.

Wasson, R.G. (1971). Soma: Divine Mushroom of Immortality. Harcourt Brace Jovanovich, New York.

Wu, Q.P., Xie, Y.Z., Deng, Z., Li, X.M., Yang, W., Jiao, C.W., Fang, L., Li, S.Z., Pan, H.H., Yee, A.J. and Lee, D.Y. (2012). Ergosterol peroxide isolated from *Ganoderma lucidum* abolishes microRNA miR-378-mediated tumor cells on chemoresistance. PloS ONE, 7(8): e44579.

Ying, J., Mao, X., Ma, Q., Zong, Y. and Wen, H. (1987). Icones of Medical Fungi from China. Science Press, Beijing, China.

Yogeeswari, P. and Sriram, D. (2005). Betulinic acid and its derivatives: A review on their biological properties. Curr. Med. Chem., 12(6): 657–666.

Zwolińska, K. (2004). Evaluation of anticancer activity of extracts from birch polypore *Piptoporus betulinus* (Bull. ex Fr.) P. Karst. Dissertation, Maria Curie-Skłodowska University, Lublin, Poland.

Therapeutics

6

Medicinal Potential of Entomopathogenic *Cordyceps*

S Shishupala

1. INTRODUCTION

Fungi are capable of producing an array of compounds and their derivatives which are useful as potential medicines to treat several human ailments. Metabolites produced by different fungi are of significant value in traditional as well as modern medicinal products. Various ethnic and tribal groups across the world have discovered medicinal properties of different fungi. Among those, mushrooms occupy a prominent place in traditional foods (as nutraceuticals) and to cure many ailments (as medicines). Many genera of fungi are exploited from the wild for such therapeutic potential. One such group of fungi is the entomopathogenic fungi capable of growing on insects to fulfil completion of their life cycles. Currently over 50 entomopathogenic fungal genera are identified as promising biocontrol agents against many insect pests (Chiu et al., 2016). Such fungi are extremely vital in maintenance of natural balance of insect population. Some of the entomopathogenic fungi are used as bioinsecticides or biopesticides to control insect pests in agriculture and these include *Beauvaria, Coelomymyces, Cordyceps, Entomophthora, Hirsutella, Lecanicillium, Metarhizium, Nomuraea, Paecilomyces* and so on. These fungi infect a wide group of insects, ranging from butterflies, flies, bugs, beetles, wasps, grasshoppers, spiders and mites (Deacon, 2006; Zhou et al., 2009; Ali, 2012). The present chapter deals with various aspects of medicinal properties of entomopathogenic *Cordyceps* and allied fungi, including their metabolites, biological activities and biotechnological applications.

Department of Microbiology, Davangere University, Shivagangothri, Davangere 577007, Karnataka, India.
Email: ssdumb@gmail.com

2. *Cordyceps*

Cordyceps belongs to the division Ascomycotina, Class Sordariomycetes, Order Hypocrales and Family Cordycipitaceae. Most of the species are entomopathogens which infect mainly the lepidopteran group of insects. *Cordyceps* is one of the promising genera used widely in Chinese medicine. The life cycle and growth of *Cordyceps* is highly complex. They are common in grasslands of Himalayas or the highlands of northern Nepal ranging from 3,000 m asl. More than 27 mountain districts of Nepal have been reported to have diverse species of *Cordyceps*. These fungi are capable of infecting *Thitarodes* caterpillars and more than 57 potential host species have been identified (Baral et al., 2015). The fungal spores infect the caterpillars of the host insects and proliferate, inducing the host to hibernate during winter. The normal life cycle may range from two to six years, depending on several environmental factors. Mycelial growth occurs in the insect body, leading to mummification (Fig. 1a). The insect larva gets killed by 15–25 days, but the fungus persist inside the exoskeleton. The fruit bodies (stroma) erupt from the dead host up to a length of 16 cm (Fig. 1b). These fruit bodies on maturity release spores when disturbed (Chen et al., 2013; Baral et al., 2015; Holliday, 2017). The fruit bodies along with the host are used for traditional Chinese medicine. An ever-increasing market for *Cordyceps* and scarcity of the product makes the cost of this fungus reach as high as US$ 100,000 per kilogram. Hence, a lot of research has gone into the artificial cultivation by many companies. The leading company Aloha

Fig. 1. Natural and cultivated *Cordyceps*. (a) Mummified insect due to colonisation of *Cordyceps* (https://allthatsinteresting.com/cordyceps-killer-fungus#7). (b) Emergence of stroma of *Cordyceps militaris* from the insect body in soil (https://hackspirit.com/cordyceps/). (c) Cultivation of *Cordyceps* in an industry (https://blog.freshcapmushrooms.com/learn/cordyceps-mushroom-health/).

Medicinals in US has been constantly involved in developing large-scale cultivation and production which is more than 50 per cent of the world market (Holliday, 2017). Some companies are trying to select useful species of *Cordyceps* for nutraceutical and pharmaceutical purposes. The requirement of unique ecological and host factors add up to the complexity of artificial cultivation (Fig. 1c).

In the Himalayan range (India, Nepal and Tibet), *Cordyceps* spp. are found at high altitudes (3500–5000 m asl) and they are called *Keeda Jadi* (meaning, insect plant). Traditionally, they were used for the treatment of heart, kidney and liver disorders (Seth et al., 2014; Yan et al., 2014). The annual estimate of the yield is about 60 tons from the Tibet autonomous region of the Himalayan plateau. Other places of its occurrence include Nepal, Bhutan and the north-eastern state of Sikkim in India (Holliday, 2017). Various local names are being designated for the medicinal fungus being used (Table 1). In China, the term 'Dong Chong Xia Cao' is being collectively used for *Ophiocordyceps* and *Cordyceps*. Health foods as well as crude medicines originate from caterpillar-shaped fruit body and mycelia from submerged culture of *Ophiocordyceps* and *Hirsutella sinensis*. So also traditional Chinese medicine is formulated from fungus-infected larvae of Hepialidae (*Thitarodes armoricanus* and *Thitarodes* spp.) (Tsim and Shao, 2005; Lo et al., 2013). Hence, what is being used as crude extract or consortium is chemically highly complex. However, considering the potential medicinal values, investigations have been carried on the crude extracts as well as purified compounds.

Table 1. Regional names of *Cordyceps.*

Traditional name	Meaning	Uses
Dong Chong Xia Cao (China)	Chinese caterpillar fungus/ vegetable wasps and plant worms Winter-insect Summer-grass	Traditional Chinese medicine for various ailments
Tochukaso (Japan)	Winter-insect Summer-grass (fungi on insects)	Treatment of different types of cancers, kidney disorders, diabetes and blood pressure
Yartsa Gunbu (Tibet and Bhutan)	Yartsa-Grass in summer Gunbu-Worm in winter	Herbal treatment for various disorders
Yarsagumba or *Yarchagumbaor Bu* or *Bhu-Sanjivani* or *Jivan Buti* or Jingani or *Kira Chhyau* or *Kira Jhar* or *Saram Buti Jadi* or *Saram Buti* (Nepal)	Insect fungus or herbal medicine	Herbal medicine
Kheeda Jadi or *Kheeda Ghaas* (India)	Insect plant	Traditional treatment of disorders in Himalayan range

2.1 Diversity and Entomopathogenic Potential of Cordyceps

Diversity in *Cordyceps* is well known with respect to morphological features and pharmacological compounds. Hence, it is important to study mycological characters as diagnostic features required for authentic identification. Use of such features will also help medical applications by avoiding similar species without biological activities.

Sung et al. (2007) provided classification of *Cordyceps* and other Clavicipitaceous fungi. *Cordyceps* and related genera are classified conventionally based on arrangement of perithecia, ascospore morphology and pathogenicity against insects. Phylogenetic relationships of 162 taxa were analysed at molecular level to describe new species. The study provided detailed analysis of this group of fungi by complementing molecular data with texture, pigmentation and stromal characteristics. It has emphasised the complexity associated with diversity of *Cordyceps* and related fungi. Most of the species were categorised into *Cordyceps*, *Metacordyceps*, *Elaphocordyceps* and *Ophiocordyceps*. It is important to explore diagnostic markers for medicinal potential of different species or strains. Liu et al. (2011) demonstrated macroscopic and microscopic differences among six *Cordyceps* spp. Transverse sections of stroma and larvae were also considered as important characteristic features. Morphological characters of stroma like stout and rough in case of *C. gunnii*, thread like in *C. liangshanesi* and stroma was absent in *C. gracilis*, while the stroma of *C. barnesii* was devoid of perithesia. Such differences were useful in distinguishing the species. However, for medicinal purposes generally *Cordyceps* is most preferred. In another taxonomic study, comparison of *Cordyceps* was made and it was found that *C. cicadae* produce asexual fruit bodies with conidiospores instead of meiotic ascospores (Lu et al., 2017).

The diversity in *Cordyceps* is highly complex and the prototype species is actually *Ophiocordyceps*. More than 600 species of *Cordyceps* are known in traditional medicine. Holliday (2017) found over thousand types of *Cordyceps* in Peru, while others include *Elaphocordyceps* and *Metacordyceps*. Even the species of much medicinal value is *Ophiocordyceps sinensis*, but it has been referred as *Cordyceps sinensis* in literature. From different parts of world, different species are reported, but for the purpose of medicinal value all of them are considered as *Cordyceps*. Most of the medicinal value and type species is from the *Cordyceps militaris*. Pathania and Sagar (2014) made a detailed analysis of *Cordyceps militaris* from the Himalayan range. For identification, shape, size and colour of the stipe and stroma were considered along with the host larvae. Light and scanning electron microscopic observations of mycelium and spores were also carried out. Growth requirements and chemical constituents were analysed. In solid media, pH 7.5 was congenial for getting maximum yield of biomass, whereas pH 5.5 was efficient in liquid medium. Biologically active components, like cordycepin and D-mannitol (being referred as cordycepic acid) were also detected apart from other nutritional factors.

Interesting analysis of microbial communities existing with the host moth (*Thitarodes*) of *Ophiocordyceps sinensis* have been made. This fungus, being used in Chinese medicine, is obligatory parasitising the larvae of lepidopteran moth. Large-scale cultivation of this fungus met with limited success because of the presence of various other microorganisms. Using molecular methods, a total of 348 bacterial genera and 289 fungal genera were identified as associated microbes with the larvae. The study revealed a significant role of these microorganisms in determining the efficient exploitation of Chinese *Cordyceps*. Complexity of diverse groups of organisms associated with the host moth may affect large-scale cultivation and extraction of therapeutic compounds from medicinal *Cordyceps* (Liang et al., 2019). Apart from the medicinal values, *Cordyceps* is also known to produce value-added enzymes. *Cordyceps farinose* was identified to produce amylases. Structural and functional analysis provided information on the novel binding site in C-terminal domain; thus, such studies expand the applications of this fungus (Roth et al., 2019). In another exciting study, entomopathogenic capacity of *Cordyceps javanica* was enhanced by combination with insect parasitoid *Eretmocerus hayati* in the management of whitefly *Bemisia tabaci*. Dose-dependent combined efficiency was found with individual biocontrol agents (Ou et al., 2019). Extensive diversity and compatibility of *Cordyceps* with other biocontrol agents provide the opportunity for successful use as biopesticide. Diversity and biological activities are an integral part of *Cordyceps*-based medicines. Sun et al. (2020) reported increased growth in eggplant and reduction in whitefly attack after seed treatment with *Cordyceps fumosorosea*. This endophyte was an effective coloniser of eggplant, preventing the attack of insect *Bemisia tabaci*. Hence, biological control potential along with plant growth promotion was practically achieved. Biopesticide formulation of *Beauveria bassiana* and *Cordyceps javanica* in combination with biosurfactant (rhamno lipid) was developed against the moth, *Bemisia tabaci*. Though *C. javanica* was effective in causing pest mortality, its efficiency was lesser than *B. bassiana* (Silva et al., 2019). Biocontrol efficiency of entomopathogenic fungi was tested by considering the development of a fluid-bed coating process for soil granule-based formulation (Stephan et al., 2020). Hence, the entomopathogenic fungi, including *Cordyceps*, are most useful candidates as biopesticides.

Cordyceps militaris is also being considered as feed additive in livestock management. With rich availability of glycans, polysaccharides, polyphenols, triterpenes, ergosterol, laccases and so on, this fungus is known to enhance the body weight and egg production in poultry. Hence, *C. militaris* along with other beneficial fungi may be used as probiotics of livestock (Chuang et al., 2020). Occurrence of many species and differential medicinal significance in *Cordyceps* and related fungi is a matter of intense study. A comparative survey of published literature on different species, such as *C. cicadae*, *C. militaris* and *C. sinensis* was made. Similar biologically active compounds were detected in all the three species. Although the *C. militaris* and *C. sinensis* are widely considered for medicinal value, it is also important to consider *C. cicadae* too owing to its appreciable bioactive capabilities. Moreover, this species is known to occur in different geographical locations (Nxumala et al., 2020). It is also important from the conservation point of view to investigate biochemical and molecular variations in the available species.

Multiple benefits have been attributed to *Cordyceps*. Fan et al. (2020) discussed the use of traditional Chinese medicine in sepsis management. This herbal formulation containing various bioactive compounds is being advised even for patients in intensive care units. This traditional medicine is also recommended against COVID-19 infection caused by the corona virus. The fungus is eaten along with the host insect directly or made in the form of a powder. Complexities of the range of compounds synergistically act in providing health benefits.

2.2 Traditional Medicines from Cordyceps

Ethnic groups and tribals in the countries of Himalayan range realised the health benefits of *Cordyceps* centuries ago. They continued to harvest and use locally available material at will. As the information on claimed health benefits passed across, it made the Western world to show interest in these mysterious fungi. Dong et al. (2013) reported the production of cordyxanthin from fruit bodies of *Cordyceps militaris*. Medicinal values of *Cordyceps sinensis* have attracted the Asian countries since many centuries. This fungus infects and grows in the body of Himalayan ghost moth (*Thitarodes* spp.) or bat moth (*Hepialis armoricanus*), mummifies it and produces fruit bodies. Such fungus-insect admixture is used in Chinese traditional medicines. It is also a favourable medicinal herbal drug in countries like Nepal, Bhutan and Tibet. *Cordyceps sinensis* has become biotechnologically significant owing to the production of a wide range of biologically active compounds (Chiu et al., 2016; Choda, 2017). Chinese and Tibetan medicinal practices involve the use of *Cordyceps* from nature. In the lifecycle of *Cordyceps*, during the summer the fungus infects the insect larvae, colonises inside the larval body and provides stiff appearance to the insect by winter; hence called as 'winter worm'. By the following summer, during the reproductive phase the fungus emerges as its stroma from the head of larvae present in the soil; then it is known as 'summer grass' (Zhou et al., 2009). Such field observations had made people to think this organism was able to change from plant in the summer and worm in the winter season due to the complex life cycle of the fungus. Historical records suggest that the available written document from China dates back to AD 620 (Bensky et al., 2004). The medicinal value was documented in the *Compendium Materia Medica* in 1694 (Winkler, 2008). Even in India, the entomopathogenic *C. sinensis* is being used in traditional medicine by the local name of *Keeda Jadi*. This fungus occurs in the high altitudes of the Himalayan mountain ranges. Bioprospecting of this fungus for its biological activities and pharmacological potentials is the immediate necessity in view of the extensive medicinal attributes (Seth et al., 2014).

In medical applications, the fungal components of insect pathogen are highlighted. In reality, the fungus growing on larvae and the insect larvae are known to have medicinal components. In traditional medicine, the larvae infected with *Cordyceps* are consumed whole. Mummified larvae would be fully colonised by the fungus and chitinaceous cell wall of insect larvae would hold all the extracellular metabolites of the fungus. A majority of the biological activity is detected in extracellular components. Hence, in traditional medicine, the whole component has been considered (Holliday, 2017). This traditional knowledge was later translated into commercial purposes. It is imperative that the medically active compounds are produced due to interactions than the fungus alone. The insect larvae may provide a lot of precursors for the metabolites of the fungus.

2.3 *Biological Activities of Cordyceps*

Extensive literature is available on the biological activities of *Cordyceps*. High value medical applications warrant evaluation of biological activities in cellular as well as secretory metabolites of fungi. The potential of *C. sinensis* in showing an array of therapeutic properties has been realised. Bioactive components of this fungus include 3'-deoxyadenosine, cordycepic acid and polysaccharides. Several ailments like night sweating, kidney problems, hyposexuality, hyperlipidaemia and other disorders (heart, liver and respiratory) are being tackled in relation to the components of *Cordyceps*. A wide range of pharmacological activities are attributed to bioactive compounds found in these entomopathogenic fungi (Zhou et al., 2019). Exopolysaccharides and intracellular polysaccharides are the important biologically active components of *C. sinensis*. A comprehensive review with respect to production, extraction, isolation, purification and physicochemical characterisation of polysaccharides from *Cordyceps* was made by Yan et al. (2014). Their study revealed the complexity, biological activity and structure of fungal polysaccharides. Optimisation of factors influencing the production of biologically-active polysaccharides is essential for improving the market value. Exopolysaccharides of *Cordyceps* are known for their structural diversity and biological functions. Medicinal properties of *Cordyceps* are partly assigned to polysaccharide complexity in terms of structure and function. Anabolic pathways of polysaccharides also contribute to the variations in the composition (Yang et al., 2020c; Ying et al., 2020). *Cordyceps cicadae* polysaccharides showed important pharmacological properties. *In vitro* antioxidant property using *Drosophila* model indicated that the polysaccharide fraction CP70 increased the catalase and glutathione peroxidase activities. The results suggested that CP70 fraction was involved in prolonging the life span of *Drosophila* and significant antioxidant and antiageing properties were found in *C. cicadae* (Zhu et al., 2020). The mechanism of action of *Cordyceps* polysaccharides on acute liver failure in rats was investigated by Gu et al. (2020). The polysaccharide was able to prevent hepatocyte apoptosis. The mechanism involved were effective in suppression of caspase, interleukin-18 and interleukin-10 expression along with simultaneous increased expression of vascular endothelial growth factor and stromal cell-derived factor 1α as tested by Western blot. The result clearly demonstrated a reduction of apoptosis by balancing pro-inflammatory and anti-inflammatory factors by *Cordyceps* polysaccharides.

In an interesting study by Boontiam et al. (2019), the effect of feeding pigs with *C. militaris*-grown substrate altered the growth and haemotological features. Pigs were provided with 2 g/kg of *C. militaris* spent-material increased the final body weight as well as the rate of weight gain on a daily basis. Increased concentration of immunoglobulins, antioxidant capacity and glutathione peroxidase activity were also noticed. Decreased cholesterol was found in pigs fed with mushroom spent-material. The study indicated the secretion of bioactive compounds into the substrate, which could be used as animal feed supplement. *Cordyceps militaris* enhanced the meat tenderness and antioxidant properties when mixed with broiler chickens. A higher amount of mixing also increased the flavour and taste-related compounds, proving nutraceutical value of this fungus (Barido et al., 2020). Use of *Cordyceps* as animal feed additives was recommended by Chuang et al. (2020). Nutraceutical and therapeutic potential was also proved in the case of using cordycepin for health and well-being (Ashraf et al., 2020). A wide range of biological activities have been addressed in different components of *Cordyceps* is given in Table 2.

Bioactive compounds of *Cordyceps nidus* showed an increase in laccase production in another mushroom, *Pleurotus ostreatus*. Ligninolytic enzymes, such as laccases are having significant applications in lignocellulosic biomass utilisation. Bioactive compounds of *C. nidus* on mixing with the substrate resulted in high quantity of enzyme production (Duran-Aranguren et al., 2020). This study extended

Table 2. Biological activities attributed to *Cordyceps*.

Component/Metabolite	Biological activity
Fruit body	Hepatoprotection
Exopolysaccharides	Immunomodulatory, anticancer and antioxidant properties
Cellular polysaccharides	Immunostimulatory, anticancer, antioxidant, hypoglycemic, renal-protective, cholesterol reducing and antiaging properties
Cordycepin	Immunomodulatory, anticancer, antiviral, cordiac hypertrophy, antileukemic, antimetastatic, antimicrobial, antidiabetic and hepatoprotective activities
Adenosine and guanosine	Immunomodulatory and coronary ailments
Cordymin	Antioxidant and anti-inflammatory activities
Lovastatin	Cholesterol reducing potential
Ergosterol, Sitosterol, sterol derivatives, and Cordyceamide A and B	Cytotoxic, vitamin D biosynthesis and treatments of autoimmune disorders
Myriocin	Immune response inhibitor and antibiotic
Melanin	Antioxidant potential
Cordysinin A to E	Anti-inflammatory potential
Cyclosporine	Immunosuppressive potential
Cyclodepsipeptides	Antimicrobial, insecticidal and cytotoxic activities
Pentostatin	Immune suppressive, inhibitor of adenosine deaminase and antineoplastic potential
Militarinones	Antimicrobial and cytotoxic potential
Fumosorinone	Inhibitor of tyrosine phosphatase, activation of insulin pathway and antidiabetic potential
Farinosones	Cytotoxic potential
Oosporein	Immunosuppressive, antimicrobial and cytotoxic potential
Serine protease	Fibrinolytic activity
Endo β-N-acetylglucosaminidase	Hydrolyse human IgG

the usefulness of *Cordyceps* spp. other than medicinal properties. Attempts are being made to use *Cordyceps* also as functional foods. In one such approach, Song (2020) made *Cordyceps* coffee to improve the quality and functionalities. Without affecting the original aroma of coffee, *C. militaris* was used. The biologically active compounds, like cordycepin and β-glucan, were functional in this mixture, providing opportunity to blend medicinal properties in functional foods (or beverages).

2.4 Therapeutic Metabolites and Chemical Diversity of Cordyceps

Fungal metabolites are having extremely special implications in various aspects of human life (Calvo et al., 2002). Chemical diversity and therapeutic metabolites are the essence of medicinal property of *Cordyceps* (Yue et al., 2012; Chiu et al., 2016). Most of the biological activities of components from *Cordyceps* have been shown to have medicinal value. Hence, research focuses on isolating and testing the whole extract as well as individual metabolite to find the efficacy of therapeutic potential. Selected bioactive metabolites, like cordycepin, D-manniotol (cordycepic acid), ergosterol, polysaccharides, nucleosides and peptides have been extracted from *Cordyceps*. These extracts possess a wide range of pharmacological activities, like antiinflammatory, antitumor, antihyperglycemic, antioxidant, antiapoptosis, immunomodulatory, nephroprotective and hepatoprotective activities.

Polysaccharide fraction of *Cordyceps sinensis* induced reduction of lung cancer cell viability by decreasing expression of vascular endothelial growth factor and basic fibroblast growth factor (Ji et al., 2011). Biological activity of cordycepin of *Cordyceps militaris* in reducing liver cancer was studied at molecular level (Guo et al., 2020). In liver cancer patients, high expression of chemokine receptor type 4 (CXCR4) promotes migration and invasion capacities of cancerous cells. Cordycepin was found to be effective in reducing expression of this receptor which was dose-dependent. The cordycepin treatment also reduced the chemotactic migration of liver cancer cells contributing to prevention of spread of cancer cells. *Cordyceps militaris* extracts were able to induce apoptosis in ovarian cancer cells. Dose-dependent induction was achieved through activation of tumor necrosis factor-reliant mechanisms (Jo et al., 2020). The results clearly demonstrated the presence of anti-cancerous metabolites in the extracts. Inhibition of lung cancer cell proliferation and apoptosis induction were associated with functional aspect of *Cordyceps militaris* (Luo et al., 2019). Anticancer metabolites in *C. sinensis* were determined by HPLC (Sang et al., 2020). Various issues related to the use of cordycepin as anticancer agent, including the mechanisms of anticancer property, were discussed by Khan and Tania (2020). Therapeutic potential of cordycepin was attributed to its ability to induce apoptosis and cause DNA damage in tumor cells. Further, the compound can also induce autophagy, exhibited immunomodulatory effects and inhibited tumor metastasis. Anticancer property of *C. militaris* extract showed reduction in viability and morphological disruption of lung cancer cells. Molecular mechanisms included induction of apoptosis by the fungal extract and downregulating of the gene expression level (Jo et al., 2020). Another monosaccharide-xylitol from *C. militaris* has the capability to induce apoptosis in cancer cells (Tomonobu et al., 2020). Induction of glutathione-degrading enzyme is the main mechanism involved in cancer-selective activity of xylitol. It was also observed that sensitisation of cancer cells by xylitol makes the cells amenable to chemotherapeutic drugs. Oh et al. (2020) analysed solvent-soluble extracts of *C. militaris* for anticancer and antifatigue activities. Ethyl acetate extracts were found to be beneficial as they showed tumor-growth inhibitory activity. In combination with adjuvant sorafenib, the extract was made into useful therapeutic product against hepatocellular carcinoma. Promising results were obtained with cancer-related fatigue conditions. The metabolomics approach was employed by Oh et al. (2019) to analyse bioactive compounds from *C. militaris* and proved effective in inhibiting hepatic cancer cell proliferation. Using mass spectrometry and NMR techniques for ethanol extracts of fruit bodies, 44 metabolites were identified. Among them, 16 amino acids and 10 organic acids were distinguished. Again the period of fruit body development was correlated with the higher amount of bioactive metabolites, like cordycepin and cordycepic acid.

In mouse model, diabetic nephropathy was effectively managed by oral administration of the fungus *Hirsutella sinensis* (anamorph of *Cordyceps sinensis*).

Pathological and biochemical analysis of samples of *H. sinensis*-administered mice revealed reduction in fast blood glucose level as well as the ratio of urinary albumin/creatinine. Increased creatinine clearance was also noticed. The results suggested renal protection by metabolic modulation through the fungal bioactive components (Lu et al., 2019). Immunomodulatory and antioxidant properties of polysaccharides from *Cordyceps kyushuensis* was also investigated (Su et al., 2020). Corbrin capsule is a patented Chinese medicine made up of powder of *Cordyceps*. Therapeutic value of this capsule has been proved against renal insufficiency, pulmonary disorders, pulmonary fibrosis and bronchitis. Wu et al. (2020) tested this capsule against acute cerebral ischemia using a mouse model. The corbrin capsule was capable of increasing ATP concentration and showed anticerebral ischemic effects when continuously used for seven days. Prolonged treatment induced anti-inflammatory response in brain tissues. Neuroprotective and therapeutic properties of *C. militaris* revealed improvement in memory impairments caused by cerebral ischemia. Experiments conducted with rats provided ample evidence in reduction of neuronal injury, memory alterations and learning ability. This study clearly demonstrated the potential of *C. militaris* in treatment against dementia and neuroinflammatory disorders (Kim et al., 2019). Extract of *C. sinensis* and bioactive metabolite cordycepin were tested for their therapeutic property against pulmonary hypertension. Both extract and metabolite significantly reduced the proliferation of human pulmonary artery's smooth muscle cells. In murine lungs, only the extract was effective for vasodialation but not the cordycepin. These results suggested the possible presence of other bioactive metabolites in the extract, showing therapeutic potential (Luitel et al., 2020).

Metabolic diseases are due a to sedentary lifestyle and excessive calorie intake in humans. Potential efficiency of *Cordyceps* extracts and metabolites were found to have specific compounds for effective therapy of metabolic disorders. Cao et al. (2020) discussed the possible biological activities in treatment of metabolic disorders. Serious illnesses, like acute pancreatitis and inflammatory response resulting in tissue necrosis, could be managed by cordycepin. This metabolite obtained from *Cordyceps militaris* showed immunomodulatory properties under *in vitro* and *in vivo* studies (Yang et al., 2020b). Using mice model, Chen et al. (2020b) showed anti-inflammatory response induced by *Cordyceps sinensis* mycelial extract. The treatment with extract showed suppression of nasal symptoms and other complications associated with allergic rhinitis and asthma. Reduction in interleukin, IgE and eosinophil peroxidase levels were responsible for the curative effects. This extract proved to be useful in reducing oxidative stress, like reactive oxygen species generation and intracellular hydrogen peroxide content. The DNA repair function in human keratinocytes was also observed. The study suggested the use of *Cordyceps* extract for topical application in order to prevent UV-induced adverse effects (He et al., 2020).

Depending on the purpose and activity, it is recommended to use either the whole powder or individual metabolites of *Cordyceps*. Yang et al. (2020a) reviewed the current trends in research on bioactive metabolite cordycepin from *Cordyceps kyushuensis* and *C. militaris*. Many biological activities have been assigned to this compound. Biotechnological approaches for the production of cordycepin include chemical synthesis, fermentation and biosynthesis. Understanding the biosynthetic pathways and genes involved in regulation may provide useful strategy for commercial production of cordycepin.

3. Production of Medicinally-valued Compounds from *Cordyceps*

Artificial cultivation of *Cordyceps* involves highly complex physicochemical and microecological factors. The soil factors, like organic carbon, oxidisable organic carbon, humin carbon, humic acid, pH, water content, available potassium and phosphorus influence the production of *Cordyceps*. Interestingly, soil microbial diversity also influences the growth of *Cordyceps* in commercial production. According to Shao et al. (2019), bacterial communities in soil may enhance *Cordyceps* cultivation, whereas the fungal communities may suppress it. Suparmin et al. (2019) investigated the difference in cordycepin production between aerial and submerged mycelia in liquid cultures of *C. militaris*. The metabolite quantity produced in liquid surface culture was higher than in submerged culture. Transcriptomic analyses at regular intervals provided information on regulation of metabolite biosynthetic pathways. Expression of cluster genes for cordycepin was not significantly different under both culture conditions. Lee et al. (2019) improved cordycepin production using casein hydrolysate in submerged cultivation.

Luo et al. (2020) developed a cost-effective liquid fermentation medium for the cultivation of *C. militaris*. Traditionally used nitrogen source (peptone) was replaced with pupa powder and wheat bran resulting in 30 per cent yield enhancement and 50 per cent cost reduction. High level of purity of bioactive cordycepin was achieved with hydrothermal reflux extraction and macroporus resin adsorption. Liu et al. (2020b) also studied the effect of pH on polysaccharide production and antioxidant activity of *Cordyceps*. A higher level of β-glucan content and increased antioxidant activity were found when *C. militaris* was grown with initial pH 8–9. Analysis of RNA sequence showed differential expression of 1088 genes under different growth pH conditions. Cultured *C. sinensis* polysaccharides were effective in modulating intestinal mucosal immunity. Specific fermentation conditions necessary for cultivation of *Cordyceps* were achieved (Tang et al., 2015; Zhang et al., 2016; Suparmin et al., 2019; Xu et al., 2019). These studies clearly demonstrate the impact of external factors on cultivation as well as biological activity of *Cordyceps*.

Tao et al. (2020) compared different strains of *C. militaris* for the production of fruit bodies and bioactive compounds based on various substrates. Among the six strains tested, the strain #CM3 was efficient in developing fruit bodies on rice and wheat as substrates. In addition, this strain showed high biological efficiency. However, another strain, #CM9 was found to have higher efficiency when grown on pupae. Hence, the production of bioactive compounds and biomass were dependent on the strain and substrate used for cultivation. Lin et al. (2020) have developed multisensor incubators for cultivation of *C. militaris*. These attempts provided further prospects to develop suitable methods of artificial cultivation. Endemic locations, micro-ecological factors and seasonal nature of the fungus need to be considered in developing the methods to build fungal biomass as well as desired metabolite. Considering the prospective therapeutic importance of *Cordyceps* metabolites, investigations were performed to enhance the bioactive potential through various means. Extended storage of *C. sinensis* is one of the vital steps in retaining biological activity of the product. Wu et al. (2015) studied the effect of heat treatment against the biological activity of *C. sinensis*. Heat treatment up to 100°C did not result in reduction of DNase activity and polysaccharide content. However, heat treatment at 60°C for 60 min resulted in a significant increase in dissolution of cordycepin. Similarly, the metabolic responses in the production of carotenoid and cordycepin under exposure to light have been addressed by Thananusak et al. (2020). Hence, the recommendation is to provide heat treatment for safe storage of *Cordyceps* without loss or alteration of biological potential.

The complex nature of insects vs. *Ophiocordyceps sinensis* interaction revealed several insights. The fungus shows blastospores-hyphae dimorphism transition in the host hemolymph, which is a critical factor in virulence. Liu et al. (2020a) studied the effect of inoculum density, fungal nutrients, fungal metabolites, quorum sensing molecules and insect hormones on the dimorphism of *O. sinensis in vitro*. The results revealed that addition of N-acetylglucosamine is necessary for hyphal growth and to inhibit budding. Interestingly, the insect hormone 20-hydroxyecdysone was involved in hyphal formation. These results are also essential in considering cultivation of *Cordyceps*.

3.1 Biotechnology of Commercial Products of Cordyceps

Extensive medicinal uses and highly acclaimed metabolites of *Cordyceps* provide ample scope for biotechnological applications, innovations and development of commercial products. Elaborate efforts are being made in understanding and mass production of useful metabolites of *Cordyceps*. Industrial fermentation, genome analysis and gene cloning strategies are envisaged in *Cordyceps* biotechnology. Optimisation of cultivation conditions with C/N ratio (8:1) and biosynthesis of cordycepin were achieved along with experimental support (Raethong et al., 2020).

Genome sequencing and metabolic analysis of *Cordyceps cicadae* in comparison with other *Cordyceps* spp. indicated that proteases and chitinases were similar. It was also found that *C. cicadae* genome encodes a series of gene clusters for secondary metabolism (Lu et al., 2017). For biotechnological consideration to grow *C. militaris* and cordycepin production, a high quality genome-scale metabolic model has been developed by Raethong et al. (2020). This model includes 1,171 metabolites along with 1,821 biochemical reactions and 1,329 genes. Comparative genomic approach was used to identify 85 sugar transporter genes in *C. militaris* for assessing the growth requirements of sugars. Analysis of gene-expression pattern for utilisation of sucrose, glucose and xylose was also achieved. Key amino acids, like phenylalanine and tryptophan interactions contributed to xylose transport functions (Sirithep et al., 2020). This study revealed molecular basis of sugar metabolism and provided useful insights into the production of bioactive metabolites, especially cordycepin.

In a recent study, Wang et al. (2020) followed the genome-mining approach for the synthesis of bioactive compounds in *Cordyceps militaris*. Biosynthetic gene cluster of the acetyl CoA and cholesterol acyltransferase inhibitor beauveriolides was found in the genome of *C. militaris*. The study identified enzyme resources for the production of cyclodepsi peptide and other molecules required to treat atherosclerosis and Alzheimer's disease. Development of MALDI-TOF-MS techniques is available for detection of peptaibol and other metabolites in *Trichoderma* and fungal cell analyses by mass spectroscopy. Such rapid and sensitive methods of detection could be employed for the detection and characterisation of molecules in *Cordyceps* (Shishupala, 2008; 2009). *Agrobacterium*-mediated transformation of *C. militaris* was achieved by Zheng et al. (2011). Insertional mutagenesis carried out in *C. militaris* generated mutant library and the study provided the requirements of functional genetic analyses.

Fibrinolytic enzymes produced by *C. militaris* are important as thromobolytic agents for blood clotting disorders associated with cardiovascular diseases. Therapeutic applications and structure-function relationships could be better approaches to understand recombinant fibrinolytic enzymes. The gene CmFE from *C. militaris* coding fibrinolytic enzyme was successfully expressed in the yeast, *Pichia pastoris*, resulting in the production of 28 kDa extracellular enzyme. Cloning, expression and structure prediction of medicinally important enzyme in *C. militaris* was achieved by Katrolia et al. (2020). This study presented biotechnological approaches in the production of therapeutic enzymes of *Cordyceps*. Marslin et al. (2020) developed cordycepin-loaded polynanoparticles (lactic-co-glycolic acid) in order to provide sustainable release of the drug. A high level of cell uptake and cytotoxicity were exhibited by this nanoparticle against human breast cancer cells (MCF7). Low hemotoxicity to the host was an added advantage over free cordycepin. Clinical efficiency of cordycepin was enhanced by converting it into nanoparticle-based preparation. This provided authentic evidence for nanobiotechnological applications of *Cordyceps* metabolites.

Varieties of commercial products (powder or capsules) are available and they are claimed to increase individual potential and cure a number of ailments (Fig. 2). The commercial products found in the market are being considered for its nutraceutical and therapeutic properties. These products are recommended to treat fatigue, kidney diseases and low sex drive. The two species used extensively in commercial production among *Cordyceps* are *C. militaris* and *C. sinensis*. Six potential benefits have been attributed in such products, including boosting exercise performance-delivering energy to muscles and hence considered in sports medicine. Antiaging properties are being targeted for middle-aged people. Proven antitumor activity makes it a natural choice in cancer therapy. Researchers have shown antidiabetic values of these products. Ever-increasing diabetic patients and long-term requirements anticipate commercial success. Cardiovascular diseases are being treated and provide a wide market. Antiinflammatory properties are essential to treat various respiratory disorders with a dosage ranging between 1000–3000 mg per day (Walle, 2018).

4. Limitations in Use of *Cordyceps*

Safety concerns are the most important factors in any innovation of human medicine. In case of regular medicines, purity of the compounds and several clinical trials are performed before licencing the product for trade. However, such a rigorous exercise

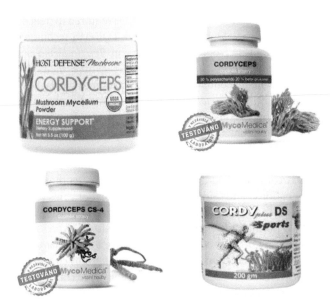

Fig. 2. *Cordyceps*-based products in the market (https://www.mycomedica.eu/eshop-cordyceps-cs-4. html?mobile=1; https://homeopathic.com/product/cordyceps-mushroom-powder-energy-support-3-5-oz/; https://www.flipkart.com/cordyceps-india-cordy-plus-ds-sports/p/itmbdf409439a635).

may not happen in the case of complex traditional medicines, like *Cordyceps* extracts. Hence, it is also most essential to gain safety data for use of such natural extracts as medicines and functional foods or additives.

A comparative study on drug safety data with respect to Chinese patent medicine—*Cordyceps sinensis* extracts—was carried out by Hu et al. (2019). This analysis showed adverse drug reactions, including intestinal disorders like nausea, diarrhoea and vomiting. Multiple data sources from clinical trial results are essential for safety concern in the use of these traditional medicines. Direct consumption of *Cordyceps* spp. is extensively followed in traditional medicine. Li et al. (2019) detected arsenic content in *C. sinensis* using HPLC and mass spectrometry. Arsenic was mainly associated with alkali-soluble proteins of the fungus and probably has potential toxicity. The health risk involved in arsenic residue concomitant with *Cordyceps* spp. limits extraordinary health benefits. Regular monitoring measures are required to detect and quantify arsenic and other toxic components present in field-sampled *Cordyceps* spp. Due consideration has to be given to possible toxicity and over-dosage complications associated with the products, especially when consumed directly.

Extensive use of *Cordyceps* spp. for human consumption as food supplements (or additives) and culture substrates for animal feed opened up many benefits. However, analysis of a series of biosynthetic gene clusters in *Cordyceps* spp. provided clues on the presence of possible mycotoxin analogs, like PR-toxin and trichothecenes (Chen et al., 2020a). This study raised the biosafety concern in allowing direct consumption of *Cordyceps* spp. by humans. Further explorations of *Cordyceps* for therapeutic purpose need careful consideration of mycotoxin contamination. Extensive demand for *Cordyceps*-based medicines allow duplicates as well as mimic products to find place in the markets. In order to avoid such spurious materials and authenticate natural *Cordyceps* in the markets, duplex PCR method was developed by Zhang et al. (2020). Chinese *Cordyceps* was precisely identified by using ITS amplicon from *Ophiocordyceps sinensis* and cytochrome oxidase C subunit amplicon. Considering the biotechnological potential of this medicinal fungus, the technique developed appears to be highly useful. Safety assessment of *C. cicadae* mycelium administered to human beings was carried out by Tsai et al. (2020). Hydroxyethyl adenosine-enriched *C. cicadae* mycelium did not show any side effects on humans. In a period of three months, no adverse effects were observed; hence consumption of cultivated mycelium seems to be safe and could also be used in functional foods.

High-value medicines and economic importance of *Cordyceps* resulted in unprecedented collections, which may affect the life cycle of the fungus. Extensive dependence of the rural population on chasing such mushrooms for their economic improvement may put the desired fungal species at risk of extinction. Baral et al. (2015) made an effort to suggest the remedial measures for *in situ* and *ex situ* conservation. Sustainable resource-management practices should include historical attributes, harvesting techniques, understanding genetic diversity, ecological niches, socio-economic concerns, regulatory and mycological considerations.

5. Conclusions and Outlook

Medicinal values of *Cordyceps* and allied species have provided a rich potential market. Most of the pharmacological investigations have led to the conclusion of exploiting such rare fungi for human benefits. It is extremely important to avoid overexploitation of such exclusive and endangered species. Standardisation of cultivation methods and use of appropriate metabolite production strategies are the present-day requirements. Everlasting research on *Cordyceps* has unravelled medicinal mysteries associated with such fungi. In the era of the significance of COVID-19, it is most imperative to look for natural medicines with therapeutic products. Search for medicinally-valuable compounds in nature is always a 'random walk in a random forest'. The ethnic knowledge along with the modern approaches should find further natural sources of bioactive compounds. The metabolites of *Cordyceps* already characterised should be used as lead molecules in chemical synthesis. The present biotechnological tools along with combinatorial chemistry should provide ample scope to produce medically-valued compounds on an industrial scale. Microbiome associated with *Cordyceps* has shown several hundreds of bacteria and fungi. It would be interesting to explore these organisms either for precursor compounds or for bioconversion processes. It is also pertinent to understand the basic biological phenomena underlying fungus-insect interactions. This information will also provide the advantages for safe and authentic commercial production. Present commercial value of *Cordyceps* clearly depends on biopesticide and the medicinal potential. Proper judicial and sustainable uses of these fungi are essential in order to prevent extinction of *Cordyceps* in nature owing to over-exploitation. Firm biological principles should be laid on priority basis for biotechnological applications of *Cordyceps*. Genome mapping of *Cordyceps* and its host may also provide imperative insights into exploration for medicinal values. It is hoped that the traditional knowledge combined with modern methodology will pave the way for tangible utilisation of *Cordyceps* spp. in the future. The techniques of metabolomics, metagenomics and reverse genetics may be highly useful in understanding *Cordyceps* along with conventional microbiological methods. It is envisioned that the elusive group of fungal products should be available to mankind at much lesser cost. Several challenges lie ahead in order to cultivate such fungi and exploit their metabolites for commercial purposes with strong ethical and scientific basis.

Acknowledgements

The author is grateful to Davangere University for providing the necessary facilities and also to the Indian Institute of Science, Challkere Campus, Chitradurga, Karnataka, India for providing library facilities for literature survey. Special thanks are due to Prof. M.S. Hegde, Convener, Talent Development Centre, Indian Institute of Science, Bangalore, India for the whole-hearted support extended.

References

Ali, M.L. (2012). The caterpillar fungus *Cordyceps sinensis* as a natural source of bioactive compounds. J. Pharma. Bio. Sci., 1: 41–43.

Ashraf, S.A., Elkhalifa, A.E.O., Siddiqui, A.J., Patel, M., Awadelkareem, A.M. et al. (2020). Cordycepin for health and wellbeing: A potent bioactive metabolite of an entomopathogenic medicinal fungus *Cordyceps* with its nutraceutical and therapeutic potential. Molecules, 25: 2735–2755.

Baral, B., Shrestha, B. and Silva, J.A.T. (2015). A review of Chinese *Cordyceps* with special reference to Nepal, focusing on conservation. Environ. Exp. Biol., 13: 61–73.

Barido, F.H., Jang, A., Pak, J.I., Kim, D.Y. and Lee, S.K. (2020). Investigation of taste-related compounds and antioxidative profiles of retorted samgyetang made from fresh and dried *Cordyceps militaris* mushrooms. Food Sci. Anim. Resour., 40: 772–784.

Bensky, D., Clavey, S. and Stoger, E. (2004). Chinese Herbal Medicine: Metriamedica. 3rd ed., Eastland Press, Seattle, p. 770.

Boontiam, W., Wachirapakorn, C. and Wattanachal, S. (2019). Growth performance and hematological changes in growing pigs treated with *Cordyceps militaris* spent mushroom substrate. Vet. World, 13: 768–773.

Calvo, A.M., Wilson, R.A., Bok, J.W. and Keller, N.P. (2002). Relationship between secondary metabolism and fungal development. Microbiol. Mol. Biol. Rev., 66: 447–459.

Cao, C., Yang, S. and Zhou, Z. (2020). The potential application of *Cordyceps* in metabolic-related disorders. Phytother. Res., 34: 295–305.

Chen, B., Sun, Y., Luo, F. and Wang, C. (2020a). Bioactive metabolites and potential mycotoxins produced by *Cordyceps* fungi: A review of safety. Toxins, 12: 410–413.

Chen, J., Chan, W.M., Leung, H.Y., Leong, P.K., Yan, C.T.M. and Ko, K.M. (2020b). Anti-inflammatory effects of a *Cordyceps sinensis* mycelium culture extract (CS-4) on rodent models of allergic rhinitis and asthma. Molecules, 25: 4051–4069.

Chen, P.X., Wang, S., Nie, S. and Marcone, M. (2013). Properties of *Cordyceps sinensis*: A review. J. Func. Foods, 5: 550–569.

Chiu, C.-P., Hwang, T.-L., Chan, Y., El-Shazly, M., Wu, T.-Y. et al. (2016). Research and development of *Cordyceps* in Taiwan. Food Sci. Human Wellness, 5: 177–185.

Choda, U. (2017). Medicinal value of *Cordyceps sinensis*. Transl. Biomed., 8: 132–136.

Chuang, W.Y., Hsieh, Y.C. and Lee, T. (2020). The effects of fungal feed additives in animals: A review. Animals, 10: 805–820.

Deacon, J. (2006). Fungal Biology. Blackwell Publishing, Oxford, UK, p. 371.

Dong, J.Z., Wang, S.H., Ai, X.R., Yao, L., Sun, Z.W. et al. (2013). Composition and characterisation of cordyxanthins from *Cordyceps militaris* fruit bodies. J. Funct. Foods, 5: 1450–1455.

Duran-Aranguren, D., Chirivi-Salomon, J.S., Anaya, L., Duran-Sequeda, D., Cruz, L.J. et al. 2020. Effect of bioactive compounds extracted from *Cordyceps nidus* ANDES-F1080 on laccase activity of *Pleurotusostreatus* ANDES-F515. Biotechnol. Rep., 26: 10.1016/j.btre.2020.e00466.

Fan, T., Cheng, B., Fang, X., Chen, Y. and Su, F. (2020). Application of Chinese medicine in the management of critical conditions: A review on sepsis. Am. J. Chin. Med., 48: 1315–1330.

Gu, L., Yu, T., Liu, J. and Lu, Y. (2020). Evaluation of the mechanism of *Cordyceps* polysaccharide action on rat acute liver failure. Arch. Med. Sci., 16: 1218–1225.

Guo, Z., Chen, W., Dai, G. and Huang, Y. (2020). Cordycepin suppresses the migration and invasion of human liver cancer cells by downregulating the expression of CXCR4. Int. J. Mol. Med., 45: 141–150.

He, H., Tang, J., Ru, D., Shu, X., Li, W. et al. (2020). Protective effects of *Cordyceps* extract against UVB-induced damage and prediction of application prospects in the topical administration: An experimental validation and network pharmacology study. Biomed. Pharmacother., 121: 10.1016/j. biopha.2019.109600.

Holliday, J. (2017). *Cordyceps*: A highly coveted medicinal mushroom. pp. 59–91. *In*: Agrawal, D.C. et al. (eds.). Medicinal Plants and Fungi: Recent Advances in Research and Development, Medicinal and Aromatic Plants of the World. Springer Nature, Singapore Pte. Ltd.

Hu, R., Golder, S., Yang, G., Li, X., Wang, D. et al. (2019). Comparison of drug safety data obtained from the monitoring system, literature and social media: An empirical proof from a Chinese patent medicine. PLoS ONE, 14: 10.1371/journal.pone.0222077.

Ji, N.-F., Yao, L.-S, Li, Y., He, W., Yi, K.-S. and Huang, M. (2011). Polysaccharide of *Cordyceps sinensis* enhances cisplatin cytotoxicity in on-small cell lung cancer H157 cell line. Integr. Canc. Therp., 10: 359–367.

Jo, E., Jang, H.-J., Yang, K.E., Jang, M.S., Huh, Y.H. et al. (2020). *Cordyceps militaris* induces apoptosis in ovarian cancer cells through TNF-α/TNFR1-mediated inhibition of NF-κB phosphorylation. BMC Comple. Med. Therp., 20: 1–11.

Katrolia, P., Liu, X., Zhao, Y., Kopparapu. N.K. and Zheng, X. (2020). Gene cloning, expression and homology modelling of first fibrinolytic enzyme from mushroom (*Cordyceps militaris*). Int. J. Biol. Macromol., 146: 897–906.

Khan, M.A. and Tania, M. (2020). Cordycepin in anticancer research: molecular mechanism of therapeutic effects. Curr. Med. Chem., 27: 983–996.

Kim, Y.O., Kim, H.J., Abu-Taweel, G.M., Oh, J. and Sung, G-H. (2019). Neuroprotective and therapeutic effect of *Cordyceps militaris* on ischemia-induced neuronal death and cognitive impairments. Saudi J. Biol. Sci., 26: 1352–1357.

Lee, S.K., Lee, J.H., Kim, H.R., Chun, Y., Lee, J.H. et al. (2019). Improved Cordycepin production by *Cordyceps militaris* KYL05 using casein hydrolysate in submerged conditions. Biomoleules, 9: 461–471.

Li, Y., Liu, Y., Han, X., Jin, H. and Ma, S. (2019). Arsenic species in *Cordyceps sinensis* and its potential health risks. Front. Pharmacol., 10: 1471–1480.

Liang, Y., Hong, Y., Mai, Z., Zhu, Q. and Guo, L. (2019). Internal and external microbial community of the Thitarodes moth, the host of *Ophiocordyceps sinensis*. Microorganisms, 7: 517–536.

Lin, J.-Y., Tsai, H.-L. and Sang, W.-C. (2020). Implementation and performance evaluation of integrated wireless multisensor module for aseptic incubator of *Cordyceps militaris*. Sensors, 20: 4272–4284.

Liu, G., Cao, Li., Qiu, X. and Han, R. (2020a). Quorum sensing activity and hyphal growth by external stimuli in the entomopathogenic fungus *Ophiocordyceps sinensis*. Insects, 11: 205–217.

Liu, H.-J., Hu, H.-B., Chu, C., Li, Q. and Li, P. (2011). Morphological and microscopic identification studies of *Cordyceps* and its counterfeits. Acta Pharm. Sinica B, 1: 189–195.

Liu, Y., Li, Y., Zhang, H., Li, C., Zhang, Z. et al. (2020b). Polysaccharides from *Cordyceps militaris* cultured at different pH, sugar composition and antioxidant activity. Int. J. Biol. Macromol., 162: 349–358.

Lo, H.-C., Hsieh, C., Lin, F.-Y. and Hsu, H. (2013). Systematic review of mysterious caterpillar fungus *Ophiocordyceps sinensis* in Dong Chong Xia Cao and related bioactive ingredients. J. Trad. Comple. Med., 3: 16–32.

Lu, Y., Luo, F., Cen, K., Xiao, G., Yin, Y., Li, C., Li, Z., Zhan, S., Zhang, H. and Wang, C. (2017). Omics data reveal the unusual asexual fruiting nature and secondary metabolic potentials of the medicinal fungus *Cordyceps cicadae*. BMC Genomics, 18: 668–682.

Lu, Z., Li, S., Sun, R., Jia, X., Xu, C. et al. (2019). *Hirsutella sinensis* treatment shown protective effects on renal injury and metabolic modulation in db/db mice. Evid. Based Compl. Alt. Med., 10.1155/2019/4732858.

Luitel, H., Novoyatleva, T., Sydykov, A., Petrovic, A., Mamazhakypov, A. et al. (2020). Yarsagumba is a promising therapeutic option for treatment of pulmonary hypertension due to the potent anti-proliferative and vasorelaxant properties. Medicina, 56: 131–138.

Luo, L., Ran, R., Yao, J., Zhang, F., Xing, M. et al. (2019). Se-enriched *Cordyceps militaris* inhibits cell proliferation, induces cell apoptosis, and causes G2/M phase arrest in human non-small cell lung cancer cells. Onco Targets Ther., 12: 8751–8763.

Luo, Q., Cao, H., Liu, S., Wu, M., Li, S. et al. (2020). Novel liquid fermentation medium of *Cordyceps militaris* and optimisation of hydrothermal reflux extraction of cordycepin. J. Asian Nat. Prod. Res., 22: 167–178.

Marslin, G., Khandelwal, V. and Franklin, G. (2020). *Cordycep innano* encapsulated in poly (lactic-co-glycolic acid) exhibits better cytotoxicity and lower hemotoxicity than free drug. Nanotech. Sci. Appl., 13: 37–45.

Nxumalo, W., Elateeq, A.A. and Sun, Y. (2020). Can *Cordyceps cicadae* be used as an alternative to *Cordyceps militaris* and *Cordyceps sinensis*?—A review. J. Ethnopharmacol., 257: 112879. 10.1016/j.jep.2020.112879.

Oh, J., Choi, E., Yoon, D.-H., Park, T.-Y., Shrestha, B. et al. (2019). H-NMR-based metabolic profiling of *Cordyceps militaris* to correlate the development process and anti-cancer effect. J. Microbiol. Biotechnol., 29: 1212–1220.

Oh, J., Choi, E., Kim, J., Kim, H., Lee, S. and Sung, G.-H. (2020). Efficacy of ethyl acetate fraction of *Cordyceps militaris* for cancer-related fatigue in blood biochemical and H-nuclear magnetic resonance metabolomics analyses. Int. Can. Therap., 19: 1–12.

Ou, D., Ren, L.-M., Yuan-Liu, Ali, S., Wang, X.-M. et al. (2019). Compatibility and efficacy of the parasitoid *Eremocerushayati* and the entomopathogenic fungus *Cordyceps javanica* for biological control of whitefly Bemisatabaci. Insects, 10: 425–434.

Pathania, P. and Sagar, A. (2014). Studies on the biology of *Cordyceps militaris*: A medicinal mushroom from northwest Himalaya. KAVAKA, 43: 35–40.

Raethong, N., Wang, H., Nielsen, J. and Vongsangnak, W. (2020). Optimising cultivation of *Cordyceps militaris* for fast growth Cf and cordycepin overproduction using rational design of synthetic media. Comput. Struct. Biotechnol. J., 18: 1–8.

Roth, C., Moroz, O.V., Turkenburg, J.P., Blagova, E., Waterman, J. et al. (2019). Structural and functional characterisation of three novel fungal amylases with enhanced stability and pH tolerance. Int. J. Mol. Sci., 20: 4902–4916.

Sang, Q., Pan, Y., Jiang, Z., Wang, Y., Zhang, H. and Hu, P. (2020). HPLC determination of massoia lactone in fermented *Cordyceps sinensis* mycelium CS-4 and its anticancer activity *in vitro*. J. Food Biochem., 44: 10.1111/jfbc.13336.

Seth, R., Haider, S.Z. and Mohan, M. (2014). Pharmacology, phytochemistry and traditional uses of *Cordyceps sinensis* (Berk.) Sacc: A recent update for future prospects. Ind. J. Trad. Know., 13: 551–556.

Shao, J.-L., Lai, B., Jiang, W., Wang, J.-T., Hong, Y.-H. et al. (2019). Diversity and co-occurrence patterns of soil bacterial and fungal communities of Chinese *Cordyceps* habitats at Shergyla mountain, Tibet: Implications for the occurrence. Microorganisms, 7: 284–304.

Shishupala, S. (2008). Biochemical analysis of fungi using matrix-assisted laser desorption/ionisation time-of-flight mass spectrometry (MALDI-TOF-MS). Fungal Diversity Research Series, 20: 327–372.

Shishupala, S. (2009). Bioactive peptaibol antibiotics from *Trichoderma*. pp. 300–320. *In*: Sridhar, K.R. (ed.). Frontiers in Fungal Ecology, Diversity and Metabolites, I.K. International Publishing House Pvt. Ltd., New Delhi.

Silva, J.N., Mascarin, G.M., Castro, R.P.V., Castilho, L.R. and Freire, D.M. (2019). Novel combination of a biosurfactant with entomopathogenic fungi enhances efficacy against *Bemisia* whitefly. Pest Manag. Sci., 75: 2882–2891.

Sirithep, K., Xiao, F., Raethong, N., Zhang, Y., Laoteng, K. et al. (2020). Probing carbon utilization of *Cordyceps militaris* by sugar transportome and protein structural analysis. Cells, 9: 401–419.

Song, H.-N. (2020). Functional *Cordyceps* coffee containing cordycepin and β-glucan. Prev. Natr. Food. Sci., 25: 184–193.

Stephan, D., Bernhardt, T., Buranjadze, M., Seib, C., Schäfer, J. et al. (2020). Development of a fluid-bed coating process for soil granule-based formulations of *Metarhiziumbrunneum*, *Cordyceps fumosorosea* or *Beauveria bassiana*. J. Appl. Microbiol., 10.1111/jam.14826.

Su, J., Sun, J., Jian, T., Zhang, G. and Ling, J. (2020). Immunomodulatory and antioxidant effects of polysaccharides from the parasitic fungus *Cordyceps kyushuensis*. Biomed. Res. Int., 10.1155/2020/8257847.

Sun, T., Shen, Z., Shaukat, M., Du, C. and Ali, S. (2020). Endophytic isolates of *Cordyceps fumosorosea* to enhance the growth of *Solanum melongena* and reduce the survival of whitefly (*Bemisiatabaci*). Insects, 11: 78–89.

Sung, G.-H., Hywel-Jones, N.L., Sung, J.M., Luangsa-ard, J.J., Shreestha, B. and Spatafora, J.W. (2007). Phylogenetic classification of *Cordyceps* and the clavicipitaceous fungi. Stu. Mycol., 57: 5–59.

Suparmin, A., Kato, T., Takemoto, H. and Park, E.Y. (2019). Metabolic comparison of aerial and submerged mycelia formed in the liquid surface culture of *Cordyceps militaris*. Microbiologyopen, 8: doi.org/10.1002/mbo3.836.

Tang, J., Qian, Z. and Zhu, L. (2015). Two-step shake-static fermentation to enhance Cordycepin production by *Cordyceps militaris*. Chem. Eng. Trans., 46: 19–24.

Tao, S., Xue, D., Lu, Z. and Huang, H. (2020). Effects of substrates on the production of fruiting bodies and the bioactive components by different *Cordyceps militaris* strains (Ascomycetes). Int. J. Med. Mush., 22: 55–63.

Thananusak, R., Laoteng, K., Raethong, N., Zhang, Y. and Vongsangnak, W. (2020). Metabolic responses of carotenoid and cordycepin biosynthetic pathways in *Cordyceps militaris* under light-programming exposure through genome-wide transcriptional analysis. Biology, 9: E242. 10.3390/biology9090242.

Tomonobu, N., Komalasari, N.L.G.Y., Sumardika, I.W., Jiang, F., Chen, Y. et al. (2020). Xylitol acts as an anticancer monosaccharide to induce selective cancer death via regulation of the glutathione level. Chem. Biol. Inter., 324: 10.1016/j.cbi.2020.109085.

Tsai, Y., Hsu, J., Lin, D.P., Chang, H., Chang, W. et al. (2020). Safety assessment of HEA-enriched *Cordycepscicadae* mycelium: A randomised clinical trial. J. Am. Coll. Nutr., 23: 1–6.

Tsim, K.W.K. and Li, S.P. (2005). *Cordyceps sinensis*—A traditional Chinese medicine known as winter-worm summer-grass. Alt. Med., 9: 1160–1164.

Walle, G.V.D. (2018). www.healthline.com/nutrition/cordyceps-benefits#.

Wang, X., Gao, Y., Zhang, M., Zhang, M., Huang, J. and Li, L. (2020). Genome mining and biosynthesis of the Acyl-CoA: Cholesterol acyltransferase inhibitor beauveriolide I and III in *Cordyceps militaris*. J. Biotechnol., 309: 85–91.

Winkler, D. (2008). Yartsa Gumba (*Cordyceps sinensis*) and the fungal commodification of Tibet's rural economy. Econ. Bot., 62: 291–305.

Wu, J., Yan, W., Wu, X., Hong, D., Lu, X. and Rao, Y. (2020). Protective effects of corbrin capsule against permanent cerebral ischemia in mice. Biomed. Pharmacother., 121: 10.1016/j.biopha.2019.109646.

Wu, P., Tao, Z., Liu, H., Jiang, G., Ma, C. et al. (2015). Effects of heat on the biological activity of wild *Cordyceps sinensis*. J. Trad. Chinese Med. Sci., 2: 32–38.

Xu, L., Wang, F., Zhang, Z. and Terry, N. (2019). Optimisation of polysaccharide production from *Cordyceps militaris* by solid-state fermentation on rice and its antioxidant activities. Foods, 8: 590–599.

Yan, J., Wang, W. and Wu. (2014). Recent advances in *Cordyceps sinensis* polysaccharides: Mycelial fermentation, isolation, structure and bioactivities: A review. J. Func. Foods, 6: 33–47.

Yang, J., Zhou, Y. and Shi, J. (2020b). Cordycepin protects against acute pancreatitis by modulating NF-κB and NLRP3 inflammasome activation via AMPK. Life Sci., 251: 10.1016/j.lfs.2020.117645.

Yang, L., Li, G., Chai, Z., Gong, Q. and Guo, J. (2020a). Synthesis of cordycepin: Current scenario and future perspectives. Fungal Genet. Biol., 143: 10.1016/j.fgb.2020.103431.

Yang, S., Yang, X. and Zhang, H. (2020c). Extracellular polysaccharide biosynthesis in *Cordyceps*. Crit. Rev. Microbiol., 46: 359–380.

Ying, M., Yu, Q., Zheng, B., Wang, H., Wang, J. et al. (2020). Cultured *Cordyceps sinensis* polysaccharides modulate intestinal mucosal immunity and gut microbiota in cyclophosphamide-treated mice. Carbohydr. Polym., 235: 10.1016/j.carbpol.2020.115957.

Yue, K., Ye, M., Zhou, Z., Sun, W. and Lin, X. (2012). The genus *Cordyceps*: A chemical and pharmacological review. J. Pharma. Pharmacol., 65: 474–493.

Zhang, F.-L., Yang, X.-F., Wang, D., Lei, S.-R., Guo, L.-A. et al. (2020). A simple and effective method to discern the true commercial Chinese *Cordyceps* from counterfeits. Sci. Rep., 10: 2974.10.1038/s41598-020-59900-9.

Zhang, Q., Liu, Y., Di, Z., Han, C.C. and Liu, Z. (2016). The strategies for increasing Cordycepin of *Cordyceps militaris* by liquid fermentation. Fungal Gen. Biol., 6: 1–4.

Zheng, Z., Huang, C., Cao, L., Xie, C. and Han, R. (2011). *Agrobacterium tumefaciens*-medicated transformation as a tool for insertional mutagenesis in medicinal fungus *Cordyceps militaris*. Fungal Biol., 115: 265–274.

Zhou, X., Gong, Z., Su, Y., Lin, J. and Tang, K. (2009). *Cordyceps* fungi: Natural products, pharmacological functions and developmental products. J. Phar. Pharmacol., 61: 279–291.

Zhou, Y., Wang, M., Zhang, H., Huang, Z. and Ma, J. (2019). Comparative study of the composition of cultivated, naturally grown *Cordyceps sinensis* and stiff worms across different sampling years. PLoS ONE, 14: 1–15.

Zhu, Y., Yu, X., Ge, Q., Li, J., Wang, D. et al. (2020). Antioxidant and antiaging activities of polysaccharides from *Cordyceps cicadae*. Int. J. Biol. Macromol., 157: 394–400.

7

Medical Mushrooms in Neurodegenerative Disorder (Alzheimer's Disease)

Manoj Govindarajulu,[1,] Sindhu Ramesh,[1]
Grace McKerley,[1] Mary Fabbrini,[1] Anna Solomonik,[1]
Rishi M Nadar,[1] Satyanarayana Pondugula,[2]
Timothy Moore[1] and
Muralikrishnan Dhanasekaran[1,]**

1. INTRODUCTION

Neurodegenerative diseases, such as Alzheimer's disease (AD), are a growing health concern worldwide. Alzheimer's disease is the most common cause of dementia in the elderly. Dementia in AD is associated with a progressive disability during the entire course of illness and death occurring within five to twelve years of the disease onset. Macroscopically, the AD can be identified by shrinkage of the brain and cortical atrophy/thinning (Chen and Mobley, 2019). The German scientist Alois Alzheimer, after whom the disease was named, was the first to discover that AD has a "distinct and recognisable neuropathological substrate" (Perl, 2010). There is much to be discovered pertaining to this disease and its effects, but it has been characteristically defined by the presence of specific neuropathologies-extracellular deposition of amyloid plaques and intraneuronal neurofibrillary tangles. Amyloid plaques consist of the deposition of β-amyloid (Aβ), which is formed by abnormal processing of amyloid precursor protein (APP). Neurofibrillary tangles are an

---

[1] Department of Drug Discovery and Development, Harrison School of Pharmacy, Auburn University, Auburn, AL, 36849, USA.
[2] Department of Anatomy, Physiology and Pharmacology, Auburn University, Auburn, AL 36849, USA.
* Corresponding authors: myg0003@auburn.edu; dhanamu@auburn.edu

accumulation of abnormally phosphorylated tau within the perikaryal cytoplasm of certain neurons (Chen and Mobley, 2019). Furthermore, several other pathologies such as mitochondrial dysfunction (Ramesh et al., 2018), oxidative stress (Tönnies and Trushina, 2017), neuroinflammation (Hampel et al., 2020), excitotoxicity and calcium dyshomeostasis (Hynd et al., 2004), brain energy dysfunction (Yin et al., 2016) and so on are associated with AD. Currently approved medications to treat AD offer only symptomatic relief and are associated with several adverse effects. Despite extensive research, there have been limited therapeutic interventions to treat cognitive disorders, such as AD. Interestingly, complementary and alternative medicine, especially functional foods, offer several benefits in treating aging-related diseases, such as AD, which has attracted wide consideration.

Mushrooms have been shown to possess therapeutic potential in mitigating the pathologies of AD due to its wide range of chemical constituents and their different biological activities. Furthermore, several constituents in mushroom have been shown to exhibit antitumor, antioxidant, anti-inflammatory, antidiabetic, and cardioprotective activities (Roupas et al., 2012). Despite their therapeutic potential, mushrooms have been less explored and are in the early stages of research when compared to herbal medicines. In neurodegenerative diseases such as AD, disease prevention is the best possible approach as it is difficult to reverse the degenerative process or the neuronal death (Crous-Bou et al., 2017). This chapter focuses on the therapeutic potential of various mushrooms with respect to AD. Furthermore, we summarise the most important mushrooms and their biological actions and/or molecular mechanisms in several cell culture and animal models of AD.

2. Mushrooms with Neuroprotective Effects against Alzheimer's Disease

The estimated number of mushroom species is 140,000, although only 10 per cent (~14,000) are known (Wasser, 2002). Among the species of mushroom identified, some are currently being studied for their benefits to neuronal health, specifically as a potential therapeutic approach to neurodegenerative diseases, such as AD. The pathological hallmarks of AD include the deposition of amyloid plaques and neurofibrillary tangles. Additionally, other pathologies, such as neuroinflammation, oxidative stress, excitotoxicity have been reported as previously described. The currently available treatment for AD provides only symptomatic relief and is associated with adverse effects. Hence, there is a growing interest in natural products, especially dietary supplements and functional foods in AD. The subsequent sections discuss the role of important mushrooms in AD.

2.1 *Hericium erinaceus*

Hericium erinaceus is a well-known culinary and medicinal mushroom in Japan and China. It, also known as 'Yamabushitake' in Japan, 'Houtou' in China, or 'Lion's mane' in Western countries, is a mushroom that typically grows on old or dead broadleaf trees (Mori et al., 2008; Wang et al., 2004). *Hericium erinaceus* has been shown to display antitumor, antimutagenic, antioxidant, hypolipidemic and immunomodulatory properties (Friedman, 2015). Furthermore, several studies have reported that *H. erinaceus* have beneficial effects on neurodegenerative diseases. The major components of *H. erinaceus* mycelium are erinacine A (HE-A), C (HE-C), and S (HE-S), which either belong to cyanthinditerpenoid (both HE-A and HE-C) or sesterterpene (HE-S).

A double-blind placebo-controlled trial to study the efficacy and safety of erinacine A-Enriched *H. erinaceus* mycelia capsules in mild AD patients showed a higher cognitive function scale and achieved better contrast sensitivity. The benefit of EAHE in reducing cognitive decline has been attributed to improved blood biomarkers, such as calcium, albumin, Hb, Hcy, SOD, BDNF, APOE4 and α-ACT, as well as reduced structural deterioration in the arcuate fasciculus (ARC) and parahippocampal cingulum (PHC) regions of patients with mild AD. Additional studies to determine the molecular mechanisms of the action of EAHE must be performed (Li et al., 2020). Another double-blind placebo-controlled trial in MCI patients showed a modest increase in cognitive function scale in the *H. erinaceus*-treated group when compared to the placebo group (Mori et al., 2009). *Hericium erinaceus* is being investigated for its therapeutic potential as an inducer of neuronal differentiation and for its neuroprotective properties (Mori et al., 2009).

Hericium erinaceus has been reported to improve spatial short-term and visual recognition memory in an Aβ-induced mouse model (Mori et al., 2011). A two-month oral supplementation with *H. erinaceus* reversed the aging-associated decline of recognition memory, possibly through an increase in hippocampal neurogenesis (Ratto et al., 2019). Similarly, in an AlCl3 combined with d-galactose-induced Alzheimer's disease mouse model, *H. erinaceus* mycelium polysaccharide-enriched aqueous extract administration improved behavioural deficits and central cholinergic system function. Specifically, it enhanced the acetylcholine and choline acetyltransferase (ChAT) in the serum and hypothalamus (Zhang et al., 2016). Erinacine A-enriched *H. erinaceus* mycelia (HE-My) and its ethanol extracts (HE-Et) attenuated cerebral Aβ plaque burden through the upregulation of insulin-degrading enzyme, which plays a key role in Aβ clearance. Furthermore, the number of plaque-activated glial cells (microglia and astrocytes) in the cerebral cortex and hippocampus were diminished, the ratio of nerve growth factor (NGF) to NGF precursor (proNGF) was increased and hippocampal neurogenesis was noted in APPswe/PS1dE9 transgenic mice (Tsai-Teng et al., 2016). Treatment with erinacine A and S in AD mice showed reduced cerebral Aβ plaques, decreased glial activation, and promoted hippocampal neurogenesis. Specifically, erinacine A showed a reduction in the level of insoluble amyloid β and C-terminal fragment of amyloid precursor protein and improved behavioural deficits in AD mice (Tzeng et al., 2018). Most of the studies described above have documented that the extracts of *H. erinaceus* improve cognition through induction of nerve differentiation.

Mechanistic insights into understanding the role of the specific action of *H. erinaceus* have been determined. 3-Hydroxyhericenone F (3HF), extracted from the fruit body of *H. erinaceus*, exhibited potent BACE1 inhibitory activity. Furthermore, improvement in the decreasing level of mitochondrial respiratory chain complexes, the calcium ion levels ($[Ca^{2+}]$), the inhibition in the production of ROS, the increase in the mitochondrial membrane potential and ATP levels, and the regulation of the expression levels of the genes encoding for the p21, COX I, COX-II, PARP1 and NF-κB proteins have been noted (Diling et al., 2017).

2.2 Neurotrophic Factors

Nerve Growth Factor (NGF) have been found to play an important role in the differentiation, survival, and maintenance of neuronal cells. Several studies indicate that decreased neurotrophins are a contributing factor in neurodegenerative diseases (Kant Tiwari and Chaturvedi, 2014). Extracts from *H. erinaceus,* such as erinacines A, B, and C, are known to stimulate NGF synthesis (Kawagishi et al., 1994; Mori et al., 2008). Erinacine A has recently been shown to cause direct neuritogenic activity through activation of TrkA and Erk1/2 dependent pathways (Zhang et al., 2017; Hwang et al., 2020). A recent study by Martinez-Marmol et al. showed N-de phenylethylisohericerin (NDPIH), an isoindoline compound along with its hydrophobic derivative hericene A, induced axon outgrowth, neurite branching and enlarged growth cones, which were attributed to brain-derived neurotrophic factor (BDNF)-like activity. Additionally, *in vivo* treatment with *H. erinaceus* crude extract and hericene A significantly improved BDNF and its downstream signalling pathway and improved learning and memory in the novel object-recognition memory test, indicating that hericene A can promote BDNF-like activity in neurons *in vitro* and *in vivo*, thereby enhancing recognition memory (Martínez-Mármol et al., 2020).

Hericium erinaceus has been found to significantly increase lipoxin A4 (LXA4), a protein known to have anti-inflammatory properties in most areas of the brain, along with the increased expression of cytoprotective proteins, such as heme oxygenase-1 (HO-1), heat shock protein 70 (Hsp70) and thioredoxin (TRX) (Trovato et al., 2016). ER stress, however, is known to be caused by sustained Ca^{2+} depletion and a study by Ueda et al. has found that *H. erinaceus* contained compounds that protect the neuron from ER stress, which reduces stress-induced neuronal cell death (Ueda et al., 2008). These studies indicate the therapeutic potential of *H. erinaceus* in mitigating the pathologies of several neurodegenerative diseases.

2.3 Ganoderma lucidum

Ganoderma lucidum, or Lingzhi, is an oriental medicinal mushroom that has a long history of promoting health and longevity in countries like China, Japan, and other Asian countries (Wachtel-Galor et al., 2011). As a widely accepted medicinal mushroom, there are several studies researching the potential benefits of *G. lucidum* for various cancers (breast and prostate) and neurodegenerative disorders, along with the molecular mechanisms of mentioned effects. So far, *G. lucidum* has been noted for its neuroprotective properties.

Ganoderma lucidum extracts have been shown to improve behavioural deficits, reduce Aβ plaques, neurofibrillary tangles, and amyloid angiopathy in transgenic AD mice (Qin et al., 2017a). Mechanistically, increased expression of ApoA1, ApoE, ABCA1, and Syt1 involved in the clearance of Aβ were noted (Qin et al., 2017b). Aqueous extract of *G. lucidum* attenuated Aβ-induced synaptotoxicity and phosphorylation of p38 MAPK, ERK and JNK in primary neuronal cultures (Lai et al., 2008). Hot water extracts of *G. lucidum* ameliorated spatial learning and memory deficits in rats infused with Aβ infused intracerebroventrically (Rahman et al., 2020).

Polysaccharides from *G. lucidum* attenuate microglia-mediated neuro-inflammation and modulate microglial phagocytosis in BV2 microglia and primary mouse microglial cells. *Ganoderma lucidum* polysaccharide treatment downregulated LPS or Aβ-induced pro-inflammatory cytokines and promoted anti-inflammatory cytokine expression. Morphological changes in microglia were associated with MCP-1 and C1q expression changes, indicating that GLP can act as a potent anti-inflammatory agent (Cai et al., 2017). A recent study by Hilliard et al. showed that *G. lucidum* extracts (GLE) downregulated G-CSF, IL1α, MCP-5, MIP3α, and RANTES expression in BV-2 cells stimulated by LPS. Furthermore, GLE modulated the expression of genes associated with NFκB and MAPK signalling, thereby attenuating neuroinflammation (Hilliard et al., 2020).

Triterpenoids from *G. lucidum* (GLT) improved cognitive deficits, inhibited oxidative damage and apoptosis through inhibition of the ROCK signalling pathway. *In vitro* GLTs improved antioxidant superoxide dismutase (SOD), inhibited malondialdehyde and lactic dehydrogenase expression in Aβ-treated hippocampal neurons, indicating that GLTs could play a protective role in AD (Yu et al., 2020).

Age-related changes in methylation have been reported to be strongly associated with AD. Alcohol extracts of *G. lucidum* treatment in mice model of aging and AD showed upregulation of methylation regulators, such as Histone H3, DNMT3A, and DNMT3B in the brain tissues. Furthermore, specific components of *G. lucidum,* such as ganoderic acid Mk, ganoderic acid C6 and lucidone A were reported to be actual constituents regulating methylation, thereby delaying AD progression (Lai et al., 2019). Inhibition of acetylcholinesterase activity is one of the main ways to reduce neuronal cell death and is a candidate for the treatment of neurodegenerative disorders, such as Alzheimer's disease. Two lanostane triterpenes named methyl ganoderate A acetonide and n-butyl ganoderate H were isolated from the fruit bodies of *G. lucidum*, as they were found to be preferential inhibitors of acetylcholinesterase (AchE) activity (Lee et al., 2011).

The water extract of *G. lucidum* induced neuronal differentiation and neuritic outgrowth in a neuronal *in vitro* cell culture model (Cheung et al., 2000). *Ganoderma lucidum* polysaccharides and its water extracts improved cognitive deficits and promoted hippocampal neurogenesis in a mice model of AD. Mechanistically, activation of fibroblast growth factor receptor 1 (FGFR1) and downstream extracellular signal-regulated kinase (ERK) and AKT signalling, which promoted self-renewal of neural progenitor cells, were noted (Huang et al., 2017). Treatment with spore powder of *G. lucidum* in AD patients showed no promising efficacy after a six-week treatment. This might be due to relatively short-term intervention and indicates the need to perform long-term studies to determine its therapeutic benefit (Wang et al., 2018).

2.4 *Antrodia camphorata*

Antrodia camphorata is a medicinal mushroom used as a traditional medicine in Taiwan for various health-related conditions. *Antrodia camphorata* has been approved to be safe for human use in the clinical trial (ClinicalTrials.gov Identifier:NCT01007656) and is frequently used as an herbal remedy for drug intoxication, diarhoea, abnormal pains, hypertension, and itchy skin due to its immunostimulatory and anti-inflammatory effects (Chien et al., 2008). In addition, *A. camphorata* has also been utilised in the treatment of various liver diseases, such as fatty liver, hepatitis and hepatocellular carcinoma (Hsiao et al., 2003; Ao et al., 2009). The pharmacological effects of *A. camphorata* and its bioactive compounds are reviewed in (Geethangili and Tzeng, 2011).

Antrodia camphorata has been shown to have utility in various neurological diseases. *A. camphorata* has antioxidant and anti-inflammatory effects, both of which can be effective in the treatment of neurodegenerative disorders. One study found that *A. camphorata* possessed strong antioxidant and anti-inflammatory abilities for inhibiting neurotoxicity in an AD animal model, in addition to suppressing hyperphosphorylated tau (*p*-tau) protein expression, which is known as an important AD risk factor (Wang et al., 2012). Additional to its antioxidant and anti-inflammatory effects, one study found that *A. camphorata* has a neuroprotective effect, preventing serum-deprived apoptosis via a protein kinase A (PKA)/CREB-dependent pathway (Huang et al., 2005). Antroquinonol, a ubiquinone derivative isolated from *A. camphorata*, has been shown to decrease oxidative stress and inflammatory cytokines via activating the nuclear transcription factor erythroid-2-related factor 2 (Nrf2) pathway, which is downregulated in AD. Furthermore, antroquionol improved learning and memory, decreased hippocampal Aβ and reduced the astrogliosis through activation of the Nrf2 pathway and decreasing histone deacetylase 2 (HDaC2) levels in a mouse model of AD (Chang et al., 2015). Finally, the decrease in ROS by *A. camphorata* could be attributed to increased activity and expression of various antioxidant mechanisms as studies have demonstrated *A. camphorata* to have potent antioxidant effects through upregulation of glutathione S-transferase, regulating normal GSH.GSSH ratio and scavenging of ROS (Tsai et al., 2007).

3. Other Mushrooms with Potential Effects on Neurodegenerative Disorders

Currently, there are many mushrooms that are being investigated for their health benefits, often due to their long history on being used for medicinal purposes. However, these mushrooms have yet to be studied extensively, unlike the mushrooms listed above that have had several studies to investigate their benefits and the molecular mechanisms. The less commonly studied mushrooms in AD are summarised in Table 1.

4. Conclusion

In this review, we have recapitulated mushrooms that have been stated to indicate therapeutic effects in neurons, with importance on either crude extracts or isolated metabolites. Overall, these medicinal mushrooms possess neuroprotective properties and improving neuronal survival and neurite extension, including enhanced neuronal recovery and performance, both *in vitro* and *in vivo*. Therefore, based on the

Table 1. Other medical mushrooms useful to treat Alzheimer's disease.

	Specific constituent	Model	Signalling mechanism	Reference
Cortinarius brunneus	Brunnein-A	PC3 cells	AChE inhibiting activity	(Teichert et al., 2007)
Cortinarius Infractus	Infractopicrin and 10-hydroxy-infractopicrin	*In vitro* cell culture model	AChE inhibiting activity Inhibition of Aβ-peptides self-aggregation	(Geissler et al., 2010)
Dictyophora indusiata	Dictyoquinazols A, B and C	Primary mouse cortical neurons	Prevent glutamate- and NMDA-induced excitotoxicities	(Lee et al., 2002)
	Dictyophorine A and B	Rat astroglial cells	Increase the synthesis of NGF	(Kawagishi et al., 1997)
Lignosus rhinocerotis	Sclerotium-hot aqueous and ethanolic extracts	*In-vitro* rat pheochromocytoma (PC-12) cells	Neuritogenic activity through MEK and ERK1/2 signalling pathway	(Seow et al., 2015)
	Sclerotium-hot aqueous and ethanolic extracts	Lipopolysaccharide (LPS)-stimulated BV2 microglia	Inhibition of nitric oxide production	(Seow et al., 2017)
Mycoleptodonoides aitchisonii	Fruit bodies	Thapsigargin (TG)-induced ER stress	Regulating calcium homeostasis	(Choi et al., 2014)
	Aqueous extract	Fourteen days pregnant Wistar strain rats	Increased NGF in newborn rats during lactation period	(Okuyama et al., 2004)
Pleurotus giganteus	Aqueous and ethanolic extracts of fruit bodies	*In-vitro* rat pheochromocytoma (PC-12) cells	Neuritogenic activity through MEK/ERK and PI3K/Akt signaling pathway	(Phan et al., 2012)
	Uridine	Differentiating neuroblastoma (N2a) cells	Neurite outgrowth by MEK/ERK and PI3K-Akt-mTOR pathway	(Phan et al., 2015)
	Ethanol extract	LPS and H_2O_2 stimulated RAW 264.7 macrophages	Inhibition of nitric oxide production and suppression of STAT3 and COX-2 pathway	(Baskaran et al., 2017)
Sarcodons Cabrosus	Scabronine A	*In vitro* rat pheochromocytoma (PC-12) cells	Increased NGF	(Obara et al., 1999)
Termitomyces albuminosus	Termitomycesphins A-D, G and H	PC12 cells	Neuritogenic effect	(Qu et al., 2012; Qi et al., 2000)
Tremella fuciformis	Fruit bodies	*In vitro* rat pheochromocytoma (PC-12) cells	Increase neurite outgrowth	(Park et al., 2012)
	Hot water extract of *T. fuciformis*	PC12 cells	Neurite outgrowth, prevent Aβ induced cytotoxicity	(Park et al., 2007)
	Fruit bodies	TMT induced impairment of memory in rats	Increased ChAT and CREB reactivity in the hippocampus, improved brain glucose metabolism	(Park et al., 2012)

various studies considered in this review, medicinal mushrooms may have potential therapeutic values to treat neurodegenerative diseases. Nevertheless, any such effort involving clinical trials must be carried out with great vigilance since the adverse pharmacological effects of these mushrooms are not well established even though many of these mushrooms are edible. Therefore, further research on medicinal mushrooms in the neurological clinical field must be attempted with a long-term objective of developing future effective therapies for alleviating neurodegenerative diseases.

References

Ao, Z.-H., Xu, Z.-H., Lu, Z.-M., Xu, H.-Y., Zhang, X.-M. and Dou, W.-F. (2009). Niuchangchih (*Antrodia camphorata*) and its potential in treating liver diseases. J. Ethnopharmacol., 121(2): 194–212.

Baskaran, A., Chua, K.H., Sabaratnam, V., Ravishankar Ram, M. and Kuppusamy, U.R. (2017). *Pleurotus giganteus* (Berk. Karun & Hyde), the giant oyster mushroom inhibits NO production in LPS/H_2O_2 stimulated RAW 264.7 cells via STAT 3 and COX-2 pathways. BMC Compl. Alt. Med., 17(1): 40. 10.1186/s12906-016-1546-6.

Cai, Q., Li, Y. and Pei, G. (2017). Polysaccharides from *Ganoderma lucidum* attenuate microglia-mediated neuroinflammation and modulate microglial phagocytosis and behavioural response. J. Neuroinflam., 14(1): 63. 10.1186/s12974-017-0839-0.

Chang, W.-H., Chen, M.C. and Cheng, I.H. (2015). Antroquinonol lowers brain amyloid-β levels and improves spatial learning and memory in a transgenic mouse model of Alzheimer's disease. Sci. Rep., 5(1): 15067. 10.1038/srep15067.

Chen, X.-Q. and Mobley, W.C. (2019). Alzheimer's disease pathogenesis: Insights from molecular and cellular biology studies of oligomeric Aβ and Tau species. Front. Neurosci., 13(659). 10.3389/fnins.2019.00659.

Cheung, W.M.W., Hui, W.S., Chu, P.W.K., Chiu, S.W. and Ip, N.Y. (2000). *Ganoderma* extract activates MAP kinases and induces the neuronal differentiation of rat pheochromocytoma PC12 cells. FEBS Lett., 486(3): 291–296. 10.1016/s0014-5793(00)02317-6.

Chien, S.-C., Chen, M.-L., Kuo, H.-T., Tsai, Y.-C., Lin, B.-F. and Kuo, Y.-H. (2008). Anti-inflammatory activities of new succinic and maleic derivatives from the fruiting body of *Antrodia camphorata*. J. Agric. Food Chem., 56(16): 7017–7022.

Choi, J.-H., Suzuki, T., Okumura, H., Noguchi, K., Kondo, M. et al. (2014). Endoplasmic reticulum stress suppressive compounds from the edible mushroom *Mycoleptodonoides aitchisonii*. J. Nat. Prod., 77(7): 1729–1733.

Crous-Bou, M., Minguillón, C., Gramunt, N. and Molinuevo, J.L. (2017). Alzheimer's disease prevention: From risk factors to early intervention. Alz. Res. Ther., 9(1): 71–71. 10.1186/s13195-017-0297-z.

Diling, C., Tianqiao, Y., Jian, Y., Chaoqun, Z., Ou, S. and Yizhen, X. (2017). Docking studies and biological evaluation of a potential β-secretase inhibitor of 3-hydroxyhericenone from *Hericium erinaceus*, Front. Pharmacol., 8(219). 10.3389/fphar.2017.00219.

Friedman, M. (2015). Chemistry, nutrition and health-promoting properties of *Hericium erinaceus* (lion's mane) mushroom fruiting bodies and mycelia and their bioactive compounds. J. Agric. Food Chem., 63(32): 7108–7123. 10.1021/acs.jafc.5b02914.

Geethangili, M. and Tzeng, Y.-M. (2011). Review of pharmacological effects of <i>*Antrodia camphorata*</i> and its bioactive compounds. Evid. Based Compl. Alt. Med., 2011: 212641. 10.1093/ecam/nep108.

Geissler, T., Brandt, W., Porzel, A., Schlenzig, D., Kehlen, A. et al. (2010). Acetylcholinesterase inhibitors from the toadstool *Cortinarius infractus*. Bioorg, Med. Chem., 18(6): 2173–2177.

Hampel, H., Caraci, F., Cuello, A.C., Caruso, G., Nisticò, R. et al. (2020). A path toward precision medicine for neuroinflammatory mechanisms in Alzheimer's disease. Front. Immunol., 11: 456–456. 10.3389/fimmu.2020.00456.

Hilliard, A., Mendonca, P. and Soliman, K.F. (2020). Involvement of NFƙB and MAPK signaling pathways in the preventive effects of *Ganoderma lucidum* on the inflammation of BV-2 microglial cells induced by LPS. J. Neuroimmunol., 345: 577269. 10.1016/j.jneuroim.2020.577269.

Hsiao, G., Shen, M.-Y., Lin, K.-H., Lan, M.-H., Wu, L.-Y. et al. (2003). Antioxidative and hepatoprotective effects of *Antrodia camphorata* extract. J. Agric. Food Chem., 51(11): 3302–3308.

Huang, S., Mao, J., Ding, K., Zhou, Y., Zeng, X. et al. (2017). Polysaccharides from *Ganoderma lucidum* promote cognitive function and neural progenitor proliferation in mouse model of Alzheimer's disease. Stem Cell Rep., 8(1): 84–94.

Hwang, J.-H., Chen, C.-C., Lee, L.-Y., Chiang, H.-T., Wang, M.-F. and Chan, Y.-C. (2020). *Hericium erinaceus* enhances neurotrophic factors and prevents cochlear cell apoptosis in senescence accelerated mice. J. Fun. Foods, 66: 103832. 10.1016/j.jff.2020.103832.

Hynd, M.R., Scott, H.L. and Dodd, P.R. (2004). Glutamate-mediated excitotoxicity and neurodegeneration in Alzheimer's disease. Neurochem. Int., 45(5): 583–595. 10.1016/j.neuint.2004.03.007.

Kant Tiwari, S.K. and Chaturvedi, R. (2014). Peptide therapeutics in neurodegenerative disorders. Curr. Med. Chem., 21(23): 2610–2631.

Kawagishi, H., Ishiyama, D., Mori, H., Sakamoto, H., Ishiguro, Y., Furukawa, S. and Li, J. (1997). Dictyophorines A and B, two stimulators of NGF-synthesis from the mushroom *Dictyophora indusiata*. Phytochemistry, 45(6): 1203–1205.

Lai, C.S.-W., Yu, M.-S., Yuen, W.-H., So, K.-F., Zee, S.-Y. and Chang, R.C.-C. (2008). Antagonizing β-amyloid peptide neurotoxicity of the anti-aging fungus *Ganoderma lucidum*. Brain Res., 1190: 215–224.

Lai, G., Guo, Y., Chen, D., Tang, X., Shuai, O. et al. (2019). Alcohol extracts from *Ganoderma lucidum* delay the progress of Alzheimer's disease by regulating DNA methylation in rodents. Front. Pharmacol., 10(272). 10.3389/fphar.2019.00272.

Lee, I.-K., Yun, B.-S., Han, G., Cho, D.-H., Kim, Y.-H. and Yoo, I.-D. (2002). Dictyoquinazols A, B, and C, new neuroprotective compounds from the mushroom *Dictyophora indusiata*. J. Nat. Prod., 65(12): 1769–1772.

Li, I.-C., Chang, H.-H., Lin, C.-H., Chen, W.-P., Lu, T.-H. et al. (2020). Prevention of early Alzheimer's disease by erinacine A-enriched *Hericium erinaceus* mycelia pilot double-blind placebo-controlled study. Front. Aging Neurosci., 12(155). 10.3389/fnagi.2020.00155.

Martínez-Mármol, R., Chai, Y., Khan, Z., Kim, S.B., Hong, S.M. et al. (2020). Hericerin derivatives from *Hericium erinaceus* exert BDNF-like neurotrophic activity in central hippocampal neurons and enhance memory. bioRxiv:2020.2008.2028.271676. 10.1101/2020.08.28.271676.

Mori, K., Obara, Y., Hirota, M., Azumi, Y., Kinugasa, S. et al. (2008). Nerve growth factor-inducing activity of *Hericium erinaceus* in 1321N1 human astrocytoma cells. Biol. Pharm. Bull., 31(9): 1727–1732. 10.1248/bpb.31.1727.

Mori, K., Inatomi, S., Ouchi, K., Azumi, Y. and Tuchida, T. (2009). Improving effects of the mushroom Yamabushitake (*Hericium erinaceus*) on mild cognitive impairment: A double-blind placebo-controlled clinical trial. Phytother. Res., 23(3): 367–372. 10.1002/ptr.2634.

Mori, K., Obara, Y., Moriya, T., Inatomi, S. and Nakahata, N. (2011). Effects of *Hericium erinaceus* on amyloid β(25-35) peptide-induced learning and memory deficits in mice. Biomed. Res., 32(1): 67–72. 10.2220/biomedres.32.67.

Obara, Y., Nakahata, N., Kita, T., Takaya, Y., Kobayashi, H. et al. (1999). Stimulation of neurotrophic factor secretion from 1321N1 human astrocytoma cells by novel diterpenoids, scabronines A and G. Eur. J. Pharmacol., 370(1): 79–84.

Okuyama, S., Lam, N.V., Hatakeyama, T., Terashima, T., Yamagata, K. and Yokogoshi, H. (2004). *Mycoleptodonoides aitchisonii* affects brain nerve growth factor concentration in newborn rats. Nutr. Neurosci., 7(5-6): 341–349.

Park, H.-J., Shim, H.S., Ahn, Y.H., Kim, K.S., Park, K.J. et al. (2012). *Tremella fuciformis* enhances the neurite outgrowth of PC12 cells and restores trimethyltin-induced impairment of memory in rats via activation of CREB transcription and cholinergic systems. Behav. Brain Res., 229(1): 82–90.

Park, K.J., Lee, S.-Y., Kim, H.-S., Yamazaki, M., Chiba, K. and Ha, H.-C. (2007). The neuroprotective and neurotrophic effects of *Tremella fuciformis* in PC12h cells. Mycobiology, 35(1): 11–15. 10.4489/MYCO.2007.35.1.011.

Perl, D.P. (2010). Neuropathology of Alzheimer's disease. Mt. Sinai J. Med., 77(1): 32–42. 10.1002/msj.20157.

Phan, C.-W., Wong, W.-L., David, P., Naidu, M. and Sabaratnam, V. (2012). *Pleurotus giganteus* (Berk.) Karunarathna & K.D. Hyde: Nutritional value and *in vitro* neurite outgrowth activity in rat pheochromocytoma cells. BMC Compl. Alt. Med., 12(1): 102.10.1186/1472-6882-12-102.

Phan, C.-W., David, P., Wong, K.-H., Naidu, M. and Sabaratnam, V. (2015). Uridine from *Pleurotus giganteus* and its neurite outgrowth stimulatory effects with underlying mechanism. PLoS ONE, 10(11): e0143004.

Qi, J, Ojika, M. and Sakagami, Y. (2000). Termitomycesphins A–D, novel neuritogenic cerebrosides from the edible Chinese mushroom *Termitomyces albuminosus*. Tetrahedron, 56(32): 5835–5841.

Qin, C., Wu, S.-Q., Chen, B.-S., Wu, X.-X., Qu, K.-Y. et al. (2017a). Pathological changes in APP/PS-1 transgenic mouse models of Alzheimer's disease treated with *Ganoderma lucidum* preparation. Zhongguo Yi Xue Ke Xue Yuan Xue Bao, 39(4): 552–561. 10.3881/j.issn.1000-503x.2017.04.015.

Qin, C., Wu, S., Chen, B., Wu, X. and Qu, K. et al. (2017b). Effect of *Ganoderma lucidum* preparation on the behaviour, biochemistry and autoimmune parameters of mouse models of APP/PS1 double transgenic Alzheimer's disease. Zhongguo Yi Xue Ke Xue Yuan Xue Bao, 39(3): 330–335. 10.3881/j.issn.1000-503x.2017.03.006.

Qu, Y., Sun, K., Gao, L., Sakagami, Y., Kawagishi, H. et al. (2012). Termitomycesphins G and H, additional cerebrosides from the edible Chinese mushroom *Termitomyces albuminosus*. Biosci. Biotechnol. Biochem., 76(4): 791–793.

Rahman, M.A., Hossain, S., Abdullah, N. and Aminudin, N. (2020). Lingzhi or Reishi medicinal mushroom, *Ganoderma lucidum* (Agaricomycetes) ameliorates spatial learning and memory deficits in rats with hypercholesterolemia and Alzheimer's disease. Int. J. Med. Mush., 22(1).

Ramesh, S., Govindarajulu, M., Jones, E., Suppiramaniam, V., Moore, T. and Dhanasekaran, M. (2018). Mitochondrial dysfunction and the role of mitophagy in Alzheimer's disease. Alzheimer's Disease & Treatment, MedDocs Publishers, LLC.

Ratto, D., Corana, F., Mannucci, B., Priori, E.C., Cobelli, F. et al. (2019). *Hericium erinaceus* improves recognition memory and induces hippocampal and cerebellar neurogenesis in frail mice during aging. Nutrients, 11(4): 715.10.3390/nu11040715.

Roupas, P., Keogh, J., Noakes, M., Margetts, C. and Taylor, P. (2012). The role of edible mushrooms in health: Evaluation of the evidence. J. Fun. Foods, 4(4): 687–709.

Seow, S.L.-S., Eik, L.-F., Naidu, M., David, P., Wong, K.-H. and Sabaratnam, V. (2015). *Lignosus rhinocerotis* (Cooke) Ryvarden mimics the neuritogenic activity of nerve growth factor via MEK/ERK1/2 signalling pathway in PC-12 cells. Sci. Rep., 5(1): 16349. 10.1038/srep16349.

Seow, S.L.-S., Naidu, M., Sabaratnam, V., Vidyadaran, S. and Wong, K.-H. (2017). Tiger's milk medicinal mushroom, *Lignosus rhinocerotis* (Agaricomycetes) sclerotium inhibits nitric oxide production in LPS-stimulated BV2 microglia. Int. J. Med. Mush., 19(5): 405–418. 10.1615/IntJMedMushrooms.v19.i5.30.

Teichert, A., Schmidt, J., Porzel, A., Arnold, N. and Wessjohann, L. (2007). Brunneins A-C, beta-carboline alkaloids from *Cortinarius brunneus*. J. Nat. Prod., 70(9): 1529–1531. 10.1021/np070259w.

Tönnies, E. and Trushina, E. (2017). Oxidative stress, synaptic dysfunction and Alzheimer's disease. J. Alz. Dis., 57(4): 1105–1121. 10.3233/JAD-161088.

Trovato, A., Siracusa, R., Di Paola, R., Scuto, M., Fronte, V., Koverech, G., Luca, M., Serra, A., Toscano, M. and Petralia, A. (2016). Redox modulation of cellular stress response and lipoxin A4 expression by *Coriolus versicolor* in rat brain: relevance to Alzheimer's disease pathogenesis. Neurotoxicology 53: 350–358.

Tsai-Teng, T., Chin-Chu, C., Li-Ya, L., Wan-Ping, C., Chung-Kuang, L. et al. (2016). Erinacine A-enriched *Hericium erinaceus* mycelium ameliorates Alzheimer's disease-related pathologies in APPswe/PS1dE9 transgenic mice. J. Biomed. Sci., 23(1): 49. 10.1186/s12929-016-0266-z.

Tsai, M.-C., Song, T.-Y., Shih, P.-H. and Yen, G.-C. (2007). Antioxidant properties of water-soluble polysaccharides from *Antrodia cinnamomea* in submerged culture. Food Chem., 104(3): 1115–1122.

Tzeng, T.-T., Chen, C.-C., Chen, C.-C., Tsay, H.-J., Lee, L.-Y. et al. (2018). The cyanthin diterpenoid and sesterterpene constituents of *Hericium erinaceus* mycelium ameliorate Alzheimer's disease-related pathologies in APP/PS1 transgenic mice. Int. J. Mol. Sci., 19(2): 598. 10.3390/ijms19020598.

Ueda, K., Tsujimori, M., Kodani, S., Chiba, A., Kubo, M., Masuno, K., Sekiya, A., Nagai, K. and Kawagishi, H. (2008). An endoplasmic reticulum (ER) stress-suppressive compound and its analogues from the mushroom *Hericium erinaceum*. Bioorganic & Medicinal Chemistry 16(21): 9467–9470.

Wachtel-Galor, S., Yuen, J., Buswell, J.A. and Benzie, I.F. (2011). *Ganoderma lucidum* (Lingzhi or Reishi). *In*: Herbal Medicine: Biomolecular and Clinical Aspects. 2nd edition. CRC Press/Taylor & Francis.

Wang, G.H., Wang, L.H., Wang, C. and Qin, L.H. (2018). Spore powder of *Ganoderma lucidum* for the treatment of Alzheimer disease: A pilot study. Medicine (Baltimore), 97(19): e0636. 10.1097/md.0000000000010636.

Wang, L.-C., Wang, S.-E., Wang, J.-J., Tsai, T.-Y., Lin, C.-H. et al. (2012). *In vitro* and *in vivo* comparisons of the effects of the fruiting body and mycelium of *Antrodia camphorata* against amyloid β-protein-induced neurotoxicity and memory impairment. Appl. Microbiol. Biotechnol., 94(6): 1505–1519.

Wang, Z., Luo, D. and Liang, Z. (2004). Structure of polysaccharides from the fruiting body of *Hericium erinaceus* Pers. Carbohyd. Polym., 57(3): 241–247.

Wasser, S.P. (2002). Medicinal mushrooms as a source of antitumor and immunomodulating polysaccharides. Appl. Microbiol. Biotechnol., 60(3): 258–274. doi:10.1007/s00253-002-1076-7.

Yin, F., Sancheti, H., Patil, I. and Cadenas, E. (2016). Energy metabolism and inflammation in brain aging and Alzheimer's disease. Free Radic. Biol. Med., 100: 108–122. 10.1016/j.freeradbiomed.2016.04.200.

Yu, N., Huang, Y., Jiang, Y., Zou, L., Liu, X. et al. (2020). *Ganoderma lucidum* triterpenoids (GLTs) reduce neuronal apoptosis via inhibition of ROCK signal pathway in APP/PS1 transgenic Alzheimer's disease mice. Oxid. Med. Cell. Long., 2020: 9894037. 10.1155/2020/9894037.

Zhang, C.-C., Cao, C.-Y., Kubo, M., Harada, K., Yan, X.-T. et al. (2017). Chemical constituents from *Hericium erinaceus* promote neuronal survival and potentiate neurite outgrowth via the TrkA/Erk1/2 pathway. Int. J. Mol. Sci., 18(8): 1659.

Zhang, J., An, S., Hu, W., Teng, M., Wang, X. et al. (2016). The neuroprotective properties of *Hericium erinaceus* in glutamate-damaged differentiated PC12 cells and an Alzheimer's disease mouse model. Int. J. Mol. Sci., 17(11): 1810.

8

Neuroprotective Attributes of *Cordyceps*

Lekshmi R,[1] Rajakrishnan R,[2] Benil PB,[3]
Naif Abdullah Al-Dhabi,[2,]*
Savarimuthu Ignacimuthu,[4] Ameer Khusro,[5]
Young Ock Kim,[6] Hak-Jae Kim[6] and
Mariadhas Valan Arasu[2,4,]*

1. INTRODUCTION

Neuroprotection is now one of the potential areas of research among investigators as they are more focused on finding suitable approaches to protect neurons against diverse pathological factors in neurodegenerative diseases. These approaches aim to find a solution either before the beginning of the illness so that the neurons cannot be affected badly or during the advancement of ailment to avoid spreading of damage from one neuron to the others. Therefore, neuroprotection could be considered as a disease-modifying method to suspend or even stop progressive neurodegeneration.

Neuroprotection refers to the approaches and mechanisms that protect the central nervous system against neuronal damage due to acute (e.g., stroke or

[1] Department of Botany and Biotechnology, MSM College, Kayamkulam, Kerala, India.
[2] Department of Botany and Microbiology, College of Science, King Saud University, PO Box 2455, Riyadh 11451, Saudi Arabia.
[3] Department of Agadatantra, Vaidyaratnam PS Varier Ayurveda College, PO Edarikode, Kottakkal, Kerala, India.
[4] Xavier Research Foundation, St. Xavier's College, Palayamkottai, Thirunelveli, Tamil Nadu, India.
[5] Research Department of Plant Biology and Biotechnology, Loyola College, Nungambakkam, Chennai, India.
[6] Department of Clinical Pharmacology, College of Medicine, Soonchunhyang University, Cheonan, Republic of Korea.
* Corresponding authors: naldhabi@ksu.edu.sa; mvalanarasu@gmail.com

trauma) and chronic neurodegenerative disorders (e.g., Alzheimer's and Parkinson's diseases) (Iriti et al., 2010). Despite many decades of research into the identification and translational developments of neuroprotective compounds, only a few strategies have progressed into appropriately designed unbiased, randomised and placebo-controlled clinical trials (Lapchak and Boitano, 2017). Among these strategies, herbal medicine may represent a valuable resource in prevention rather than in therapy of diseases pertaining to the central nervous system (CNS) in association with a healthy lifestyle, including appropriate dietary habits and moderate physical activities. As a complementary and alternative therapy, phytotherapy refers to the medical use of plant parts for their therapeutic properties. Phytomedicine usually possesses complex blends of bioactive components termed as phytochemicals, which include alkaloids, flavonoids, terpenoids, phenolics and so on. Herbal drug therapy is becoming popular owing to its relatively minor side effects, therapeutic efficacy, ease of availability and less expensive compared to the modern therapeutics (Malik, 2017). Another strategy is nutritional therapy, which is based on functional foods and nutraceuticals as therapeutics. This complementary therapy depends on the assumption that food is not only a source of nutrients and energy, but can also provide health benefits (Zhao, 2007). Consumed as part of a normal diet, plant foods are thus not only a source of nutrients and energy, but may additionally provide health benefits beyond basic nutritional functions by virtue of their dietary therapeutics (Esposito et al., 2002).

Many species of fungi have been valued as important resources of food as well as medicine since time immemorial. They can be regarded as functional foods because of their ability to improve human health and quality of life. In recent past, several studies were conducted on specific pharmacological activities of edible and medicinal fungi, which provided insights into their neuroprotective effects—anti-oxidant, anti-neuroinflammatory, cholinesterase inhibitory properties and their ability to prevent neuronal death (Wong et al., 2017). The number of fungal species evaluated for neuro-health studies are very much limited as compared to the total edible/medicinal species identified so far. There are reports that some species of marine fungi, such as *Aspergillus glaucus* and *Tacromyces islandicus* produce polyketides with antioxidant and neuroprotective properties (Sun et al., 2013; Li et al., 2016).

The genus *Cordyceps* consists of insect-inhabiting ascomycetous fungi belonging to the family Clavicipitaceae, which is regarded as a rich reservoir of valuable natural products with several biological activities and those have been used as a health-supplement for sub-health patients, especially the elderly people in many Asian countries. This chapter highlights the neuroprotective attributes of *Cordyceps*.

2. Neurotoxicity and Neuroprotection

In a broader sense, neurotoxins are exogenous and endogenous substances or metabolites resulting in neuronal cell death by inducing apoptotic and necrotic or necroptoptic events (Segura-Aguilar et al., 2006). Exogenous neurotoxicants are now discoursed under those producing neurodegenerative disorders, psychiatric disorders and substance-abuse disorders (Segura-Aguilar et al., 2004). Several metabolites endogenously derived from the metabolism of food articles, like *Lathyrus* and toxic metabolites of amino acid tryptophan, such as quinolinic acid, are potent neurotoxins (Guillemin et al., 2005). Health impacts of neurotoxicity are now attributed to a plethora of neurodegenerative diseases, like Amyotropic Lateral Sclerosis (ALS), Parkinsonism, Alzhiemer's disease, Huntington's disease, myasthenia gravis and so on. Several mechanisms have been elucidated for the mechanism of neurotoxicity by these agents (Palomo et al., 2004). Gene-environment interplay by creating mediators like nitric oxide and superoxides capable of producing devastating neuronal damage coupled with genes providing the basic attributes of enzymes necessary for the biochemical pathways are considered a key event in neurotoxicity. The β-amyloid protein aggregates signalling axonal death is another major pathway of neurodegeneration (Ferrer, 2004). Mediators, like TNF-α, exert neurotoxicity in regions of the brain, where their receptors are absent by producing profound escalation of oxidative stress (Sriram et al., 2006). Schwann cell dysregulation resulting from neuronal damage, especially of the peripheral nerves, is also attributed with neurotoxic endpoints (Boyle et al., 2005). Bacterial lipopolysaccharide endotoxin-induced neuroinflammation also claims a sizeable percentile in the mechanism of neurotoxicity (Johansson et al., 2005). Several kinds of biological processes have also been associated in the development and pathogenesis of neurodegenerative disorders, which may include oxidative stress, neuro-inflammation, excitotoxicity, cell death and mitochondrial dysfunction (Chitinis and Weiner, 2017; Mattson, 2017; Singh et al., 2019).

Among the different mechanisms of toxicity cited above, the disturbance in cerebral redox homeostasis is the primary triggering event. Several experimental models inconclusively demonstrated the 'primo loco' status of oxidative stress in neuronal cell death. The high concentration of reactive oxygen and nitrogen species produced by oxidative stress damages the proteins, lipids and DNA of the neurons. Several synthetic compounds have been established for curbing the damages produced by these free radicals. Compounds like edaravone produced significant free radical-scavenging of hydroxyl radical, leading to the prevention of lipid peroxidation. Antibiotic Minocycline, a semi-synthetic derivative of tetracycline, is also used clinically as a free radical scavenger and an inhibitor of microglial activity and caspase-1-dependent apoptosis. Although it has been effectively used in treatment, it possesses severe setbacks in the form of adverse events and toxicities. Hence, there is a high demand for compounds derived from natural sources to exhibit neuroprotective accomplishments. A multi-pronged approach has been adopted while employing natural agents for neuroprotection. The chief event being oxidative stress, primary investigation has been in its direction. Several compounds derived from natural sources showed significant neuroprotection by increasing the levels of antioxidant enzymes, like SOD and catalase, restoration of GSH levels and decreased MDA levels, thereby decreasing the lipid peroxidation.

Natural products have been recognised and employed for their healing properties since olden times. During the past few years, several research studies have published the therapeutic effects of natural products, such as higher plants, algae, fungi, lichens and their biologically active molecules against almost all human diseases, including neurodegenerative disorders. Nowadays natural products are considered as potential neuroprotective agents for the treatment of neurodegenerative diseases (Sairazi and Sirajuddin, 2020). Many medical practitioners around the world are showing interest in complementary and alternative medicine, especially dietary supplements and functional foods in delaying the onset of age-associated neurodegenerative diseases. The search for more effective neuroprotective agents has transgressed the realms of phytomedicine and evaded the use of microscopic and macroscopic lower life forms. Mushrooms have long been used not only as food but also for the treatment of various ailments. There are a number of edible mushrooms available, which possess many rare and exotic components that exhibit positive effects on the brain cells *in vitro* as well as *in vivo*. When compared to phytomedicines, which are broadly explored and more advanced, the brain and cognitive health impacts of fungi are in the preliminary stages of research. The *in vitro* toxicological evaluation of various mushroom extracts on embryonic fibroblast and neuroblastoma cell lines revealed the safety of the extracts even at high doses and therefore they can be considered as a dietary supplement to improve the brain and cognitive health (Phan et al., 2013). There are reports about some medicinal mushrooms, which have been used in traditional medicine to promote neurotrophic activities, especially nerve regeneration (Wong et al., 2009), neuroprotection (Zhou et al., 2010) and neurite outgrowth (Eik et al., 2012). Even though reports are scarce, certain studies recommended that culinary-therapeutic mushrooms may play an important role in the prevention of many age-related neurological dysfunctions, such as Alzheimer's and Parkinson's diseases. Such mushrooms include *Antrodia camphorata, Ganoderma lucidum, Grifola frondosa, Hericium erinaceus, Lignosus rhinocerotis, Pleurotus giganteus, Sarcodon* spp. and many more (Phan et al., 2015). It is reported that a marine fungus *Aspergillus glaucus* (HB1-19) and an endophytic fungus *Talaromyces islandicus* (EN-501) produce polyketides with efficient antioxidant and neuroprotective properties (Sun et al., 2013; Li et al., 2016). Another fungal metabolite, xyloketal B, isolated from *Xylaria* sp. Affords protection on PC12 cells against ischemia-induced cell injury and MPTP-induced neurotoxicity (Lu et al., 2010). Yurchenko et al. (2018) reported the neuroprotective activity of certain compounds, such as 6-hydroxy-N-acetyl-β-oxotryptamine, 3-methylorsellinic acid, 8-methoxy-3,5-dimethylisochroman-6-ol, candidusin A, 4″-dehydroxycandidusin A and diketopiperazinemactanamide isolated from various marine fungi. Such promising results on the neuroprotective capability of fungal metabolites is an indication of the elaborate efforts to identify more fungal species that may provide neuroprotection, memory improvement and cognition functions.

3. *Cordyceps* and their Therapeutic Potential

The genus *Cordyceps* was first recognised in 1818 and was conventionally included in the family Clavicipitaceae under Ascomycota. This genus houses species that parasitise a number of orders of insects and certain fungi, producing elongated and cylindrical or filamentous stromata with perithecioid filiform ascocarp with multi-septate ascospores (Shrestha et al., 2017). *Cordyceps* is distinguished from other members of Clavicipitaceae by its production of superficial to completely-immersed perithecia on stipitate and often clavate to capitate stromata and its ecology as a pathogen of arthropods and the fungal genus *Elaphomyces* (Sung et al., 2007). They are distributed widely among almost all terrestrial habitats except Antarctica, with maximum species diversity noted in tropics and sub-tropics, mainly in the eastern and south-east Asian regions (Kobayasi, 1941; Samson et al., 1988). Based on molecular phylogenetic and morphological studies, the genus *Cordyceps* was recently revised and divided into four genera—*Cordyceps*, *Elaphocordyceps*, *Metacordyceps* and *Ophiocordyceps*.

3.1 *Cordyceps*

- They consist of pallid to brightly coloured species that produce soft, fleshy stroma.
- Most of them attack larvae and pupae of insect orders, such as Lepidoptera and Coleoptera.
- Numerous species that produce highly reduced stromata, loosely organised hyphae or a subiculum on the host also occur in this genus.

3.2 *Elaphocordyceps*

- They include all species that parasitise *Elaphomyces* and closely related species that attack nymphs of cicadas.
- The morphology of the *Elaphomyces* parasites and the cicada pathogens are remarkably similar and attest to the recent history of inter-kingdom host-jumps in a common subterranean environment.

3.3 *Metacordyceps*

- They include only a limited number of described species, of which all but one are known from East Asia.
- The stromatal colour of fresh specimen ranges from white to lilac, purple or green and the darker pigments are almost black in dried specimens.
- The texture of the stromata is fibrous and not fleshy like *Cordyceps s.s.* and the hosts are almost always buried in soil.

3.4 *Ophiocordyceps*

- This is the largest genus of arthropod-pathogenic fungi.
- Many species are darkly pigmented and occur at immature stages of hosts buried in the soil or in the decaying wood.
- Stromatal morphology is diverse, ranging from filiform and wiry to clavate and fibrous. Many species produce their perithecia in non-terminal regions of the stroma either distinctly superficial or in broad irregular patches or in lateral pads.

Many researchers are showing interest in *Cordyceps* species since the use of entomopathogenic fungi is widely accepted due to their economic as well as environmental significance. A number of *Cordyceps* spp. has been cultivated globally considering their medicinal value (e.g., *Cordyceps cicadicola, C. liangshanesis, C. militaris, C. ophioglossoides, C. sinensis* and *C. sobolifera*) (Tuli et al., 2017). At the same time, demand is increasing because of their therapeutic properties but their distribution at about 14,000 feet altitude in the Himalayan regions of China, Nepal,

Tibet and India makes them high priced, at around US\$ 12,000/kg (Ashraf et al., 2020; Paterson, 2008). Some of the recent studies in the Western Ghats and southwest India also showed occurrence of various *Cordyceps* species and allied genera (Sridhar and Karun, 2017; Dattaraj et al., 2018). Even with so many hurdles like their distribution and procurement difficulties, they are still treated as a highly-valued fungi for their abundant natural bioactive reserves with various potent biological activities (Yue et al., 2013). The list of therapeutic compounds and biological activities of *Cordyceps* spp. is summarised in Table 1. For hundreds of years, *Cordyceps* were used as folk-food tonic, but only in recent times its potential pharmaceutical as well as nutraceutical applications have been explored, which has attracted food scientists globally (Ashraf et al., 2020).

3.5 *Cordyceps* as Neuroprotective Agents

Cordyceps spp. are a group of rare and naturally occurring entomopathogenic fungi, which are generally seen in high altitude regions, like the Himalayan plateau and are a proven therapeutic mushroom widely used in traditional Chinese medicine. *Cordyceps* are among the thousands of available mushrooms containing various bioactive components with innumerable health benefits.

Many species of *Cordyceps* are widely used in traditional medicines in China, Japan, Korea and other East Asian countries (Zhu et al., 1998). Even though extensive research has been carried out on the various therapeutic properties of *Cordyceps* spp., studies on their neuroprotective assessment are limited, but available reports clearly indicate the importance of *Cordyceps* spp. as neuroprotective agents and the need of further research. Among the various *Cordyceps* spp., neuroprotective studies were conducted mainly on *C. cicadae, C. militaris, C. ophioglossoides* and *C. sinensis*.

Cordycepin, an active constituent in *Cordyceps* as a derivative of the nucleoside adenosine is reported to have neuroprotective effects. Song et al. (2018) have conducted a detailed study on the neuroprotective ability of cordycepin. Their results confirmed that treatment with cordycepin significantly repressed the $A\beta_{25-35}$-induced cell death, reduced ROS production and the per cent cell death. The result of Ca^{2+}-dependent fluorescence and whole-cell current recording showed that abnormal calcium homeostasis induced by $A\beta_{25-35}$ is rebalanced by cordycepin. Cordycepin also reduces the over-activity of acetylcholine esterase and increased p-Tau expression. These findings suggest that cordycepin exerts notable protective effects against $A\beta_{25-35}$-induced neurotoxicity. The DPCPX, a specific A_1R antagonist, partially reduced the ability of cordycepin in preventing $A\beta_{25-35}$-induced neurotoxicity as well as the ability to attenuate neuronal death via the activation of A_1Rs. It also significantly reduced the acetylcholine esterase over-activity and Ca^{2+} influx. Thus, the neuroprotective effects of cordycepin on hippocampal neurons from $A\beta_{25-35}$-induced impairment are partially dependent on the activation of adenosine A_1 receptor.

The neuroprotective efficacy of *C. militaris* (which is a highly nutritive and edible fungus) on ischemia-induced neuronal death and cognitive impairments was reported by Kim et al. (2019). The *C. militaris* is found to decrease the neuronal cell death and spatial memory loss on Morris water maze (MWM). Evaluation

Table 1. Some important biologically active compounds reported from *Cordyceps* spp.

Compounds	Source	Biological activity	Reference
Nucleoside/Nucleoside derivatives			
Cordycepin (3-deoxyadenosine)	Cultured cells of *C. militaris* and *C. sinensis*	Anticancer	Nakamura et al., 2005; Wu et al., 2007; Jen et al., 2009
		Antimicrobial, insecticidal, antithrombotic, antirestenosis, antioxidant and immunomodulatory	Ahn et al., 2000; Kim et al., 2002; Cho et al., 2006; Chang et al., 2008; Hwang et al., 2008; Zhou et al., 2008
Adenosine	Cultured cells of *C. sinensis*	Anticonvulsant anti-inflammatory anti-tumor	Kuo et al., 2015; Liu et al., 2015; Elkhateeb et al., 2019
Guanosine	Cultured cells of *C. sinensis*	Immunomodulatory	Yu et al., 2007
Polysaccharides			
Exopolysaccharide fraction	Cultured supernatant of *C. sinensis*	Anticancer	Zhang et al., 2005
Exopolysaccharide-peptide complexes	Submerged mycelial culture of *C. sphecocephala*	Anticancer	Oh et al., 2008
Polysaccharide-enriched fraction	Cultured cells of *C. militaris*	Hypoglycemic	Zhang et al., 2006
Acidic polysaccharides	Cultured cells of *C. militaris*	Anti-influenza virus	Ohta et al., 2007
Intracellular polysaccharides (83 kDa)	Mycelia of *C. sinensis*	Immunomodulatory	Wu et al., 2006
Cordysinocan (82 kDa)	*C. sinensis*; UST 2000	Immunopotentiating	Cheung et al., 2009
CBP-1 (17 kDa)	Cultured cells of *C. militaris*	Antioxidant	Yu et al., 2009
Alkaloids			
Cordyformamide	*C. brunearubra*; BCC1395	Antimalarial	Isaka et al., 2007
Militarinone A, B and C	Mycelium of *Paecilomyces militaris*	Neurotrophic	Schmidt et al., 2003
Farinosone A	*P. farinosus*; RCEF 0101	Neurotrophic	Cheng et al., 2004
(3R,6R)-4-methyl-6-(1-methylethyl)-3-phenylmethylperhydro-1,4-oxazine-2,5-dione	Fruiting bodies of *Isaria japonica*	Anticancer	Oh et al., 2002
Cordypyridone A and B	*C. nipponica*; BCC 1389	Antimalarial	Isaka et al., 2001
P-Terphenyls			
Gliocladinin A and B	Solid cultures of Gliocladium sp. that colonizes *C. sinensis*	Anticancer	Guo et al., 2007

Table 1 Contd. ...

...Table 1 Contd.

Compounds	Source	Biological activity	Reference
Diphenyl ethers			
Cordyol A	Cultured broth of *Cordyceps* sp. BCC 1861	Antimicobacterial	Bunyapaiboonsri et al., 2007
Cordyol C	Cultured broth of *Cordyceps* sp. BCC 1861	Anticancer	Bunyapaiboonsri et al., 2007
Violaceol-I and -II	Cultured broth of *Cordyceps* sp. BCC 1861	Anticancer	Bunyapaiboonsri et al., 2007
Proteins			
Beauvericin	*C. dicadae, P. tenuipes, C. takaomanlana*	Anticancer	Calo et al., 2004
Hemagglutinin	*C. militaris*	Anticancer and anti-HIV	Wong et al., 2009
CML (lectin)	*C. militaris*	Hemagglutination and mitogenic	Jung et al., 2007
Antibiotic			
Ophiocordin	*Cordyceps ophioglossoides*	Antifungal	Kneifel et al., 1977
Sterols			
4γ-acetoxy-scirpendiol	Fruiting bodies of *C. takaomantana*	Hypoglycemic	Yoo and Lee, 2006
Ergosterol	Cultivated mycelium of *C. sinensis*	Anticancer	Matsuda et al., 2009
β-Sitosterol	*C. sinensis*	Anticancer	Wu et al., 2007
Ergosterol peroxide	*C. cicadae*	Immunomodulatory	Kuo et al., 2003

of hippocampal region and neuron count in the CA1 region of the hippocampus confirmed the neuroprotective potential of *C. militaris*. So also, it significantly reduced the activity of microglia and affords protection against ischemic brain damage by targeting inflammatory response. Because of the nutritive value as well as biological activities of *C. sinensis*, it is regarded as a functional food (Yue et al., 2008). Nallathamby et al. (2015) was successful in isolating ergosterol from *C. militaris*, which reduces LPS-induced inflammation in BV2 microglia cells.

Cordyceps ophiglossoides, is a parasite which is reported to exhibit protective activity against the Aβ-induced neuronal cell death and spatial memory loss in Aβ-treated rats (Jin et al., 2004). In this study, the methanol extracts of cultured *C. ophioglossoides* were assessed for their protective effects on the cellular model of Alzheimer's dementia (AD). It was generated by treating human neuronal cells with Aβ and also examined their potential to protect from spatial memory loss in rat model of AD using MWM method. The results proposed that the deterioration in spatial memory and learning capacity in Aβ-injected rats is mainly associated with ROS generation by accumulated Aβ in the brain and that suppression of the ROS release may mediate the protective effect of the extract on the Aβ-induced memory loss. These results further suggest that *C. ophioglossoides* mycela may have a therapeutic potential against AD.

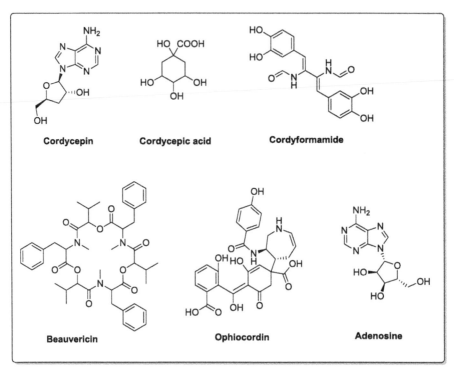

Fig. 1. Chemical structures of important biologically active compounds derived from *Cordyceps* spp. (*Source*: Pubchem - https://pubchem.ncbi.nlm.nih.gov/).

Cordyceps sinensis is an acclaimed medicinal fungus, which has been widely employed therapeutically in China for over two millennia. A number of biological properties have been documented for this fungus, but its ability to provide neuroprotection is least studied. The prophylactic and therapeutic effects of fermented *C. sinensis* powder on subcortical ischemic vascular dementia in mice models was studied by Chen et al. (2018). This study showed chronic cerebral hypoperfusion induced by rUCCAO, which caused the impairment of white matter, thereby damaging the spatial working memory. The results showed that prophylaxis with either 0.2 g/kg or 1.0 g/kg of fermented *C. sinensis* powder one week before and 28 days after modelling diminished the rUCCAO-induced cognitive impairment. But it is not clearly revealed which all components are bioactive and responsible for its benefits and whether these components act alone or synergistically. All these issues need to be addressed in future investigations.

4. Conclusions and Prospects

Investigations on natural biocompatible nutritional supplements with balanced micronutrient level are in progress. Bioactive components from natural sources are necessary for the prevention and treatment of several neurodegenerative diseases, without causing harmful side effects. Fungi being the promising source, *Cordyceps* spp. are extensively studied for their anticancer, antioxidant and anti-inflammatory activities with a few attempts made on their protective action against neurodegenerative disorders. The molecular mechanisms underlying their neuroprotective potential is not clearly elucidated. In order to develop new strategies (e.g., application of nanotechnology for the delivery of neuroprotective natural constituents to the brain) are needed for prevention and healing of neurodegenerative diseases.

References

Ahn, Y.J., Park, S.J., Lee, S.G., Shin, S.C. and Choi, D.H. (2000). Cordycepin: Selective growth inhibitor derived from liquid culture of *militaris* against Clostridium spp. J. Agric. Food Chem., 48: 2744–2748.

Ashraf, S.A., Elkhalifa, A.E.O., Siddiqui, A.J., Patel, M., Awadelkareem, A.M. et al. (2020). Cordycepin for health and wellbeing: A potent bioactive metabolite of an entomopathogenic medicinal fungus *Cordyceps* with its nutraceutical and therapeutic potential. Molecules, 25: 2735. 10.3390/molecules25122735.

Boyle, K., Azari, M.F., Profyris, C. and Petratos, S. (2005). Molecular mechanisms in Schwann cell survival and death during peripheral nerve development, injury and disease. Neurotox, Res., 7: 151–167.

Bunyapaiboonsri, T., Yoiprommarat, S., Intereya, K. and Kocharin, K. (2007). New diphenyl ethers from the insect pathogenic fungus *Cordyceps* sp. BCC 1861. Chem. Pharm. Bull., 55: 304–307.

Calò, L., Fornelli, F., Ramires, R., Nenna, S., Tursi, A. et al. (2004). Cytotoxic effects of the Mycotoxin beauvericin to human cell lines of myeloid origin. Pharmacol. Res., 49: 73–77.

Chang, W., Lim, S., Song, H., Song, B.W., Kim, H.J. et al. (2008). Cordycepin inhibits vascular smooth muscle cell proliferation. Eur. J. Pharmacol., 597: 64–69.

Cheng, Y., Schneider, B., Riese, U., Schubert, B., Li, Z. and Hamburger, M. (2004). Farinosones A-C, neurotrophic alkaloidal metabolites from the entomogenous deuteromycete *Paecilomyces farinosus*. J. Nat. Prod., 67: 1854–1858.

Cheung, J.K., Li, J., Cheung, A.W., Zhu, Y., Zhend, K.Y. et al. (2009). Cordysinocan, a polysaccharide isolated from cultured *Cordyceps*, activates immune responses in cultured T-lymphocytes and macrophages: Signalling cascade and induction of cytokines. J. Ethnopharmacol., 124(1): 61–68.

Chitnis, T. and Weiner, H.L. (2017). CNS inflammation and neurodegeneration. J. Clin. Invest., 127(10): 3577–3587.

Cho, H.J., Cho, J.Y., Rhee, M.H., Lim, C.R. and Park, H.J. (2006). Cordycepin (3'-deoxyadenosine) inhibits human platelet aggregation induced by U46619, a TXA2 analogue. J. Pharm. Pharmacol., 58: 1677–1682.

Dattaraj, H.R., Jagadish, B.R., Sridhar, K.R. and Ghate, S.D. (2018). Are the scrub jungles of southwest India potential habitats of *Cordyceps*? KAVAKA - Tr. Mycol. Soc. India, 51: 20–22.

Eik, L.F., Naidu, M., David, P., Wong, K.H., Tan, Y.S. and Sabaratnam, V. (2012). *Lignosus rhinocerus* (Cooke) Ryvarden: A medicinal mushroom that stimulates neurite outgrowth in PC-12 cells. Evid. Based Compl. Alt. Med., 2012: 320308. 10.1155/2012/320308.

Elkhateeb, W.A., Daba, G.M., Thomas, P.W. and Wen, T.C. (2019). Medicinal mushrooms as a new source of natural therapeutic bioactive compounds. Egypt. Pharm. J., 18: 88–101.

Esposito, E., Rotilio, D., Di Matteo, V., Di Giulio, C., Cacchio, M. and Algeri, S. (2002). A review of specific dietary antioxidants and the effects on biochemical mechanisms related to neurodegenerative processes. Neurobiol. Aging., 23: 719–35.

Ferrer, I. (2004). Stress kinases involved in tau phosphorylation in Alzheimer's disease, tauopathies and APP transgenic mice. Neurotox. Res., 6: 469–475.

Guillemin, G.J., Kerr, S.J. and Brew, B.J. (2005). Involvement of quinolinic acid in AIDS dementia complex. Neurotox. Res., 7: 103–123.

Guo, H., Hu, H., Liu, S., Liu, X., Zhou, Y. and Che, Y. (2007). Bioactive p-terphenyl derivatives from a *Cordyceps*-colonising isolate of *Gliocladium* sp. J. Nat. Prod., 70: 1519–1521.

Hwang, K., Lim, S.S., Yoo, K.Y., Lee, Y.S., Kim, H.G. et al. (2008). A phytochemically characterised extract of *Cordyceps militaris* and cordycepin protect hippocampal neurons from ischemic injury in berbils. Planta Med., 74: 114–119.

Iriti, M., Vitalini, S., Fico, G. and Faoro, F. (2010). Neuroprotective herbs and foods from different traditional medicines and diets. Molecules, 15: 3517–3555.

Isaka, M., Tanticharoen, M., Kongsaeree, P. and Thebtaranonth, Y. (2001). Structures of cordypyridones A-D, antimalarial Nhydroxy- and N-methoxy-2-pyridones from the insect pathogenic fungus *Cordyceps nipponica*. J. Org. Chem., 66: 4803–4808.

Isaka, M., Boonkhao, B., Rachtawee, P. and Auncharoen, P. (2007). Axanthocillin-like alkaloid from the insect pathogenic fungus *Cordyceps brunnearubra* BCC 1395. J. Nat. Prod., 70: 656–658.

Jen, C.Y., Lin, C.Y., Leu, S.F. and Huang, B.M. (2009). Cordycepin-induced MA-10 mouse Leydig tumor cell apoptosis through caspase-9 pathway. Evid. Based Compl. Alt. Med., 6: 1–10.

Jin, D.Q., Park, B.C., Lee, J.S., Choi, H.D., Lee, Y.S. et al. (2004). Mycelial extract of *Cordyceps* ophioglossoides prevents neuronal cell death and ameliorates β-amyloid peptide-induced memory deficits in rats. Biol. Pharm. Bull., 27(7): 1126–1129.

Johansson, S., Bohman, S., Radesäter, A.C., Oberg, C. and Luthman, J. (2005). Salmonella lipopolysaccharide (LPS) mediated neurodegeneration in hippocampal slice cultures. Neurotox. Res., 8(3,4): 207–220.

Jung, E.C., Kim, K.D., Bae, C.H., Kim, J.C., Kim, D.K. and Kim, H.H. (2007). A mushroom lectin from ascomycete *Cordyceps militaris*. Biochem. Biophys. Acta., 1770: 833–838.

Kim, J.R, Yeon, S.H., Kim, H.S. and Ahn, Y.J. (2002). Larvicidal activity against Plutellaxylostella of cordycepin from the fruiting body of *Cordyceps militaris*. Pest Manag. Sci., 58: 713–717.

Kim, Y.O., Kim, H.J., Abu-Taweel, G.H., Oh, J. and Sung, G.H. (2019). Neuroprotective and therapeutic effect of *Cordyceps militaris* on ischemia-induced neuronal death and cognitive impairments. Saudi J. Biol. Sci., 26: 1352–1357.

Kneifel, H., Konig, W.A., Loeffler, W. and Müller, R. (1977). Ophiocordin, an antifungal antibiotic of *Cordyceps ophioglossoides*. Arch. Microbiol., 113: 121–130.

Kobayasi, Y. (1941). The genus *Cordyceps* and its allies. Science Reports of the Tokyo Bunrika Daigaku, (Section B, # 84), 5: 53–260.

Kuo, Y.C., Weng, S.C, Chou, C.J, Chang, T.T. and Tsai, W.J. (2003). Activation and proliferation signals in primary human T lymphocytes inhibited by ergosterol peroxide isolated from *Cordyceps cicadae*. Br. J. Pharmacol., 140: 895–906.

Kuo, H.C., Huang, I.C. and Chen, T.Y. (2015). *Cordyceps* S.l. (Ascomycetes) species used as medicinal mushrooms are closely related with higher ability to produce cordycepin. Int. J. Med. Mush., 17: 1077–1085.

Lapchak, P.A. and Boitano, P.D. (2017). Reflections on neuroprotection research and the path toward clinical success. *In*: Lapchak, P. and Zhang J. (eds.). Neuroprotective Therapy for Stroke and Ischemic Disease. Springer Series in Translational Stroke Research, Springer, Cham, Switzerland. 10.1007/978-3-319-45345-3_1.

Li, H.L., Li, X.M., Liu, H., Meng, L.H. and Wang, B.G. (2016). Two new diphenylketones and a new xanthone from *Talaromyces islandicus* EN-501, an endophytic fungus derived from the marine red alga *Laurencia okamurai*. Mar. Drugs, 14(12): 233. doi:10.3390/md14120223.

Liu, Y., Wang, J., Wang, W., Zhang, H., Zhang, X. and Han, C. (2015). The chemical constituents and pharmacological actions of *Cordyceps sinensis*. Evid. Based Compl. Alt. Med., 2015: 575063. 10.1155/2015/575063.

Lu, X.L., Yao, X.L., Liu, Z., Zhang, H., Li, W. et al. (2010). Protective effects of xyloketal B against MPP+-induced neurotoxicity in *Caenorhabditis elegans* and PC12 cells. Brain Res., 1332: 110–119.

Malik, J. (2017). A new hope for Huntington's disease. pp. 213–241. *In*: Farooqui, T. and Farooqui, A.A. (eds.). Neuroprotective Effects of Phytochemicals in Neurological Disorders. John Wiley & Sons, Inc.

Matsuda, H., Akaki, J., Nakamura, S., Okazaki, Y., Kojima, H. et al. (2009). Apoptosis-inducing effects of sterols from the dried powder of cultured mycelium of *Cordyceps sinensis*. Chem. Pharm. Bull. (Tokyo), 57: 411–414.

Mattson, M.P. (2017). Excitotoxicity in neurodegeneration. pp. 37–44. *In*: Schapira, A., Wszolek, Z., Dawson, M. and Wood, N. (eds.). Neurodegeneration. Wiley Online Library, Hoboken, NJ, USA.

Nakamura, K., Konoha, K., Yoshikawa, N., Yamaguchi, Y., Kagota, S. et al. (2005). Effect of cordycepin (3'-deoxyadenosine) on hematogenic lung metastatic model mice. Vivo, 19: 137–142.

Nallathamby, N., Serma, L.G., Vidyadaran, S., Abd Maleka, S.N., Raman, J. and Sabaratnam, V. (2015). Ergosterol of *Cordyceps militaris* attenuates LPS induced inflammation in BV2 microglia cells. Nat. Prod. Comm., 10(6): 885–886.

Oh, H., Kim, T., Oh, G.S., Pae, H.O., Hong, K.H. et al. (2002). (3R,6R)-4-methyl-6-(1-methylethyl)-3-phenylmethyl-perhydro-1, 4-oxazine-2,5-dione: An apoptosis-inducer from the fruiting bodies of *Isaria japonica*. Planta Med., 68: 345–348.

Oh, J.Y., Baek, Y.M., Kim, S.W., Hwang, H.J., Huwang, H.S. et al. (eds.). (2008). Apoptosis of human hepatocarcinoma (HepG2) and neuroblastoma (SKN-SH) cells induced by polysaccharides-peptide complexes produced by submerged mycelial culture of an entomopathogenic fungus *Cordyceps sphecocephala*. J. Microbiol. Biotechnol., 18: 512–519.

Ohta, Y., Lee, J.B., Hayashi, K., Fujita, A., Park, D.K. and Hayashi, T. (2007). *In vivo* anti-influenza virus activity of an immunomodulatory acidic polysaccharide isolated from *Cordyceps militaris* grown on germinated soybeans. J. Agric. Food Chem., 55: 10194–10199.

Palomo, T., Archer, T., Kostrzewa, R.M. and Beninger, R.J. (2004). Gene-environment interplay in schizopsychotic disorders. Neurotox. Res., 6: 1–9.

Paterson, R.R. (2008). *Cordyceps*: A traditional Chinese medicine and another fungal therapeutic biofactory? Phytochem., 69: 1469–1495.

Phan, C.W., David, P., Naidu, M., Wong, K.H. and Sabaratnam, V. (2013). Neurite outgrowth stimulatory effects of culinary-medicinal mushrooms and their toxicity assessment using differentiating Neuro-2a and embryonic fibroblast BALB/3T3. BMC Compl. Alt. Med., 13: 261. 10.1186/1472-6882-13-261.

Phan, C.W., David, P., Naidu, M., Wong, K.H. and Sabaratnam, V. (2015). Therapeutic potential of culinary-medicinal mushrooms for the management of neurodegenerative diseases: Diversity, metabolite and mechanism. Crit. Rev. Biotechnol., 35(3): 355–368.

Sairazi, N.S.M. and Sirajudeen, K.N.S. (2020). Natural products and their bioactive compounds: Neuroprotective potentials against neurodegenerative diseases. Evid. Based Compl. Alt. Med., 2020: 6565396. 10.1155/2020/6565396.

Samson, R.A., Evans, H.C. and Latgé, J.P. (1988). Atlas of Entomopathogenic Fungi. Springer-Verlag, Berlin, Heidelberg, New York.

Schmidt, K., Riese, U., Li, Z. and Hamburger, M. (2003). Novel tetramic acids and pyridone alkaloids, militarinones B, C, and D, from the insect pathogenic fungus *Paecilomyces militaris*. J. Nat. Prod., 66: 378–383.

Segura-Aguilar, J. and Kostrzewa, R.M. (2004). Neurotoxins and neurotoxic species implicated in neurodegeneration. Neurotox. Res., 6: 615–630.

Segura-Aguilar, J. and Kostrzewa, R.M. (2006). Neurotoxins and neurotoxicity mechanisms. An overview. Neurotox. Res., 10(3-4): 263–285.

Shrestha, B., Sung, G.H. and Sung, J.M. (2017). Current nomenclatural changes in *Cordyceps sensulato* and its multidisciplinary impacts. Mycology, 8(4): 293–302.

Singh, A., Kukreti, R., Saso, L. and Kukreti, S. (2019). Oxidative stress: A key modulator in neurodegenerative diseases. Molecules, 24: 1583. 10.3390/molecules24081583.

Song, H., Huang, L.P., Lia, Y., Liu, C., Wang, S. et al. (2018). Neuroprotective effects of cordycepin inhibit Aβ-induced apoptosis in hippocampal neurons. Neurotoxicol., 68: 73–80.

Sridhar, K.R. and Karun, N.C. (2017). Observations on *Ophiocordyceps nutans* in the Western Ghats. J. New Biol. Rep., 6: 104–111.

Sriram, K., Matheson, J.M., Benkovic, S.A., Miller, D.B., Luster, M.I. and O'Callaghan, J.P. (2006). Deficiency of TNF receptors suppresses microglial activation and alters the susceptibility of brain regions to MPTP-induced neurotoxicity: Role of TNF-α. FASEB J., 20: 670–682.

Sun, S.W., Ji, C.Z., Gu, Q.Q., Li, D.H. and Zhu, T.J. (2013). Three new polyketides from marine-derived fungus *Aspergillus glaucus* HB1-19. J. Asian Nat. Prod. Res., 15(9): 956–961.

Sung, G.H., Hywel-Jones, N.L., Sung, J.M., Luangsaard, J.J., Shrestha, B. and Spatafora, W. (2007). Phylogenetic classification of *Cordyceps* and the clavicipitaceous fungi. Stud. Mycol., 57: 5–59.

Tuli, H.S., Kashyap, D. and Sharma, A.K. (2017). Cordycepin: A *Cordyceps* metabolite with promising therapeutic potential. pp. 761–782. *In*: Mérillon, J.-M. and Ramawat, K.G. (eds.). Fungal Metabolites. Springer International Publishing, Cham, Switzerland.

Wong, J.H., Wang, H. and Ng, T.B. (2009). A hemagglutinin from the medicinal fungus *Cordyceps militaris*. Biosci. Rep., 29: 321–327.

Wong, K.H., Naidu, M., David, R.P., Abdulla, M.A., Abdullah, N. et al. (2009). Functional recovery enhancement following injury to rodent peroneal nerve by Lion's Mane Mushroom, *Hericium erinaceus* (Bull.: Fr.) Pers. (Aphyllophoromycetideae). Int. J. Med. Mushr., 11: 225–236.

Wong, K.H., Chee Ng, C., Kanagasabapathy, G., Yow, Y.Y. and Sabaratnam, V. (2017). An overview of culinary and medicinal mushrooms in neurodegeneration and neurotrauma research. Int. J. Med. Mush., 19(3): 191–202.

Wu, J.Y., Zhang, Q.X. and Leung, P.H. (2007). Inhibitory effects of ethyl acetate extract of *Cordyceps sinensis* mycelium on various cancer cells in culture and B16 melanoma in C57BL/6 mice. Phytomedicine, 14: 43–49.

Wu, W.C., Hsiao, J.R., Lian, Y.Y., Lin, C.Y. and Huang, B.M. (2007). The apoptotic effect of cordycepin on human OEC-M1 oral cancer cell line. Canc. Chemother. Pharmacol., 60: 103–111.

Wu, Y., Sun, H., Qin, F., Pan, Y. and Sun, C. (2006). Effect of various extracts and a polysaccharide from the edible mycelia of *Cordyceps sinensis* on cellular and humoral immune response against ovalbumin in mice. Phytother. Res., 20: 646–652.

Yoo, O. and Lee, D.H. (2006). Inhibition of sodium glucose cotransporter-I expressed in *Xenopuslaevis* oocytes by 4-acetoxyscirpendiol from *Cordyceps takaomantana* (anamorph = *Paecilomyces tenuipes*). Med. Mycol., 44: 79–85.

Yu, L., Zhao, J., Zhu, Q. and Li, S.P. (2007). Macrophage biospecific extraction and high performance liquid chromatography for hypothesis of immunological active components in *Cordyceps sinensis*. J. Pharm. Biomed. Anal., 44: 439–443.

Yu, R., Yin, Y., Yang, W., Ma, W., Yang, L. et al. (2009). Structural elucidation and biological activity of a novel polysaccharide by alkaline extraction from cultured *Cordyceps militaris*. Carbohydr. Polym., 75: 166–171.

Yue, G.G., Lau, C.B., Fung, K.P., Leung, P.C. and Ko, W.H. (2008). Effects of *Cordyceps sinensis*, *Cordyceps militaris* and their isolated compounds on ion transport in calu-3 human airway epithelial cells. J. Ethnopharmacol., 117: 92–101.

Yue, K., Ye, M., Zhou, Z., Sun, W. and Lin, X. (2013). The genus *Cordyceps*: A chemical and pharmacological review. J. Pharm. Pharmacol., 65: 474–493.

Yurchenko, E.A., Menchinskaya, E.S., Pislyagin, E.A., Trinh, P.T.H., Ivanets, E.V. et al. (2018). Neuroprotective activity of some marine fungal metabolites in the 6-hydroxydopamin- and paraquat-induced Parkinson's disease models. Mar. Drugs, 16(11): 457. 10.3390/md16110457.

Zhang, G.Q., Huang, Y.D., Bian, Y., Wong, J.H., Ng, T.B. and Wang, H.X. (2006). Hypoglycemic activity of the fungi *Cordyceps militaris, Cordyceps sinensis, Tricholoma mongolicum*, and *Ompalia lapidescens* in streptozotocin-induced diabetic rats. Appl. Microbiol. Biotechnol., 72: 1152–1156.

Zhang, W., Yang, J., Chen, J., Hou, Y. and Han, X. (2005). Immunomodulatory and antitumour effects of an exopolysaccharide fraction from cultivated *Cordyceps sinensis* (Chinese caterpillar fungus) on tumour-bearing mice. Biotechnol. Appl. Biochem., 42: 9–15.

Zhao, J. (2007). Nutraceuticals, nutritional therapy, phytonutrients and phytotherapy for improvement of human health: A perspective on plant biotechnology application. Rec. Pat. Biotech., 1: 75–97.

Zhou, X., Luo, L., Dressel, W. et al. (2008). Cordycepin is an immunoregulatory active ingredient of *Cordyceps sinensis*. Am. J. Chin. Med., 36: 967–80.

Zhou, Z.Y., Tang, Y.P., Xiang, J., Wua, P., Jin, H.P. et al. (2010). Neuroprotective effects of water-soluble *Ganoderma lucidum* polysaccharides on cerebral ischemic injury in rats. J. Ethnopharmacol., 131: 154–164.

Zhu, J.S., Halpern, G.M. and Jones, K. (1998). The scientific rediscovery of an ancient Chinese herbal medicine: *Cordyceps sinensis* Part I. J. Alt. Compl. Med., 4: 289–303.

9

Neurological and Related Adverse Events Associated with Pharmacokinetic Interactions of Illicit Substances of Fungal Origin with Clinical Drugs

Julia M Salamat,[1,#] Kodye L Abbott,[1,#]
Patrick C Flannery,[1,#] Kristina S Gill,[1,#]
Muralikrishnan Dhanasekaran[2] and
Satyanarayana R Pondugula[1,]*

1. INTRODUCTION

Illicit substances refer to the compounds that are either stimulatory or inhibitory to the central nervous system. These compounds cause hallucinogenic effects, hence, their use has been forbidden (Smelser and Baltes, 2001). This chapter emphasises the psychoactive illicit substances of fungal origin. According to the National Institute on Drug Abuse (NIDA, 2018), the illicit substances of fungal origin include psilocybin, psilocin and lysergic acid diethylamide (LSD) (Fig. 1).

[1] Department of Anatomy, Physiology and Pharmacology, Auburn University, Auburn, AL 36849, USA.
[2] Department of Drug Discovery and Development, Harrison School of Pharmacy, Auburn University, AL 36849, USA.
[#] Authors contributed equally.
[*] Corresponding author: srp0010@auburn.edu

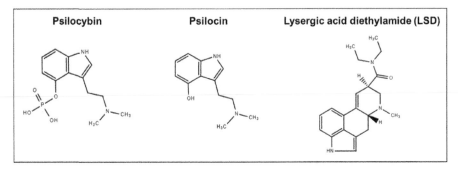

Fig. 1. Chemical structures of the illicit substances of fungal origin.

Table 1. Illicit substances and their fungal origin.

Illicit substance	Fungal origin
Psilocybin and psilocin	*Conocybe*
	Gymnopilus
	Panaeolus
	Pluteus
	Psilocybe
	Psilocybe cubensis
	Psilocybe baeocystis
	Psilocybe semilanceata
	(Stamets, Beug et al., 1980)
Lysergic acid diethylamide (LSD)	*Claviceps purpuria*
	(Bujarski and Sperling, 2012)

Psilocybin is present in hallucinogenic mushrooms which are also called 'magic mushrooms'. Although it is reported that there are about a hundred 'magic mushroom' species, some specific psychedelic mushrooms are in the genus *Conocybe, Gymnopilus, Panaeolus, Pluteus*, and *Psilocybe. Psilocybe cubensis* is considered one of the most popular ones as well as *Psilocybe semilanceata* and *Psilocybe baeocystis* (Stamets et al., 1980) (Table 1). The LSD is present in ergot. Ergot fungi refer to a genus of fungus called *Claviceps*, which contain about 40 species (Schiff, 2006). The most popular species is named *Claviceps purpurea* or rye ergot fungus (Schardl et al., 2006) (Table 1).

2. Therapeutic Relevance of Illicit Substances of Fungal Origin

Lately, the therapeutic prospective of illicit substances has gained significance and several studies have looked at the potential therapeutic benefits of illicit substances for a variety of medical conditions. For example, cannabis has been shown to have several therapeutic benefits. This substance has been studied for its use in various conditions, such as nausea, migraines, sleep deprivation, anxiety, and pain (Borgelt et al., 2013; Piper et al., 2017; Russo et al., 2007). In addition, products derived from cannabis have been approved as prescription therapies, for example, Dronabinol, derived from tetrahydrocannabinol, a substance found in cannabis, and Nabilone, which mimics tetrahydrocannabinol, are prescription drugs that are approved by the

FDA. Both drugs are used for treatment of nausea and vomiting in humans undergoing chemotherapy as well as in treatment of AIDS-associated anorexia (Bedi et al., 2013). Lately, the FDA also approved Epidiolex®, a cannabidiol drug, for treatment of epileptic seizures (FDA, 2018). Additionally, Sativex®, an extract derived from cannabis, has been approved in several countries for treatment of multiple sclerosis-related symptoms (Fallon et al., 2017; Lichtman et al., 2018; Markova et al., 2019; Turri et al., 2018).

Similar to other illicit substances, the beneficial effects of illicit substances of fungal origin has gained attention and several studies have explored the potential therapeutic benefits of illicit substances of fungal origin for a variety of diseases. For example, the psychedelic psilocybin, which is found in magic mushrooms, was revealed to be an effective treatment against post-traumatic stress disorder (PTSD), anxiety and depression (Byock, 2018). Several clinical trials are being conducted to explore the ability of psilocybin in treating many medical conditions, such as obsessive-compulsive disorder, cancer-associated anxiety, smoking cessation, migraine and depression (NIH).

Similarly, LSD found in ergot has recently been explored for its use as a therapy for addiction and inflammation as well as for mental distress in cancer patients (Nichols et al., 2017). Clinical trials are also being carried out to ascertain the therapeutic effects of LSD in patients with a variety of medical conditions, such as anxiety and depression (NIH). While psilocybin and LSD alone can cause stimulation or hallucinogenic effects (Graziano et al., 2017), the stimulation and hallucinogenic effects caused by these illicit substances are further exasperated with concomitant use of clinical drugs. This is because of altered response of these illicit substances or the co-administered clinical drugs. Indeed, clinically relevant adverse neurological interactions have been observed to accompany the simultaneous usage of these substances with clinical drugs (Bonson and Murphy, 1996; Vollenweider et al., 1998; Baumeister et al., 2015).

2.1 *Mechanisms of Adverse Pharmacokinetic Interactions*

The absorption, distribution, metabolism and elimination of clinical drugs or illicit substances is mainly determined by drug-metabolising enzymes and drug-transporters (Fig. 2). Cytochromes P450 (CYPs), UDP-glucuronosyltransferases (UGTs) and sulfotransferases (SULTs) are predominant drug-metabolising enzymes. ATP-binding cassette transporters and organic anion and cation transporters are the major drug-transporters (Fig. 2). A combinatorial regimen of clinical drugs and illicit substances of fungal origin can result in alterations in the expression/activity of drug-metabolising enzymes and/or drug-transporters. These changes can result in interactions at the pharmacokinetic level, eventually leading to neurological and related adverse events (Fig. 2).

In this chapter, we discuss the neurological and related adverse events associated with pharmacokinetic interactions between illicit substances of fungal origin and clinical drugs in humans. This chapter also highlights insights into the future studies to further understand the underlying pharmacokinetic mechanisms of adverse neurological and related adverse events.

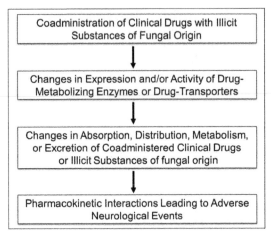

Fig. 2. Mechanisms of pharmacokinetic interactions between illicit substances of fungal origin and clinical drugs.

3. Illicit Substances of Fungal Origin Associated with Neurological and Related Adverse Drug Interactions

3.1 Psilocybin and Psilocin

Psilocybin is found in certain mushrooms, such as those from genus *Conocybe, Gymnopilus, Panaeolus, Pluteus*, and *Psilocybe* (Stamets et al., 1980; Johnson and Griffiths, 2017). It is either eaten or consumed as a tea (NIDA, 2018). The mechanism behind its hallucinogenic and euphoric properties seems to occur due to partial agonism of serotonin receptors (Geiger et al., 2018).

Psilocybin is a prodrug. Hence, after ingestion, psilocybin is converted into its active metabolite—psilocin, through dephosphorylation by alkaline phosphatase. Psilocin has psychotropic effects analogous to LSD (van Amsterdam et al., 2011). Psilocin is further metabolised by the enzymes—monoamine oxidase (MAO) or aldehyde dehydrogenase (ALDH) (Dinis-Oliveira, 2017). Psilocin is also metabolised by UGT1A9 and UGT1A10 (Manevski et al., 2010).

Inhibitors of MAO or ALDH may increase the psychedelic effects of psilocin. Indeed, as seen in human patients, the dopamine antagonist haloperidol amplified the psychotomimetic actions of psilocybin and psilocin (Vollenweider et al., 1998). This can be attributed to the ability of haloperidol to inhibit MAO (Fang et al., 1995). It is currently unknown whether co-administered clinical drugs that can induce or inhibit UGT1A9/UGT1A10 can contribute to psilocin-mediated adverse neurological events (Fig. 2).

3.2 Lysergic Acid Diethylamide (LSD)

LSD is a hallucinogenic substance that is typically consumed orally (NIDA, 2018). The mechanism behind LSD's hallucinogenic effects is its role as a partial agonist of D2 dopaminergic receptor (Marona-Lewicka et al., 2005). LSD is metabolised by several CYPs, such as CYP1A2, CYP2C9, CYP2C19, CYP2D6, CYP2E1, and CYP3A4, as shown by several *in vitro* studies (Wagmann et al., 2019; Luethi et al., 2019; Libanio Osorio Marta, 2019). Therefore, it is possible that a co-administered clinical drug that can induce or inhibit CYPs could potentially result in adverse drug interactions (Fig. 2).

It is a possibility that pharmacokinetic interactions can occur during the concomitant usage of LSD with clinical drugs; however, there are no reported studies of such instances. Nevertheless, other mechanisms of adverse interactions have been discovered. For instance, imipramine, desipramine and clomipramine, considered as the anti-depressant tricyclics, were shown to intensify the psychological and hallucinatory effects of LSD (Bonson and Murphy, 1996). The mechanism behind this adverse interaction might be attributed to changes in both serotonin and dopamine systems by the anti-depressants, imipramine, desipramine or clomipramine (Bonson and Murphy, 1996).

4. Conclusions and Future Directions

Although there is a significant potential regarding the therapeutic benefits of illicit substances of fungal origin, their possible pharmacokinetic interactions with clinical drugs, that can result in adverse neurological events, haven't received much attention. Despite the awareness that simultaneous administration of illicit substances of fungal origin with clinical drugs can lead to neurological adverse events, the pharmacokinetic mechanisms driving these interactions are not thoroughly recognised. Therefore, there is a need to comprehensively understand and elucidate these neurological pharmacokinetic interactions between these illicit substances of fungal origin and clinical drugs.

Similar to psilocybin and LSD, clinical drugs are also metabolised by drug-metabolising enzymes. Hence, it is important to understand how illicit substances of fungal origin may influence the therapeutic response of clinical drugs during the combinatorial usage. While some progress was made in determining the enzymes involved in the metabolism of psilocybin and LSD, no published reporters are available regarding the role of transporters. Drug-transporters can impact the absorption, distribution, metabolism and excretion of illicit substances of fungal origin. More importantly, it is known that drug-transporters also contribute to adverse pharmacokinetic drug interactions (Marquez and Van Bambeke, 2011). Hence, understanding the impact drug-transporters on psilocybin and LSD movement should facilitate prediction of adverse pharmacokinetic interactions of illicit substances of fungal origin with clinical drugs.

It is known that illicit substances, such as cannabis, can induce or inhibit drug-metabolising enzymes and drug-transporters, thereby contributing to pharmacokinetics-based adverse neurological events (Abbott et al., 2020). While it is possible that psilocin and LSD alter the expression or function of drug-metabolising enzymes, it remains to be examined whether psilocybin and LSD can induce or inhibit drug-metabolising enzymes and/or drug-transporters. Several *in vivo, in vitro* and *in silico* models can be employed to study and predict adverse pharmacokinetic interactions. Scientific investigators have been employing various animal models for preclinical *in vivo* evaluation of clinical drugs or illicit substances. However, it is possible that the illicit substances of fungal origin and clinical drugs can modulate the drug-metabolising enzymes and drug-transporters in a species-dependent manner. Similarly, the drug-metabolising enzymes and drug-transporters can display differential species-specificity for illicit substances of fungal origin and clinical drugs. These species-specific aspects can greatly complicate the extrapolation of preclinical evaluation of pharmacokinetic interactions from animal models to humans.

Because of the notable interspecies differences, humanised animal models could be useful to address the species-specific pharmacokinetic interactions involving the drug-metabolising enzymes and drug-transporters. Humanised mice models are currently available for major drug-metabolising enzymes and drug-transporters (Cheng et al., 2009; Hasegawa et al., 2012; Holmstock et al., 2013; Fujiwara et al., 2018; Yamasaki et al., 2018). *In vitro* models (Issa et al., 2017), such as engineered cell lines (Pondugula et al., 2015; Abbott et al., 2019), primary cells (Pondugula et al., 2015; Abbott et al., 2019), 3-D cultures, tissue slices and organ cultures are also useful preclinical tools for studying pharmacokinetic interactions. For instance, blood-brain barrier models of rodent, porcine and human origin are available to study drug-metabolising enzymes and drug-transporters (Torres-Vergara et al., 2020). These models include cell lines, primary cultures, isolated capillaries and so on. These models have already been validated for studying drug-metabolising enzymes and drug-transporters (Torres-Vergara et al., 2020). *In silico* tools can also be useful to predict metabolism and transport of illicit substances and clinical drugs (Tyzack and Kirchmair, 2019). *In silico* models may be employed to predict illicit substances of fungal origin and clinical drugs with the potential to serve as substrates and inhibitors of drug-metabolising enzymes and drug-transporters (Demel et al., 2008; Martiny et al., 2015). Moreover, these tools can be useful to predict pharmacokinetic interactions with drug-metabolising enzymes and drug transporters (Ecker et al., 2008; Estrada-Tejedor and Ecker, 2018). Taken together, *in vivo*, *in vitro* and computer models can be used to understand the underlying mechanisms of pharmacokinetics-based interactions between clinical drugs and illicit substances of fungal origin.

References

Abbott, K.L., Chaudhury, C.S., Chandran, A., Vishveshwara, S., Dvorak, Z. et al. (2019). Belinostat, at its clinically relevant concentrations, inhibits rifampicin-induced CYP3A4 and MDR1 gene expression. Mol. Pharmacol., 95(3): 324–334. 10.1124/mol.118.114587.

Abbott, K.L., Flannery, P.C., Gill, K.S., Boothe, D.M., Dhanasekaran, M. et al. (2020). Adverse pharmacokinetic interactions between illicit substances and clinical drugs. Drug Metab. Rev., 52(1): 44–65. 10.1080/03602532.2019.1697283.

Baumeister, D., Tojo, L.M. and Tracy, D.K. (2015). Legal highs: Staying on top of the flood of novel psychoactive substances. Ther. Adv. Psychopharmacol., 5(2): 97–132. 10.1177/2045125314559539.

Bedi, G., Cooper, Z.D. and Haney, M. (2013). Subjective, cognitive and cardiovascular dose-effect profile of nabilone and dronabinol in marijuana smokers. Addict Biol., 18(5): 872–881. 10.1111/j.1369-1600.2011.00427.x.

Bonson, K.R. and Murphy, D.L. (1996). Alterations in responses to LSD in humans associated with chronic administration of tricyclic antidepressants, monoamine oxidase inhibitors or lithium. Behav. Brain Res., 73(1-2): 229–233.

Borgelt, L.M., Franson, K.L., Nussbaum, A.M. and Wang, G.S. (2013). The pharmacologic and clinical effects of medical cannabis. Pharmacotherapy, 33(2): 195–209. 10.1002/phar.1187.

Bujarski, K.A. and Sperling, M.R. (2012). Hallucinations. Encyclopedia of Human Behaviour, 283–289.

Byock, I. (2018). Taking Psychedelics seriously. J. Palliative Med., 21(4): 417–421. 10.1089/jpm.2017.0684.

Cheng, J., Ma, X., Krausz, K.W., Idle, J.R. and Gonzalez, F.J. (2009). Rifampicin-activated human pregnane X receptor and CYP3A4 induction enhance acetaminophen-induced toxicity. Drug Metab. Dispos., 37(8): 1611–1621. 10.1124/dmd.109.027565.

Demel, M.A., Schwaham, R., Kramerm, O., Ettmayer, P., Haaksma, E.E. and Ecker, G.F. (2008). *In silico* prediction of substrate properties for ABC-multidrug transporters. Exp. Opin., Drug Met. Toxicol., 4(9): 1167–1180. 10.1517/17425255.4.9.1167.

Dinis-Oliveira, R.J. (2017). Metabolism of psilocybin and psilocin: Clinical and forensic toxicological relevance. Drug Met. Rev., 49(1): 84–91. 10.1080/03602532.2016.1278228.

Ecker, G.F., Stockner, T. and Chiba, P. (2008). Computational models for prediction of interactions with ABC-transporters. Drug Discovery Today, 13(7-8): 311–317. 10.1016/j.drudis.2007.12.012.

Estrada-Tejedor, R. and Ecker, G.F. (2018). Predicting drug resistance related to ABC transporters using unsupervised consensus self-Organising maps. Sci. Rep., 8(1): 6803. 10.1038/s41598-018-25235-9.

Fallon, M.T., Albert Lux, E., McQuade, R., Rossetti, S., Sanchez, R. et al. (2017). Sativex oromucosal spray as adjunctive therapy in advanced cancer patients with chronic pain unalleviated by optimised opioid therapy: Two double-blind, randomised, placebo-controlled phase 3 studies. Br. J. Pain, 11(3): 119–133. 10.1177/2049463717710042.

Fang, J., Yu, P.H., Gorrod, J.W. and Boulton, A.A. (1995). Inhibition of monoamine oxidases by haloperidol and its metabolites: Pharmacological implications for the chemotherapy of schizophrenia. Psychopharmacology, 118(2): 206–212.

FDA. (2018). Epidiolex (https://www.accessdata.fda.gov/drugsatfda_docs/label/2018/210365lbl.pdf).

Fujiwara, R., Yoda, E. and Tukey, R.H. (2018). Species differences in drug glucuronidation: Humanised UDP-glucuronosyltransferase 1 mice and their application for predicting drug glucuronidation and drug-induced toxicity in humans. Drug Mer. Pharmacokin., 33(1): 9–16. 10.1016/j.dmpk.2017.10.002.

Geiger, H.A., Wurst, M.G. and Daniels, R.N. (2018). DARK Classics in chemical neuroscience: Psilocybin. ACS Chem. Neurosci., 9(10): 2438–2447. 10.1021/acschemneuro.8b00186.

Graziano, S., Orsolini, L., Rotolo, M.C., Tittarelli, R., Schifano, F. and Pichini, S. (2017). Herbal highs: Review on Psychoactive effects and neuropharmacology. Curr. Neuropharmacol., 15(5): 750–761. 10.2174/1570159X14666161031144427.

Hasegawa, M., Tahara, H., Inoue, R., Kakuni, M., Tateno, C. and Ushiki, J. (2012). Investigation of drug-drug interactions caused by human pregnane X receptor-mediated induction of CYP3A4 and CYP2C subfamilies in chimeric mice with a humanized liver. Drug. Metab. Dispos, 40(3): 474–480. 10.1124/dmd.111.042754.

Holmstock, N., Gonzalez, F.J., Baes, M., Annaert, P. and Augustijns, P. (2013). PXR/CYP3A4-humanized mice for studying drug-drug interactions involving intestinal P-glycoprotein. Mol. Pharm., 10(3): 1056–1062. 10.1021/mp300512r.

Issa, N.T., Wathieu, H., Ojo, A., Byers, S.W. and Dakshanamurthy, S. (2017). Drug metabolism in preclinical drug development: A survey of the discovery process, toxicology and computational tools. Curr. Drug Metab., 18(6): 556–565. 10.2174/1389200218666170316093301.

Johnson, M.W. and Griffiths, R.R. (2017). Potential therapeutic effects of Psilocybin neurotherapeutics. J. Am. Soc. Exp. NeuroTherapeut., 14(3): 734–740. 10.1007/s13311-017-0542-y.

Libanio Osorio Marta, R.F. (2019). Metabolism of lysergic acid diethylamide (LSD): An update. Drug Metabolism Reviews, 51(3): 378–387. 10.1080/03602532.2019.1638931.

Lichtman, A.H., Lux, E.A., McQuade, R., Rossetti, S., Sanchez, R. et al. (2018). Results of a double-blind, randomized, placebo-controlled study of nabiximols oromucosal spray as an adjunctive therapy in advanced cancer patients with chronic uncontrolled pain. J. Pain Symptom Manage., 55(2): 179–188 e171. 10.1016/j.jpainsymman.2017.09.001.

Luethi, D., Hoener, M.C., Krahenbuhl, S., Liechti, M.E. and Duthaler, U. (2019). Cytochrome P450 enzymes contribute to the metabolism of LSD to nor-LSD and 2-oxo-3-hydroxy-LSD: Implications for clinical LSD use. Biochem. Pharmacol., 164: 129–138. 10.1016/j.bcp.2019.04.013.

Manevski, N., Kurkela, M., Hoglund, C., Mauriala, T., Court, M.H. et al. (2010). Glucuronidation of psilocin and 4-hydroxyindole by the human UDP-glucuronosyltransferases. Drug Metab. Dispos., 38(3): 386–395. 10.1124/dmd.109.031138.

Markova, J., Essner, U., Akmaz, B., Marinelli, M., Trompke, C. et al. (2019). (Sativex (R)) as add-on therapy vs. further optimised first-line ANTispastics (SAVANT) in resistant multiple sclerosis spasticity: a double-blind, placebo-controlled randomised clinical trial. Int. J. Neurosci., 129(2): 119–128. 10.1080/00207454.2018.1481066.

Marona-Lewicka, D., Thisted, R.A. and Nichols, D.E. (2005). Distinct temporal phases in the behavioral pharmacology of LSD: Dopamine D2 receptor-mediated effects in the rat and implications for psychosis. Psychopharmacology, 180(3): 427–435. 10.1007/s00213-005-2183-9.

Marquez, B. and Van Bambeke, F. (2011). ABC multidrug transporters: Target for modulation of drug pharmacokinetics and drug-drug interactions. Curr. Drug Targets, 12(5): 600–620.

Martiny, V.Y., Carbonell, P., Chevillard, F., Moroy, G., Nicot, A.B. et al. (2015). Integrated structure- and ligand-based *in silico* approach to predict inhibition of cytochrome P450 2D6. Bioinformatics, 31(24): 3930–3937. 10.1093/bioinformatics/btv486.

Nichols, D.E., Johnson, M.W. and Nichols, C.D. (2017). Psychedelics as medicines: An emerging new paradigm. Clin. Pharmacol. Therapeut., 101(2): 209–219. 10.1002/cpt.557.

NIDA. (2018). Commonly Abused Drugs Charts. https://www.drugabuse.gov/drugs-abuse/commonly-abused-drugs-charts.

NIH. ClinicalTrials. gov. https://clinicaltrials.gov/ct2/home.

Piper, B.J., DeKeuster, R.M., Beals, M.L., Cobb, C.M., Burchman, C.A. et al. (2017). Substitution of medical cannabis for pharmaceutical agents for pain, anxiety, and sleep. J. Psychopharmacol., 31(5): 569–575. 10.1177/0269881117699616.

Pondugula, S.R., Ferniany, G., Ashraf, F., Abbott, K.L., Smith, B.F. et al. (2015). Stearidonic acid, a plant-based dietary fatty acid, enhances the chemosensitivity of canine lymphoid tumor cells. Biochem. Biophy. Res. Comm., 460(4): 1002–1007. 10.1016/j.bbrc.2015.03.141.

Russo, E.B., Guy, G.W. and Robson, P.J. (2007). Cannabis, pain, and sleep: Lessons from therapeutic clinical trials of Sativex, a cannabis-based medicine. Chem. Biodivers, 4(8): 1729–1743. 10.1002/cbdv.200790150.

Schardl, C.L., Panaccione, D.G. and Tudzynski, P. (2006). Ergot alkaloids—Biology and molecular biology. Alkaloids Chem. Biol., 63: 45–86. 10.1016/s1099-4831(06)63002-2.

Schiff, P.L. (2006). Ergot and its alkaloids. Am. J. Pharm. Edu., 70(5): 98. 10.5688/aj700598.

Smelser, N.J. and Baltes, P.B. (2001). International encyclopedia of the Social and Behavioural Sciences, International Encyclopedia of the Social and Behavioural Sciences.

Stamets, P.E., Beug, M.W., Bigwood, J.E. and Guzman, G. (1980). A new species and a new variety of psilocybe from North-America. Mycotaxon., 11(2): 476–484.

Torres-Vergara, P., Ho, Y.S., Espinoza, F., Nualart, F., Escudero, C. and Penny, J. (2020). The constitutive androstane receptor and pregnane X receptor in the brain. Br. J. Pharmacol., 177(12): 2666–2682. 10.1111/bph.15055.

Turri, M., Teatini, F., Donato, F., Zanette, G., Tugnoli, V. et al. (2018). Pain modulation after oromucosal cannabinoid spray (SATIVEX((R))) in patients with multiple sclerosis: a study with quantitative sensory testing and laser-evoked potentials. Medicines (Basel), 5(3): 10.3390/medicines5030059.

Tyzack, J.D. and Kirchmair, J. (2019). Computational methods and tools to predict cytochrome P450 metabolism for drug discovery. Chem. Biol. Drug Des., 93(4): 377–386. 10.1111/cbdd.13445.

van Amsterdam, J., Opperhuizen, A. and van den Brink, W. (2011). Harm potential of magic mushroom use: A review. Reg. Toxicol. Pharmacol., 59(3): 423–429. 10.1016/j.yrtph.2011.01.006.

Vollenweider, F.X., Vollenweider-Scherpenhuyzen, M.F., Babler, A., Vogel, H. and Hell, D. (1998). Psilocybin induces schizophrenia-like psychosis in humans via a serotonin-2 agonist action. Neuroreport., 9(17): 3897–3902.

Wagmann, L., Richter, L.H.J., Kehl, T., Wack, F., Bergstrand, M.P. et al. (2019). *In vitro* metabolic fate of nine LSD-based new psychoactive substances and their analytical detectability in different urinary screening procedures. Anal. Bioanal. Chem., 411(19): 4751–4763. 10.1007/s00216-018-1558-9.

Yamasaki, Y., Kobayashi, K., Okuya, F., Kajitani, N., Kazuki, K. et al. (2018). Characterisation of P-Glycoprotein humanized mice generated by chromosome engineering technology: its utility for prediction of drug distribution to the brain in humans. Drug Metab. Dispos., 46(11): 1756–1766. 10.1124/dmd.118.081216.

10

Application of Selected Species of the Genus *Xylaria* in Traditional Medicine

Sunil K Deshmukh,[1,] Kandikere R Sridhar[2,3] and Manish K Gupta[4]*

1. INTRODUCTION

Traditional Chinese medicine and traditional Indian medicine (Ayurveda) are the most established forms of ethnic medicine in Asia (Kim et al., 2011). The former is dominant in the eastern Asia and the Far East (China, Korea, Japan and Vietnam) while the latter mainly in the Indian Subcontinent (India, Pakistan and Tibet) (Koller, 2007). Ascomycetous genera *Cordyceps* and *Xylaria* are commonly used in traditional Chinese, Tibetan and Indian medicines (Zhou et al., 2009; Panda and Swain, 2011; Rogers, 2011). *Xylaria* Hill ex Schrank is the largest genus of the family Xylariaceae Tul. and C. Tul. (Xylariales, Sordariomycetes) and presently includes ca. 300 accepted species of stromatic pyrenomycetes (Kirk et al., 2008). However, a list of over 800 records are found in Index Fungorum (accessed on 10 September, 2020). Several members of *Xylariales* are well-known prolific producers of secondary metabolites (Helaly et al., 2018). They are commonly called dead man's finger and are used in traditional Chinese medicine as also in Indian traditional medicine, Ayurveda (Rogers, 2011; Latha et al., 2015). In ethnic Chinese medicine, *X. nigripes* serves to

[1] TERI-Deakin Nano Biotechnology Centre, The Energy and Resources Institute, Darbari Seth Block, IHC Complex, Lodhi Road, New Delhi 110003, India.
[2] Department of Biosciences, Mangalore University, Mangalagangotri, Mangalore, Karnataka, India.
[3] Centre for Environmental Studies, Yenepoya (Deemed to be) University, Mangalore, Karnataka, India.
[4] SGT College of Pharmacy, SGT University, Gurugram 122505, Haryana, India.
* Corresponding author: sunil.deshmukh1958@gmail.com

cure many human ailments (e.g., insomnia, trauma and convulsions (Rogers, 2011; Lin et al., 2013)). Similarly, *X. polymorpha* in Ayurveda is useful in promoting milk flow during a woman's post-natal period (Rogers, 2011). Interestingly, the hypogeal *Xylaria acuminatilongissima* (grow in abandoned termite nests) is also used by traditional healers and tribals in Kerala state (India) to cure gastrointestinal ailments (Shortt, 1867; Balakrishnan and Kumar, 2001). As many as 20 species of *Xylaria* are known to exist in association with termite nests (Rogers et al., 2005; Ju and Hsieh, 2007). It is interesting to note that the *Xylaria* spp. associated with termite mounds possess medicinal potential (e.g., Ko et al., 2011; Song et al., 2011; Zhao et al., 2014). The widespread fungus, *X. polymorpha* prefers to grow on wooden logs. Similar to *X. nigripes*, the *X. polymorpha* also possesses high medicinal potential and is used traditionally to treat several human ailments. In this paper, traditional uses of two *Xylaria* spp. (*X. nigripes* and *X. polymorpha*) are discussed along with their bioactive compounds.

1.1 *Xylaria nigripes*

Xylaria nigripes, a high-value medicinal fungus, is known to develop underground (~ 30 cm below), in dark, damp and warm conditions and is usually associated with termite nests, specifically the nest of the black-winged termite, *Odontotermes formosanus* (Fig. 1a, b). Nine species of *Xylaria* including *X. nigripes* were found in termite mounds of *O. formosanus* in Taiwan (Ju and Hsieh, 2007). It is also an inhabitant in the gut of another termite host species, *Odontotermes horni* (Sreerama and Veerabhadrappa, 1993). It is not surprising because many *Xylaria* spp. are endophytic with seeds as well as aerial parts of host plants (e.g., Bayman et al., 1998; Davis et al., 2003). It is one of the common mushrooms in termite mounds seen in the scrublands of south-west India (Karun and Sridhar, 2015). In some of these termite mounds, dual association of *X. nigripes* or *X. escharoidea* with an edible termitomycete, *Termitomyces umkowaan,* is seen (Karun and Sridhar, unpub. obs.).

In Chinese traditional medicine, sclerotia of *X. nigripes* are used to eliminate dampness, reduce convulsions in infants and curb heart palpitations (Rogers, 2011). It serves as a sedative, diuretic, promotes lactation and loss of blood through the nose and the lungs following childbirth. Traditionally, *X. nigripes* is used to cure insomnia and trauma and serves as a nerve tonic (Lin et al., 2013). *In vivo* and *in vitro* studies suggest that *X. nigripes* possesses antioxidants (Ko et al., 2009; Ma et al., 2013) and serves as immunomodulatory, anti-inflammatory (Ko et al., 2011), hepatoprotective

Fig. 1. *Xylaria nigripes*, immature (a) and mature (b) fruit bodies grown with termite mound in scrubland of southwest India.

(Song et al., 2011), anti-tumor (Ma et al., 2013), anti-spatial memory impairment (Zhao et al., 2014) and as an antidepressant in epileptic patients (Peng et al., 2015) besides enhancing insulin sensitivity (Chen et al., 2015) and neuroprotective activities (Xiong et al., 2016). 'Wuling Capsule' is one of the products derived from *X. nigripes* and which is used to treat Major Depressive Disorder (MDD) (Zheng et al., 2016). These capsules are produced by *X. nigripes* through a liquid fermentation technology (Wang et al., 2020).

Xylaria nigripies is a rich source of amino acids and the most exciting amino acid in this medical mushroom is gamma-aminobutyric acid (GABA). GABA is a neurotransmitter and a chemical that acts as a messenger between the brain and the nervous system, taking messages from the brain to the nerve cells to slow down their activity. The GABA-rich Zylaria™ product contains glutamic acid and glutamate decarboxylase, which have a natural, and more importantly, gentle tranquillising effect. The *X. nigripes* supplements have been proven to help reset the natural rhythm of the human body, allowing sleep patterns to improve and sufferers of insomnia to get more restorative and peaceful sleep, take less time to fall asleep and remain asleep for longer periods (https://nulivscience.com/blog/zylaria-fungal-answer-good-nights-sleep). Zylaria™ also serves as a supplement to promote restful sleep and does not have side-effects like 'brain fog' and grogginess that some natural sleeping aids produce, Zylaria™ is known to help patients suffering from depression too. It has a slower and milder effect than more conventional antidepressant medicines and, being a natural substance, can be better tolerated with fewer side effects. Generally people with depression tend to have lower levels of GABA. Therefore, topping up the levels of GABA with GABA-rich supplements of Zylaria™ in foods seems to have positive impacts. GABA in Zylaria™ also helps in relieving anxiety and panic attacks owing to its effect on reducing the excitability of nerve cells. The same is seen in children and young adults with Attention Deficit Hyperactivity Disorder (ADHD), where it has a calming effect and increases focus and concentration, specifically as it relates to sleep quality, insomnia, anxiety and general wellness. The Nrf2/HO-1 pathway mediated the neuroprotective effect of *X. nigripes* following an ischemic condition in the brain (Sun et al., 2017). Some of the bioactive compounds responsible for bioactivity are reported and depicted in Fig. 1.

The presence of 15 amino acids (aspartic acid, glutamic acid, serine, histidine, glycine, arginine, proline, alanine, γ-aminobutyric acid, threonine, methionine, valine, phenylalanine, isoleucine and leucine) was confirmed by the HPLC analysis of Wuling capsules (He et al., 2010). Nucleosides (adenosine, uridine and guanosine) (Lu et al., 2011) and 5-methylmellein (1) , 5-hydroxymellein (2), 5-carboxylmellein (3), genistein (4) (Fig. 2a) (Chen et al., 2012; Lu et al., 2014) were also purified from *X. nigripes*.

Compound 5,8-dihydroxy-3-methyl-3,4-dihydroisocoumarin (5) (Fig. 2a) was isolated from *X. nigripes* and displayed antioxidant activity in DPPH radical-scavenging assay, 1.67 times as that of vitamin C or 2.10 times as that of Vitamin E at the concentration of 20 μM/L (Wu, 2001). Two new naturally-occurring spirocyclic pyrrole alkaloids, such as xylapyrrosides A (6) and B (7) and already-isolated metabolites, pollenopyrrosides A (8) and B (9) (Fig. 2a), were extracted from *X. nigripes*. Compounds (6–9) displayed an average scavenging activity by preventing the oxidative stress-induced cytotoxicity of A7r5 rat vascular smooth muscle cells (VSMCs) (Li et al., 2015).

New compounds named nigriterpenes A–F (10–15), 2-hydroxymethyl-3-pentylphenol (16) along with known metabolites, fomannoxin alcohol (17), 3-butyl-7-hydroxyphthalide (18), scytalone (19) and fomannoxin (20) (Fig. 2a) were obtained from *X. nigripes* (YMJ653). The nigriterpene C (12) exhibited inhibition on

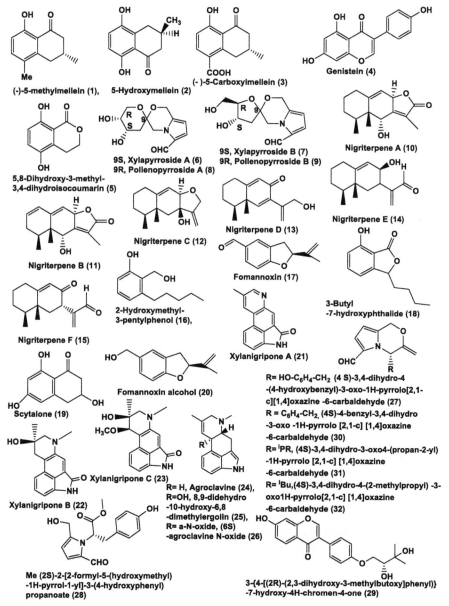

Fig. 2a. Structures of metabolites obtained from *X. nigripes* (1–32).

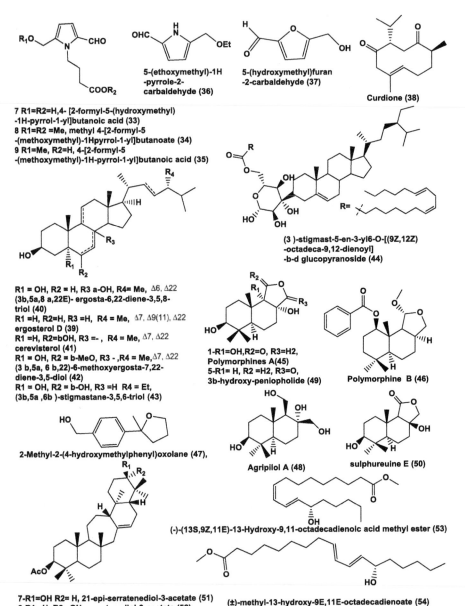

Fig. 2b. Structures of metabolites obtained from *X. nigripes* (33–54).

NO production and iNOS and COX-2 expression with IC50 values of 21.7, 8.1 and 16.6 μM, respectively. Fomannoxin alcohol (17) also inhibited NO production and iNOS and COX-2 expression with IC50 values of 33.8, 11.6 and 12.7 μM, respectively. The positive-control curcumin showed inhibition of NO production at IC50, 4.5 μM (Chang et al., 2017).

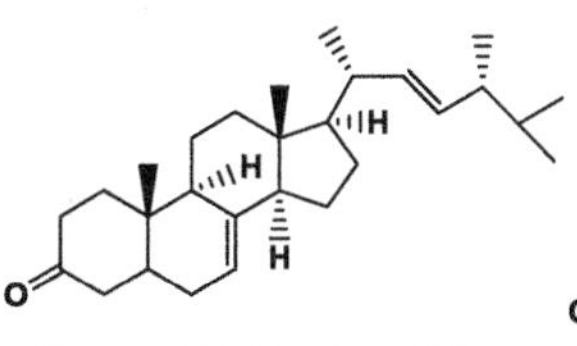

Fig. 2c. Structures of metabolites obtained from *X. nigripes* (55–69).

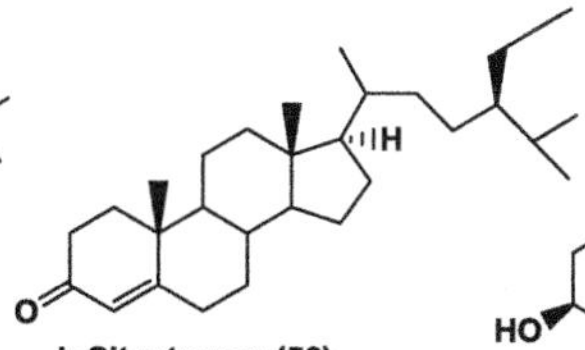

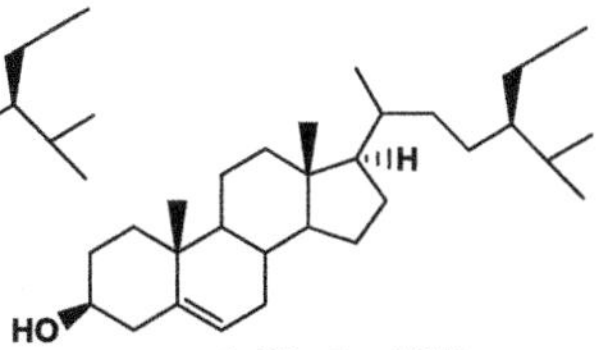

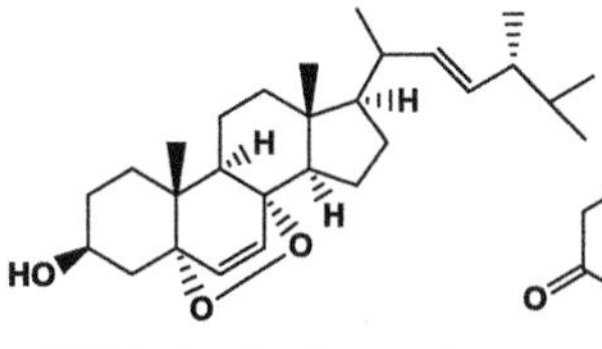

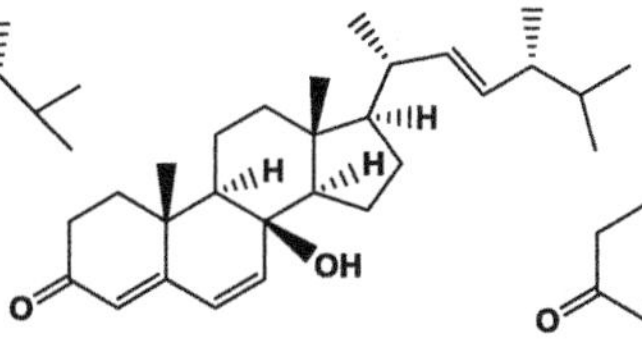

Ergot alkaloids, namely xylanigripones A (21) and B (22) C (23) alongwith three previously reported compounds, agroclavine (24), 8,9-didehydro-10-hydroxy-6,8-dimethylergolin (25) and (6S)-agroclavine N-oxide (26) (Fig. 2a) were purified from *X. nigripes* collected from Chuxiong county in Yunnan province of China. The compound (23) could inhibit the AChE activity up to 11.9 per cent at the concentration of 50 μM, and compound (22) (with the concentration of 50 μM) could inhibit the activity of AChE up to 38.1 per cent as compared with 45.4 per

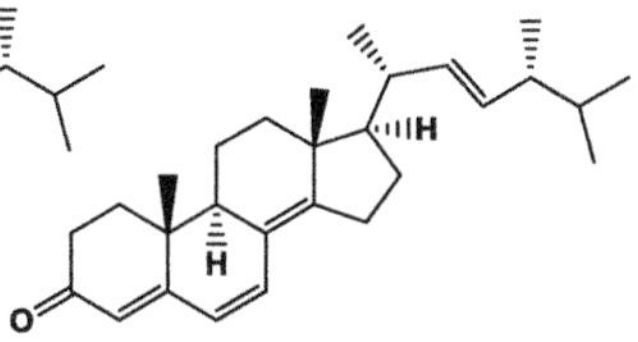

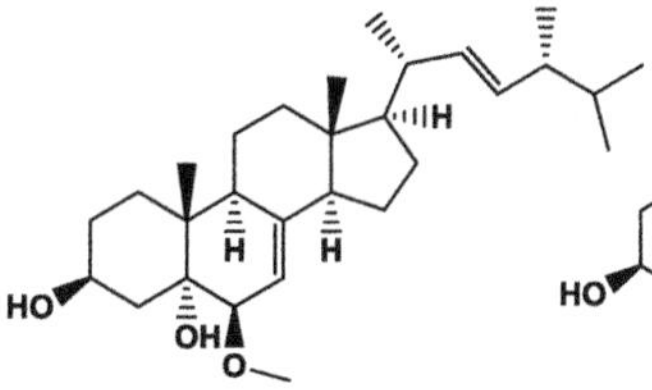

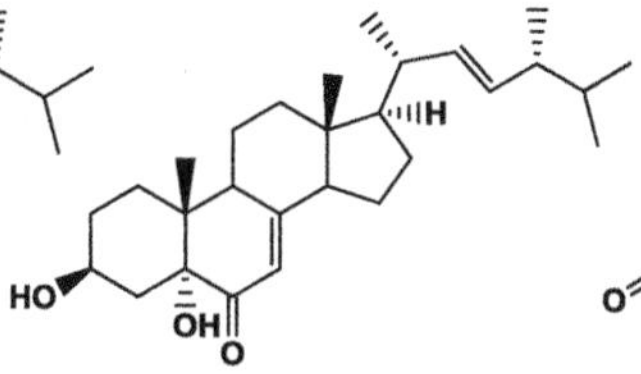

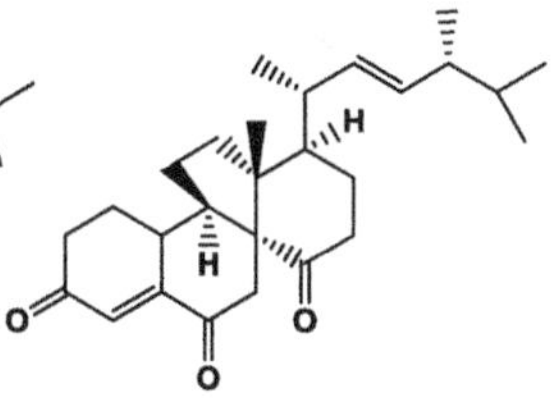

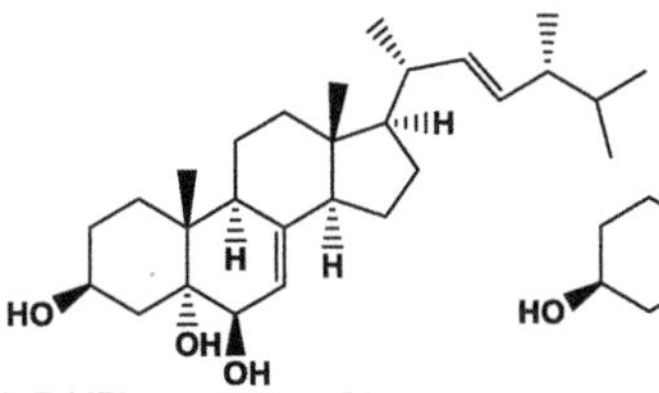

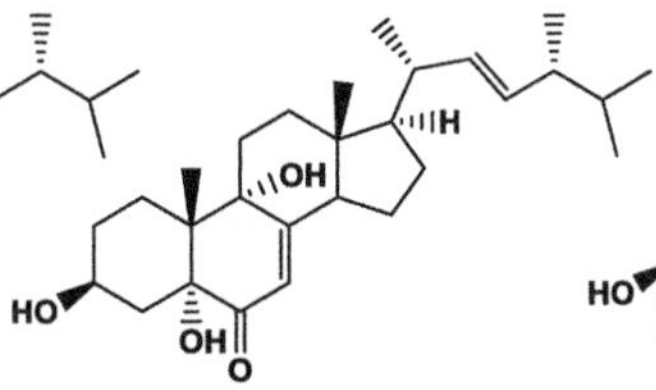

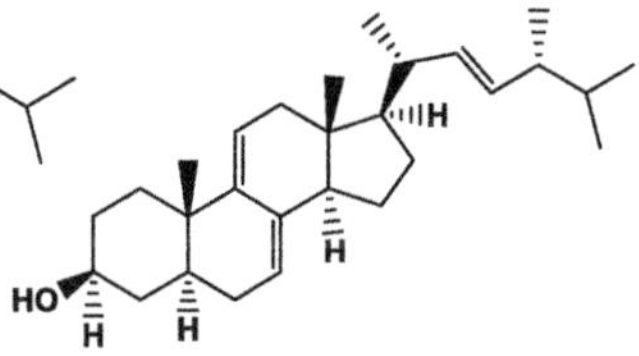

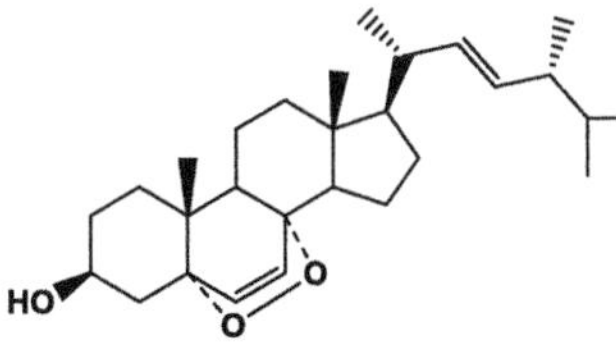

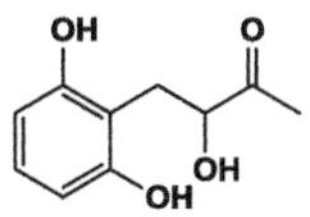

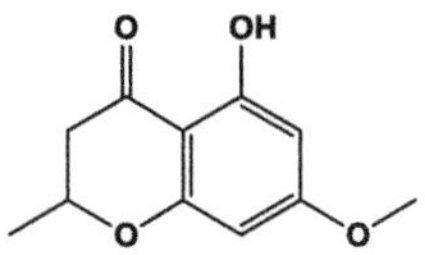

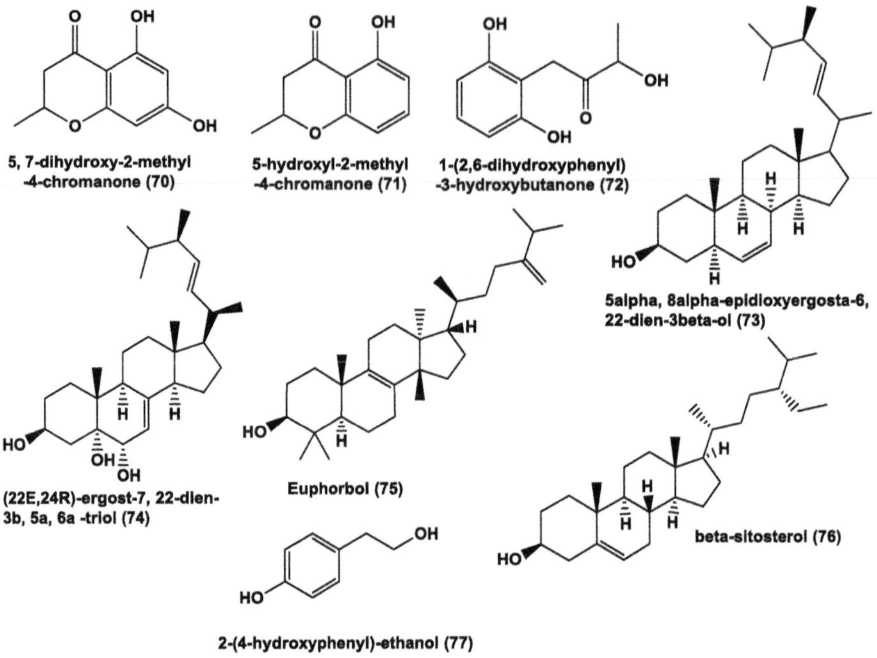

5, 7-dihydroxy-2-methyl -4-chromanone (70)

5-hydroxyl-2-methyl -4-chromanone (71)

1-(2,6-dihydroxyphenyl) -3-hydroxybutanone (72)

5alpha, 8alpha-epidioxyergosta-6, 22-dien-3beta-ol (73)

(22E,24R)-ergost-7, 22-dien-3b, 5a, 6a -triol (74)

Euphorbol (75)

beta-sitosterol (76)

2-(4-hydroxyphenyl)-ethanol (77)

Fig. 2d. Structures of metabolites obtained from *X. nigripes* (70–77).

cent inhibition rate of the positive-control tacrine. The compounds (21, 24, 25, 26) exhibited different levels of inhibition of Cholesteryl Ester Transfer Protein (CETP) activity with inhibition rates of 49 per cent, 73 per cent, 36.50 per cent and 58.50 per cent when compared with blank control (Hu and Li, 2017).

New compounds (4S)-3,4-dihydro-4-(4-hydroxybenzyl)-3-oxo-1H-pyrrolo[2,1-c][1,4]oxazine-6-carbaldehyde (27), Me (2S)-2-[2-formyl-5-(hydroxymethyl)-1H-pyrrol-1-yl]-3-(4-hydroxyphenyl)propanoate (28) and 3-{4-[(2R)-(2,3-dihydroxy-3-methylbutoxy]phenyl)}-7-hydroxy-4H-chromen-4-one (29), along with 16 known compounds (4S)-4-benzyl-3,4-dihydro-3-oxo-1H-pyrrolo[2,1-c][1,4]oxazine-6-carbaldehyde (30), (4S)-3,4-dihydro-3-oxo4-(propan-2-yl)-1H-pyrrolo[2,1-c] [1,4]oxazine-6-carbaldehyde (31), (4S)-3,4-dihydro-4-(2-methylpropyl)-3-oxo1H-pyrrolo[2,1-c] [1,4]oxazine-6-carbaldehyde (32) (Fig. 2a), 4-[2-formyl-5-(hydroxymethyl)-1H-pyrrol-1-yl] butanoic acid (33), methyl 4-[2-formyl-5-(methoxymethyl)-1Hpyrrol-1-yl] butanoate (34), 4-[2-formyl-5-(methoxymethyl)-1H-pyrrol-1-yl]butanoic acid (35), 5-(ethoxymethyl)-1H-pyrrole-2-carbaldehyde (36), 5-(hydroxymethyl)furan-2-carbaldehyde (37), curdione (38) (3β,22E)-ergosta-7,9(11),22-trien-3-ol,ergosterol D (39), (3β,5α,8α,22E)-ergosta-6,22-diene-3,5,8-triol (40), cerevisterol (41), (3β,5α,6β,22E)-6-methoxyergosta-7,22-diene-3,5-diol (42), (3β,5α,6β)-stigmastane-3,5,6-triol (43), (3β)-stigmast-5-en-3-yl6-O-[(9Z,12Z)-octadeca-9,12-dienoyl]-b-d glucopyranoside (44) (Fig. 2b), andgenistein (4), were obtained from *X. nigripes*. Compound (3β,5α,6β,22E)-6-methoxyergosta-7,22-diene-3,5-diol (42) displayed a good neuroprotective effect by attenuating α-amyloid 25–35 (Ab$_{25–35}$)-induced cell damage in the human SH-SY5Y neuroblastoma cells (20.9 per cent of increase in cell viability) at the concentration of 10 μm. (3β,5α,8α,22E)-ergosta-6,22-diene-3,5,8-triol (40) could significantly decrease NO production (IC$_{50}$, 27.6 μm), which was comparable to that of the positive control, NG-monomethyl-l-arginine (L-NMMA, IC$_{50}$ = 14.4 μm). Steroids (39, 42, 43) displayed potent cytotoxicity against U2OS cell lines (IC$_{50}$, 0.93, 6.0 and 13.2 μm, respectively). Additionally, compound (42) also displayed cytotoxicity against A549 cell line (IC$_{50}$, 11.9 μm). Staurosporine was used as a positive control which displayed cytotoxicity with IC$_{50}$, 0.01 and 0.003 μm against U2OS and A549 cell lines, respectively (Xiong et al., 2016).

New sesquiterpenoids named polymorphines A, B (45, 46), a new phenyloxolane 2-methyl-2-(4-hydroxymethylphenyl) oxolane (47) and 18 known compounds like agripilol A (48), 3β-hydroxy-peniopholide (49), sulphureuine E (50), 21-epi-serratenediol-3-acetate (51), serratenediol-3-acetate (52), (-)-(13S,9Z,11E)-13-Hydroxy-9,11-octadecadienoic acid methyl ester (53), (±)-methyl-13-hydroxy-9E,11E-octadecadienoate (54) (Fig. 2b), ergosta-7,22-Dien-3-one (55), β-sitosterone (56), β-sitosterol (57), (22E)-5α,8α-episodioxergosta-6,22-diene-3β-ol (58), isocyathisterol (59), ergosta-4,6,8 (14),22-tetraen-3-one (16) (60), (3β,5α,6β,22E)-6-methoxyergosta-7,22-diene-3,5-diAlcohol(17) (61), 3β,5α-Dihydroxy-(22E,24R)-ergosta-7,22-diene-6-Ketone (62), dankasterone A (63), (22E,24R)-ergosta-7,22-Diene-3β,5α,6β-triol (64) and (22E,24R)-ergosta-7,22-di En-3β,5α,9α-triol-6-one (65) (Fig. 2c) were purified from *X. polymorpha*. The compound (46) displayed anti-acetylcholinesterase and α-glucosidase inhibitory activities. Moreover, the compound (47) showed moderate inhibitory activity against the nematode *Panagrellus redivivus* with mortality ratio of 59.6 per cent at 2.5 mg/mL (Yang et al., 2017a).

Ergostarien-3β-ol (66) and ergosterol peroxide (67) (Fig. 2c) were purified from hexane fraction obtained from fruit bodies of *X. nigripes* and displayed potent anti-inflammatory effects on NO, TNF-α, IL-1β, IL-6, and prostaglandin E$_2$ production in LPS-stimulated macrophages XN-CP2 along with strong suppressive effects on iNOS and COX-2 expression and NF-κB activation. It is also reported that ergosterol peroxide (67) is responsible for potent anti-inflammatory activity of *X. nigripes* and the anti-inflammatory property is due to inhibition of iNOS and COX-2 expression via the NF-κB signalling pathway (Liaw et al., 2017).

Other compounds reported include (2,6-dihydroxyphenyl)-3-hydroxybutan-one (68) purified from fermental mycelium of *X. nigripes* and displayed strong DPPH radical-scavenging ability and reducing capacity (Gong et al., 2008a). The compounds, 5-hydroxy-7-methoxy-2-methyl-4-chromanone (69) (Fig. 2c), 5, 7-dihydroxy-2-methyl-4-chromanone (70), 5-hydroxyl-2-methyl-4-chromanone (71), 1-(2,6-dihydroxyphenyl)-3-hydroxybutanone (72), 5α,8α-epidioxyergosta-6,22-dien-3β-ol (73), (22E,24R)-ergost-7,22-dien-3β,5α,6α-triol (74), euphorbol (75), β-sitosterol (76) and 2-(4-hydroxyphenyl)-ethanol (77) (Fig. 2d) were obtained from *X. nigripes* but no activity was reported (Gong et al., 2008b).

1.2 *Xylaria polymorpha*

Xylaria polymorpha is common and widespread near the base of rotting stumps (Fig. 3a, b). The candle snuff is named for its shape, which is like a tool used for putting out the candle flame. Other authors suggest that the name is related to the

Fig. 3. *Xylaria polymorpha*, immature (a) and mature (b) fruit bodies grown on the dead logs of *Pongamia pinnata* in scrubland of southwest India.

way its spores are carried away by the current of the wind. This fungus is common throughout the year in the Pacific northwest and has been reported to possess luminous mycelia. Another name, staghorn, is related to its antler-like shape, as seen in *Xylaria longipes*, but slimmer than *X. polymorpha* are found on hardwood trees across North America. It is very common on wood logs of *Pongamia pinnata* in the scrublands of southwest India (Karun and Sridhar, 2015).

In Indian Ayurvedic folk-medicine, *X. polymorpha* is used to promote milk flow after birth and is known as '*phoot doodh*' (meaning, 'to gush milk') (Rai et al., 1993; Rogers, 2011). The fruit body is ground into a powder and mixed with sugar (1:1) to prepare pills of pea-size. These pills are consumed with cow's milk daily, prior to taking the meal up to five days. Some of the bioactive compounds responsible for bioactivity of *Xylaria polymorpha* are reported and depicted in Fig. 4.

The chemical analysis of fruiting bodies of *Xylaria polymorpha*, contain a hydroxyphthalide derivative, xylaral (78) (Fig. 4a), which exhibits a violet colour reaction with aqueous ammonia (Gunawan et al., 1990). Optimal conditions to produce *X. polymorpha* polysaccharides by growing in liquid culture (Yang and Huaan, 2004).

Xylaria polymorpha contains about 6 per cent mannitol (by dry weight), a sugar used as a diuretic agent (Snatzke and Wolff, 1987). The piliformic acid (79) (Fig. 4a) is a major metabolite obtained from *X. polymorpha, X. longipes, X. mali, X. hypoxylon* (Anderson et al., 1985). Later on, it was purified from *Halorosellinia oceanica* BCC 5149 with average cytotoxicity against KB and BC-1 cell lines with IC 50 values of 13.0 and 5.0 µg/ml (Chinworrungsee et al., 2001).

Xylarinic acids A (80) and B (81) (Fig. 4a) were purified from the fruit bodies of *X. polymorpha* collected from Gwangneung Forest in Gyeonggi province of Korea. The compounds, (80) and (81) displayed good antifungal activity with clear inhibition zone, 16–20 mm diameter against *Pythium ultinum* and *Magnaporthe grisea*. Both the compounds showed average activity with clear inhibition zone 11–15 mm diameter against *Alternaria panax, Aspergillus niger* and *Fusarium oxysporium* and poor activity against *Alternaria mali, A. porri, Botrytis cinerea, Cylindrocarpon destructans, Fulviafulva, Phytophthora capsici* and *Rhizoctonia solani* (Jang et al., 2007).

Xylarinic acid A (80) (Fig. 4a) isolated from *X. polymorpha* inhibited NO production and mRNA expressions of inducible nitric oxide synthase (iNOS),

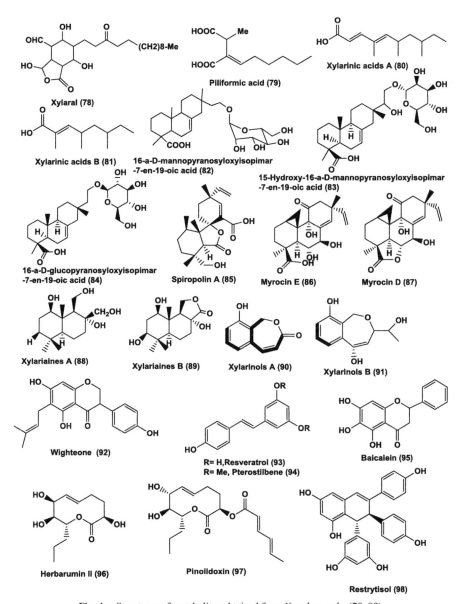

Fig. 4a. Structures of metabolites obtained from *X. polymorpha* (78–98).

cyclooxygenase-2 (COX-2), interleukin-1β (IL-1β) and interleukin-6 (IL-6) in a concentration-dependent pattern without cytotoxic effects (Kim et al., 2010).

The compounds 16-α-D-mannopyranosyloxyisopimar-7-en-19-oic acid (82), 15-hydroxy-16-α-D-mannopyranosyloxyisopimar-7-en-19-oic acid (83) and 16-α-D-glucopyranosyloxyisopimar-7-en-19-oic acid (84) (Fig. 4a) were purified from the fruit bodies of *X. polymorpha.* Compound (82) exhibited cytotoxicity against HL60, K562, HeLa and LNCaP cell lines (IC$_{50}$, 165, 143, 235; 215 μM, respectively). The compound (84) exhibited cytotoxicity against HL60, K562, HeLa

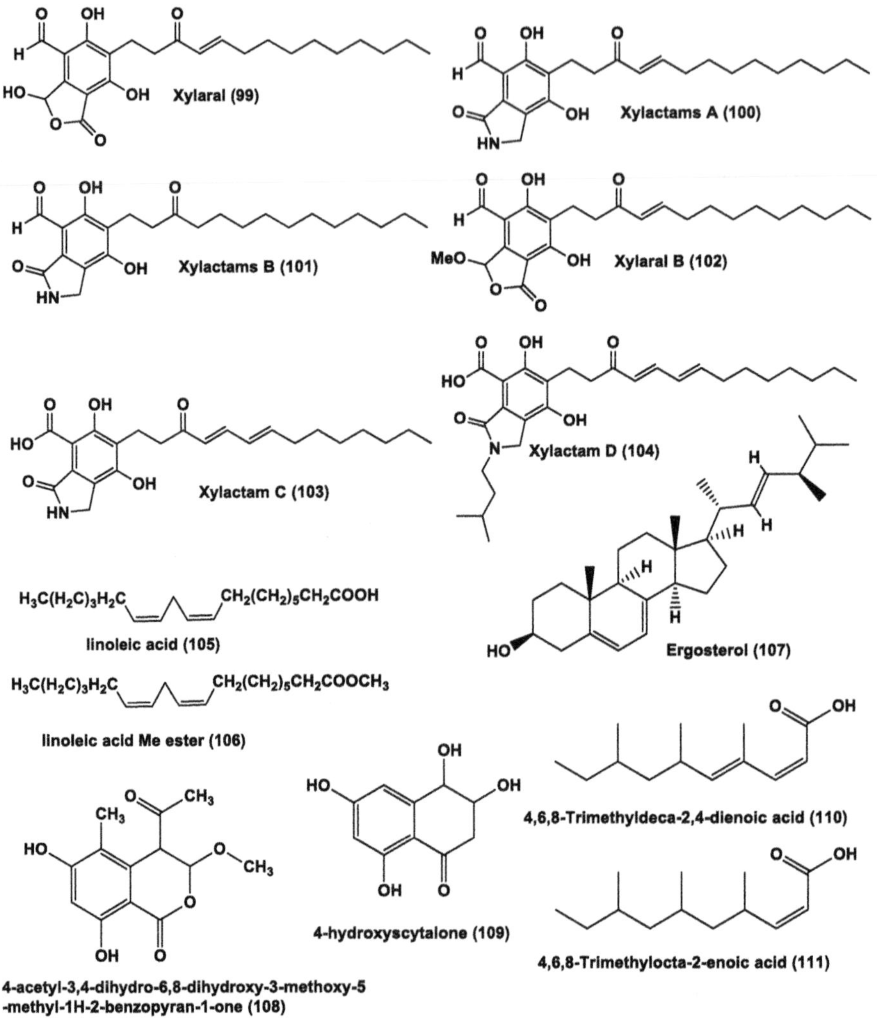

Fig. 4b. Structures of metabolites obtained from *X. polymorpha* (99–111).

and LNCaP cell lines (327, 390, 288 and 607 µM, respectively). The compound (84) exhibited cytotoxicity against HL60, K562, HeLa, and LNCaP cell lines (132, 71, 75 and 112 µM respectively). The positive control camptothecin displayed cytotoxicity against HL60, K562, HeLa and LNCaP cell lines (0.02, 0.10, 0.90 and 0.07 µM, respectively). The compounds (82) and (84) induced apoptosis along with typical DNA fragmentation in HL60 cells (Shiono et al., 2009).

Two new isopimarane-type diterpenes, spiropolin A (85) and myrocin E (86) and known compound, myrocin D (87) (Fig. 4a) were purified from the fruit bodies of *X. polymorpha*. Spiropolin A (85) restored the growth inhibition caused by the hyperactivated Ca-signalling in mutant yeast (Shiono et al., 2013).

Sesquiterpenoids, namely xylariaines A–B (88–89) (Fig. 4a), were purified from *X. polymorpha*. The site of collection was the valley of Yinggeling in the Hainan province of China. The compounds (88) and (89) displayed weakly inhibitory AChE activities with inhibition ratios of 12.4 per cent and 18.0 per cent, respectively at a concentration of 50 µg/mL (Tacrine, as positive control showed inhibitory rate of 56.7 per cent) (Yang et al., 2017b).

Xylarinols A (90) and B (91) (Fig. 4a) were found in the fruit bodies of *X. polymorpha*. These compounds were found to exhibit moderate ABTS radical scavenging activity at 100 µM concentration with 40 and 45 per cent inhibition, respectively (Lee et al., 2009).

The compounds, such as wighteone (92), resveratrol (93), pterostilbene (94), baicalein (95), herbarumin II (96), pinolidoxin (97), restrytisol (98) (Fig. 4a), xylaral (99), xylactams A (100), and B (101), xylaral B (102), xylactam C (103) and xylactam D (104) (Fig. 4b), were purified from fruit bodies of *Xylaria polymorpha* (Brown et al., 2018).

The compounds linoleic acid (105), linoleic acid Me ester (106), ergosterol (107), 4-acetyl-3,4-dihydro-6,8-dihydroxy-3-methoxy-5-methyl-1H-2-benzopyran-1-one (108) and 4-hydroxyscytalone (109) (Fig. 4b) were purified from the fruit bodies of *X. polymorpha* (Jang et al., 2009).

The compounds 4, 6, 8-Trimethyldeca-2,4-dienoic acid (110) and 4,6,8-trimethylocta-2-enoic acid (111) (Fig. 4a) were purified from *Xylaria polymorpha* and were found to have inhibitory activities against plant pathogenic fungus (Yoon et al., 2009).

Conclusion and Outlook

Although about 300 species of *Xylaria* have been reported, only a few of them have been explored in traditional medicine, especially in Chinese medicine and Indian Ayurveda. They have been reported from a wide ecological niche, like termite nests, rotten wood, marine habitats and in association with live plants, like endophytic fungi. There is an urgent need of extensive research on this genus as their potency have been already proved for application in traditional medicine. In the present study, nearly 77 and 34 compounds are reported from *X. nigripes* and *X. polymorpha*, respectively. These compounds belong to the diverse chemical classes, including pyrrol, furan, lactam, banzopyran, steroid, chromanone, glycoside, etc. Among the identified compounds, only a limited number of compounds are screened for various biological activities. To screen these compounds, the cultures of species of *Xylaria* used in traditional medicine are raised and screened against selected targets. It is also necessary to study the anamorph and teleomorph stages separately for the bioactive potency. Bioactive compounds should be obtained by using activity-guided isolation. One of the bottlenecks in the traditional screening methods is the downregulation or inactivation of secondary metabolite production under axenic laboratory culture conditions. Techniques like optimisation using OSMAC (one strain-many compounds) strategy, co-cultivation and epigenetic modifications should be used for optimisation of yield and to explore the chemical diversity. Secondary metabolites are produced from the multi-enzyme complexes encoded by the secondary metabolite biosynthetic gene clusters (smBGCs). Genome mining approaches will be knowledge-based, repurposing of smBGCs to produce different metabolites by *Xylaria* sp. Application of UPLC-MS technique can accelerate the screening processes. The studies highlighted in this chapter revealed the potential of *Xylaria* in traditional medicine, which can be used for exploring chemical diversity to develop molecules of pharmaceutical and agricultural interest.

References

Anderson, J.R., Edwards, R.L. and Whalley, A.J. (1985). Metabolites of the higher fungi, Part 22, 2-butyl-3-methylsuccinic acid and 2-hexylidene-3-methylsuccinic acid from xylariaceous fungi. J. Chem. Soc. Perkin Trans. I., 1985: 1481–1485.

Balakrishnan, V. and Kumar, N.A. (2001). 'Nilamanga' (*Sclerotium stipitatum?*)—a rare termite fungal sclerotia with medicinal properties known among the tribal and rural communities of Kerala. Ethnobot., 13: 9–14.

Bayman, P., Angulo-Sandoval, P., Baez-Ortiz, Z. and Lodge, J. (1998). Distribution and dispersal of *Xylaria* endophytes in two tree species in Puerto Rico. Mycol. Res., 102: 944–948.

Brown, C.E., Liscombe, D.K. and McNulty, J. (2018). Three new polyketides from fruiting bodies of the endophytic ascomycete *Xylaria polymorpha*. Nat. Prod. Res., 32(20): 2408–2417.

Chang, J.C., Hsiao, G., Lin, R.K., Kuo, Y.H., Ju, Y.M. and Lee, T.H. (2017). Bioactive constituents from the termite nest-derived medicinal fungus *Xylaria nigripes*. J. Nat. Prod., 80(1): 38–44.

Chen, Y., Lu, J., Zhu, M. and Luo, L. (2012). Determination of multiple constituents in wuling capsules by HPLC simultaneously. Zhongguo Zhong Yao Za Zhi., 37(2): 218–221 (in Chinese).

Chen, Y.I., Tzeng, C.Y., Cheng, Y.W. et al. (2015). The involvement of serotonin in the hypoglycemic effects produced by administration of the aqueous extract of *Xylaria nigripes* with steroid-induced insulin-resistant rats. Phytother. Res., 29: 770–776.

Chinworrungsee, M., Kittakoop, P., Isaka, M., Rungrod, A., Tanticharoen, M. and Thebtaranonth, Y. (2001). Antimalarial halorosellinic acid from the marine fungus *Halorosellinia oceanica*. Bioorg. Med. Chem. Lett., 11(15): 1965–1969.

Davis, E.C., Franklin, J.B., Shaw, J.A. and Vilgalys, R. (2003). Endophytic *Xylaria* (Xylariaceae) among liverworts and angiosperms: Phylogenetics, distribution and symbiosis. Am. J. Bot., 90: 1661–1667.

Gong, Q.F., Wu, S.H., Tan, N.H. and Ch, Z.H. (2008a). Study on compounds with antioxidant and antitumor activity from fermental mycelium of *Xylaria nigripes*. Food Sci. Technol., 12: 28–31.

Gong, Q.F., Zhang, Y.M., Tan, N.H. and Chen, Z.H. (2008b). Chemical constituents in fermented mycelium of *Xylaria nigripes*. Zhongguo Zhong Yao Za Zhi., 33(11): 1269–1272.

Gunawan, S., Steffan, B. and Steglich, W. (1990). Xylaral, a hydroxyphthalide derivative from fruiting bodies of *Xylaria polymorpha* (Ascomycetes). Liebigs Annalen Der Chemie., 1990(8): 825–827.

Helaly, S.E., Thongbai, B. and Stadler, M. (2018). Diversity of biologically active secondary metabolites from endophytic and saprotrophic fungi of the ascomycete order *Xylariales*. Nat. Prod. Rep., 35: 992–1014.

He, X.R. and Liu, P. (2010). Determination of 14 kinds of amino acids in Wuling capsule by HPLC-FLD. Chin. Trad. Pat. Med., 32(8): 1358–361 (in Chinese).

Hu, D. and Li, M. (2017). Three new ergot alkaloids from the fruiting bodies of *Xylaria nigripes* (Kl.). Sacc. Chem. Biodivers., 14(1): e1600173. 10.1002/cbdv.201600173.

Jang, Y.W., Lee, I.K., Kim, Y.S., Lee, S., Lee, H.J. et al. (2007). Xylarinic acids A and B, new antifungal polypropionates from the fruiting body of *Xylaria polymorpha*. J. Antibiot., 60(11): 696–699.

Jang, Y.W., Lee, I.K., Kim, Y.S., Seok, S.J., Yu, S.H. and Yun, B.S. (2009). Chemical constituents of the fruiting body of *Xylaria polymorpha*. Mycobiology, 37: 207–10.

Ju, Y.M. and Hsieh, H.-M. (2007). *Xylaria* species associated with nests of *Odonotermes formosanus* in Taiwan. Mycologia, 99(6): 936–957.

Karun, N.C. and Sridhar, K.R. (2015). *Xylaria* complex in the south Western India. Pl. Pathol. Quarant., 5: 83–96.

Kim, C.G., Kim, T.W., Endale, M., Yayeh, T., Seo, G.S. et al. (2010). Inhibition of nitric oxide production and mRNA expressions of pro-inflammatory mediators by xylarinic acid A in RAW 264.7 cells. J. Med. Plants Res., 4(22): 2370–2378.

Kim, J.Y., Pham, D.D. and Koh, B.H. (2011). Comparison of sasang constitutional medicine, traditional Chinese medicine and ayurveda. Evid. Based Compl. Alt. Med., 2011: 239659. 10.1093/ecam/neq052.

Kirk, P.F., Cannon, P.F., Minter, D.W. and Stalpers, J.A. (2008). Dictionary of the Fungi. 10th ed., CAB International, Egham, UK.

Ko, H.-J., Song, A., Lai, M.-N. et al. (2009). Antioxidant and antiradical activities of Wu Ling Shen in a cell free system. Am. J. Chin. Med., 37: 815–828.

Ko, H.-J., Song, A., Lai, M.-N. et al. (2011). Immunomodulatory properties of *Xylaria nigripes* in peritoneal macrophage cells of Balb/c mice. J. Ethnopharmacol., 138: 762–768. 10.1016/j.jep.2011.10.022.

Koller, J.M. (2007). Asian Philosophies, Pearson Education, New Jersey, USA.

Latha, K.P.D., Veluthoor, S. and Manimohan, P. (2015). On the taxonomic identity of a fungal morph used in traditional medicine in Kerala State, India. Phytotaxa, 201(4): 287–295.

Lee, I.K., Jang, Y.W., Kim, Y.S., Yu, S.H., Lee, K.J. et al. (2009). Xylarinols A and B, two new 2-benzoxepin derivatives from the fruiting bodies of *Xylaria polymorpha*. J. Antibiot., 62(3): 163–165.

Li, M., Xiong, J., Huang, Y., Wang, L.J., Tang, Y. et al. (2015). Xylapyrrosides A and B, two rare sugar-morpholinespiroketal pyrrole-derived alkaloids from *Xylaria nigripes*: Isolation, complete structure elucidation, and total syntheses. Tetrahedron, 71(33): 5285–5295.

Liaw, C.C., Wu, S.J., Chen, C.F., Lai, M.N. and Ng, L.T. (2017). Anti-inflammatory activity and bioactive constituents of cultivated fruiting bodies of *Xylaria nigripes* (Ascomycetes), a Chinese medicinal fungus. Int. J. Med. Mushrooms, 19(10): 915–924.

Lin, Y., Wang, X.Y., Ye, R., Hu, W.-H., Sun, S.C. et al. (2013). Efficacy and safety of Wuling capsule, a single herbal formula, in Chinese subjects with insomnia: A multicenter, randomised, double-blind, placebo-controlled trial. J. Ethnopharmacol., 145: 320–327.

Lu, J.X., Lei, L., Chen, Y., Chen, J. and Zhu, M. (2014). Chemical constituents of wuling fermentative powder. Zhongguo Xian Dai Ying Yong Yao Xue, 31(5): 541–543 (in Chinese).

Lu, J.X., Zhu, M., Chen, Y., Zhang, P. and Fang, L. (2011). HPLC specific chromatogram of chemical constituents of Wuling capsules. Yao Wu Fen Xi Za Zhi., 4: 764–767 (in Chinese).

Ma, Y.P., Mao, D.B., Geng, L.J. et al. (2013). Production optimisation, molecular characterisation and biological activities of exopolysaccharides from *Xylaria nigripes*. Chem. Biochem. Eng. Quart., 27: 177–184.

Panda, A.K. and Swain, K.C. (2011). Traditional uses and medicinal potential of *Cordyceps sinensis* of Sikkim. J. Ayur. Integr. Med., 2(1): 9–13.

Peng, W.-F., Wang, X., Hong, Z. et al. (2015). The antidepression effect of *Xylaria nigripes* in patients with epilepsy: A multicenter randomised double-blind study. Seizure, 29: 26–33.

Rai, B.K., Ayachi, S.S. and Rai, A.A. (1993). Note on ethno-myco-medicines from central India. Mycologist, 7(4): 192–193.

Rogers, J.D., Ju, Y.-M. and Lehmann, J. (2005). Some *Xylaria* species on termite nests. Mycologia, 97: 914–923.

Rogers, R. (2011). The Fungal Pharmacy. North Atlantic Books, Berkeley, California.

Shiono, Y., Matsui, N., Imaizumi, T., Koseki, T., Murayama, T. et al. (2016). An unusual spirocyclic isopimarane diterpenoid and other isopimarane diterpenoids from fruiting bodies of *Xylaria polymorpha*. Phytochem. Lett., 6(3): 439–443.

Shiono, Y., Motoki, S., Koseki, T., Murayama, T., Tojima, M. and Kimura, K.I. (2009). Isopimaranediterpene glycosides, apoptosis inducers, obtained from fruiting bodies of the ascomycete *Xylaria polymorpha*, Phytochemistry, 70(7): 935–939.

Shortt, J. (1867). An account of *Sclerotium stipitatum* (Berk. et. Curr.) of Southern India. J. Linn. Soc., 9: 417–419.

Snatzke, G. and Wolff, H.P. (1987). Mannitol from *Xylaria polymorpha*. Zeitschrift fürMykologie, 53(1): 137–138.

Song, A., Ko, H.J., Lai, M.N., Ng, L.-T. et al. (2011). Protective effects of Wu-Ling-Shen (*Xylaria nigripes*) on carbon tetrachloride-induced hepatotoxicity in mice. Immunopharmacol. Immunotoxicol., 33: 454–460.

Sreerama, L. and Veerabhadrappa, P.S. (1993). Isolation and properties of carboxylesterases of the termite gut associated fungus, *Xylaria nigripes* K., and the identity from the host termite, *Odontotermes horni* W., midgut carboxylesterases. Int. J. Biochem., 25: 1637–1651.

Sun, R., Zhang, Y., He, S., Li, R., Xue, F. et al. (2017). *Xylaria nigripes* protects mice against cerebral ischemic injury by activating Nrf2/HO-1 pathway. Int. J. Clin. Exp. Med., 10(8): 11636–11645.

Wang, H., Chen, H., Gao, Y., Wang, S., Wang, X. et al. (2020). The effect of Wuling capsule on depression in type 2 diabetic patents. Biosci. Rep., 40: BSR20191260. 10.1042/BSR20191260.

Wu, G. (2001). A study on DPPH free-radical scavengers from *Xylaria nigripes*. Acta Mycol. Sinica, 41(3): 363–366.

Xiong, J., Huang, Y., Wu, X.-Y. et al. (2016). Chemical constituents from the fermented mycelia of the medicinal fungus *Xylaria nigripes*. Helv. Chim. Acta, 99: 83–89.

Yang, L. and Huaan, W. (2004). Cultivation and polysaccharide extraction of *Xylaria polymorpha*. Mycosystema, 23(4): 536–547.

Yang, N., Kong, F., Ma, Q., Xie, Q., Luo, D. et al. (2017a). Chemical constituents from the cultures of fungus *Xylaria polymorpha*. Chinese J. Org. Chem., 37(4): 1033–1039.

Yang, N.N., Kong, F.D., Ma, Q.Y., Huang, S.Z., Luo, D.Q. et al. (2017b). Drimane-type sesquiterpenoids from cultures of the fungus *Xylaria polymorpha*. Phytochem. Lett., 20: 13–16.

Yoon, B.S., Lee, I.G., Jang, Y.U., Kim, Y.S. and Jung, J.Y. (2009). *Xylaria polymorpha* extracts with antifungal activity. Republic of Korea, KR2009052011 A 2009-05-25.

Zhao, Z., Li, Y., Chen, H. et al. (2014). *Xylaria nigripes* mitigates spatial memory impairment induced by rapid eye movement sleep deprivation. Int. J. Clin. Exp. Med., 7: 356–362.

Zheng, W., Zhang, Y.F., Zhong, H.Q., Mai, S.M., Yand, X.H. and Xiang, Y.T. (2016). Wuling capsule for major depressive disorder: A meta-analysis of randomised controlled trials. East Asian Psych., 26: 87–97.

Zhou, X., Gong, Z., Su, Y., Lin, J. and Tang, K. (2009). *Cordyceps* fungi: Natural products, pharmacological functions and developmental products. J. Pharm. Pharmacol., 61(3): 279–291.

Nutraceuticals

11

Mushrooms as Functional Foods

János Vetter

1. INTRODUCTION

Mushrooms, mainly through their fruit bodies, have been part of human nutrition since prehistoric times. During their collection, our ancestors noticed that mushroom fruit bodies in nature possessed a high water content, striking colours and often a pleasant aroma and which could be consumed raw after processing. Perhaps it had also been noticed that dried fruit bodies could later be used to produce good taste in foods. The possible bad taste of the mushrooms or bad experience, harmful or pronounced toxic effect after consumption, shaped the developing knowledge on mushroom utilisation. The so-called ethnomycological knowledge provided and is still providing the basis for mushroom consumption and utilisation. Natural ecological conditions have developed a larger selection of fungi in some places and only a smaller selection in others. A large but uncertain time and space selection of wild mushrooms dates back to the end of the 20th century, but especially at the beginning of the 21st century it was supplemented by the rapidly developing mushroom industries. Nowadays, more and more wild mushrooms have become domesticated and mushroom cultivation as a local food production is becoming more and more popular in different countries. Today, the mushroom industry of the world stands on three 'legs' as three species of mushrooms (*Agaricus bisporus* = button mushroom or champignon, *Pleurotus ostreatus* = oyster mushroom and *Lentinula edodes* = shiitake) account for the majority of cultivation, although other mushrooms are growing and domestication of new species is ongoing.

Department of Botany, University of Veterinary Science, Budapest, Hungary.
Email: Vetter.Janos@univet.hu

The biological value of mushrooms is also indicated by the names that mushrooms have recently earned, such as 'super foods' or 'functional foods'. We can legitimately apply the term 'functional food' to mushrooms as their general composition and properties (minerals and proteins, dietary fibre, polysaccharides, glucans and other biologically active components) make them suitable for consumption, maintain human health and useful in industrial applications (Blumfield et al., 2020; Kozarski et al., 2020; Waktola and Temesgen, 2020). It is possible that the mushroom food product could be enriched with other ingredients, such as vitamin D content of champignon slices by irradiation (Szabó et al., 2012) or by growing a fungus enriched in mineral elements (e.g., Se and Li) (da Silva et al., 2012; de Assuncao et al., 2012; Kora, 2020). The aim of this chapter is to evaluate the biochemical composition of various mushrooms and their most important properties in consumption or application as functional foods.

2. Inorganic Components

2.1 *Water*

The fruit bodies of fresh mushrooms are biological objects with a very high water content, on an average 85–90 per cent, of water. This may be even higher, for example, in inky mushrooms (*Coprinus* species) but for some fruiting body types of tinder (*Fomes fomentarius* or *Ganoderma lucidum*), it is only around 50–60 per cent. The mushroom with a high water content loses a lot of water quickly after harvesting, or one can produce air-dried mushroom preparations by targeted drying. When measuring the amount of a chemical component in a mushroom, it is evaluated to refer to the stable dried state (dry matter, DM), from which it can then be counted back to the unit of fresh mushroom. Let us not forget that the two comparison modes are very different; the difference is 10 times at 90 per cent water content. The water in fungi is an essential solvent, a medium for biochemical processes essential at all stages of development (spore germination, mycelial growth and fruiting).

2.2 *Ash*

If the organic matter of air-dried fungi is removed by incineration, the remaining material, the so-called crude ash content, is a parameter of the total mineral content. The crude ash content of mushroom fruit bodies is usually between 4–19 per cent (based on DM). This relatively easy-to-measure parameter is influenced by many factors, including the mineral content of fungal habitats and substrates, the nutrition mode of the fungus (for example, wood-destroying species have already lower mineral content, as the woody materials).

The crude ash content of some wild-growing mushrooms is documented in Table 1, and the ash content of some cultivated species is shown in Table 2. A comparison of the two tables suggests that samples of wild mushrooms are characterised by higher while the samples of cultivated species are characterised by lower ash contents. The phenomenon is explained by the greater variability of the mineral elements in samples from nature.

Table 1. Average crude ash content of common wild mushrooms from Hungary (Arithmetical Mean ± SD) (* Different samples collected from different habitats and time; Vetter, 1993a) (** Dry Matter).

Wild mushrooms*	Number of independent samples	Crude ash (% DM**)
Armillaria mellea (Honey fungus)	6	15.15 ± 1.44
Boletus edulis (Porcino)	10	7.45 ± 0.68
Cantharellus cibarius (Golden cantharelle)	8	19.96 ± 4.60
Clitocybe nebularis (Clouded funnel)	6	10.89 ± 1.09
Craterellus cornucopioides (Black trumpet)	6	16.90 ± 8.10
Leccinum scabrum (Rough-stemmed bolete)	9	11.53 ± 1.22
Marasmius oreades (Fairy ring mushroom)	3	11.71 ± 0.51
Xerocomus subtomentosus (Soude bolete)	7	10.54 ± 0.15

Table 2. Average ash content of cultivated mushrooms (Lelley and Vetter, 2005).

Mushroom varieties	Crude ash (% DM)
Agaricus bisporus (Sylvan A-15)	11.86 ± 0.06
Agaricus bisporus (Sylvan 608)	11.95 ± 0.09
Agaricus bisporus (Le Lion C-9)	12.23 ± 0.02
Agaricus bisporus (Le Lion C-9) (open cap)	10.53 ± 0.04
Pleurotus ostreatus (Somycel, HK-35)	7.60 ± 0.62
Pleurotus ostreatus (Amycel 3015)	9.45 ± 0.07
Pleurotus eryngii	10.05 ± 0.01
Lentinula edodes (Sylvan 4087)	7.41 ± 0.01

Another interesting question can be answered by our previously unpublished data set (Table 3), which compares the crude ash contents of the fruit body fractions of several cultivated species. Based on whole fruit-body samples, three varieties of button mushrooms were found to have crude ash ranging between 8.71–10.64 per cent, two varieties of oyster mushroom possess 13 per cent and one variety has only 5.38 per cent, while shiitake varieties range between 12.1–17.6 per cent (Tables 3–4). There are significant differences not only between species, but there may be significant, consistent differences between certain morphological parts of the fruit body.

The lowest ash content was formed in each case in the stipe and the highest values were in the peel of the cap and in the inner cap. All this can also be quantified by forming the ratio of the ash content of the inner cap and the stipe; the obtained data are: 1.40, 1.18 and 1.15 (for *Agaricus bisporus* varieties); 1.39, 0.74 and 1.30 (for *Pleurotus ostreatus* varieties); and 1.53, 1.03; 0.95 (for *Lentinula edodes* varieties). The resulting trend indicates that the pileus (cap) usually contains 20–30 per cent more raw ash than stipe. The raw ash value of the gills is slightly lower than that of the pileus, but is clearly higher than stipe. The fruit bodies are not homogeneous in the raw ash content as the data in Tables 3–4 clearly indicate, but follow a characteristic distribution (not previously known in the literature). The pileus (including the peel of the cap, representing the outer protection zone), the

Table 3. Ash content in different parts of most important cultivated *Agaricus bisporus* and *Pleurotus ostreatus* varieties (J. Vetter, unpublished data).

Mushroom variety	Part of fruit body	Crude ash content (% DM)	Mushroom and variety	Part of fruit body	Crude ash content (% DM)
Agaricus bisporus 'S-800'	Peel	9.48 ± 0.05	*Pleurotus ostreatus* 'G-24'	Peel	8.19 ± 0.05
	Inner cap	10.66 ± 0.14		Inner cap	6.58 ± 0.01
	Gills	9.94 ± 0.11		Gills	7.66 ± 0.02
	Stipe	7.58 ± 0.01		Stipe	4.73 ± 0.15
	Whole fruit body	9.40 ± 0.05		Whole fruit body	5.38 ± 0,21
	Primordium	9.70 ± 0.10		Primordium	7.89 ± 0.12
Agaricus bisporus 'Brown'	Peel	11.29 ± 0.08	*Pleurotus ostreatus* 'HD-35'	Peel	14.16 ± 0.20
	Inner cap	12.6 ± 1.22		Inner cap	10.24 ± 0.34
	Gills	10.18 ± 0.58		Gills	12.75 ± 0.05
	Stipe	10.64 ± 0.03		Stipe	13.80 ± 0.80
	Whole fruit body	11.40 ± 0.06		Whole fruit body	13.04 ± 0.03
	Primordium	11.77 ± 0.09		Primordium	13.54 ± 1.19
Agaricus bisporus 'A-15'	Peel	9.30 ± 0.001	*Pleurotus ostreatus* 'H-7'	Peel	17.86 ± 0.57
	Inner cap	10.43 ± 0.18		Inner cap	11.78 ± 1.02
	Gills	10.08 ± 0.55		Gills	14.10 ± 0.19
	Stipe	9.02 ± 0.07		Stipe	9.02 ± 0.24
	Whole fruit body	8.71 ± 0.06		Whole fruit body	13.09 ± 0.33
	Primordium	9.26 ± 0.24		Primordium	13.07 ± 0.21

inner cap and the whole fruit bodies have a higher ash, i.e., total mineral content than does the stipes. A likely explanation for this fact may be the increased protection of the site of spore-producing processes. Further data in Table 3 also suggest that the ash content of very young, developing fruit bodies (primordia) calculated for DM is almost the same (the difference is only 0.2–0.6 per cent) as the ash content of the mature fruit body. The tendency of loss of minerals while cooking the fruit bodies cannot be ruled out. If an edible mushroom has to be cooked, we should select the method that leads to least loss or no loss of minerals.

2.3 *Macroelements*

Concentration ranges of the most abundant macronutrients are shown in Table 5, with examples of some important edible mushrooms. Potassium is considered to be the main mineral constituent, varying between 30,000–33,000 mg/kg DM. The fruiting

Table 4. Ash content in whole fruit bodies and different parts of fruit bodies of *Lentinula edodes* varieties.

	Part of fruit body	Crude ash (% DM)
Lentinula edodes (40–80)	Peel	14.43 ± 0,01
	Inner cap	12.92 ± 0.14
	Gills	12.31 ± 0.09
	Stipe	8.43 ± 0.10
	Whole fruit bodies	12.10 ± 0.18
	Primordium	14.53 ± 0.08
Lentinula edodes (KST-67)	Peel	15.12 ± 0.15
	Inner cap	13.19 ± 0.22
	Gills	14.55 ± 0.30
	Stipe	12.8 ± 0.13
	Whole fruit bodies	13.80 ± 0.10
	Primordium	14.19 ± 0.10
Lentinula edodes (KST-76)	Peel	18.58 ± 0.12
	Inner cap	18.93 ± 0.38
	Gills	20.02 ± 0.29
	Stipe	19.56 ± 0.10
	Whole fruit bodies	17.57 ± 0.11
	Primordium	16.86 ± 0.15

Table 5. Macroelements in selected cultivated mushrooms (Vetter, 2010a) (* Based Györfi et al., 2010; ** based on Krüzselyi et al., 2016).

Mushrooms	Potassium (mg/kg DM)	Phosphorus (mg/kg DM)	Calcium (mg/kg DM)	Magnesium (mg/kg DM)
Agaricus bisporus	43–51000	9–11000	700–800	1200–1500
*Agaricus subrufescens**	27–32800	6–12000	570–2600	700–1500
Cyclocybe aegerita (*Agrocybe aegerita*)	29000	10300	1130	1317
Ganoderma lucidum	4–5000	1480	1800–5000	500
Grifola frondosa	23–33000	5400–6900	600–1000	1000–1230
Lentinula edodes (Shiitake)	13–25000	6000	800	1600
*Pleurotus eryngii***	1600–3200	6500–9900	760–1050	700–1595
Pleurotus ostreatus	29–34000	6–7000	500–700	1300–1600
Average of wild mushrooms (n = 625)	34865	7798	1464	1443

bodies of *Ganoderma lucidum* have only 4000–5000 mg/kg DM, while *Pleurotus eryngii* possesses 1600–3200 mg/kg DM. The bottom row of Table 5 shows the comparative average values of our previous studies of 625 mushroom samples. The biological role of high and balanced potassium content is multifaceted, as potassium is an activator of many fungal enzymes and is a key regulator of osmotic conditions in the hyphae, mycelium and the fruit body. In mushrooms, potassium levels are high and differences between taxa are quite narrow. The optimal nature of the above-mentioned two functions is ensured by the mentioned potassium interval (with some exception, lower K-levels are seen in wood-destroying mushrooms or slightly higher concentrations in Amanitaceae, that is, about 42–45,000 mg/kg DM) (Vetter, 2005). It follows from the significant concentrations of potassium generally experienced that mushroom diets are also among the more potassium-rich foods for human nutrition. In the group of wood-destroying fungi, significantly lower potassium levels were found as in the case of *Stereum* species, 2600–6000 mg/kg DM or in *Trametes* species, 1400–9000 mg/kg DM.

Phosphorus concentrations in mushrooms show a much wider concentration range. Wild mushrooms have an average phosphorus content of 7800 mg/kg DM (Table 5), but many species, genera, or even distinct fungal groups of different nutrition types have very different P contents. *Agaricus, Clitocybe, Laccaria* species have higher P levels, for example, a sample of *Lepista irina* has 25,254 mg/kg DM and *Lepista nuda* has 30,000 mg/kg DM. Interestingly, the P contents of mycorrhizal *Amanita* and *Boletus* are below average (*Amanita* spp.: 6800 mg/kg DM; *Boletus* spp.: 6065 mg/kg DM), while the P contents of wood-destroying taxa are very low (1000–3000 mg/kg DM) (Vetter, 2003a). Our Table 5 for cultivated species also reflects this as well: *Lentinula edodes* has 6000 mg/kg DM; *Pleurotus ostreatus* has 6–7000 mg/kg DM, but *Ganoderma lucidum* contains only 1480 mg/kg DM. Our previous study documented important P contents in pileus and stipes of the three most important cultivated species (Vetter et al., 2005). According to the data in Table 6, the two wood-destroying fungi (*Pleurotus ostreatus* and *Lentinula edodes*) possess low P contents, but the proportions of P in the pileus and stipes are very different. In the group of wood-destroying species, the P level of pileus is twice as high than that in stipes. For the saprotrophic *Agaricus bisporus* varieties, the P level of pileus is 33 per cent higher than in the stipes. The biological role of phosphorus in fungi is very diverse, as indicated by the large number and variety of phosphorus-containing compounds, like nucleic acids, triphosphates (ATP, GTP, CTP, and UTP) and phospholipids. Mushrooms are thus important not only as a source of potassium but also as a source of phosphorus.

Mushrooms have lower levels of calcium and magnesium, with an average of 1464 mg/kg DM in the wild-growing mushrooms and with an equivalent amount of magnesium (1443 mg/kg DM) (Table 5). Cultivated species have a lower Ca content (1000 mg/kg DM), although the Ca content of *Ganoderma lucidum* is several times higher. Regarding the distribution of calcium, it is interesting that Ca_{pileus}/Ca_{stipe} proportions are smaller than 1.0 for the three most important mushrooms; the Ca

Table 6. Phosphorus content (mg/kg DM) in Pileus (P) and Stipe (S) of the most important cultivated mushrooms (Vetter et al., 2005).

	Pileus (P)	Stipe (S)	Ratio (P/S)
Agaricus bisporus varieties (n = 18)	12151 ± 2119	9129 ± 1384	1.33
Pleurotus ostreatus varieties (n = 5)	6983 ± 2985	2950 ± 1849	2.37
Lentinula edodes varieties (n = 5)	7412 ± 1622	3772 ± 1297	1.97

content of stipe is higher than that of the pileus for *Agaricus bisporus*. The situation is similar for the increasingly important varieties of *Agaricus subrufescens* (Györfi et al., 2010). The amount of magnesium in wild mushrooms is on an average very close to the amounts of Ca, but is higher in cultivated species than the average Mg level in wild-growing ones. In terms of the fruit body distribution, Mg_{pileus}/Mg_{stipe} proportion is 1.30 for *Agaricus bisporus*, 1.47 for *Pleurotus ostreatus*, but only 0.94 for *Lentinula edodes*.

The average occurrence of macroelements is characteristic of mushroom fruit bodies. The four elements—K, P, Ca and Mg—account for 97–98 per cent (Fig. 1) while all other elements provide only 2–3 per cent of the minerals. This feature does not diminish the importance of the other elements, but it indicates the peculiarity of the mineral element distribution of fungi.

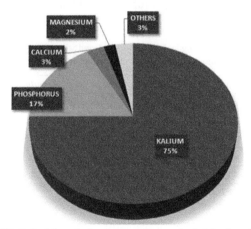

Fig. 1. Distribution of K, P, Ca, Mg and other mineral elements in fruit bodies of cultivated mushrooms (Vetter, 2010a).

2.4 Microelements

Quantities of most important microelements in pileus and stipes of some cultivated mushrooms are given in Table 5. Sodium is most essential because it is a component of K/Na pumps in cell membranes. In macrofungi, such a role is absent; therefore, the sodium concentration is low. The average in wild mushrooms is around 325 mg/kg DM. According to our studies, sodium in pileus and stipes, is about 700–942 mg/kg DM in button mushrooms and 450 mg/kg DM in the shiitake, while lowest in pileus and stipe of oyster mushrooms (260–270 mg/kg DM). In today's diets of plant origin, but even more so of animal origin, the sodium level may be higher (even tens of thousands of mg/kg DM). Apart from the sodium added during cooking or preserving fruit bodies, mushrooms are generally low-Na foods. Consumption of 100 g of fresh oyster mushrooms provides only 2.6–2.7 mg of sodium. Considering the high potassium content of fungi (*see* Table 5), the K/Na ratio reaches or significantly exceeds 100. This proportion (i.e., high K and very low Na) is highly advantageous, according to the principles of modern medicine. The Na content of edible mushrooms ranges between 160–390 mg/kg DM (Vetter, 2003b), which is found to be relatively constant, not being too dependent on nutrition type and the taxonomic position of mushrooms. Sodium-accumulating mushroom species have not been described so far.

2.4.1 Copper

The average copper content of wild mushrooms is 58.6 mg/kg DM and the values measured in pileus and stipes of cultivated mushrooms are significantly lower (Table 7). According to the copper content, the following order is seen in three mushrooms: *Agaricus bisporus* > *Pleurotus ostreatus* > *Lentinula edodes*. The pileus parts (except of oyster mushroom) appear to contain more copper than stipes. The Cu_{pileus}/Cu_{stipe} ratios are 1.35; 1.63 and 0.93 for *Agaricus bisporus*, *Lentinula edodes* and *Pleurotus ostreatus*, respectively. In the wild-growing mushrooms, a significant bioaccumulation of copper (up to 100–200 mg/kg DM) is a characteristic of *Macrolepiota* spp.

Table 7. Contents of microelements (mg/kg DM) in Pileus (P) and Stipe (S) of the most important cultivated mushrooms (Vetter et al., 2005) and the averages of element contents in wild growing mushrooms.

	Agaricus bisporus			*Pleurotus ostreatus*			*Lentinula edodes*			**Average in fruit bodies**
	Pileus	Stipe	P/S	Pileus	Stipe	P/S	Pileus	Stipe	P/S	
Boron	22.6	20.8	1.09	2.85	3.50	0.80	26.2	12.5	2.09	16.4
Copper	39.5	29.2	1.35	18.7	20.2	0.93	13.2	8.1	1.63	58.6
Manganese	8.30	6.47	1.28	9.62	6.23	1.54	21.4	16.9	1.27	43.2
Sodium	699	942	0.74	269	265	1.02	439	454	0.97	325
Zinc	64.7	55.7	1.16	76.6	36.5	2.10	82.2	49.9	1.65	115

2.4.2 Zinc

From the nutritional point of view, zinc is absolutely necessary. The average zinc level of wild mushrooms is 115 mg/kg DM (Table 7), while some taxa (e.g., *Russula atropurpurea*) may contain up to 10 times of this level. Examples of cultivated mushrooms show that the amount of zinc reaches half to two-thirds of the average in wild mushrooms. The pileus contains more zinc than stipes; this ratio is 2.1, for oyster mushrooms. As zinc is involved in enzyme activators, it is essential for metabolism of consumers.

2.4.3 Manganese

The manganese content of fruit bodies is 17–21 mg/kg DM for *Lentinula edodes* while in all parts of other cultivated mushrooms it will be below 10 mg/kg DM (the average for wild mushrooms is 43.2 mg/kg DM) (Table 7). In white decaying mushrooms, the manganese ions play a major in role in lignin decomposition. In the woody substrate such activity of fungi shows an increased Mn content.

2.4.4 Boron

In the case of boron, compared to the average in wild mushrooms (16.4 mg/kg DM), cultivated mushrooms have a higher quantity (20–23 mg/kg DM) while oyster mushrooms have a significantly lower (2.8–3.5 mg/kg DM) boron content (Table 7).

There are some mushrooms which accumulate extreme boron contents (e.g., *Mycena pura*, 300–600 mg/kg DM; *M. rosea*, 75–200 mg/kg DM) and where hundreds of mg/kg DM values indicate that it is due to bioaccumulation, although the nature of boron-binding molecule is not yet known.

2.4.5 Selenium

In most mushroom species, the selenium content is higher than the detection limit, but significant accumulation can be demonstrated in some taxa. Selenium level is 3–5 mg/kg DM in some *Amanita* (*A. pantherina*, *A. strobiliformis*, *A. muscaria*) and in *Lepista* (*L. irina*, *L. luscina*, *L. flaccida*) species, while it is higher in *Macrolepiota* species (*M. rhacodes*, 12 mg/kg; *M. mastoidea*, 10.4 mg/kg; *M. procera*, 3–8 mg/kg DM). Significant bioaccumulation was observed in *Boletus edulis* (10–30 mg/kg DM). Our previous studies (Vetter and Lelley, 2004) revealed that some varieties of cultivated button mushrooms (marked '158', 'K-23', 'K-7') contain 3–5 mg/kg DM selenium, while its content in 'D-13' variety is only 0.4–0.7 mg/kg DM. The selenium level of oyster mushrooms or even more than shiitake is only a few tenths of 1 mg/kg DM. Selenium is one of the biologically active microelements, but the ecosystems in Europe are fundamentally and inherently deficient and such deficiency was seen in foods too. Under these conditions, consumption of 100 g of a given mushroom variety means intake of circa 50 µg selenium and an uptake of a similar amount of wild-growing *Boletus* species means about 250 µg.

2.4.6 Iodine

We have very little data on the iodine content of fungi. According to earlier investigations (Vetter, 2010b), the average iodine level of the 48 tested wild mushrooms was 0.28 mg/kg DM, with the highest value found in *Macrolepiota procera* (0.92 mg/kg DM) and lowest in *Fistulina hepatica* (0.012 mg/kg DM). Button mushrooms, oyster mushrooms and shiitake have 0.17, 0.18 and 0.15 mg/kg DM iodine, respectively, which does not differ significantly from the average in wild mushrooms. Comparing the mushrooms, the most important factor is the substrate and the mode of nutrition. The iodine content of wood-destroying mushrooms is significantly lower as compared to other mushrooms.

2.5 Bioaccumulation of Minerals

Bioaccumulation is an important phenomenon related to the mineral uptake and constituents of mushrooms. In the 60s and 70s of the 20th century, it was noticed that some taxa of ascomycetous and basidiomycetous fungi possess strikingly high concentrations of some elements. This phenomenon was later renamed 'bioaccumulation', which is due to the uptake and storage of an element through a special taxon-dependent compound encoded by genetic factors. A classic example is the 100–150 mg/kg DM vanadium content in the fruit body of *Amanita muscaria*. Vanadium is a trace element present in very small amounts, for instance, fruit bodies of mushrooms contain an average not more than 0.1–0.2 mg/kg DM. However, *A. muscaria* is able to synthesise amavadin, a compound that absorbs vanadium with high affinity and 'binds' it; also one V atom is incorporated into its structure

(Fig. 2). It is an octa-coordinated vanadium (IV) complex, which is bonded to two ligands (derived from N-hydroxyimino-2,2'-dipropionic acid). The ligands coordinate through the nitrogen and three oxygen centres. Such a taxon-dependent phenomenon has been known for several other elements and the binding compounds which are summarised in Table 8. Another question persists and that is the nutritional and possibly toxicological effects of element enrichment as a result of bioaccumulation. In fact, for some elements and mushrooms, it may be absolutely necessary to control the amount of that element (both for growing substrates and for fruit bodies) (*see* Table 8).

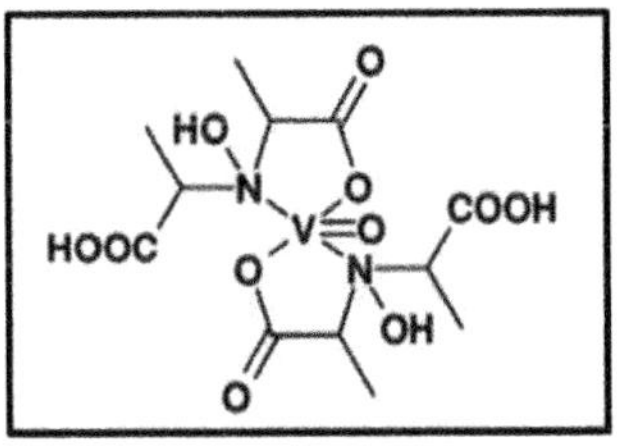

Fig. 2. Chemical structure of the vanadium-binding amavadine molecule.

Table 8. Bioaccumulation of selected elements, affected taxonomic category and the chemical nature of the binder substance.

Elements	Affected taxonomic category	Binder substance
Arsenic (As)	*Agaricus, Laccaria* and *Sarcoscypha* genera	Smaller molecule of ionic character
Boron (B)	*Lycoperdon* spp. and *Mycena pura*	?
Cadmium (Cd)	*Agaricus* spp.	Cd-mycophosphatide (a glucoproteid)
Cupper (Cu)	*Macrolepiota, Agaricus* and *Grifola* spp.	Probable peptides
Mercury (Hg)	*Agaricus, Lepista* and *Macrolepiota* spp.	Proteins? Peptides?
Selenium (Se)	*Boletus* spp. and *Agaricus bisporus* varieties	?
Vanadium (V)	*Amanita muscaria*, 2–3 other *Amanita* spp.	Amavadin
Zinc (Zn)	*Russula atropurpurea*	?

Great care should be taken about the cadmium content in button mushrooms and their varieties (due to the ability to synthesise the Cd-mycophosphatide molecule) or even the mercury levels.

3. Organic Components

The vast majority of the air-dried mushroom fruit bodies are composed of various organic ingredients. Using the crude ash data presented earlier (Tables 1–4), it can be well demonstrated that the 100 minus crude ash content's numerical value can characterise the total amount of organic matter. This usually ranges between 80–90 per cent. The following subheadings review the most important organic components.

3.1 Proteins

The protein content in fungi, like other organisms, is most often measured by the total nitrogen content. The classical analytical method multiplies the N content by a factor of 6.25, but this has proved to be excessive for fungi due to the presence of nitrogenous and non-proteinaceous substances. The most important factor (chemical compound) is chitin, i.e., the content of non-protein nitrogen (NPN) is considerably high. Sincerom the late 1980s, a factor reduced by 30 per cent (4.38), which indicates a much more realistic situation. Sometimes it is confusing but we still have to check some data in the literature or correct these data as corrective measures.

The crude protein content can be influenced by several factors, such as genetic ones (taxonomic position), nutritional factors and the developmental state of the fungus. But the pileus and stipes have different protein levels. The literature reported a large, almost opaque mass of data on crude protein contents of mushrooms. However, the correct comparison is hampered by several factors: unknown habitats of the fungi, unknown developmental phases of the fruit bodies, preparation (sometimes determination) of samples and the variety of analytical methods employed.

Table 9 demonstrates the crude protein contents of some of the wild edible mushrooms. Intra-species variability of crude protein content is very high; for example, 18.6 and 38.9 per cent DM were measured in the common species, *Agaricus campestris*, but the data for *Boletus edulis* were 15–53 per cent, for *Lepista nuda* 19.8–59.4 per cent or for *Macrolepiota procera*, the data were 7.6–24.2 per cent, thus showing high deviations. The crude protein content of the most common edible and mainly basidiomycetous mushrooms is an average ranging between 25–35 per cent DM, which corresponds to concentrations of 2.5–3.5 per cent based on fresh mass (FM). Table 10 shows the crude protein content of cultivated mushrooms with high deviations. Significant changes in the crude protein content of *Agaricus bisporus* or *Lentinula edodes* was 14.0–36.3 and 4.4–20.5 per cent DM, respectively is influenced not only by the very different cultivation conditions, but also by the role of the increasing variety selection. Protein content of varieties is genetically controlled and by their breeding. The wood-destroying mushrooms like *Ganoderma lucidum*, *Auricularia auricula-judae*, *Pleurotus* species and *Lentinula edodes* have lower crude protein levels than that of *Agaricus* species or *Volvariella volvacea*.

3.1.1 Pileus and Stipes

The crude protein contents measured in pileus and stipe show a characteristic distribution in the three most important cultivated mushrooms. Based on previous comparison (Vetter, 2000a), the crude protein contents of the pileus of *Agaricus bisporus* were 25.33 per cent DM (for 'K-23' variety), 28.8 per cent DM (for 'K-7' variety), and 24.1 per cent DM (for '158' variety), respectively. Similar values for the stipes were 24.15 per cent DM (for 'K-23'), 24.32% DM (for 'K-7') and 20.03 per cent DM (for '158'). In all *Agaricus* varieties, the crude protein level of the pileus is higher (often significantly) than that of the stipe. Thus the crude protein in pileus/ stipe ratios stands between 1.04 and 1.20. There is a similar but stronger correlation between the pileus and the stipe of oyster mushroom (24.77/7.42 = 2.34) and of *Lentinula edodes* (24.77/16.64 = 1.48). In these two mushrooms, the crude protein content shifted strongly in favour of pileus.

Table 9. Crude protein content of selected wild mushrooms.

	Crude protein (% DM)	Reference
Agaricus campestris	18.6	Pereira et al., 2012
	38.9	Beluhan and Ranogajec, 2011
Armillaria mellea	7.6	Ayaz et al., 2011a
	17.3	Akata et al., 2012
Boletus edulis	16.4	Fernandes et al., 2014
	22.8	Ayaz et al., 2011a
	26.5	Ouzouni and Riganakos, 2007
	36.9	Beluhan and Ranogajec, 2011
Cantharellus cibarius	15.1	Ouzouni et al., 2009
	30.9	Beluhan and Ranogajec, 2011
	53.7	Barros et al., 2008
Coprinus comatus	11.8	Stojkovic et al., 2013
	15.7	Vaz et al., 2011
	29.5	Akata et al., 2012
Lepista nuda	19.8	Ouzouni and Riganakos, 2007
	24.1	Ouzouni et al., 2009
	59.4	Barros et al., 2008
Macrolepiota procera	7.6	Barros et al., 2007b
	18.4	Ayaz et al., 2011a
	19.0	Fernandes et al., 2014
	24.2	Beluhan and Ranogajec, 2011
Pleurotus ostreatus	13.2	Akata et al., 2012
	24.9	Beluhan and Ranogajec, 2011
Tuber aestivum	19.1	Krüzselyi and Vetter, 2014

Table 10. Crude protein contents of selected cultivated mushrooms (Ulzijargal and Mau, 2011; Hung and Nhi, 2012; Kalač, 2013; Pardo-Gimenez et al., 2013; Krüzselyi and Vetter, 2014; Krüzselyi et al., 2016).

	Crude protein (% DM)
Agaricus bisporus	14.0–36.3
Agaricus subrufescens	26.2–39.3
Agrocybe cylindracea	16.5–22.2
Auricularia auricula-judae	5.7–15.5
Coprinus comatus	10.9
Ganoderma lucidum	13.3
Flammulina velutipes	3.9–26.6
Lentinula edodes	4.4–20.5
Pleurotus eryngii	11.0–22.0
Pleurotus ostreatus	7.0–23.8
Tuber aestivum	17–20
Volvariella volvacea	36.5

3.1.2 Effect of Development

During the development of fruit bodies, the characteristic changes in crude protein content of oyster mushrooms were observed in previous studies (Vetter and Rimóczi, 1993). According to the data, in the second developmental phase (pileus size 5–8 cm), the highest protein content was measured in pileus (25.4 per cent DM). This value is only 8.25 per cent in stipe, while the proportion of protein in pileus and stipe is 3.07. Thus, the change in crude protein content follows a maximum curve and the ratio of crude protein contents in pileus and stipes alters similarly.

3.1.3 Amino Acids

The composition of amino acids is the determinant factor in the biological value of mushroom proteins. In general, their amino acid composition is closer to that of animal proteins and thus the latter have a higher biological value than plant proteins. Our data on some older varieties of button mushroom, like *Agaricus bisporus*, are given in Table 11 which showed the total amino acid content at 25.85 per cent DM (average of three varieties) (Vetter, 1993b). The proportion of essential amino acids is 32.9 per cent on an average, i.e., one-third of the total amino acid content. Occurrence and ratio of sulphur-containing and aromatic amino acids are relatively low. The amino acid composition of 11 common edible mushrooms from the Black Sea region of Turkey was studied by Ayaz et al. (2011a). The average total amino acid content of the species was 148 mg/g DM, the average essential amino acid content was 58.2 mg/g DM (39.6 per cent of the total amino acid content) and the non-essential amino acid level was 90.0 mg/g DM (i.e., 60.4 per cent of the total amino acid content). The combined amount of the two aromatic amino acids (Phe + Tyr) was 10.9 mg/g DM, the sum of the sulphur-containing ones (Met + Cys) being 5.5 mg/g DM. In general, glutamine, aspartic acid and leucine have the highest concentrations (glutamic acid, between 10.9 and 37.6 mg/kg DM, with an average of 14.7 per cent of the total amino acid content). Studies by many authors (e.g., Kim et al., 2009) confirm this observation. Rare data reported on mushrooms from habitats of Korea show that they are rich in alanine, such as *Agaricus bisporus, A. blazei* and *Pleurotus eryngii* (Kim et al., 2009). In other studies, leucine (*Lactarius quieticolor*), methionine (*Gomphus flococcus* and *Russula integra*) or even lysine (*Russula brevispora*) were found in significant amounts in some Indian mushrooms (Agrahar-Murugkar and Subbulakahmi, 2005). Most of the amino acids were present in *Agaricus arvensis*

Table 11. Total essential, sulphur-containing and aromatic amino acid composition of three *Agaricus bisporus* varieties (% DM) (Vetter, 2010a) (Per cent total amino acid content in parenthesis).

	Agaricus bisporus 'D-15'	*Agaricus bisporus* 'Pc-1'	*Agaricus bisporus* 'Pc-17'	Average
Total amino acids	21.92 (100)	27.33 (100)	26.82 (100)	25.86 (100)
Essential amino acids	8.37 (38)	8.87 (32.4)	8.31 (30.9)	8.51 (32.9)
Sulphur-containing amino acids	0.40 (1.8)	0.30 (1.0)	0.40 (1.4)	0.36 (1.4)
Aromatic amino acids	2.08 (9.4)	2.25 (8.2)	1.94 (7.2)	2.09 (8.0)

(essential amino acids, EAA: 86.8 mg/g DM; non-essential amino acids, NEAA: 143 mg/g DM), while the lowest content was found in *Cantharellus cibarius* (EAA, 39.7 mg/g DM; NEAA, 56.1 mg/g DM). The biological value of mushrooms collected and consumed in the given region of Turkey is significant and as their cultivation is still largely impossible today, it is advisable to pay more attention to their collection, distribution and preservation. A Korean working group (Lee et al., 2011) studied an undiscovered species of the genus *Agrocybe*, namely *A. chanxingu*. Its total amino acid content was 8.01 per cent DM (EAA, 2.70 per cent DM; NEAA, 5.31 per cent DM). *A. chanxingu* is better than the amino acid distributions in *Pleurotus ostreatus* or in *Flammulina velutipes*.

3.1.3.1 Free Amino Acids

The proportion of free amino acids present in mushrooms is generally very low. The total free amino acid content of wild mushrooms varies between 0.15–7.2 per cent DM (Kalač, 2013). In Portuguese studies, Ribeiro et al. (2008) found the lowest free amino acid content in *Tricholomopsis rutilans* (0.46 per cent DM) and the highest in *Boletus edulis* (2.2 per cent DM). The average free amino acid content of five most important cultivated species is 12.2 per cent DM, the leading contents being glutamic acid, ornithine and alanine (Kim et al., 2009).

3.1.4 Differences between Flushes of Cultivation

There are interesting differences in the composition of flushes during the cultivation of button mushroom (*Agaricus bisporus*) (Vetter, 2000a). The crude protein contents in pileus of the first, second and third flushes were: 25.3 per cent DM, 22.27 per cent DM and 22.8 per cent DM, respectively. The highest in first flush was followed by two flushes of significantly lower crude protein contents. The explanation for this phenomenon may be that the requirements of the second (and of additional) flushes were not satisfied totally by the substrate, so the crude protein contents of the fruiting bodies slightly declined.

3.1.5 Biological Evaluations

Biological evaluations of mushroom proteins can be approximated in a complex way by taking into account several aspects and chemical facts. An important parameter in the biological evaluation of a given nutrient is its solubility. American biochemist Thomas B. Osborne reported firstly (more than 100 years ago in 1905) his method based on fractionation of four major groups of proteins. This method or its improved versions are still in use today. The Osborne protein groups are the water-soluble albumins, the globulins soluble in dilute saline, the alcohol-soluble prolamines and the glutelins obtainable by acidic and alkaline extractions. The residues after extraction of the above fractions form the non-protein nitrogen (NPN) group. This grouping is important because the decomposition (digestion) and subsequent absorption of food proteins show a logical relationship with their solubility, which is easier or heavier in nature. The higher the combined ratio of albumins and globulins the better is the expected bioavailability of protein. According to our recent studies (Krüzselyi et al., 2016), the sum of albumin and globulin protein groups is approximately 80 per cent for *Pleurotus eryngii* and 60 per cent for *Pleurotus ostreatus*. According to the data obtained from six Hungarian samples of *Tuber aestivum* (summer truffle), the average amount of albumins is only 40.9 per cent and that of globulins it is 5.9 per cent (summarised 46.8 per cent), of total protein content (Krüzselyi and Vetter, 2014).

3.1.6 Other Evaluation Options

There are several approaches and methodologies for the qualitative evaluation of fungal proteins and for their comparison with other proteins. For example, it is possible to compare the utilisation parameters, such as protein efficiency ratio (PER) or net protein utilisation (NPU) in animal feeding experiments (as in the work of Darbour and Takruri, 2002). However, the extrapolation of the obtained values for humans is questionable, even debatable, in several respects. There are *in vitro* digestion methods that break down proteins at a given temperature over a period of time, under the influence of added digestive enzymes (pepsin and trypsin or enzyme mixtures) and then the digestion rate by determining the amount of residual (undegraded) proteins is calculated. It is clear that the result obtained can at the best roughly characterise the *in vivo* digestibility. It is worth considering the results of Cuptapun et al. (2010), who examined the *in vitro* protein digestibility (IVPD) and *in vivo* digestibility of some cultivated fungal species. The IVPD values of 62–63 per cent for *Pleurotus sajor-caju*, *Pleurotus ostreatus* and *Lentinula edodes* were 43.4 per cent, 47.2 per cent and 52.16 per cent *in vivo* digestibilities, respectively. Thus, *in vitro* values were 10–20 per cent higher than the *in vivo* results. It should also be noted that for the reference casein protein chosen, the *in vitro* protein digestibility was 84 per cent and the *in vivo* protein digestibility was 87 per cent. The *in vivo* digestibility of the best mushroom (*L. edodes*) is 35 per cent lower than the similar parameter of casein.

3.2 Lipids

The crude fat content in fungi is the sum of compounds that can be dissolved by a standard method using organic solvents (this parameter is also called total lipid content). The crude fat content of mushroom fruiting bodies is low, most often 1–3 per cent of the DM (Table 12). As indicated by data, we can rarely find in the literature a fat content higher than 6 per cent (*Lactarius deliciosus*, 8.0 per cent; *Armillaria mellea*, 5.5–6.5 per cent) (Kalač, 2013) and these can only be considered as outstanding sporadic data. One hundred grams of fresh mushrooms contain mostly 0.1–0.3 per cent crude fat content. This low value is also one of the reasons for the very low calorific value of mushroom fruit bodies and in general, of mushroom foods.

Another important issue of the total lipid content of fungi is the qualitative side, i.e., which compounds are typically present in lipid fractions (Bengu, 2020). A logical division is to separate the groups of neutral and polar lipids. Neutral lipids account for an average of 34.4 per cent of crude fat content, while polar ones account for 65.6 per cent of wild mushrooms in Canada (Pedneault et al., 2008), although published values range widely. Wild mushrooms in Turkey contained three most common fatty acids, which were linoleic acid, oleic acid and palmitic acid (Ayaz et al., 2011b).

Table 12. Crude fat, total carbohydrate contents and energy value of the most frequently cultivated mushrooms (Vetter, 2010a) (ND, No Data).

	Crude fat (% DM)	Total carbohydrate (% DM)	Energy value (kcal/100 g DM)
Agaricus bisporus	0.8–2.5	50.9–74.0	303–325
Agaricus subrufescens	0.9–4.0	39–64.0	340–360
Auricularia auricula-judae	0.4–4.5	77–91	97–140
Coprinus comatus	2.0	76.5	368
Cyclocybe cylindracea	3.5–3.6	ND	130–179
Flammulina velutipes	2.9–9.2	56.6–86	467
Ganoderma lucidum	3.0	82.3	ND
Lentinula edodes	1.7–6.3	67–87	386
Pleurotus eryngii	1.45–1.57	70.5–81.4	421
Pleurotus ostreatus	0.5–5.4	51.9–85	416
Tuber aestivum	2.3	48.9	293
Volvariella volvacea	2.2	52.3	ND

The proportions of linoleic, oleic and the saturated palmitic acids in total fat content vary between 5 and 76.5 per cent; 1.3 per cent and 58.1 per cent; 4.6 and 24.7 per cent, respectively. This wide range of data also indicates that there may be significant differences in fatty acid metabolism of different taxa. The more or less leading role of linoleic acid is also related to the fact that this molecule is a precursor to 'fungal alcohol', the 1-octen-3-ol molecule, which is a characteristic aromatic substance of fungi. Unsaturated fatty acids are dominant in the examined wild mushrooms of Turkey (55.7–87.1 per cent of total fatty acid content). Average composition of all tested species was: saturated fatty acids, 15.9 per cent; unsaturated fatty acids, 81.1 per cent; monounsaturated fatty acids, 36.4 per cent, polyunsaturated fatty acids, 44.7 per cent. In the summarised table of Kalač (2016) for the 16 cultivated mushrooms, saturated fatty acids occur in 21.9 per cent, monounsaturated fatty acids in 14.4 per cent, while polyunsaturated fatty acids occur in 62.6 per cent. Thus, in the case of cultivated species, the proportion of polyunsaturated fatty acids is approximately two-thirds of the total fat content. The ratio of physiologically important ω-3 and ω-6 fatty acids in mushrooms is not close to the proportion 1:5 which is considered optimal today (Kalač, 2016).

3.3 *Carbohydrates*

Carbohydrates make up the bulk of dry matter in the mushroom fruit bodies. Determination of the total carbohydrate content is actually based on calculations, the principle of which is to subtract from 100 the sum of moisture, crude protein, crude fat and crude ash contents per 100 units: Carbohydrate = 100 – (moisture in g + crude protein in g + crude fat in g + crude ash in g) (Manzi et al., 2004).

The calculated parameter indicates a very wide group of substances, where various monosaccharides and their derivatives (e.g., sugar alcohols), oligosaccharides (trehalose) and polysaccharides (glycogen, chitin and glucans) are present. A review and grouping of mushroom carbohydrates is presented in Fig. 3.

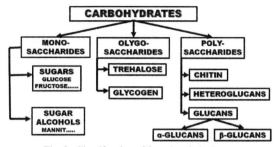

Fig. 3. Classification of fungal carbohydrates.

The total carbohydrate contents of cultivated mushrooms are between 50–91 per cent DM (Table 12). They represent not only a large fraction but also a wide concentration range for a given species. For both *Agaricus* species, the width of concentration interval is 24 per cent, while the values published for *Pleurotus ostreatus* range between 51.9–85 per cent. The highest numerical value (91 per cent) was determined for *Auricularia* species. This is explained by the fact that crude protein, crude fat and ash contents of the mushrooms are low and therefore, the carbohydrate content calculated on the basis of the difference is very high.

3.3.1 Monosaccharides

The sugars present in fungi (mainly C5 or C6) are mainly arabinose, glucose, fructose in free form and can be measured in low or very low concentrations. Another issue is that glucose, for example, is the most important essential substrate for biochemical respiration and energy production. Thus the free glucose concentration at a given moment is very low. The molecules are continuously metabolised (used) or involved in sugar-requiring biosynthetic processes (e.g., glucan formation).

3.3.2 Sugar Alcohols

The most important representative is mannitol, which is synthesised from fructose in a two-step biochemical process. Mannitol has several roles but it mainly regulates the formation and maturation process of fruit bodies. Less mannitol can be measured in the immature fruit bodies and is significantly more in matured ones. The mannitol content in mature fruit bodies used for our nutrition was significant, i.e., up to 40 per cent of dry matter. *Agaricus bisporus* showed 40–50 per cent DM mannitol; 54 per cent occurs in *Pleurotus ostreatus*; 60 per cent in *P. eryngii*; but *Flammulina velutipes* has only 9.7 per cent mannitol (Reis et al., 2012). Consumption of fruit bodies with high concentrations of mannitol (even for diabetic patients) is not a problem. The sweetness of mannitol is moderate and as it is absorbed poorly, it does not elicit a significant insulin response in the human body. Considerable mannitol content of fruit bodies (on DM basis) is a specific fact of mushroom chemistry.

3.3.3 Oligosaccharides

Sucrose (=saccharose) is not a characteristic of fungi, although it can be detected in some species. The concentration of trehalose, a disaccharide composed of two glucose units, is ranged widely. For six cultivated mushrooms, the values ranging from 0.16–8.0 per cent to FM were measured by Reis et al. (2012). The two button mushrooms (white and brown *Agaricus bisporus*) contained the least (0.16–0.2 per cent FM); two oyster mushrooms contained the highest quantities (*Pleurotus ostreatus*, 4.42 per cent; *P. eryngii*, 8.01 per cent). The difference between trehalose contents of button mushrooms and oyster mushrooms was 22–40 times. Barros et al. (2007a) found the trehalose content of some edible mushrooms collected in Portugal to be between 0.02–1.46 per cent. The authors also measured the concentration of mannitol in fruit bodies. Comparing the contents of two sugar derivatives, it appears that trehalose concentrations are low with higher mannitol levels and vice versa. This tendency is confirmed in the studies by Heleno et al. (2009) on other fungal species. The question of the function of the molecule is interesting in connection with trehalose concentration, which can be measured in significant proportions. Recent studies by Liu et al. (2019) suggest that trehalose can be a protective molecule against abiotic stress. Heat stress increases the intracellular trehalose accumulation in mushrooms. In the case of heat-sensitive strains of mushrooms, the enzyme background (enzyme activities) of trehalose metabolism help in the accumulation of trehalose. Concentrations of both sugar types decrease during cooking (e.g., in genera *Lactarius* and *Macrolepiota*). In the case of *P. eryngii*, the measured high trehalose concentration remains essentially unchanged during various preservation procedures (Li et al., 2015). Inadequate degradation of trehalose can also cause abdominal symptoms in some individuals, which is reminiscent of the problem of intolerance of lactose degradation. Underlying the problem is the too low trehalase activity in the small intestine (Kalač, 2016).

3.3.4 *Polysaccharides*

Chitins are water-insoluble, highly resistant and cell wall-forming polysaccharides occurring in many groups of true fungi. Chemically, chitin is a macromolecule composed of 1,4-N-acetyl-D-glucosamine monomers (Fig. 4). Partial removal of the acetyl group from its molecules produces chitosan. The biosynthesis of chitin fibres takes place in chitosomes, the special cellular organelles derived from ER which contain all the necessary molecules for biosynthesis (precursors and enzymes). Our previous studies allowed several findings on wild and cultivated mushrooms (Vetter and Siller, 1991; Vetter, 2007). In group of wild mushrooms, their taxonomic affiliation was a less important factor, but their mode of life seems to be an important aspect. Wood-destroying species (tinders, Judas ear fungus or even oyster mushrooms) always have lower chitin content than those of mycorrhizal and litter-inhabiting mushrooms. The average chitin content of xylophagous species is 2.45 per cent DM, the mycorrhizal fungi is 5.79 per cent DM and litter decomposing species,

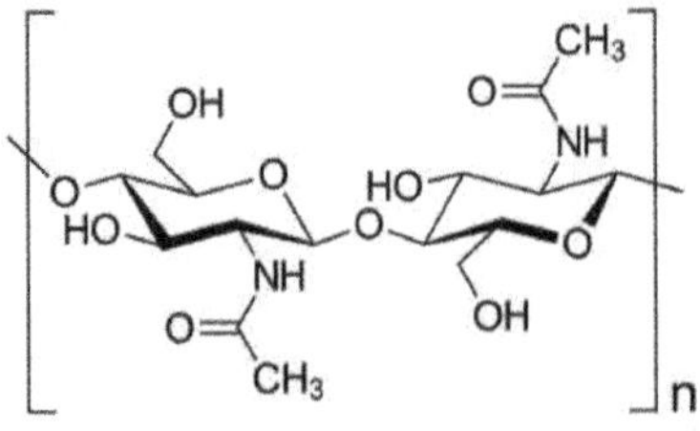

Fig. 4. Chemical composition of chitin polysaccharide (monomers: 1,4-N-acetyl-D-glucosamin).

5.5 per cent DM. Wood-destroying species are known to live on very low nitrogen-containing substrates (on various woods) but their protein content is not very low. The distribution of available nitrogen allows relatively low chitin contents.

The cultivated button mushroom (*A. bisporus*) varieties have a chitin content of 6–8 per cent DM, the pileus/stipe ratio is less than 1.0 (0.8–0.9). Chitin content of oyster mushrooms is significantly lower (from 2.16 to 5 per cent DM), while the pileus/stipe ratio is between 1.25 and 1.36. Thus the pileus parts contain significantly more chitin than the stipes. A similar relationship (showing a surplus in the pileus) occurs for *Lentinula edodes* (shiitake) (Vetter, 2007). This distribution of chitin may be explained by the fact that the process of spore formation, protected by pileus, takes place on the fruiting layer. Here the relatively large amount of chitin serves to strengthen the cell wall.

The actual chitin content of a fruit body, especially at higher concentrations, is also an important factor in the digestibility of mushrooms. The degradability and digestibility of chitin molecules are low, as the chitinase activity in the spectrum of animal-human digestive enzymes is very low (with a few exceptions). Another issue is that chitin is an important part of the dietary fibre fraction (TDF) and has a role as a fibre producer in normal digestive processes. The older experiments have shown that with increasing chitin administration, the protein digestibility in Wistar rats decreases as chitin reduces liver triacylglyceride and cholesterol levels (Zacour et al., 1992).

Chitin molecules have an increasing role (as a functional biomaterial) in certain foods, agriculture, medicine, pharmaceutics, paper industry and cosmetics. Chitin extracted from fungi is much purer, contains very few minerals and can be isolated with better efficiency than from the animal source. A recent study isolated chitin from pileus, stipes and gills of button mushrooms by an alkaline method and then examined it by advanced methods (thermal analysis, X-ray diffraction, infrared spectroscopy and solid state 13C NMR) (Hassainia et al., 2018). The extracted chitin proved to be α-chitin which is the most stable form of chitin.

3.3.5 Glycogen

Among the polysaccharides, the amount of glycogen (otherwise classified as animal starch) ranges in fruit bodies of mushrooms between 5–15 per cent DM (Dikeman et al., 2005). Incidentally, the glycogen molecules can be a carbon source for developing high-carbon-requiring fruit bodies. From the consumer's point of view, the 0.5–1.5 per cent DM glycogen in fresh mushrooms has no particular role.

3.3.6 Glucans

Polysaccharides composed of glucose units are called glucans. Despite recovery from the same units, there may be fine chemical differences like anomeric configurations of glucose units, nature of the glycosidic bonds, type and extent of branches and their molecular weights. Their most important types are the α- or β-glucans according to the anomeric configuration of glucose and the groups of 1,3-, 1,4-, and 1,6-glucans according to the position of glycosidic bonds. They are the basic constituents of the cell walls of fungi that protect the cells from environmental influences. The β-glucans are linked covalently to chitin molecules while the location and nature of α-glucans are more liquid.

The (1→3)-α-D-glucans are present in significant amounts in some taxa of basidiomycetous mushrooms, like *Piptoporus* (or *Fomitopsis*) *betulina* has 44–54 per cent, or fruit bodies of *Laetiporus sulphureus* have 75–88 per cent (Zlotko et al., 2019). They occur in some ascomycetous mushrooms in low percentages, but are generally absent in yeasts, expcept in *Histoplasma capsulatum* (46.5 per cent). They are present in many species of edible *Pleurotus* mushrooms (*P. citrinopileatus, P. djamor, P. eryngii, P. ostreatus*) where their quantity is usually between 2–6 per cent. Unlike β-glucans, α-glucans are still not fully known even today, although their structure is targeted by many studies. The study of α-glucans is significantly complicated due to their insolubility in water. In fact, in α-glucan molecules, the glucose units are mainly linked by 1,3 glycosidic bonds, but there are molecules containing 1,4-bonds and even those where 1,3 and 1,4 bonds are present as alternates. In the well-known and valuable *Boletus edulis*, for example, the glucan polymer has about 67 per cent of α-(1→3)-D-glucans and contains 28 per cent of α-(1→3)-D-mannan. In general, α-glucans are insoluble in water due to their strong hydrogen bonds and this makes it difficult to determine the mass or conformation of the molecule. The α-glucan of *Lentinula edodes* has a molecular weight of 500 kDa which is determined by an advanced method. Its spatial structure can be a flexible chain or a random coil (the latter formed after the breakdown of hydrogen bridges). The α-glucans occur in different crystalline forms, depending on the mushroom, on its part studied, on the preparation's conditions and so on. According to X-ray diffraction studies, α-glucan of mushrooms has three possible forms: the first is obtained in the natural form (*L. sulphureus* and *P. betulinus*); the second is obtained by precipitation from an alkaline solution and the third can be produced by drying (at 60–90°C). Interestingly, the linear α-glucan of the microfungus *Aspergillus wentii*, for instance, has a molecular weight of 850 kDa; the glucan consists of 25 subunits and each subunit has 200 α-D-glucose moieties (Choma et al., 2013).

The function of α-glucans within the mushroom is to increase the strength of the cell wall against the environmental effects and at the same time to maintain a kind of flexibility that allows growth. Undoubtedly, the most important biological effect of mushroom glucans is the anti-cancer effect which is caused mainly by β-glucans. Due to their insolubility, α-glucans do not have such an effect, but their derivatives (carboxymethylated or sulfonated) are already water-soluble and thus they have antitumor activity (Zlotko et al., 2019). Interestingly, a relationship has been demonstrated in the ecosystem between the presence of α-glucans and the virulence of a given fungal species. In the case of many fungal pathogens, the presence of α-glucans damages and reduces the defence system of the attacked plant. In other words, the balance between plant resistance and virulence of the invading fungus is disturbed by the α-glucans leading to plant disease development. Recent studies have also found that α-glucans protect the invading fungus (both spatially and functionally) against the plant defence reactions. It is also generally possible that these molecules inhibit the enzymatic degradation of β-glucans in both fungal-plant and fungal-mammalian relationships.

The β-glucans are very significant members of the fungal cell wall polysaccharides. In these molecules, D-glucose units are linked by β-1,3-bonds to form the polysaccharide chain but they also contain varying degrees of β-1-6 branches. This means that many different types of β-glucans can be formed in different fungi. The main function of β-glucans is the protection against dehydration and many other harmful factors. Glucans in the cell on the other hand can also be energy stores for the mushrooms at certain stages of development. The possibility of a glucan-chitin relationship may protect the cell wall against glucanase enzymes. Indirectly, glucans contribute to the preservation and osmotic stability of the cell wall proteins (Dalonso et al., 2015). From a nutritional point of view, fungal β-glucans are part of the dietary fibre (DF) fraction and have a wide range of biological activities. Their newer name is intended to express the diversity of their biological effects as Biological Response Modifiers (BRMs) (Bulam et al., 2018).

The glucan concentration in fruit bodies was sporadically studied. Only the work of Sari et al. (2017) can be considered. The authors determined the β- and total glucan contents for 23 wild mushrooms (for pileus and stipes) and for seven fungal fruit bodies devoid of pileus and stipe. Wild mushrooms are characterised by a total glucan content of 26 per cent with a very wide variance, most of the glucans are in fruit bodies of tinder species (*Trametes versicolor*, 61.2 per cent; *Piptoporus betulinus*, 54.2 per cent; glucan content is high in *Laetiporus sulphurous* at 52.6 per cent; *Auricularia auricula-judae*, 42.1 per cent). Proportion of β-glucans is 83–99 per cent of the total glucan content and on an average of 92.9 per cent. The glucan content varies between 10–14 per cent in cultivated button mushroom varieties, 18–25 per cent in *Pleurotus* and 20–26 per cent in shiitake varieties. Glucan levels measured in wood-destroying mushrooms are generally higher as compared to the non-wood-destroying ones. On examining mushroom pileus and stipes separately, in most species the glucan content of stipes is 20–30 per cent higher than that of pileus. The molecular weights of β-glucans vary widely, ranging from a few times to a thousand kDa. Their spatial structure is very diverse, the best example being the 'famous' β-glucan of *Lentinula edodes* (lentinan). The spatial structure of this molecule can vary between random skeins, single, double or triple helix and aggregate or spherical structure. In aqueous solution, the structure is triple helical with many H-bridges. From a pharmacological point of view, it is very important that the helical glucans are able to bind to immunoglobulins in the human blood serum. The solubility of glucans, which is also a fundamental issue in their biological action, depends on several factors (e.g., structure, origin, even their temperature).

The various β-glucans are named in general after the genus names of the fungus (*Lentinula edodes*, lentinan; *Pleurotus ostreatus*, pleuran; *Grifola frondosa*, grifolan and so on). Most important biological effect of β-glucans seems to be the modification of the immune system. Their molecular biological activity appears to be based on interaction with specific β-glucopyranose receptors on leukocytes. This relationship is influenced by many factors [(e.g., spatial structure, water solubility, number of branches of the glucose backbone (it is best if the ratio of branches is 0.2–0.3) and molecular weight (the higher is more efficient)]. The β-glucans are relatively resistant to gastric acid; they enter slowly into the duodenum and bind to the macrophage receptors of the intestinal wall. This is followed by several biochemical events (activation, cytokine production and induction of immune system in a specific way). The study of different β-glucans of mushrooms and in some cases the utilisation of some extracted and purified preparations as drugs, the presentation and evaluation of the most important anticarcinogenic effects are the topics of different disciplines (mainly in oncology and immunology). A series of such interesting literature reviews are available (Rop et al., 2009; Vannucci et al., 2017; Vetvicka and Vetvickova, 2018).

3.4 *Vitamins*

Vitamins are substances that are essential for the body and their uptake is usually from external sources. There are two main groups—water-soluble and fat-(lipid) soluble vitamins. The characteristic nature of vitamins can be seen in the world of mushrooms.

3.4.1 *Water-soluble Vitamins*

Vitamin C (ascorbic acid) is present in cultivated mushrooms in the concentration of 70–350 mg/kg DM (Table 13). One hundred grams of fresh mushrooms consumed ($\sim$ 10 g DM) have about 0.7–5.3 mg of vitamin C. The vitamin content can decrease further during the gastronomical preparations (cooking, heating and chemical treatments) and it is possible that the mushroom dish no longer contains vitamin C at all.

Among the members of the vitamin B group, the amount of B_3 (= niacin) varies between 100 and 650 mg/kg DM. By consuming 100 g of fresh mushrooms, one can have 1–6.5 mg of vitamin, which can be up to one-third of the daily requirement. Vitamins B_1 and B_2 are also present in mushrooms, but the intake of mushrooms at 100 g is not significant to meet the vitamin requirement (Table 13). It should be noted that the quantities of members from vitamin B group decrease during storage and preservation of mushrooms. There is a lot of data in the literature on vitamin B_{12} in mushrooms samples. This vitamin is not produced by metabolical processes of mushrooms through bacterial fungal samples contain some of it. Due to the above, we do not cite such data, as these molecules can actually enter the fungal sample as 'contaminants'. They do not characterise the mushroom species.

Table 13. Contents of vitamin B group and of vitamin C in the most important cultivated mushrooms (Bernas et al., 2006; Kalač, 2016).

	Agaricus bisporus	*Lentinula edodes*	*Pleurotus ostreatus*
Vitamin B_1 (thiamine) (mg/kg DM)	6–10	6	3–42
Vitamin B_2 (riboflavin) (mg/kg DM)	39–55	18	16–51
Vitamin B_3 (niacin) (mg/kg DM)	240–530	310	100–650
Vitamin B_9 (folic acid) (μg/kg DM)	4500–5900	3000–6600	3000–7000
Vitamin C (ascorbic acid) (mg/kg DM)	70–140	250	59–350

3.4.2 *Fat-soluble Vitamins*

The α-tocopherol (vitamin E) is the biologically most active form of vitamin and its main role is in the protection of the cell membranes against lipid peroxidation. The tocopherol contents in edible mushrooms are very low (0.02–200 μg/100 g DM) (Kozarski et al., 2015) and 0.2–1.0 μg/100 g DM values were measured in most cultivated mushrooms (Kalač, 2016).

The significance of vitamin D can be determined by two factors: the first one is the well-known and increasing importance of biological roles of vitamin D (anti-carcinogenic effect and positive effect against a variety of chronic diseases like cancer and diabetes); and the second is the level of desirable vitamin D concentrations and the number of patients suffering from vitamin D deficiency. Based on this, mushrooms have also come to the forefront in the supply of vitamin D. The level of vitamin D_2 in fresh (or dried) mushrooms can be significantly stimulated by UV irradiation. Edible mushrooms contain ergosterol in 100–650 mg/100 g DM (~ 10–65 mg/100 g FM) (Kalač, 2016). Vitamin D_2 (= ergocalciferol) molecules are formed from some of the ergosterol molecules by UV (mainly UV-B) irradiation followed by a heat-dependent isomerisation reaction. The majority of ergosterol is located in the gills of fruit bodies where most of the vitamin D_2 is formed. In oyster mushrooms, the original 5.2 μg/g DM concentration will be nearly four times higher (22.8 μg/g DM) (Cardwell et al., 2018). The sliced oyster mushrooms on exposure to UV-B for 60 minutes provide up to 140 μg/g of vitamins. Treatment up to 90 min at 28°C with energy of 1.14 W/m² increases the vitamin level up to 240 μg/g (Wu and Ahn, 2014). Various mushroom companies enrich the fresh mushrooms with UV radiation to a minimum vitamin content of 10 μg/100 g, which can meet 50–100 per cent of the minimum daily vitamin requirements (Cardwell et al., 2018). An increase in vitamin D_2 levels is also possible in dried mushrooms, although this method is rare and uncommon.

3.5 *Odorous, Aromatic Substances*

The odorous, aromatic substances belong to the secondary metabolites of fungi and occur in different mushrooms. Despite their low concentration, their effect may be mainly nutritional. Their beneficial influence may lead to increased excretion of digestive enzymes, absorption and utilisation of nutrients in the consumer. Their effects can be the opposite in other cases. It can be alarming or even disgusting, leading to chances of decreased consumption. Aroma of fungal taxa can often help in identifying the mushroom in question. Modern and sensitive analytical methods have isolated many substances responsible for odour and aroma (Moliszewska, 2014). The most typical fragrance compounds are C_8 volatile molecules (e.g., 1-octen-3-ol and oct-1-en-3-ol) and the derivatives. *Agaricus bisporus* contains more than 150 fragrant compounds; 70 molecules are known from *Boletus edulis* and 130 from *Lentinula edodes,* including 18 sulphur-containing compounds. In the case of *Tuber aestivum*, the number of compounds detected is already in the order of hundreds. In four wild mushrooms of northern Europe, mainly saturated and unsaturated aldehydes and ketones form the characteristic odour (Aisala et al., 2019). Cultivated button mushrooms contain mainly C_8 volatiles, but in addition to these, various aromatic compounds (benzyl alcohol, benzaldehyde, p-anisaldehyde, benzyl acetate and others) are dominant (Fratz and Zorn, 2010). Interestingly, the 1-octen-3-ol molecule, which is a main component of many mushrooms, is an R isomer (= (-)-oct-1-en-3-ol) while the other S isomer has a much weaker odour. The latter is found in moulds but also appears in some plants (Fratz and Zorn, 2010). Another fragrance is derived from the molecule oct-1-en-3-one and is described as a fragrance which is characteristic of cooked mushrooms. Nucleotides, amino acids and carbohydrates are present in the fragrance composition of *A. bisporus*.

In addition to the fact that their chemical background is complex, perception is also a multifactorial process and contains subjective elements. The mushroom scents described in the mycological literature do not always match, sometimes there are different scents for the same species. The contents of fragrance compounds increases with maturity of fruit bodies. It is important for the mature fruit body to function as a rich source of fragrance to attract the spore spreading fauna. The composition of fragrance of *Tuber* (truffle) species, for example, is closely related to not simple task of spore propagation from underground fruiting bodies. Some mushrooms produce unpleasant (for humans) odours, especially when indole and 3-chloroindole-like substances are formed. These compounds also have biological significance. For instance, the strong and disgusting odour released during the maturation of the octopus stinkhorn (or devil's fingers) *Clathrus archeri* attracts insects required for successful spore distribution.

3.6 Antioxidants

The normal, healthy function of human body is characterised by the balance of oxidative and antioxidant factors an oxidative homeostasis exists. Various causes like formation or entry of free radicals and weakening of the antioxidant system can lead to imbalance. The *Free-Radical Theory of Aging* published by Harman (Kozarski et al., 2015) in the mid-1950s was the origin of these medicinal and biochemical relations. It is well known that oxidative predominance in this case is a determining factor in many metabolic disorders and chronic diseases (cardiovascular, neurodegenerative, inflammations and different cancers). External factors leading to oxidant predominance may include high temperature, various radiations and chemical agents or toxins of biological origin. Among the antioxidant factors, the diversity and roles of antioxidants of foods should be emphasised. At the same time, the importance of all substances and nutraceuticals, which can help in maintaining and restoring the oxidative homeostasis in the body, is growing. The antioxidant molecules of mushrooms are located in mycelia as well as in the fruit bodies.

3.6.1 Phenolics

It is a great group of aromatic, hydroxylated molecules with one or more aromatic rings and hydroxyl groups. The main characteristic of these molecules is their antioxidant activity, which entails inhibition of free radicals, decomposition of peroxide and scavenging of oxygen. The number of such scientific data and investigations is very high, but the methodology of this problem and mainly the evaluation of chemical data is very complicated: (1) qualitative and quantitative analysis of such molecules in fruit bodies is only an analytical question, due to the new and sensitive methods of separation (HPLC and others); (2) the second more difficult problem is the determination of antioxidant activity of a mushroom. The common method involves production of different extracts (in different solvents) of fruit bodies and *in vitro* determination of the antioxidant activity. The main questions that arise are how does the antioxidant system of mushroom work and what is the relationship between *in vitro* and *in vivo* function?

The total phenolic content of wild and cultivated mushrooms was studied with various extraction solvents (acetone, ethanol, water and hot water) (Wang and Xu, 2014). Most phenolics are extracted by aqueous treatment in the range of 1–10 mg/g DM, that is, with hot water. The measurable total phenolic content increases significantly in hot water in some species (*Coprinus comatus*), while in others (*Pleurotus citrinopileatus*, it is 9.32–2.96; *Lentinula edodes*, 4.26–1.82) the levels are significantly lower. The background of the mentioned methodological problems is that the different phenolics obviously cannot be measured exactly in a single or strictly by a specific reaction. The nature and ratio of 'free' and 'bound' phenolics are different in each species, which need to be extracted differently. The total amount of phenolics is usually given in gallic acid equivalents (i.e., the amount of phenolics measured this way is roughly approximate). Kalač (2016) concluded that the published data are between 1 and 6 mg/g DM and the phenolic level rarely exceeds 10 mg/g DM. The data by Hung and Nhi (2012) on cultivated mushrooms indicate clearly that in addition to the 'free' (easily extractable phenolic content) and 'bound' (difficult to extract) phenolic fraction can also be measured (Table 14). The latter accounts for 2.6–28 per cent of the total phenolic content (the smallest in *Pleurotus ostreatus* at 2.6 per cent; the largest in *Ganoderma lucidum*, at 28.1 per cent). It is also important to underline that the classical Folin-Ciocalteu's reagent definitely overestimates the amount of phenolics because it reacts with other molecules of non-phenolics (e.g., ascorbic acid and amino acids). The main components of phenolics in a given fungus have to be determined. Among the hydroxybenzoic acid derivatives, protocatechuic, gentisic, vanillic and syryngic acids are the most important components, while the most frequent cinnamon derivatives are o- and p-cumaric, caffeic, ferulic, synapic acids and their derivatives. The main effect of the above compounds is the scavenging of free radicals. The consequence of the previous studies is that there is generally a linear, sometimes tighter and sometimes less close, correlation between the phenolic concentration and free radical scavenging ability.

Table 14. Free and bound phenolic contents (mg/g DM) of selected mushrooms (Hung and Nhi, 2012).

	Total free phenolics	Total bound phenolics	Free + bound phenolics
Auricularia polytricha	0.47	0.07	0.54
Ganoderma lucidum	2.35	0.67	3.02
Lentinula edodes	2.87	0.29	3.16
Pleurotus ostreatus	2.60	0.07	2.67
Volvariella volvacea	4.12	0.19	4.31

3.6.2 Flavonoids

As a subgroup of phenolics, although present in only a low quantity, flavonoids contribute significantly to the antioxidant activity. In comparative studies by Krüzselyi (2018), the phenolic content of *Cyclocybe cylindracea* was 3.62 mg/g DM and the brown variety of button mushroom has 2.38 mg/g DM, while *Pleurotus ostreatus* contained only 0.84 mg/g DM. In Krüzselyi's (2018) studies, for the first time in the literature, fungal fruit bodies were partly divided in morphologically (peel, inner cap, gills and stipe) and partly in developmental terms (primordium, immature, 1–2 cm fruit bodies and fully developed fruit bodies). It was found most of the phenolics in button mushroom and oyster mushroom varieties are distributed in the peel of the cap and in the gills, while in *Lentinula edodes*, the differences are similar but smaller. There is little difference in phenolic levels between immature and mature mushroom fruit bodies. This also means that the entire phenolic stock of the mature fruit body is already (in primordia) established.

3.6.3 Other Antioxidants

The polysaccharide glucans and their different biological effects (mainly the immunostimulatory and carcinogenic properties) are important. Several studies suggest that many antioxidant polysaccharides with antioxidant functions are also found in mushrooms (e.g., Kozarski et al., 2015). Antioxidant characteristic of these molecules are due to their radical scavenging, reducing and chelating properties and holding back of lipid peroxidation. The most important antioxidant activity may be due to the presence of β-glucans. The free radical scavenging ability of these molecules is caused by the presence of hydrogen from certain monosaccharide units.

Ascorbic acid (vitamin C) is undoubtedly an effective compound with its free radical-scavenging properties. It is present in the fruit body of several mushrooms in small amounts and from which it can enter the body only in insignificant amounts with the mushroom food. The situation is similar for vitamin E (tocopherols) as its intake by the mushroom food is insignificant in terms of the antioxidant effect. Carotenoids are not uncommon components in fruit bodies, although they are usually pigments present in fairly small amounts. In fact, several carotenoid derivatives are involved in the yellowish-orange-reddish colour formation of fruit bodies. However, nowadays, the focus is mostly on the lycopene as an antioxidant. This molecule may be an antioxidant due to its unsaturated nature and on the other hand, being a substance that can cross the blood-brain barrier, it is also present in the central nervous system albeit in low concentrations.

The ergosterol, which is present in significant amounts in many mushrooms, can produce vitamin D_2 (ergochalciferol) under the influence of UV radiation. The vitamins D_2 and D_3 (and some of their active derivatives) act as membrane antioxidants in the human body (Wiseman, 1993).

Ergothioneine is a sulphur-containing derivative of the amino acid called histidine (Fig. 5), which is unknown in either the animal or the plant kingdom. Without it, the human cells are more sensitive to oxidative stress, with increased potential for mitochondrial DNA damage, levels of protein oxidation and lipid peroxidation. The ergothioneine molecules are concentrated in the mitochondria, where they play a protective role against the oxygen radicals ($\cdot O_2^-$) formed here. Since ergothioneine can

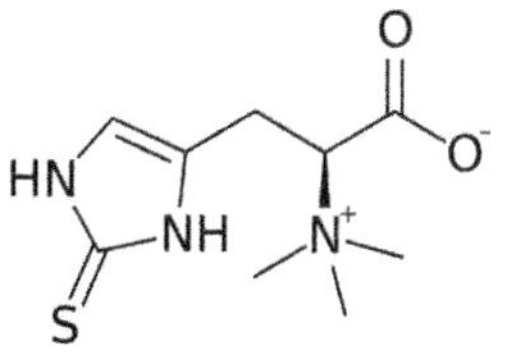

Fig. 5. Structure of the ergothioneine.

only enter with nutrients or its lack can cause problems, it is actually a new vitamin. The ergothioneine content is distributed over a very wide concentration range. Kalač (2016) mentions the concentrations of 0.2 and 1000 mg/100 g DM, while Chen et al. (2012) demonstrated, in each of the 29 species of mushrooms studied, the highest concentrations in *Pleurotus* spp. (1245 and 2850 mg/kg DM).

3.7 Total Dietary Fibre

The concept of dietary fibre (DF) is very important on the field of mushroom carbohydrates too. The notion means, the sum of indigestible carbohydrates (for duodenum), the determining component of which is the chitin. The total dietary fibre (TDF) consists of soluble and mainly insoluble fractions. Soluble constituents absorb water, form a gel-like structure and increase the viscosity of foods and chyme. The insoluble fraction, on the other hand, softens stools and increases stool bulk and generally provides a favourable colonic condition (Kalač, 2016). Dietary fibre also has other beneficial physiological effects (laxation, keeping blood sugar and cholesterol content at optimum physiological levels). Many physiological studies have previously demonstrated that adequate DF care is beneficial for maintaining health, controlling diabetes and body weight (Cheung, 2013). Average data for 13 mushrooms grown in the Far East was TDF, 31.1 per cent DM; insoluble fraction, 29.2 per cent DM; soluble fraction, 1.9 per cent DM (i.e., it is only 6.1 per cent of the total fibre fraction).

The TDF values determined for fruit bodies vary within wide limits, approximately between 5–40 per cent DM, with an average of 26 per cent (Wang et al., 2014). Majority of TDF fraction is the insoluble fibre. The evaluation and comparison of data in the literature is complicated by the different analytical methods used to determine the fibre. Cooking increases the amount of TDF in the mushrooms, as it extracts more material, thus increasing the relative proportion of fibre materials.

Other important components of TDF fraction are the various heteroglucans (i.e., polysaccharides), where mainly but not only glucose units are found. The previous studies indicate that the amount of TDF in shiitake pileus is 32 per cent and in stipes 40 per cent (Cheung, 1996). Regarding its sugar constituents, in addition to the crucial glucose (70–73 per cent), 14–15 per cent of glucosamine, 3–7 per cent of uronic acid, 1.7–3.3 per cent of galactose, 5.2 per cent of mannose, 1.5–1.9 per cent xylose and even rhamnose in stipe (0.6 per cent) were present. In *Pleurotus sajor-caju* and *Volvariella volvacea*, the composition of TDF is essentially the same.

3.8 Calorific Value

The energy content of mushrooms is usually evaluated by using the following classical formula (Kalač, 2016): Energy value (kcal for 100 g DM) = 4 × (crude protein in g + carbohydrate in g) + 9 × crude fat in g. The kcal value can be converted to kJ as needed (by multiplication factor, 4.168). We use in calculation three slightly overestimated parameters, so the obtained energy content is also high; therefore it would need to be reduced by one-third (Kalač, 2016). The need for reduction is mainly supported by the fact that the largest parameter of the formula is the carbohydrate content, which provides a significant part of insoluble dietary fibre and is not metabolised, so that it does not provide energy for the body.

As shown in Table 12, the energy values of frequently grown mushrooms are generally between 300 and 500 kcal/100 g DM (i.e., the energy content is theoretically 30–50 kcal per 100 g of fresh mushrooms). The large literature table by Kalač (2016) on the nutritional value of mushrooms contains some data which significantly exceed the limit values shown above. The reality of energy contents above 500 kcal/100 g DM, is somewhat uncertain as the 50–60 per cent crude protein contents on which the calculation is based are also in doubt. Mushrooms with a higher energy level should only be those where the crude fat content is unusually high as much as more than 20–25 per cent.

There are some substances in mushrooms which have negative, harmful or detrimental biological effects. We need to fix several things, especially the concept of edibility of mushrooms presupposes that it does not contain toxic metabolites. Many mushrooms known in the world contain poisonous molecules to humans (and also to animals). There is a wealth of interesting toxicological and mycological literature available on the toxicoses (or mycetisms).

3.9 Antinutritional Components

Among the so-called antinutritive substances (those that slow down the uptake, breakdown and utilisation of food), trypsin inhibitors come first. Many types of substances are known from the plant kingdom, where mainly the fruits (seeds) of species from some plant families (Fabaceae, Solanaceae and Poaceae) may contain such substances. It is often advisable to reduce their amount (to reduce their biological effects) with certain treatments (e.g., heat treatment).

3.9.1 Trypsin Inhibitors

We compared the trypsin inhibitory activity (TIU) of aqueous extracts of wild, edible and common mushrooms in our previous study (Vetter, 2000b). We found that the trypsin inhibitory activity of these species ranged between 0.36–10.4 TIU/mg DM. The activity found is generally higher than that of cereals, but lower than that in the seeds of fabaceous plants. The highest TIU values were measured in *Laccaria, Hygrophorus, Craterellus cornucopioides, Gomphidius glutinosus, Macrolepiota rhacodes* and *Hydnum repandum*. The measured activities are undoubtedly antinutritive, but it should be noted that raw mushrooms are very rarely consumed and various gastronomic operations will largely or completely degrade the proteinaceous antinutritive molecules.

3.9.2 Nitrate Ions in Fruit Bodies

Occurrence of NO_3-ions in fruit bodies can cause different problems. It is well known that nitrate ion is also an important inorganic nitrogen source for many edible mushrooms. Products of enzymatic reduction of nitrate ions ($NO_3 \rightarrow NO_2 \rightarrow NH_3$) are important for amino acid metabolism (especially for glutamate). However, the question is to find how the actual nitrate level of a fruit body develops and whether its uptake (consumption) is important in a toxicological sense. There may also be a possibility of nitrate to nitrite conversion and of methaemoglobinaemia. In our studies, we analysed the nitrate content of samples of wild and most important cultivated mushrooms representing different nutrition types (Bóbics et al., 2016). Low and relatively constant nitrate levels were characteristic for mycorrhizal and wood-destroying mushrooms (216 mg/kg DM and 228 mg/kg DM, respectively), while the variability was wider in the saprotrophic group (150–12750 mg/kg DM). This indicates that the inherent nitrate levels of fungal substrates may be an important factor. For some mushrooms, the possibility of nitrate accumulation also arose (*Clitocybe nebularis, C. odora, Lepista* and *Macrolepiota* species). Among the cultivated fungi, *Agaricus bisporus* is characterised by an average nitrate level of 562 mg/kg DM; *Pleurotus ostreatus* has 221 mg/kg DM and *Lentinula edodes* has 154 mg/kg DM. For the wild-growing 'accumulating' mushrooms the nitrate intake with 100 g of fresh mushroom can range from a few per cent to 22–28 per cent of the daily tolerable nitrate content. In principle, the consumption of one such mushroom can deplete a person's nitrate load.

3.9.3 Agaritin

A molecule of controversial significance was isolated from different *Agaricus* species in the early 1960s. This substance is agaritin (β-N-(γ-L-(+)-glutamyl-4-(hydroxymethyl)-phenylhydrazine), which is in fact a weakly mutagenic compound. The 4-(hydroxymethyl) phenylhydrazine eventually produced the 4-(hydroxymethyl)-benzenediazonium cation, which induced tumors in the gastrointestinal tract and bladder in experimental animals (rats and mice). This molecule appears mainly in *Agaricus* species (including *A. bisporus*) (Schulzová et al., 2009). Agaritin contents in some mushrooms are below 100 mg/kg FM; a large group of mushrooms possess 100–1000 mg/kg FM and a significant group of mushrooms has an average of 1000–3995 mg/kg FM. In addition to *Agaricus*, lower agaritin contents were measured in some *Leucoagaricus* and *Macrolepiota* species. The average agaritin concentration in cultivated mushrooms was between 200–500 mg/kg FM. The agaritin content of fresh mushrooms decreases very significantly during the different gastronomical steps (storage, freezing, drying and cooking or other treatments). Its concentration in preserved *A. bisporus* can only be measured at 15–20 mg/kg FM; it is a nearly ten-fold decrease (Andersson et al., 1999). Thus, further investigation of human biological effects of the agaritin molecule is absolutely necessary. Breeders' efforts to encourage the cultivation of varieties with the lowest possible agaritin content from the current variety selections should be supported. It is reassuring however that (assuming the actual carcinogenicity of agaritin-derived products) the amount of agaritin remaining during storage and preparation of the mushrooms is only a small part of the original. We also think that only a long-term and continuous agaritin uptake may be risky.

3.9.4 *Coprine*

In some respects, the coprine molecule can also be considered as carcinogen (procarcinogen); in some *Coprinus* species it is ~ 150 mg coprine/kg FM, while significantly lower levels have been measured for some *Clitocybe* or *Pholiota squarrosa*. This molecule is a non-proteinogenic amino acid and cyclopropanone is converted in humans to cyclopropanal hydrate. When a person consumes alcohol along with a coprine-containing mushroom, it leads to development of a characteristic group of symptoms, which actually is acetaldehyde poisoning. Breakdown of alcohol is stopped by coprine and acetaldehyde accumulates. Fortunately, the number of coprine-containing mushrooms is low and symptoms occur only when consumption is associated with alcohol; more insight are found in Kalač (2016) and Jo et al. (2014).

3.9.5 *Allergies*

Allergic reactions caused by spores of certain mushrooms have long been known to occur mainly among the growers of given species of mushroom or the pickers of the mushrooms (Senti et al., 2000). The mushroom allergens and the problems caused due to mushroom consumption are important. Very little data suggests such a possibility, but it can occur rarely. An 18-year-old female patient suffered from the age of eight from the effects of *Lentinula edodes*, *Grifola frondosa* or Shimeji mushroom (*Hypsizigus ulmarius*), which manifested in the form of oedema and urticaria after consuming the listed fungi. In the study, the patient showed clear skin reactions to extracts of all the three fungi (Ito et al., 2020). Further studies on mushroom allergens throw more light on their importance and prevention of such threats.

4. Conclusion

Mushrooms are of increasing physiological significance and value in our diet. They are derived partly from mushrooms collected from nature and partly from increasing quantity and variety of fungi produced by the mushroom industry. Beyond the historical and sociological background of mushrooms, some nations are more mushroom lovers (mycophiles) while others are less so (mycophobes). Nutritional and physiological benefits of mushrooms as food are shown in the Fig. 6. Mineral composition of fruit bodies is important due to their high (and almost constant) potassium level, their significant phosphorus content and some microelements that occur through bioaccumulation (mainly Cu, Zn, Se). Sodium content of fruit bodies is modest which, together with the high K content, is very beneficial for patients of cardiovascular problems. Their protein content is moderate, their amino acid composition is similar to animal proteins and the sum of albumin + globulin-type proteins is favourable. Mushrooms have low fat concentration, which is very important for the favourable low-energy level of mushroom foods. Another advantage is that most fatty acids are unsaturated. The total carbohydrate content is very high in the total dry weight. It is a very heterogeneous group of substances, characterised by high levels of mannitol sugar alcohol (with low glycaemic index) or the presence of insoluble polysaccharides. The latter, also known as dietary fibre, consists of the cell wall-forming chitin and mainly of β-glucans. The dietary fibre fraction may be the main contributor to digestive processes and in the case of β-glucans, other beneficial

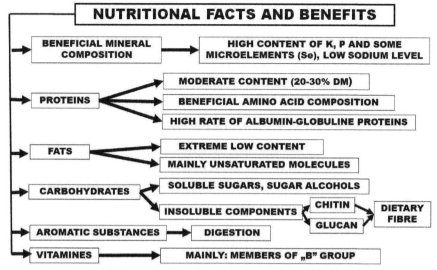

Fig. 6. Nutritional facts and benefits of mushroom fruit bodies.

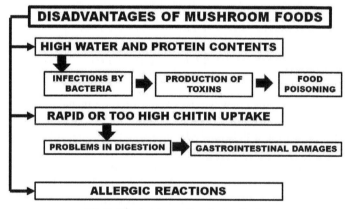

Fig. 7. Disadvantages of mushroom foods.

preventive and therapeutic anti-cancer effects have become increasingly important. Vitamins of mushrooms include those of the vitamin B group (B_1 and B_3) which are important to meet the daily vitamins requirements of the consumer.

Are there any disadvantages, or properties which must be taken into account during use? Obviously yes (*see* Fig. 7)! The two basic properties of mushrooms are their high water level and significant protein content. It follows that they can easily be a source of spoilage processes (caused mainly by bacteria), where food poisoning associated with toxin production can occur easily (e.g., salmonellosis). Statistics are often a bit misleading because the poisoned human actually consumed fungal food when the poisonous agents were not produced by a mushroom but by the proliferating bacteria. It is important to keep this in mind during storage as they have a short shelf-life. The higher chitin content of the mushroom, especially after consuming it in large quantities, temporarily can overcharge the digestive system,

Table 15. The medicinal effects of the higher mushrooms.

	Molecular basis	Mechanism of action	Examples
Anticarcinogenic	Polysaccharides: β-glucans, Complexes of polysaccharides and proteins; molecules of triterpene character	Stimulation of immune system and effect on carcinogenesis or on enzymes of cancer tissue (aromatase)	Lentinan (*Lentinula edodes*); Schizophyllan (*Schizophyllum commune*)
Antidiabetic	Multi-component effects (dietary fibre, antioxidants and polysaccharides)	Reduction of blood glucose; inhibition of α-glucosidase and stimulation of pancreatic cells for insulin synthesis; inhibition of gluconeogenesis	*Pleurotus ostreatus, Coprinus comatus* (comatin) and *Ganoderma lucidum* (ganopoly) decrease of serum glucose by hepatic pyruvate carboxikinase
Antioxidant	Phenolics (including flavonoids), polysaccharides (β-glucans) and vitamins (ascorbic acid, tocopherols and carotenoids)	Restoration of oxidant-antioxidant balance scavenging of free radicals: slowing the development of chronic diseases; stabilisation and improvement of the condition of chronic patients	*Agaricus subrufescens Agaricus bisporus Lentinula edodes* and *Pleurotus ostreatus*
Antimicrobial	Terpenoids; β-glucans and laccase enzyme	Antiviral effects (against-HIV); Lentinan: immune system hepatitis C virus	*Ganoderma lucidum Pleurotus ostreatus* and *Lentinula edodes*

causing mild gastrointestinal damage. It is very rare for a fungal mushroom dish to cause allergic reactions (*see* Fig. 7).

In addition to the physiological effects of nutrition, mushrooms have many other medicinal and biological effects. Most important of these are summarised in Table 15. The first of these effects, which is becoming increasingly important is the fact that a number of large mushrooms have anticarcinogenic effects. The molecular basis of these effects are mainly polysaccharides or polysaccharide-protein complexes which attack the immune system. The anti-diabetic effects are due to several groups of substances (dietary fibre and antioxidants) and these effects are aimed at lowering blood sugar by stimulating the pancreatic function or by inhibiting gluconeogenesis. Antioxidant factors are becoming increasingly important. Their chemical backgrounds are mainly phenolics (including flavonoids), some fungal polysaccharides and vitamins (tocopherols, carotenoids, ascorbic acid). The main effect of this complex is to restore the oxidative balance and bind the generated or absorbed free radicals. It follows that they can play an important role in improving chronic diseases due to oxidant predominance by stabilising the patient's condition or in a preventive way by slowing down the development of different diseases. The active ingredients of mushrooms presented in this chapter have little or no side effects. From a chemical point of view, fungal terpenoids, certain polysaccharides, or even enzymes (laccase) can induce antibacterial, antiviral or even fungicidal effects. The history of active ingredients of mushrooms began a long ago but it is coming to limelight now with the discovery and validation of more and more substances that contribute to human nutrition and health.

References

Agrahar-Murugkar, D. and Subbulakshmi, G. (2005). Nutritional value of wild edible wild mushrooms collected from the Khasi Hills of Meghalaya. Food Chem., 89: 599–603.

Aisala, H., Sola, J., Hopia, A., Linderberg, K.M. and Sandell, M. (2019). Odour-contributing volatile compounds of wild edible Nordic mushrooms analysed with HS-SPME-GC-MS and HS-SPME-GC-O/FID. Food Chem., 283: 566–578.

Akata, I., Ergönül, B. and Kalyoncu, F. (2012). Chemical composition and antioxidant activities of 16 wild edible mushroom species grown in Anatolia. Int. J. Pharmacol., 8: 134–138.

Andersson, H.C., Hajslova, J., Schulzova, V., Panouska, Z., Hajkova, L. and Gry, J. (1999). Agaritin content in processed foods containing the cultivated mushroom (*Agaricus bisporus*) on the Nordic and Czech market. Food Addit. Contam., 16: 439–446.

Assuncao, de L.S., da Luz, J.M.R., da Silva, M.C.S., Vieira, P.A.F., Bazzolli, D.M.S. et al. (2012). Enrichment of mushrooms: An interesting strategy for the acquisition of lithium. Food Chem., 134: 1123–1127.

Ayaz, F.A., Torun, H., Colak, A., Sesli, E., Millson, M. and Glew, R.H. (2011a). Macro- and microelement contents of fruiting bodies of wild-edible mushrooms growing in the East Black Sea region of Turkey. Food Nutr. Sci., 2: 53–59.

Ayaz, F.A., Chuang, L.T., Colak, A., Sesli, E., Presley, J. et al. (2011b). Fatty acid and amino acid composition of selected wild edible mushrooms consumed in Turkey. Int. J. Food Sci. Nutr., 62: 328–335.

Barros, L., Baptista, P., Correira, D.M., Casal, S., Oliveira, B. and Ferreira, I.C.F.R. (2007a). Fatty acid and sugar composition and nutritional value of five wild edible mushrooms from Northeast Portugal. Food Chem., 105: 140–145.

Barros, L., Baptista, P., Correira, D.M., Morais, J.S. and Ferreira, I.C.F.R. (2007b). Effects of conservation treatment and cooking on the chemical composition and antioxidant activity of Portuguese wild edible mushrooms. J. Agric. Food Chem., 55: 4781–4788.

Barros, L., Ventuizini, B.A., Baptista, P., Estevinho, L.M. and Ferreira, L.C.F.R. (2008). Chemical composition and biological properties of Portuguese wild mushrooms: A comprehensive study. J. Agric. Food Chem., 56: 3856–3862.

Beluhan, S. and Ranogajec, A. (2011). Chemical composition and non-volatile components of Croatian wild edible mushrooms. Food Chem., 124: 1076–1082.

Bengu, A.S. (2020). The fatty acid composition in some economic and wild edible mushrooms in Turkey. Progress in Nutrition, 22(1): 185–192.

Bernas, E., Jaworska, G. and Lisiewska, L. (2006). Edible mushrooms as a source of valuable nutritive constituents. Acta Sci. Pol. Technol. Aliment., 5(1): 5–20.

Blumfield, M., Abbott, K., Duve, E., Cassettari, T., Marshall, S. and Fayet-Moore, F. (2020). Examining the health effects and bioactive components in *Agaricus bisporus* mushrooms: a scoping review. J. of Nutritional Biochemistry, 84.

Bóbics, R., Krüzselyi, D. and Vetter, J. (2016). Nitrate content in a collection of higher mushrooms. J. Sci. Food Agric., 96: 430–436.

Bulam, S., Üstün, N.S. and Peksen, A. (2018). β-glucans: An important bioactive molecule of edible and medicinal mushrooms. Intern. Technol. Sci. Design Symp., 27–29 June, Giresun, Turkey, 1242–1258.

Cardwell, G., Bornmann, J.F., James, A.P. and Black, LK.J. (2018). A review of mushrooms as a potential source of dietary vitamin D. Nutrients, 10(10): 1498. 10.3390/nu10101498.

Chen, S.Y., Ho, K.J., Hsich, Y.J., Wang, L.T. and Mau, J.L. (2012). Contents of lovastatin, γ-aminobutyric acid and ergothioneine in mushroom fruiting bodies and mycelia. LWT Food Sci. Technol., 47: 274–278.

Cheung, P.C.K. (1996). Dietary fibre content and composition of some cultivated edible mushroom fruiting bodies and mycelia. J. Agric. Food Chem., 44: 468–371.

Cheung, P.C.K. (2013). Mini-review on edible mushrooms as a source of dietary fibre: Preparation and health benefits. Food Sci. Hum. Wellness, 2: 136–166.

Choma, A., Wiater, A., Komaniecka, I., Paduch, R., Pleszczynska, M. and Szczodrak, J. (2013). Chemical characterisation of a water insoluble (1→3)-α-d-glucan from an alkaline extract of *Aspergillus wentii*. Carbohydr. Polym., 91: 603–608.

Cuptapun, Y., Hengsawadi, D., Mesomya, W. and Yaiciam, S. (2010). Quality and quantity of protein in certain kinds of edible mushrooms in Thailand. Kasetsart J. (Nat. Sci.), 44: 664–670.

Dalonso, N., Goldman, G.H. and Gera, R.M.M. (2015). β-(1-3), (1-6)-Glucans: Medicinal activities, characterisation, biosynthesis and new horizons. App. Microbiol. Biotechnol., 99: 7893–7906.

Darboure, I.R. and Takruri, H.R. (2002). Protein quality of four types of edible mushrooms founds in Jordan. Pl. Foods Hum. Nutr., 57: 1–11.

Dikeman, C.L., Bauer, L.L., Flickinger, E.A. and Fahey, Jr. G.C. (2005). Effects of stage of maturity and cooking on the chemical composition of select mushroom varieties. J. Agric. Food Chem., 53: 1130–1138.

Fernandes, A., Berreira, J.C.M., Antonio, A.L., Oliveira, M.B.P.P., Martins, A. and Ferreira, I.C.E.R. (2014). Feasibility of electron-beam irradiation to preserve wild dried mushroom: Effects on chemical composition and antioxidant activity. Innov. Sci. Eng. Technol., 22: 158–166.

Fratz, M.A. and Zorn, H. (2010). Fungal flavours. *In*: Hofrichter, M. (ed.). The Mycota X – Industrial Applications. 2nd ed., Springer-Verlag, Berlin.

Győrfi, J., Geösel, A. and Vetter, J. (2010). Mineral composition of different strains of edible medicinal mushroom *Agaricus subrufescens* Peck. J. of Medicinal Food, 13(6): 1510–1514.

Hassainia, A., Satha, H. and Boufi, S. (2018). Chitin from *Agaricus bisporus*: Extraction and characterisation. J. Biol. Macromol., 117: 1334–1342.

Heleno, S., Barros, L., Sousa, M.J., Martins, A. and Ferreira, I.C.F.R. (2009). Study and characterisation of selected nutrients in wild mushrooms from Portugal by gas chromatography and high performance liquid chromatography. Microchem. J., 93(2): 195–199.

Hung, P.V. and Nhi, N.Y. (2012). Nutritional composition and antioxidant capacity of several edible mushrooms grown in the Southern Vietnam. Int. Food Res. J., 19: 611–615.

Ito, Z., Kobayashi, T., Egusa, C., Maeda, T., Abe, N. et al. (2020). A case of food allergy due to three different mushroom species. Allerg. Int., 69: 152–153.

Jo, W.S., Hossain, M.A. and Park, S.C. (2014). Toxicological profiles of posionous, edible and medicinal mushrooms. Mycobiology, 42(3): 215–220.

Kalač, P. (2013). A review of chemical composition and nutritional value of wild-growing and cultivated mushrooms. J. Sci. Food Agric., 93: 209–218.

Kalač, P. (2016). Edible Mushrooms: Chemical Composition and Nutritional Value. Academic Press, Amsterdam.

Kim, M.Y., Chung, I.M., Lee, S.J., Ahn, J.K.J., Kim, E.H. et al. (2009). Comparison of free amino acid, carbohydrates concentrations in Korean edible and medicinal mushrooms. Food Chem., 113: 386–393.

Kora, A.J. (2020). Nutritional and antioxidant significance of selenium-enriched mushrooms. Bulletin of the National Research Centre, 44: 34.

Kozarski, M., Klaus, A., Jakovljevic, D., Todorovic, N., Vunduk, J., Petrovic, P., Niksic, M., Vrvic, M.M. and Griensven, van, L. (2015). Antioxidants of edible mushrooms. Molecules, 20: 19489–19525.

Kozarski, M., Klaus, A., Vunduk, J., Jakovljevic, D. Jadranin, M. and Niksic, M. (2020). Health impact of the commercially cultivated mushroom *Agaricus bisporus* and the wild-growing *Ganoderma resinaceum*—A comparative overview. J. of the Serbian Chemical Society, 85(6): 721–735.

Krüzselyi, D. and Vetter, J. (2014). Complex chemical evaluation of summer truffle (*Tuber aestivum* Vittadini). J. Appl. Bot. Food Qual., 87: 291–295.

Krüzselyi, D., Kovács, D. and Vetter, J. (2016). Chemical analysis of king oyster mushroom (*Pleurotus eryngii*) fruit bodies. Acta Aliment., 45(1): 20–27.

Krüzselyi, D. (2018). Bioactive Antioxidants of Basidiomycetous Mushrooms. Ph.D. dissertation, Szent István University, Gödöllő (in Hungarian).

Lee, K.J., Yun, I.J., Kim, K.H., Lim, S.H. and Ham, H.J. (2011). Amino acid and fatty acid composition of *Agrocybe chaxingu*, an edible mushroom. J. Food Comp. Anal., 24: 175–178.

Lelley, J.I. and Vetter, J. (2005). The possible role of mushrooms in maintaining good health preventing diseases *Acta Edulis*. Fungi, 12: 412–419.

Li, X., Feng, T., Zhou, F., Zhou, S., Liu, Y. et al. (2015). Effects of drying methods on the tasty compounds of *Pleurotus eryngii*. Food Chem., 166: 358–364.

Liu, X.M., Wu, X. L., Gao, W., Qu, J.B., Chen, Q., Huang, C.Y. and Zhang, J.X. (2019). Protective roles of trehalose in *Pleurotus pulmonarius* during heat stress response. J. Integr. Agric., 18(2): 428–437.

Manzi, P., Marconi, S., Aguzzi, A. and Pizzoferrato, L. (2004). Commercial mushrooms: nutritional quality and effect of cooking. Food Chem., 84: 201–206.

Moliszewska, E. (2014). Mushroom flavour. Folia Biol. Oceol., 10: 80–88.

Ouzouni, P.K. and Riganakos, K.A. (2007). Nutritional value and metal content of Greek wild edible fungi. Acta Aliment., 36: 99–110.

Ouzouni, P.K., Petridis, D., Koller, W.D. and Riganakos, K.A. (2009). Nutritional value and metal content of wild edible mushrooms collected from West Macedonia and Epirus Greece. Food Chem., 115: 1575–1580.

Pardo-Gimenez, A., Figueiredo, V.R., Dias, E.S., Pardo-Gonzalez, J.E., Alvarez-Orti, M. and Zied, D.C. (2013). Proximate analysis of sporophores of *Agaricus subrufescens* Peck. ITEA Inf. Tec. Econ. Ag., 109: 290–302.

Pedneault, K., Angers, P., Gossehin, A. and Twedell, R.J. (2008). Fatty acids profiles of polar and neutral lipids of ten species of higher basidiomycetes indigenous to eastern Canada. Mycol. Res., 112: 1418–1434.

Pereira, E., Barros, L., Martins, A. and Ferreira, I.C.F.R. (2012). Towards chemical and nutritional inventory of Portuguese wild edible mushrooms in different habitats. Food Chem., 130: 394–403.

Reis, F.S., Barros, L., Martins, A. and Ferreira, I.C.F.R. (2012). Chemical composition and nutritional value of the most widely appreciated cultivated mushrooms: An inter species comparative study. Food and Chemical Toxicology, 50: 191–197.

Ribeiro, B., Andrade, P.B., Silva, B.M., Baptista, P., Seabra, R.M. and Valentao, P. (2008). Comparative study on free amino acid composition of wild edible mushroom species. J. Agric. Food Chem., 56: 10973–10979.

Rop, O., Mlcek, J. and Jurikova, T. (2009). Beta-glucans in higher fungi and their health effects. Nutr. Rev., 67(11): 624–631.

Sari, M., Prange, A., Lelley, J. and Hambitzer, R. (2017). Screening of beta-glucan content in commercially cultivated and wild growing mushrooms. Food Chem., 216: 45–51.

Schulzová, V., Hajslova, J., Perautka, R., Hlavasek, J., Gry, J. and Andersson, H.C. (2009). Agaritine content of 53 *Agaricus* species collected from nature. Food Add. Contam., 26(1): 82–93.

Senti, G., Leser, C., Lundberg, M. and Wüthrich, B. (2000). Allergic asthma to shiitake and oyster mushroom. Allergy, 55: 975–976.

Silva, da M.C.S., Nozuka, J., Luz, da J.M.R., Assuncao de, L.S., Oliviero, P. et al. (2012). Enrichment of *Pleurotus ostreatus* with selenium in coffee husks. Food Chem., 131(2): 558–563.

Stojkovic, D., Reis, F.S., Barros, L., Glamoclija, J., Ciric, A. et al. (2013). Nutrients and non-nutrients composition and bioaccumulation of wild and cultivated *Coprinus comatus* (O.F. Müll.) Pers. Food Chem. Toxicol., 59: 289–296.

Szabó, A., Gyepes, A., Nagy, A., Abrankó, L. and Győrfi, J. (2012). The effects of UVB radiation on the vitamin D content of white and cream type button mushrooms (*Agaricus bisporus* Lange/Imbach) and oyster mushroom (*Pleurotus ostreatus* (Jacq.) P. Kumm). Acta Aliment., 41: 187–196.

Ulzijargal, E. and Mau, J.L. (2011). Nutrient composition of culinary-medicinal mushroom fruiting bodies and mycelia. Int. J. Med. Mush., 13: 343–349.

Vanucci, L., Sima, P., Vetvicka, V. and Krizan, J. (2017). Lentinan properties in anticancer therapy: A review on the last 12-year literature. Am. J. Immunol., 13(11): 50–61.

Vaz, J.A., Barros, L., Nartins, A., Santos-Buelga, C., Vasconselos, M.H. and Ferreira, I.C.F.R. (2011). Chemical composition of wild edible mushrooms and antioxidant properties of their water soluble polysaccharidic and ethanolic fractions. Food Chem., 126: 610–616.

Vetter, J. (1993a). Chemical composition of eight edible mushrooms. Z. für Lebensm. Unters, Forschung, 196: 224–227 (in German).

Vetter, J. (1993b). Über die Aminosäure-zusammensetzung der eßbaren Russula (Täubling) und Agaricus (Champignon)-Großpilzarten. Z. Lebensm. Unters. Forsch., 197: 381–384.

Vetter, J. (2000a). New data on chemical composition of cultivated mushroom species–II. Int. Conf. Mush. Culti. (in Hungarian), Budapest, 22–23 May 2000, 51–54.

Vetter, J. (2000b). Trypsin inhibitor activity of basidiomycetous mushrooms. Eur. Food Res. Technol., 211: 346–348.

Vetter, J. (2003a). Monographic processing of the mineral element composition of mushrooms. Final Research Report of OTKA Project # 31702 (in Hungarian).

Vetter, J. (2003b). Data on sodium content of common edible mushrooms. Food Chem., 81(4): 589–593.

Vetter, J. (2005). Mineral composition of Basidiomes of Amanita species. Mycol. Res., 109(6): 746–750.

Vetter, J. (2007). Chitin content of cultivated mushrooms *Agaricus bisporus, Pleurotus ostreatus* and *Lentinula edodes*. Food Chem., 102: 6–9.

Vetter, J. (2010a). Nutritional values of fungi. pp. 48–63. *In:* Győrfi, J. and Mezőgazda, K. (eds.). Fungal Biology—Cultivation of Fungi. Budapest (in Hungarian).

Vetter, J. (2010b). Inorganic iodine content of common edible mushrooms. Acta Aliment., 39 (4): 424–430.

Vetter, J. and Lelley, J. (2004). Selenium level of the cultivated mushroom *Agaricus bisporus*. Acta Aliment., 333(3): 297–301.

Vetter, J. and Rimóczi, I. (1993). Roh-, verdauliche und unverdauliche Fruchtkörperproteine in Austernseitlingen (*Pleurotus ostreatus*) Pilze. Z. Lebensm, Unters. Forsch., 197: 427–428.

Vetter, J. and Siller, I. (1991). Chitin content of higher fungi. Z. Lebensmittel Untersuchung und Forschung, 193: 36–38 (in German).

Vetter, J., Hajdú, Cs., Győrfi, J. and Maszlavér, P. (2005). Mineral composition of the cultivated mushrooms *Agaricus bisporus, Pleurotus ostreatus* and *Lentinula edodes*. Acta Aliment., 34(4): 441–451.

Vetvicka, V. and Vetvickova, J. (2018). Glucans and cancer: Comparison of commercially available beta-glucans. Part IV, Anticancer Research, 38(3): 1327–1333.

Waktola, G. and Temesgen, T. (2020). Pharmacological activities of Oyster mushroom (*Pleurotus ostreatus*). Novel Research in Microbiology Journal, 4(2): 688–695.

Wang, Y. and Xu, B. (2014). Distribution of antioxidant activities and total phenolic contents in acetone, ethanol, water and hot water extracts from 20 edible mushrooms via sequential extraction. Austin J. of Nutrition and Food Sciences, 2(1).

Wang, X.M., Yhang, J., Wu, L.H., Yhao, Z.L., Li, T. et al. (2014). A mini-review of chemical composition and nutritional value of edible wild-grown mushrooms from China. Food Chem., 151: 279–285.

Wiseman, H. (1993). Vitamin D is a membrane antioxidant ability to inhibit iron-dependent lipid peroxidation in liposomes compared to cholesterol, ergosterol and tamoxifen and relevance to anticancer action. FEBS Lett., 326: 285–288.

Wu, W.J. and Ahn, Y.B. (2014). Statistical optimisation of ultraviolet irradiate conditions for vitamin D_2 synthesis in oyster mushrooms (*Pleurotus ostreatus*) using response surface methodology. PLoS ONE, 9: e95359.

Zacour, A.C., Silva, M.E., Cecon, P.R., Bambira, E.A. and Vieira, E.C. (1992). Effect of dietary chitin on cholesterol absorption and metabolism in rats. J. Nutr. Sci. Vitaminol., 38: 609–613.

Zlotko, K., Wiater, A., Wasko, A., Pleszczynska, M., Paduch, R. et al. (2019). A report on fungal $(1{\rightarrow}3)$-α-D-glucans: Properties, functions and application. Molecules, 24: 3972. doi.10.3390/molecules24213972.

12

White Rot Fungi in Food and Pharmaceutical Industries

Deepak K Rahi,[1], Sonu Rahi[2] and Ekta Chaudhary[1]*

1. INTRODUCTION

White rot fungi belong to the division Eumycota or true fungi (subdivision Basidiomycotina, class Hymenomycetes and subclass Holobasidiomycetidae) (Hawksworth et al., 1995). The subclass Holobasidiomycetidae consists of almost all wood decay fungi, litter decomposer fungi and mycorrhizal fungi. White rot fungi are ubiquitous in nature, particularly in hardwood rather than in softwood forests (Blanchette et al., 1990). Specific characteristic features of white rot fungi are documented in Table 1.

White rot fungi possess the ability to degrade almost all wood components (i.e., cellulose, hemicellulose and lignin), and preferentially lignin. The former are named simultaneous or non-selective white rot fungi, while the latter as selective white rot fungi. The selective white rot fungi are of special bioindustrial interest, since they remove lignin and leave the valuable cellulose intact (Dashtban et al., 2010). There are also white rots that cause both types of attacks within one substrate (Blanchette, 1984, 1985; Adaskaveg and Gilbertson, 1986). This ability is due to the secretion of extracellular non-specific ligninolytic enzymes during their secondary metabolism, which is usually triggered by nutrient exhaustion. The non-specificity of these enzymes enables them to transform a great variety of recalcitrant and hazardous components, such as polycyclic aromatic hydrocarbons (PAHs), pesticides, fuels, alkanes, polychlorinated biphenyls (PCBs), explosives and dyes. Besides, their extracellular nature allows these fungi to access non-polar and insoluble compounds. In addition to lignocellulolytic enzymes, white rot fungi are

[1] Department of Microbiology, Panjab University, Chandigarh-160014, India.
[2] Department of Botany, Government Girls College, A.P.S. University, Rewa-486003, India.
* Corresponding author: deepakraahi10@gmail.com

Table 1. Characteristic features of white rot families.

Characteristics & Important Species
ORDER I: AGARICALES
Family 1: Inocybaceae
Fruiting body: Small, fleshy, conical, fibrous with raised central knob, brown in colour, some liliac/ purple coloured species also exist; Stipe: thin, fibrous, cylindrical; Gills: closely places, white to brown in colour; Spores: brown, oval to ellipsoid in shape, smooth; Hyphae: clam connections present. **Habitat:** Living deciduous trees **Important species:** *Crepidotus* spp.
Family 2: Marasmiaceae
Basidiocarp: Convex to plano convex, fibrrilose, colour ranging from white to pale orange; Stipe: small, cylindrical, eccentric, white in colour, tapered at apex; Spores: ellipsoid, hyaline, thin walled; clamp connections present. **Habitat:** Decaying wood **Important species:** *Clitocybula* spp., *Crinipellis* spp., *Gerronema* spp. and *Marasmius* spp.
Family 3: Physalacriaceae
Fruiting body: Plane, umbonate, with decurved margin, greyish yellow to yellowish white in colour; Hyphae: generative, monomitic hyphae with clamp connections; Basidia: club shaped; Spores: ellipsoid, cylindrical, tear drop like shape, thin walled, hyaline. **Habitat:** Decaying wood **Important species:** *Armillaria* spp., *Gloiocephala* spp. and *Hymenopellis* spp.
Family 4: Pleurotaceae
Basidiocarp: Fleshy, laterally attached with no stipe; 14–15 cm in size; Upper surface irregular, white; Margin: wavy; Gills: white, decurrently arranged, 3–4 per cm, thick texture and fimbriate margin. Hyphal system: mostly monomitic, clamp connection present; Spores: smooth, elongated. **Habitat:** Dead hardwood or conifers **Important species:** *Hohenbuehelia* spp. and *Pleurotus* spp.
Family 5: Schizophyllaceae
Basidiocarp: Fan shaped when attached to the side of the log and irregular when attached at the top, white to grey in colour, densely covered by hair, 1–4 cm long; Gills: bear longitudinally divided gills, white to grey in colour; Hyphae: hyaline, septate, with clamp connections; Spores: cylindrical to elliptical; cystidia absent. **Habitat:** Decaying hardwood **Important species:** *Auriculariopsis* spp. and *Schizophyllum* spp.
Family 6: Tricholomataceae
Basidiocarp: Covex, broadly conical, covered with scales, dry to moist; Gills: free, thick, white in colour, unattached to the stipe, widely spaced; Stipe: thick, equal in width or club shaped; Spores: oval shaped with white spore print; Clamp connections present, cystidia absent. **Habitat:** Living tree **Important species:** *Clitocybe* spp., *Melanoleuca* spp., *Resupinatis* spp. and *Tricholoma* spp.
ORDER II: HYMENOCHAETALES
Family 1: Hymenochaetaceae
Basidiocarp: Annual, effuse-reflexed or pileated, turns black in the presence of an alkali; Hyphal system: generative hyphae, lacks clamp connections; thick walled thorn shaped cystidia present; Basidiospore: cylindrical or allantoid, hyaline, thin walled, smooth. **Habitat:** Dead wood and living tree **Important species:** *Fomitiporia* spp., *Hymenochaete* spp., *Inonotus* spp. and *Phellinus* spp.

Table 1 Contd. ...

...Table 1 Contd.

Characteristics & Important Species
ORDER III: POLYPORALES
Family 1: Ganodermataceae
Basidiocarp: Annual/perennial, laccate to non laccate, woody to corky, dimidiate, 10–15 cm in diameter; upper surface ranges from white to yellow to pale brown and dark brown with distinctive base; Hymenophore: upto 10 mm long; Pore surface: turns brown on bruising, 3–4 pores per mm, spherical to ovoid; Hyphal system: trimitic, occasionally dimitic or monomitic containing thin-walled clamped, branched, generative hyphae; Basidiospore: ellipsoid, double walled, inner wall is coloured and the surface is usually ornamental.   **Habitat:** Decaying wood or living trees   **Important Species:** *Amauroderma* spp., *Elfvingia* spp. and *Ganoderma* spp.
Family 2: Grifolaceae
Fruiting body: Fruiting body is 3–14 cm in size; fan shaped, dark to pale grey brown, often with concentric zones, velvety texture with wavy margins; Stem: branched, white, non-centric; Pore surface: non-bruising, 1–3 pores pre mm ; Spores: ellipsoid, smooth, hyaline; hymenial cystidia absent; Hyphae: dimitic, generative with clamp connections.   **Habitat:** Decaying wood and living trees grow together at the base of the tree, sharing a single stem   **Important Species:** *Grifola* spp.
Family 3: Meripilaceae
Basidiocarp: Fan shaped; tan to dull brown in colour which darkens with age, concentric rings present, fibrillose surface with tiny scales; Pore surface: 3–6 pores per mm, bruises brown and black; Spores: spherical to ovoid or elliptical, hyaline, smooth, non-amyloid; Hyphae: monomitic, generative.   **Habitat:** Hardwoods, rarely on conifers   **Important Species:** *Physisporinis* spp. and *Rigidoporus* spp.
Family 4: Meruliaceae
Basidiocarp: Funnel shaped, fan shaped, semicircular, crust like; Hyphal system: monomitic, generative hyphae with clamp connections; Cystidia: present in hymenium which is covered by folds, undulations, or ridges; Spores: smooth, thin walled, hyaline   **Habitat:** Decaying wood   **Important species:** *Bjerkandera* spp., *Gloeoporus* spp., *Hyphoderma* spp., *Irpex* spp., *Phlebia* spp. and *Steccherinum* spp.
Family 5: Phanerochaetaceae
Fruiting body: Fruiting body is membranous, crust like, no definite shape; Hyphal system: monomitic, septate generative hyphae, with single or multiple clamp connections; Basidiospores: cylindrical to ellipsoid, smooth, thin walled.   **Habitat:** Dead wood (conifers and hard wood)   **Important Species:** *Antrodiella* spp., *Ceriporia* spp., *Ceriporiospsis* spp. and *Phanerochaete* spp.
Family 6: Polyporaceae
Gilled fungi: Basidiocarp: 1–3 cm wide, broady convex with navel like depression, dry, fibrillose scaly surface, incurved margin; Gills: running down the stem, crowded, serrated edges, white to creamy in colour, covered with partial veil; stipe: 4–5 cm long, slightly tapered, scaly; Spores: ellipsoid, smooth, hyaline, inamyloid, thick walled hyhae with clamp connections.   **Porous fungi:** Basidiocarp: hoof or disc shapes attached directly to the substrate with no stipe, colour ranges from grey to black, concentric rings may be present; Pore surface: pale brown, with numerous pores and brown pore tubes; Hyphae: dimitic/trimitic which can be generative, skeletal or binding type, clamp connections present: Spores: cylindrical, large, hyaline, smooth.   **Habitat:** Decaying wood   **Important species:** *Cerrena* spp., *Coriolopsis* spp., *Fomes* spp., *Lentinus* spp., *Lenzites* spp., *Panus* spp., *Polyporus* spp., *Trametes* spp. and *Tyromyces* spp.

Table 1 Contd. ...

...Table 1 Contd.

Characteristics & Important Species
ORDER IV: RUSSULALES
Family 1: Stereaceae
Basidiocarp: Shelf-like, 0.5–2 cm in size, fused, white to greyish in colour, surface covered by coarse, stiff hair; Lower surface: smooth, reddish to yellow in colour, lacks gills or tubes; clamp connections absent and produce amyloid basidiospores. **Habitat:** Decaying wood **Important species:** *Aleurodiscus* spp. and *Stereum* spp.

capable of producing a large quantity of diverse metabolites of industrial significance (e.g., exopolysaccharides, bioactive proteins, terpenoides, phenolic compounds, sterols and so on). This makes white rot fungi very appealing for application in different industrial and biotechnological adventures. The implementation of such applications would contribute to the establishment of a more sustainable industrial development for circular economy. The present chapter intends to present the importance of white rot fungi in various food and pharmaceutical industries; more specifically food and pharmaceutical industries with examples of exopolysaccharides, bioactive proteins, terpenoids, phenolic compounds and sterols.

2. Applications in Food Industry

2.1 Foods, Beverages and Baking

Laccases produced by the white rot fungi have been widely used for varied applications in the food industry. They can be used in different processes to enhance or modify the appearance, texture and taste of foods and beverages. Carbohydrates, unsaturated fatty acids, phenols and thiol-containing proteins in foods and beverages are susceptible to laccases. Their modification by laccases may lead to new functionality, quality improvement, which leads to reduction in cost of processing by other methods.

In the beverage industry, an interesting application of laccases involves elimination of undesirable phenolic compounds, which are responsible for the development of turbidity, browning and haze formation in clear fruit juice, beer and wine. Fruit juices contain naturally-occurring phenolic compounds and their oxidation products contributing to their colour as well as taste. A higher concentration of phenolic compounds and polyphenols results in enzymatic darkening, causing undesirable changes in the colour as well as in aroma (Ribeiro et al., 2010). These undesirable changes can be avoided by applying laccase in combination with filtration during the stabilisation of fruit juices. Addition of laccase results in removal of phenols and increase in the stability colour. Laccase treatment has also been found to be more effective for flavour stability as compared to conventional treatments (e.g., addition of ascorbic acid and sulphites) (Minussi et al., 2002).

Laccases have the ability to crosslink biopolymers and are currently of great interest in baking as well as in other food industries. The baking industry utilises a variety of enzymes to improve the bread texture, volume, flavour and freshness.

Addition of laccases to the dough results in an oxidising effect, which causes improved gluten structure strength, stability, machinability and reduced stickiness in dough as well as baked products (Minussi et al., 2002).

2.2 Wine and Beer Stabilisation

Wine is a mixture of various chemical compounds, such as ethanol, organic acids, salts and phenolic compounds. The polyphenols present in wine need to be removed in order to avoid any undesirable changes in the characteristics of wine. Laccases from white rot fungi are used as an alternative to chemically synthesised adsorbents for the treatment of polyphenols since they are stable in acidic conditions and lack reversible inhibition with sulphite (Minussi et al., 2002). Laccases are also used for preparing cork stoppers, which oxidatively reduce the cork taint or astringency (Tanrioven and Eksi, 2005).

Haze formation is a common phenomenon, which occurs in beer during the cooling process. The haze formation occurs by small quantities of naturally-occurring proanthocyanidins and polyphenols which lead to protein precipitation that results in haze formation (Mathiasen, 1995). Laccase can be added at the end of the process in order to remove the unwanted oxygen in the finished beer and thereby increase its storage life. A commercialised laccase preparation named 'Flavourstar', manufactured by Novozymes A/S (public limited company), is marketed for using in brewing beer to prevent the formation of off-flavour compounds (e.g., trans-2-nonenal) by scavenging the oxygen, which otherwise would react with fatty acids, amino acids, proteins and alcohol to result in off-flavour (Olempska, 2004).

2.3 Delignification of Feedstock

Fossil fuels are considered as a major cause for global warming and to mitigate this problem, the biofuels have received substantial attention in the last few decades as sustainable alternatives (Santori et al., 2012). Production of different biofuels, such as the second-generation ethanol, is considered to be one such environment-friendly alternative (Fang et al., 2015). The second generation ethanol production is based on the non-nutritional (lignocellulosic substrates) and industrial waste products (whey or glycerol) (Balat et al., 2008; Thompson and Meyer, 2013). Production of ethanol using lignocellulosic biomass as an inexpensive source, has received global attention because of its renewable and eco-friendly nature. Delignification of lignocellulose is an important and challenging step in the bioconversion of lignocellulose to ethanol. Biological method of delignification using enzymes from the white rot fungi are now being considered the best option due to its mild reaction conditions, higher product yield and low energy demand (Sánchez et al., 2011). However, the process requires a long incubation period (several weeks to months) before obtaining the product (cellulose) recovery as it is obtained with the physical and chemical pretreatment methods (Kondo et al., 2014). In order to overcome these challenges, bioconversion of lignocelluloses with lignocellulolytic enzyme systems have been suggested as an effective treatment strategy (Mukhopadhyay et al., 2011; Wang et al., 2013). The suggested lignocellulolytic enzyme systems include LiP, MnP and laccase among others and the associated merits are reaction specificity and high product yield

Table 2. Delignification of substrates by ligninolytic enzymes from white rot fungi for ethanol production.

Substrate	Fungal strain	Enzyme involved	Ethanol yield	Reference
Cotton gin trash	*Trametes versicolor*	Laccase	31.6%	Placido et al., 2013
Eucalyptus wood	*Trametes villosa*	Laccase	16.2%	Gutierrez et al., 2012
Hardwood craft pulp	*Phlebia* sp. MG-60	Laccase, MnP, LiP	71.8%	Kamei et al., 2012
Oak wood	*Phlebia* sp. MG-60	Laccase, MnP, LiP	43.9%	Kamei et al., 2012
Oil palm empty fruit Bunches	*Pleurotus floridanus*	Laccase	27.9%	Ishola et al., 2014
Poplar wood	*Trametes velutina*	MnP, LiP, Laccase	22.2%	Wang et al., 2012
Sugar cane baggase	*Pleurotus ostreatus* IBL-02	Laccase, MnP, LiP	16.3 g/l	Asgher et al., 2013
Sugar cane baggase	*Phlebia* sp. MG-60	Laccase, MnP, LiP	45%	Kondo et al., 2014
Sugarcane baggase	*Phlebia* sp. MG-60	Laccase, MnP, LiP	65.7%	Kondo et al., 2014
Waste newspaper	*Phlebia* sp. MG-60	Laccase, MnP, LiP	51.1%	Kamei et al., 2012
Wheat straw	*Ceriporiopsis Subvermispora*	Laccase, MnP, LiP	90%	Salvachua et al., 2011
Wheat straw	*Irpex lacteus*	MnP, Laccase	74%	López-Abelairas et al., 2013

occasioned by non-utilisation of products as a source of energy (Wang et al., 2013; Ma and Ruan, 2015) (Table 2). Therefore, the enzymatic conversion of renewable lignocellulosic materials is an environment-friendly and sustainable alternative to fossil-derived fuels.

2.4 Wastewater Treatment

Phenolic compounds are the main pollutants present in food-industry effluents. The presence of phenolic compounds in drinking and irrigation waters or in cultivated lands is responsible for significant health as well as environmental hazards. The government policies on pollution control is becoming more and more stringent, forcing industries to adapt more effective treatment technologies. Laccase is capable of degrading phenolic compounds and these are utilised for bioremediation of food industry effluents. The application of laccase for bioremediation of wastewater streams is particularly of interest to beer factories. Fractions of wastewater released from beer factories contain a large amount of polyphenols and dark-brown appearance. Yague et al. (2000) reported that the laccase produced by the white rot fungus *Coriolopsis gallica* is capable of degrading polyphenols that are present in wastewaters. Other research by Gonzalez et al. (2000) utilised laccase from *Trametes* sp. for the bioremediation of distillery waste waters generated from ethanol production by using fermentation of sugarcane molasses with a high content of organic matter with intense dark brown appearance.

3. Applications in Pharmaceutical Industry

Enzymes derived from white rot fungi find application in many pharmaceutical industries. Laccases have been used extensively for synthesis of several products, like anticancer drugs, antioxidants, hormones and antiviral derivatives owing to the high value of their oxidation potential. Laccases also serve as important components in various cosmetic products, which reduce the toxicity (Arora et al., 2010). The first chemical product of pharmaceutical importance that has been prepared using laccase enzyme is actinocin, which has been obtained by the action of laccase on 4-methyl-3-hydroxyanthranilic acid. Actinocin has anticancer capability due to its ability to block the transcription of DNA in tumor cells (Burton, 2003). Vinblastine, another anticancer drug obtained as a natural product from the plant *Catharanthus roseus* is used in the treatment of leukemia. Nevertheless, the product quantity of the phytochemical vinblastine is very low. To enhance the synthesis of vinblastine-containing compounds, many precursors were identified. These precursors include katarantine and vindoline of which 40 per cent are converted into final products using the laccase enzyme. Similar methods are used for the preparation of various medically important products, especially antibiotics (Yaropolov et al., 1994; Pilz, 2003). Catechins are important antioxidant compounds which help to combat cancer, inflammation and cardiovascular diseases. However, catechins have low antioxidant ability and such limitation can be solved by laccases derived from white rot fungi by oxidation of catechins into small units of tannins (these are also found in tea, herbs and vegetables) with improved antioxidant abilities (Kurisawa et al., 2003).

3.1 Cosmeceutics

Application of different enzymes in cosmetics has been continuously increasing. Enzymes are used as free radical-scavengers in sunscreen cream, toothpaste, mouthwash, hair waving and dyeing. Recently, application of ligninoytic enzymes has been extended to the development of value-added cosmeceutical and dermatological products. Most notable products, include Melanozyme™ (lignin peroxidase-based product), which is marketed as '*elure*™ skin brightening cream' and luminase. These serve as catalytic skin-tone illuminators, which are manufactured by Syneron Medical Ltd, Irvine, California, USA. These are used in treatment of hyperpigmentation (sun spots or age spots) and skin lightening. The LiP used in the development of these skin-lightening products has solely been derived from the white rot fungus, *Phanerochaete chrysosporium*. Laccases, oxidases, peroxidases and polyphenol oxidases are also important enzymes used in hair dyes (Lang et al., 2004).

3.2 Synthesis of Organic Compounds

Organic synthesis leads to construction of complex organic compounds by a series of chemical reactions. Organic synthesis of chemicals has certain drawbacks, including the high cost of chemicals, cumbersome multi-step reactions and toxicity of reagents. Fungal laccases might prove useful in synthetic chemistry, where they may be applicable for production of complex polymers and medical products. Indeed, the application of laccases and peroxidases in organic synthesis has arisen due to its range of broad substrates and the conversion of substrates into unstable free (cation) radicals that may undergo further non-enzymatic reactions, such as polymerisation or hydration. The increased consumption of polymers, the growing concern for human health and environmental safety has led to utilisation of fungal enzymes in the synthesis of biodegradable polymers. *In vitro* enzyme-catalysed synthesis of polymer is an environmentally safe process having several advantages over conventional chemical methods (Vroman and Tighzert, 2009; Kadokawa and Kobayashi, 2010). Biopolymers are environment friendly as they are synthesised from renewable carbon sources via biological processes and degrade naturally as renewable resources (Hiraishi and Taguchi, 2009). Biopolymers, especially polyesters, polycarbonates and polyphosphates are used in various biomedical applications. For instance, biopolymers are immensely useful in orthopaedic devices, tissue engineering, adhesion barriers, control drug delivery and so on (Gunatillake and Adhikari, 2003; Ulery et al., 2011).

4. Exopolysaccharides and Applications

Extracellular polysaccharides (EPS) are high-molecular weight, hydrated, long-chain polysaccharides containing branched and repeating units of sugars or sugar derivatives, such as glucose, fructose, mannose and galactose, which get secreted into the surrounding environment during microbial growth (Ismail and Nampoothiri, 2010). The EPS consists of macromolecules, such as polysaccharides, proteins, nucleic acids, humic substances, lipids and other non-polymeric constituents of low molecular weight (Bramhachari and Dubey, 2007). Carbohydrates and proteins are usually the major and best investigated components of EPS (Frolund et al., 1996). In addition, some organic matters from the medium can also be adsorbed by the EPS matrix (Nielsen and Jahn, 1999; Liu and Fang, 2003). Production of EPS is a common property of microorganisms and occurs in prokaryotes (bacteria and archea) as well as eukaryotes (algae and fungi) under non-limited oxygen conditions. The EPS-producing microorganisms are found in a number of ecological niches. Ecological niches having a medium with high carbon/nitrogen ratio, such as effluents from the sugar, paper, food industries and wastewater plants harbour microbes producing polysaccharides.

The fungal kingdom is known to produce high-value exopolysaccharides with structural and functional specificities useful in various industrial and pharmacological applications. A number of fungi, including higher Basidiomycetes, lower filamentous fungi and yeasts are known for their ability to synthesise EPS in the laboratory cultures (Rahi and Rahi, 2010; Mahapatra et al., 2013). In recent years, the demand for natural polymers in various industrial applications has led to increased attention towards EPS production, especially from fungi due to their multifaceted applications (Yadav et al., 2014). The EPS of fungal origin can serve as an alternative to plant and algal products since their properties are almost identical to the currently used gums. In addition, the microbial EPS have unusual molecular structures with interesting conformations resulting in unique potential industrial applications (Rahi et al., 2018). Fungal exopolysaccharides (EPSs) have been recognised as high-value biomacromolecules since the last two decades (Rahi, 2015). However, fungal EPS of white rot fungi have gained much importance comparatively due to various applications that not only indicate the alternative source of marketed plant or seaweed polysaccharides, but also have some new and interesting bioapplicability. Table 3 summarises various EPS produced by white rot fungi and their applications.

Exopolysaccharides of Basidiomycetes, especially the white rots have been a matter of research as therapeutic agents and several bioactive substances possess health-promoting benefits. These polysaccharides act as immune modulators by enhancing human immune responses against cancer cells, bacteria, viruses and oxidative stress via non-specific mechanisms. These polysaccharides also act as dietary fibres with hypolipidemic, hypoglycemic, antiatherogenic and prebiotic properties (Kim et al., 2006; Stachowiak and Reguła, 2012; Giavasis, 2013; Mizuno and Nishitani, 2013). Numerous fungal formulations are used in traditional Eastern medicine for prevention and treatment of diseases like migraine, hypertension, arthritis, bronchitis, asthma, diabetes, hypercholesterolemia and hepatitis (Lawal et al., 2019). The white rot fungi are known to produce high-value exopoysaccharides; however, till date, only a few of these fungi have been explored for their ability to produce exopolysaccharides (Malik et al., 2018).

4.1 *Role as Prebiotics*

The market for functional foods has been increasing at a prolific rate since they provide additional health benefits than do the conventional foods and one such category of functional foods include the prebiotics (Rahi et al., 2018). Most of the bioactive mushroom polysaccharides can serve as dietary fibre, as they cannot be digested in the human intestine (Giavasis, 2013). Fungal glucans from *Cordyceps miltaris, Cordyceps sinensis, Ganoderma lucidum, Grifolla frondosa, Lentinus edodes, Pleurotus ostreatus* and *Schizophyllum commune* are able to reduce blood sugar and serum cholesterol according to the several animal and human trials (Lakhanpal and Rana, 2005; Lindequist et al., 2005; Chen et al., 2013). Hypoglycemic effects augment via attachment of the indigestible polysaccharides to the intestinal epithelium, which decelerates glucose absorption, while hypolipidemic effects are due to the interruption of the enterohepatic circulation of bile acids, favouring their excretion in the faeces (Lakhanpal and Rana, 2005; Giavasis, 2013). Bioactive white rot fungal polysaccharides exert a hypoglycemic effect via the modulation of carbohydrate metabolism and insulin synthesis. For instance, glucans from *Ganoderma frondosa* exhibit antidiabetic and antiobesity properties (Lakhanpal and Rana, 2005). Lentinan has been used as a hypocholesterolemic agent in humans owing its capacity to reduce the blood levels of lipoproteins (LDL as well as HDL) (Lakhanpal and Rana, 2005).

4.2 *Functional Foods and Nutraceutics*

Mushrooms and their products have been used as nutraceuticals since a long time (Rahi et al., 2004, 2005). A number of functional foods and nutraceuticals have been developed recently, based on these polysaccharides (Rahi and Soni, 2007; Giavasis, 2014). One such example is the beta-D-glucan isolated from the white rot fungus *Lentinus edodes*, which can be used as a replacement for wheat flour in fibre-rich baked foods to produce a novel functional food with low calories and high fibre

Table 3. Exopolysaccharides producing white rot fungi and their applications.

Organism	Nature of polysaccharide	Applications	References
Amauroderma rugosum	Glucan	Immunostimulatory; antioxidant properties	Zhang, 2017
Cerrena unicolor	Glucan	Antioxidant; antibacterial; antiviral	Jaszek et al., 2013
Flammulina velutipes	Glucan	Antioxidant; immunomodulatory	Wu et al., 2014; Ma et al., 2015
Fomes fomentarius	Glucan	Antioxidant; anticancer	Rehman et al., 2020
Funalia trogii	Glucan	anticancer; antioxidant	Yegenoglu et al., 2011
Ganoderma applanatum	Glucan	cytostatic; antibacterial	Osinska-Jaroszuk et al., 2014; Rahi and Barwal, 2015a
Ganoderma carnosum	Glucan	antimicrobial	Demir and Yamac, 2008
Ganoderma lucidum	Glucan	Anti-ageing; inhibit ROS production; lipid peroxidation; antifungal	Nirwan et al., 2016; Wang et al., 2017
Grifola frondosa	Grifolan	Inhibit glycogen synthase kinase/hypoglycemic activity	Ishibashi et al., 2001; Ma et al., 2014
Inonotus obliquus	Glucan	Cytotoxic/cytostatic; tumor growth inhibition; topoisomerase II inhibition (growth arrest)	Zhang et al., 2007; Zheng et al., 2011
Lentinula edodes	Lentinan	Anti-aging; antioxidant; antimicrobial	Mattila et al., 2000; Ina et al., 2013; Giavasis, 2014; Sugiyama, 2016
Lentinus strigosus	Lentinan	Antidiabetic	Demir and Yamac, 2008; Ina et al., 2013; Sugiyama, 2016
Pleurotus tuber-regium Singer	Beta-D-glucan	Anti-hyperglycemic; antioxidant	Zhang et al., 2007; Huang et al., 2012
Pleurotus citrinopileatus	Galactomannan	Anti-inflammatory	Zhang et al., 2007; Minato et al., 2019
Pleurotus eryngii	Glucan	Activation of macrophages; tumor growth inhibition	Ma et al., 2014
Pleurotus florida	Glucan	Antioxidant; anticancer	Latha and Bhaskar, 2014
Pleurotus ostreatus	Pleurotan	Antioxidant activity	Majtan, 2013; Barakat and Sadik, 2014
Pleurotus pulmonarius RDM9	Glucan	Antioxidant activity	Malik et al., 2019
Pleurotus sajor caju	Glucan	Immunostimulatory; antidiabetic; Antibacterial;	Rahi and Brwal, 2015b; Sermwittayawong et al., 2018
Polyporus arcularius	Glucan		Demir and Yamac, 2008
Schizophyllum commune	Schizophyllan	Antioxidant	Yogita et al., 2011; Zong et al., 2012; Zhang et al., 2013

Table 3 Contd. ...

...Table 3 Contd.

Organism	Nature of polysaccharide	Applications	References
Trametes versicolor	Polysaccharide-Krestin (PSK); Polysaccharopeptide (PSP)	Cytotoxic/antiproliferative; tumor growth inhibition; immunostimulation	Stachowiak and Reguła, 2012; Sugiyama, 2016; Friedmann, 2016

content. It also improves the pasting parameters, batter viscosity and elasticity (Kim et al., 2011). Glucans *L. edodes* are also added to noodles as a partial wheat flour replacement to increase the fibre content, antioxidant and hypocholesterolaemic effects with improved quality characteristics (Kim et al., 2008, 2009). It can also be utilised effectively as oil barriers and texture-enhancing ingredients in frying batters. Glucans obtained from *Ganoderma* spp. act as free radical-scavangers on addition to foods, prevents lipid peroxidation and enhances interferon synthesis in blood cells on consumption (Kozarski et al., 2011). Glucans derived from other white rot species, such as *Flammulina velutipes, Ganoderma lucidium, Lentinus edodes,* have notable antioxidant potential. Thus they are being proposed as natural antioxidant additives in foods (Giavasis, 2014). Cultures of *Schizophyllum commune* can be used to ferment lactose to produce beta-glucan cheese, which is known to have significant antithrombotic effects (Okamura-Matsui et al., 2001). These polysaccharides are also being used in formulation of functional snack foods. On addition of polysaccharide produced by *Pleurotus ostreatus* to the Indian *papad* snack results in improved fibre content. Another advantage of using these polysaccharides is that they contain monosodium glutamate-like components which possess savoury taste, helping in improving the flavour of the final product (Tsai et al., 2006).

4.3 Bioemulsification

A mixture of oil and water is generally termed as emulsion, where oil serves as dispersed phase and water serves as continuous phase. When kept still for a long time, the oil droplets begin to separate from water. To prevent such separation and to establish homogeneity, emulsifiers are useful. Emulsifiers are the amphiphatic molecules consisting of a hydrophilic head (directed towards aqueous phase) and a hydrophiobic tail (directed towards oil phase). The emulsifiers help in stabilising the emulsion by forming a physical barrier, which prevents the droplets from coalescing. Emulsifiers also help to increase the uniformity of nutrients, especially fatty acids, fat-soluble vitamins and amino acids (Calvo et al., 2004). Since the chemical emulsifiers are generally toxic and often non-biodegradable, various naturally-occurring non-toxic and biodegradable emulsifiers are used. Emulsifiers of microbial origin can solve the problem to a large extent and therefore there is a need to explore their possibilities to use on a regular basis (Yadav et al., 2014; Rahi and Urvashi, 2015). Bioemulsifiers are high molecular-weight amphiphatic compounds composed of a mixture of lipopolysaccharides, heteropolysaccharides, lipoproteins and proteins. Due to their properties, such as biodegradability, low toxicity, emulsifying and high surface activity, tolerance to pH, temperature and ionic strength, bioemulsifiers have a wide range of utility in food, cosmetic, pharmaceutical, detergent, textile and petroleum industries (Nitschke and Costa, 2007; Lourenço et al., 2018).

The emulsifying potential of exopolysaccharides produced by white rot fungi is made use of in enhancing the stability of nano- or microemulsions. In a study carried out by Guo et al. (2018), it was found that the *Ganoderma luicidium*-derived polysaccharide, when integrated with coix oil, the microemulsion becomes capable of enhancing the emulsion stability and in turn, its anti-tumor activity. In another study carried out by Gallotti et al. (2020), it was found that the beta-glucan extracts of *Pleurotus ostreatus* were capable of providing suitable emulsifying properties in spray drying as well as liquid emulsions. It also provides protection of vitamin E against oxidation. According to Veverka et al. (2018), beta-glucan isolated from *P. ostreatus* was capable of forming stable emulsion gels with various natural oils (olive oil, cocoa butter, coconut oil and linolenic acid) without addition of any consolvent or surfactant.

4.4 Bioflocculation

Flocculation is one of the essential parameters employed in removal of suspended solids from a solution for which a variety of natural and synthetic flocculants are employed. Algae, bacteria, actinomycetes and fungi have been reported to produce bioflocculants composed of substances like polysaccharides, proteins, glycoproteins or nucleic acids (Lazarova and Manem, 1995; Gao et al., 2006). Flocculants can be divided into three main categories: (1) inorganic flocculants (alum); (2) synthetic organic flocculants (polyacrylamide); (3) natural flocculants (chitosan, sodium alginate, cellulose, lignin and tannin). Synthetic inorganic and organic flocculants are widely used due to their low cost and better efficiency. However, they possess inherent drawbacks as sources of carcinogenic monomers and are often non-biodegradable (Okaiyeto et al., 2016). Hence, natural polysaccharide-based bioflocculants can be replaced since they are non-toxic, biodegradable and free from secondary pollution risks. Polysaccharide-based flocculants have several wide-range usage in wastewater treatment, textile, pharmacology, cosmetology and food industries (Saeed et al., 2011; Yang et al., 2016b; Grenda et al., 2017; Kolya et al., 2017).

A number of bacteria, fungi and algae have been studied for their bioflocculating potential. However, very few reports are available on the bioflocculant production by the white rot fungi. Chitosan is one of the polysaccharide-based flocculants produced by white rot fungal genera *Ganoderma* and *Pleurotus* (Kannan et al., 2010). According to Zhang et al. (2013), flocculation effect of polysaccharide from *Phanerochaete chrysosporium* on coal slurry showed the highest flocculation (93.5 per cent). It was inferred that *Phanerochaete chrysosporium* contains a huge amount of acidic polysaccharides, which were responsible for the very high rate of flocculation.

4.5 Antioxidant Potential

Fungi are known for abundant antioxidant compounds, like exopolysaccharides, which have proven to be effective in removing reactive oxygen species (ROS), such as superoxide, anions, hydrogen peroxide and hydroxyl radicals (Rahi et al., 2018).

Polysaccharides extracted from various white rot fungi, such as *Ganoderma lucidum, Lentinus edodes, Pleurotus linteus* and *Trametes versicolor* exhibit reducing and chelating properties, which inhibit lipid oxidation and reduce the oxidative stress. The beta-glucan produced by these fungi and some phenolic compounds, such as tyrosine, are responsible for the anti-oxidant effects (Kozarski et al., 2011; Giavasis, 2013). *In vitro* studies show that the *G. lucidum* peptidoglycan can also protect the mitochondria, endoplasmic reticulum and microvilli of macrophages against chemically-induced damage and malfunction (You and Lin, 2002). In addition, methanolic extracts of *G. lucidum* and *G. tsugae* glucans and proteoglucans act as antioxidants by scavenging ROS, which are linked to oncogenesis and to lipid oxidation (Lakhanpal and Rana, 2005). Another mechanism of antioxidant activity is the ability of white rot fungal polysaccharides (e.g., *G. lucidum*) to limit the production of oxygen-free radicals and the activity of peripheral mononuclear cells in murine peritoneal macrophages, which are related to the respiratory burst and ageing process or to enhance the activity of antioxidant enzymes in blood serum (You and Lin, 2002; XiaoPing et al., 2009).

4.6 Immunostimulation and Antitumor Properties

A number of *in vitro* studies on the human cell and various clinical trials have verified the immunomodulating or immunostimulating properties of polysaccharides produced by white rot fungi. Lentinan and schizophyllan are the most well-studied polysaccharides known to have immunomodulating properties by stimulation of secretion of tumor necrosis factor-α (TNF-α) by human monocytes and activation of macrophages. On administering a combination with the conventional antitumor drugs or radiotherapy, the polysaccharides prove effective against gastric, breast, lung, cervical and colorectal cancers. They are also known to prevent metastasis and minimise the side effects of chemotherapy and radiotherapy on healthy tissues (Ikekawa, 2001; Lo et al., 2011; Chang and Wasser, 2012; Stachowiak and Reguła, 2012; Giavasis, 2013). The glucans from *Grifola frondosa* and *Trametes versicolor* have also been used in clinical trials, where they have increased the survival rate and survival time of patients suffering from different types of cancers, especially when supplied in combination with chemotherapy or on administration after surgery in order to prevent metastasis (Lindequist et al., 2005; Stachowiak and Reguła, 2012). Ganoderan is another white rot fungal biopolymer that has been used as an adjuvant in cancer therapy as it increases the cytotoxic effect of chemotherapy besides enhancing the immune responses in patients with prostate cancer (Yuen and Gohel, 2005; Mahajna et al., 2008; Vannucci et al., 2013).

4.7 Antimicrobial Activities

Mushroom polysaccharides are active against many bacterial and viral infections as they stimulate the phagocytosis of microbes by neutrophils and macrophages. Antimicrobial activity has been shown by lentinan against *Escherichia coli, Listeria monocytogenes, Salmonella enteritis, Staphylococcus aureus* including the tuberculosis bacilli (Mattila et al., 2000; Giavasis, 2013). The antiviral activity of glucans has also been reported, which is possibly due to the increased release of

IFN-γ and enhanced proliferation of peripheral blood mononuclear cells (PBMC) (Lindequist et al., 2005; Sasidhara and Thirunalasundari, 2012) as also by the polysaccharide of *Ganoderma* sp. (ganoderan) against the herpes virus. Schizophyllan is another polysaccharide produced by *Schizophyllum* sp., which is known to enhance the immune responses in hepatitis B patients (Kakumu et al., 1991; Sasidhara and Thirunalasundari, 2012). The antimicrobial activity of mushroom glucans seems to be indirectly related to immunomodulation. It affects the intestinal microflora and is thus feasible in formulating the glucan-based nutraceuticals which have prophylactic antimicrobial properties (Van Nevel et al., 2003).

5. Bioactive Proteins and Applications

The white rot fungi produce many bioactive proteins and peptides, specifically lectins, immunomodulatory proteins (FIP), ribosome inactivating proteins (RIP), antimicrobial proteins, antifungal proteins, ribonucleases and laccases, which have become popular as natural antitumor, antiviral, antimicrobial, antioxidant and immunomodulatory agents.

5.1 Lectins

Lectins are non-immune proteins or glycoproteins which bind specifically to the cell surface carbohydrates, leading to cell agglutination. Lectins play a crucial role in various biological processes, such as cellular signalling, scavenging glycoproteins from the circulatory system, cell-to-cell interactions in the immune system, differentiation and protein targeting to cellular compartments, including the host defence mechanisms, inflammation and cancer. Among all the sources of lectins, plants have been most extensively studied. Recently, fungal lectins have attracted a wide attention owing to their antitumor, antiproliferative and immunomodulatory activities. White rot fungal genera *Ganoderma* and *Pleurotus* have been mainly studied for the production of lectins. Lectins produced by *Ganoderma capense* and *Pleurotus citrinopileatus* are known to possess mitogenic activity with respect to murine splenocytes. The lectins produced by *Pleurotus citrinopileatus* and *Schizophyllum commune* are known to inhibit HIV-1 reverse transcriptase (Han et al., 2005; Li et al., 2008).

5.2 Ribosome-inactivating Proteins

The ribosome inactivating proteins (RIP) are enzymes involved in inactivating ribosomes by eliminating one or more adenosine residues from rRNA. A number of RIPs have been purified from several white rot fungi (e.g., *Lyophyllum shimeiji* and *Pleurotus tuber-regium*) (Wang and Ng, 2000, 2001; Ng et al., 2003). The RIP produced by these species exhibit HIV-1 reverse transcriptase inhibitory, antifungal and antiproliferative activities.

5.3 Immunomodulatory Proteins

Fungal immunomodulatory proteins (FIP) are a new family of bioactive proteins obtained from some of edible and medicinal mushrooms (e.g., *Flammulina velutipes* and *Ganoderma lucidum*). The FIP possess antitumor activities, including inhibition of cell growth and proliferation, induction of apoptosis, autophagy and reduction of invasion and migration. Among the white rot fungi, the genus *Ganoderma* is widely studied for the production of FIP. A number of immunomodulatory proteins, such as GMI, ling zhi-8 (LZ-8) and Se-containing protein, Se-GL-P, have been isolated from *Ganoderma* sp. and they are known to suppress tumor invasion and metastatis by inhibiting epidermal growth-factor-mediated migration (Kino et al., 1989; Du et al., 2007; Lin et al., 2010). Another protein, reLZ-8 (a recombinant immunomodulatory protein derived from *Ganoderma lucidum*), has been reported to induce IL-2 gene expression through the Sac-family protein, tyrosine kinase, reactive oxygen species and different protein kinase-dependent signalling pathways. A glycoprotein PCP-3A, isolated from the fresh fruit body of golden oyster mushroom *Pleurotus citrinopileatus*, is known to inhibit the proliferation of human tumor cell lines (Chen et al., 2009).

5.4 Other Proteins and Peptides

Table 4 summarises various bioactive proteins produced by white rot fungi and their impact. A potent antifungal protein lentin was isolated by Ngai and Ng (2003) from the fruit bodies of white rot fungus, *Lentinus edodes*. This protein is capable of inhibiting the mycelial growth of a variety of fungal species, such as *Botrytis cinerea, Mycosphaerella arachidicola* and *Physalospora piricola*. It is also known for its inhibitory activity against HIV-1 reverse transcriptase and proliferation of leukemia cells. Another protein, ganodermin, isolated from *G. lucidium* by Wang and Ng (2006b) inhibits the mycelial growth of *Botrytis cinerea* as well as *Fusarium oxysporum*. Antibacterial protein, isolated by Zheng et al. (2010) from the dried fruit bodies of *Clitocybe sinopica*, is known for its potential antibacterial activity against *Agrobacterium rhizogenes, A. tumefaciens, A. vitis, Xanthomonas oryzae* and *X. malvacearum*.

6. Terpenoids and Applications

Terpenoids are naturally-occurring hydrocarbons composed of basic five carbon units, isopentenyl diphosphate and its isomer dimethylallyl diphosphate coupled by prenyltransferase enzymes. Depending on the carbon units, terpenoids are divided into monoterpenes, sesquiterpenes, diterpenes and triterpenes. Monoterpenes are derived from geranyl diphosphate (GPP, C10), sesquiterpenes from farnesyl diphosphate (FPP, C15) and diterpenes from geranylgeranyl diphosphate (GGPP, C20) (Davis and Croteau, 2000). More than 55,000 molecules of terpenoids have been discovered till date, mainly as plant products, but a number of terpenoids have been isolated from the microbial source as well. Recent advances in genome sequencing for Basidiomycota suggest that white rot mushrooms have the potential to produce hundreds of terpenoids, which have so far not been identified precisely. Fungal terpenoids have proven to be a rich source of valuable bioactive products. Terpenoids, isolated from various white rot fungi, have been mainly studied for their therapeutic potential, like anticancer (Wang et al., 2013), antimalarial, anticholinesterase (Lee

Table 4. Bioactive proteins produced by white rot fungi.

Category	Bioactive agent	Mushroom	Impact	References
Lectins	Lectins	*Ganoderma capense, Grifola frondosa, Pleurotus citrinopileatus* and *Schizophyllum commune*	Increases insulin secretion, decrease blood glucose, improves ovulation; mitogenic activity and HIV-1 reverse transcriptase inhibitory activity	Han et al., 2005; Lakhanpal and Rana, 2005; Li et al., 2008; Suwannarach et al., 2020
FIP's (Fungal immunomodulatory proteins)	FIP-gat Fip-SJ75 FIP-gsi FIP-gts Fip-lti1, Fip-lti2	*Ganoderma atrum, Ganoderma lucidum, Ganoderma sinensis, Ganoderma tsugae* and *Lentinus tigrinus*	Stimulating lymphocyte mitogenesis; enhances transcription of IL-2; induces apoptosis via autophagy; cytokines regulation (IL-2, IL-3, IL-4); cytokines regulation (TNF-alpha, IL-beta1, and IL-6); activating macrophage M1 polarisation and initiating pro-inflammatory response; lymphocytes activator and cytokine regulation	Brown et al., 2003; Li et al., 2014; Li et al., 2015; Xu et al., 2016; Gao et al., 2019; Suwannarach et al., 2020
RIP's (Ribosome inactivating protein)	Pleuturegin	*Pleurotus tuber-regium*	HIV-1 reverse transcriptase inhibitory activity; anti-inflammatory and antiproliferative	Wang and Ng, 2000, 2001; Ng et al., 2003
	GMI, ling zhi-8 (LZ-8)	*Ganoderma* sp.	Antitumor activity	Du et al., 2007; Lin et al., 2010
Other proteins and peptides	Lentin	*Lentinus edodes*	Antifungal activity and HIV-1 reverse transcriptase inhibitory activity	Ngai and Ng, 2003
	Ganodermin	*Ganoderma lucidium*	Antifungal	Wang and Ng, 2006

et al., 2011), antiviral (Mothana et al., 2003), antibacterial (Arpha et al., 2012) and anti-inflammatory activities (Kamo et al., 2004) (Table 5).

7. Phenolic Compounds and Their Applications

Phenolic compounds are aromatic hydroxylated compounds produced by plants and fungi. They are classified into two main groups: flavonoids and non-flavinoids. Flavonoids are composed of two aromatic rings linked through an oxygen heterocycle. They can be sub-classified as flavonols, flavones, isoflavones, anthocyanins, proanthocyanidins, flavanones and so on. Flavonoids are mostly found in the form of glycoside. The main sugars to which they are linked are glucose, rhamnose, galactose and xylose. Non-flavonoids, like benzoic and cinnamic compounds, are commonly

Table 5. Terpenoids produced by white rot fungi and their biological activity.

Terpenoid	Mushroom	Compound	Activity	References
Monoterpenoids	*Pleurotus cornucopiae*	Pleurospiroketal	Cytotoxic	Wang et al., 2013
Sesquiterpinoids	*Pleurotus cornucopiae*	Pleurospiroketal A Pleurospiroketal B Pleurospiroketal C	Inhibits nitrogen oxide production; cytotoxic	Wang et al., 2013
		Pleurospiroketal D Pleurospiroketal E	Inhibits nitrogen oxide production; cytotoxic	Wang et al., 2013
Diterpenoids	*Pleurotus eryngii*	Eryngiolide A	Cytotoxic	Wang et al., 2012
	Tricholoma sp.	Tricholomalide A Tricholomalide B Tricholomalide C	Cytotoxic	Tsukamoto et al., 2003
Triterpenoids	*Ganoderma lucidum*	Methyl ganoderate A acetonide n-Butyl ganoderate H Methyl ganoderate A Ganoderic acid B Ganoderic acid E	Anticholinesterase	
		Lucidenic acid N Lucidenic acid A	Anti-invasive	Weng et al., 2007; Lee et al., 2011
		Ganoderic acid Sz, Ganoderic acid C1	Anticomplement	Seo et al., 2009
	Ganoderma amboinenese	Ganodermacetal Methyl ganoderate C Ganoderic acid X	Toxic activity against brine shrimp larvae; cytotoxic	El Dine et al., 2008; Yang et al., 2012
	Ganoderma boninense	Ganoboninketal A Ganoboninketal B Ganoboninketal C	Antiplasmodial NO inhibition; cytotoxic	Ma et al., 2014
	Ganoderma colossum	Lucidenic acid N Lucidenic acid A	Viral protease	El Dine et al., 2008
	Ganoderma tsuage	Ganoderic acid Sz, Ganoderic acid C1	Cytotoxic	Su et al., 2000

called phenolic acids. They contain an aromatic ring attached to different functional groups or esterified to organic acids. Some other phenolic compounds are stilbenes, tannins, lignins and lignans. Several properties, like colour, flavour and astringency are caused by these compounds (Gutiérrez-Grijalva et al., 2016). Phenolic acids are the main phenolic compounds found in white rot mushrooms and these can be divided into two major groups: hydroxybenzoic acids and hydroxycinnamic acids (Heleno et al., 2015) (Table 6). The phenolic compounds have been the focus due to their properties as antioxidant, anti-inflammatory, anticancer agents and so on (Puttaraju et al., 2006; Alves et al., 2013) (Table 6).

Table 6. Phenolic compounds produced by different white rot fungi and their health benefits.

Phenolic compound	Mushroom	Health benefits	References
5-O-Caffeoylquinic acid	*Phellinius lentius* and *Pleurotus ostreatus*	Antioxidant and antitumor	Kim et al., 2008
Caffeic acid	*Phellinius lentius, Pleurotus eryngii* and *Pleurotus djamor*	Antimicrobial and antioxidant	Puttaraju et al., 2006; Piazzon et al., 2012; Alves et al., 2013
Cinnamic acid	*Lentinus squarrulosus, Pleurotus djamor, Pleurotus ostreatus* and *Pleurotus sajor-caju*	Antimicrobial	Puttaraju et al., 2006; Kim et al., 2008
Chrysin	*Pleurotus ostreatus*		Puttaraju et al., 2006
Ferulic acid	*Pleurotus djamor, Pleurotus ostreatus* and *Pleurotus sajor-caju*	Antimicrobial and antioxidant	Puttaraju et al., 2006; Piazzon et al., 2012; Alves et al., 2013
Gallic acid	*Ganoderma lucidum, Pleurotus djamor, Pleurotus eryngii, Pleurotus linteus, Pleurotus ostreatus* and *Pleurotus sajor-caju*	Antineoplastic, bacteriostatic, antimelanogenic, antioxidant properties and anticancer	Puttaraju et al., 2006; Kim, 2007
Gentisic acid	*Pleurotus djamor* and *Pleurotus sajor-caju*	Anti-inflammatory, antirheumatic, analgesic activities and cytostatic agent	Ashidate et al., 2005; Puttaraju et al., 2006
Homogentisic acid	*Pleurotus ostreatus*	Antioxidant and antibacterial	Kim et al., 2008
Myricetin	*Ganoderma lucidum* and *Pleurotus ostreatus*	Antimicrobial and antioxidant	Kim et al., 2008
Naringin	*Ganoderma lucidum* and *Pleurotus ostreatus*	Antimicrobial and antioxidant	Kim et al., 2008
p-hydroxybenzoic acid	*Ganoderma lucidium, Lentinus edodes, Pleurotus eryngii, Phellinus lentius* and *Pleurotus ostreatus*	Antioxidant and anti-inflammatory	Mattila et al., 2001; Barros et al., 2008; Kim et al., 2008
Protocatechuic acid	*Ganoderma lucidum, Lentinus edodes, Lentinus squarrulosus, Lentinus sajor caju, Pleurotus ostreatus* and *Pleurotus sajor-caju*	Antioxidant, antimicrobial, cytotoxic, apoptotic and neuroprotective	Puttaraju et al., 2006; Yip et al., 2006; Kim et al., 2008; Yin et al., 2009; Alves et al., 2013
Pyrogallol	*Ganoderma lucidum*	Antioxidant and anti-inflammatory	Kim et al., 2008; Nicolis et al., 2008
Quercitin	*Ganoderma lucidium* and *Pleurotus ostreatus*	Antioxidant and anti-inflammatory	Kim et al., 2008
Vanillic acid	*Pleurotus djamor, Pleurotus eryngi* and *Pleurotus sajorcaju*	Antisicking, anthelmintic activities and suppress hepatic fibrosis in chronic liver injury	Puttaraju et al., 2006

8. Sterols and Applications

Sterols or steroid alcohols are lipid molecules occurring in plants, animals and fungi. Sterols occurring in plants are known as phytosterols, while those in animals are called zoosterol. Ergosterol is the sterol present in the cell membranes of the fungi. Ergosterol is a phytosterol consisting of ergostane with double bonds at the 5, 6, 7, 8- and 22, 23 positions and a 3-beta-hydroxy group (Fig. 1). It is the major product of sterol biosynthesis in fungi (and also in some trypanosomes) and has been found essential in the aerobic growth of most fungi. Ergosterol contents in basidiomycetous fungi are between 40–85 per cent of the total sterols (Lösel, 1988; Weete, 1989; Weete and Gandhi, 1996).

The fungal sterol, ergosterol, is abundant in mushrooms, including white rot fungi and is known to serve as pro-vitamin D_2 and is converted by ultraviolet radiation into ergocalciferol or vitamin D_2, which is a nutritional factor that promotes bone development in humans and other mammals. *Cantharellus cibarius, Craterellus tubaeformis, Cordyces sinensis, Ganoderma lucidium, Lentinus edodes* and *Pleurotus ostreatus* are a few white rot fungi previously studied for ergosterol production (Teichmann et al., 2007; Yuan et al., 2007). Ergosterol has shown to inhibit phorbol-12-myristate 13-acetate (TPA)-induced inflammatory ear oedema in mice (Yasukawa et al., 1994) and vitamin D_2 has been shown to contribute to prevention of prostate and colon cancers (Guyton et al., 2003). The peroxide of ergosterol, $5\alpha,8\alpha$-epidioxy-22E-ergosta-6, 22-dien-3β-ol (ergosterol peroxide), is common in mushrooms (Bok et al., 1999; Yaoita et al., 2002; Takei et al., 2005) and has been shown to inhibit the growth of some cancer cells and induce apoptosis of HL60 human leukaemia cells (Bok et al., 1999; Takei et al., 2005). Ergosterol peroxide inhibits TPA-induced inflammation and tumour promotion in mice (Yasukawa et al., 1994), while decreasing lipid peroxidation of rat liver microsomes and suppresses proliferation of mouse and human lymphocytes stimulated with mitogens (Fujimoto et al., 1994; Kim et al., 1999; Kuo et al., 2003). Thus, ergosterol peroxide shows antitumour, antioxidant and immunosuppressive properties, but the molecular mechanism of its antitumour action has not been clarified.

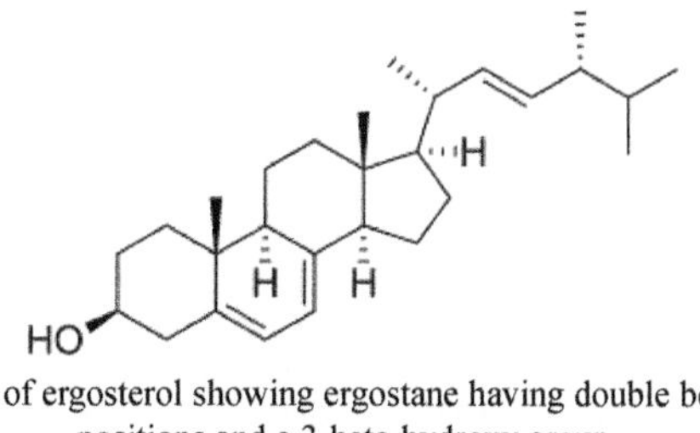

Fig. 1. Molecular structure of ergosterol showing ergostane having double bonds at 5, 6, 7, 8 and 22, 23 positions and a 3-beta-hydroxy group.

9. Conclusion

White rot fungi have immense value of nutritional as well as pharmaceutical significance. They are capable of producing a variety of enzymes, especially laccases and peroxidases of industrial importance. Besides paper and textile industries, enzymes produced by the white rot fungi are used in food and pharmaceutical industries. White rot fungal enzymes are useful in production of functional foods, beverages, value-added cosmetics and delignification of lignocellulose to produce ethanol. They are also useful in bioremediation processes, like wastewater treatment of paper and textile industries. White rot fungi produce exopolysaccharides, bioactive proteins, terpenoids, phenolic compounds and sterols. Their exopolysaccharides serve as prebiotics in functional foods, useful in bioemulsification, biofloculation, serve as antioxidants, antimicrobial agents and possess anticancer and immunomodulatory properties. Owing to their broad range of applications, white rot fungal enzymes, proteins and polysaccharides serve as essential commodities in the food and pharmaceutical industries. Their metabolites are helpful in environment-friendly bioremediation as well as in treating several human lifestyle diseases. Utilisation of white rot fungal products is still in its infancy and needs focused research attention for fullest utilisation of their potential in the food and pharmaceutical industries.

References

Adaskaveg, J.E. and Gilbertson, R.L. (1986). *In vitro* decay studies of selective delignification and simultaneous decay by the white rot fungi *Ganoderma lucidum* and *G. tsugae*. Can. J. Bot., 64(8): 1611–1619.

Alves, M.J., Ferreira, I.C., Froufe, H.J., Abreu, R.M.V., Martins, A. and Pintado, M. (2013). Antimicrobial activity of phenolic compounds identified in wild mushrooms, SAR analysis and docking studies. J. Appl. Microbiol., 115(2): 346–357.

Arora, N., Agarwal, S. and Murthy, R.S.R. (2012). Latest technology advances in cosmaceuticals. Int. J. Pharm., 4(3): 168–182.

Arora, D.S. and Sharma, R.K. (2010). Ligninolytic fungal laccases and their biotechnological applications. Appl. Biochem. Biotechnol., 160(6): 1760–1788.

Arpha, K., Phosri, C., Suwannasai, N., Mongkolthanaruk, W. and Sodngam, S. (2012). Astraodoric acids A-D: New lanostane triterpenes from edible mushroom *Astraeus odoratus* and their anti-Mycobacterium tuberculosis H37Ra and cytotoxic activity. J. Agric. Food Chem., 60(39): 9834–9841.

Asgher, M., Ahmed, N. and Iqbal, H.M.N. (2011). Hyperproductivity of extracellular enzymes from indigenous white rot fungi (*Phanerochaete chrysosporium*) by utilising agro-wastes. Bioresources, 6(4): 4454–4467.

Asgher, M., Yasmeen, Q. and Iqbal, H.M.N. (2013). Enhanced decolourisation of solar brilliant red 80 textile dye by an indigenous white rot fungus *Schizophyllum commune* IBL-06. Saudi J. Biol. Sci., 20(4): 347–352.

Ashidate, K., Kawamura, M., Mimura, D., Tohda, H., Miyazaki, S. et al. (2005). Gentisic acid, an aspirin metabolite, inhibits oxidation of low-density lipoprotein and the formation of cholesterol ester hydroperoxides in human plasma. Eur. J. Pharmacol., 513(3): 173–179.

Balat, M., Balat, H. and Öz, C. (2008). Progress in bioethanol processing. Prog. Energy Combust., 34(5): 551–573.

Barakat, O.S. and Sadik, M.W. (2014). Mycelial growth and bioactive substance production of pleurotusostreatusin submerged culture. Int. J. Curr. Microbiol. Appl. Sci., 3(4): 1073–1085.

Barros, L., Dueñas, M., Ferreira, I.C., Baptista, P. and Santos-Buelga, C. (2009). Phenolic acids determination by HPLC–DAD–ESI/MS in sixteen different Portuguese wild mushrooms species. Food Chem. Toxicol., 47(6): 1076–1079.

Blanchette, R.A. (1984). Screening wood decayed by white rot fungi for preferential lignin degradation. Appl. Environ. Microbiol., 48(3): 647–653.

Blanchette, R.A., Otjen, L., Effland, M.J. and Eslyn, W.E. (1985). Changes in structural and chemical components of wood delignified by fungi. Wood Sci. Technol., 19(1): 35–46.

Blanchette, R.A., Nilsson, T., Daniel, G. and Abad, A. (1990). Biological degradation of wood. Adv. Chem. Ser., 225: 141–174.

Bok, J.W., Lermer, L., Chilton, J., Klingeman, H.G. and Towers, G.N. (1999). Antitumor sterols from the mycelia of *Cordyceps sinensis*. Phytochemistry, 51(7): 891–898.

Bramhachari, P.V., Kishor, P.K., Ramadevi, R., Rao, B.R. and Dubey, S.K. (2007). Isolation and characterisation of mucous exopolysaccharide (EPS) produced by *Vibrio furnissii* strain VB0S3. J. Microbiol. Biotechn., 17(1): 44–51.

Brown, G.D., Herre, J., Williams, D.L., Willment, J.A., Marshall, A.S. and Gordon, S. (2003). Dectin-1 mediates the biological effects of β-glucans. J. Exp. Med., 197(9): 1119–1124.

Burton, S.G. (2003). Oxidising enzymes as biocatalysts. Tr. Biotechnol., 21(12): 543–549.

Calvo, C., Toledo, F.L., Pozo, C., Martínez-Toledo, M.V. and González-López, J. (2004). Biotechnology of bioemulsifiers produced by micro-organisms. J. Food Agric. Environ., 2(3): 238–243.

Calvo, C., Manzanera, M., Silva-Castro, G.A., Uad, I. and González-López, J. (2009). Application of bioemulsifiers in soil oil bioremediation processes: Future prospects. Sci. Total Environ., 407(12): 3634–3640.

Chang, S.T. and Buswell, J.A. (1996). Mushroom nutriceuticals. World J. Microb. Biotechnol., 12(5): 473–476.

Chang, S.T. and Wasser, S.P. (2012). The role of culinary-medicinal mushrooms on human welfare with a pyramid model for human health. Int. J. Med. Mushrooms, 14(2): 95–134.

Chen, J.N., Wang, Y.T. and Wu, J.S.B. (2009). A glycoprotein extracted from golden oyster mushroom *Pleurotus citrinopileatus* exhibiting growth inhibitory effect against U937 leukemia cells. J. Agric. Food Chem., 57(15): 6706–6711.

Chen, X., Siu, K.C., Cheung, Y.C. and Wu, J.Y. (2014). Structure and properties of a $(1 \rightarrow 3)$-β-d-glucan from ultrasound-degraded exopolysaccharides of a medicinal fungus. Carbhyd. Polm., 106: 270–275.

Chen, F., Long, X., Yu, M., Liu, Z., Liu, L. and Shao, H. (2013). Phenolics and antifungal activities analysis in industrial crop Jerusalem artichoke (*Helianthus tuberosus* L.) leaves. Ind. Crop. Prod., 47: 339–345.

Cheng, J.J., Lin, C.Y., Lur, H.S., Chen, H.P. and Lu, M.K. (2008). Properties and biological functions of polysaccharides and ethanolic extracts isolated from medicinal fungus, *Fomitopsis pinicola*. Proc. Biochem., 43(8): 829–834.

Dashtban, M., Schraft, H., Syed, T.A. and Qin, W. (2010). Fungal biodegradation and enzymatic modification of lignin. Int. J. Biochem. Mol. Biol., 1(1): 36–50.

Davis, E.M. and Croteau, R. (2000). Cyclisation enzymes in the biosynthesis of monoterpenes, sesquiterpenes, and diterpenes. pp. 53–59. *In*: Leeper, F.J. and Vederas, J.C. (eds.). Biosynthesis. Springer, Berlin and Heidelberg.

Demir, M.S. and Yamac, M. (2008). Antimicrobial activities of basidiocarp, submerged mycelium and exopolysaccharide of some native Basidiomycetes strains. J. Appl. Biol. Sci., 2(3): 89–93.

Dou, H., Chang, Y. and Zhang, L. (2019). *Coriolus versicolor* polysaccharopeptide as an immunotherapeutic in China. Prog. Mol. Biol. Transl. Sci., 163: 361–381.

Du, M., Zhao, L., Li, C., Zhao, G. and Hu, X. (2007). Purification and characterisation of a novel fungi Se-containing protein from Se-enriched *Ganoderma lucidum* mushroom and its Se-dependent radical scavenging activity. Eur. Food Res. Technol., 224(5): 659–665.

El Dine, R.S., El Halawany, A.M., Ma, C.M. and Hattori, M. 2008. Anti-HIV-1 protease activity of lanostane triterpenes from the vietnamese mushroom *Ganoderma colossum*. J. Nat. Prod., 71(6): 1022–1026.

Fang, Z., Liu, X., Chen, L., Shen, Y., Zhang, X. et al. (2015). Identification of a laccase Glac15 from *Ganoderma lucidum* 77002 and its application in bioethanol production. Biotechnol. Biofuels, 8(1): 1–12.

Friedman, M. (2016). Mushroom polysaccharides: Chemistry and antiobesity, anti-diabetes, anticancer, and antibiotic properties in cells, rodents and humans. Foods, 5(4): 80. 10.3390/foods5040080.

Frolund, B., Palmgren, R., Keiding, K. and Nielsen, P.H. (1996). Extraction of extracellular polymers from activated sludge using a cation exchange resin. Water Res., 30: 1749–1758.

Fujimoto, H., Nakayama, M., Nakayama, Y. and Yamazaki, M. (1994). Isolation and characterisation of immunosuppressive components of three mushrooms, P*isolithus tinctorius, Microporus flabelliformis* and *Lenzites betulina*. Chem. Pharm. Bull., 42(3): 694–697.

Gallotti, F., Lavelli, V. and Turchiuli, C. (2020). Application of *Pleurotus ostreatus* β-glucans for oil–in–water emulsions encapsulation in powder. Food Hydrocoll., 105841.

Gao, J., Bao, H.Y., Xin, M.X., Liu, Y.X., Li, Q. and Zhang, Y.F. (2006). Characterisation of a bioflocculant from a newly isolated *Vagococcus* sp. W31. J. Zhejiang Univ. Sci. B, 7(3): 186–192.

Gao, Y., Wáng, Y., Wāng, Y., Wu, Y., Chen, H. et al. (2019). Protective function of novel fungal immunomodulatory proteins fip-lti1 and fip-lti2 from *Lentinus tigrinus* in concanavalin a-induced liver oxidative injury. Oxid. Med. Cell. Longev., 3139689. 10.1155/2019/3139689.

Giavasis, I. (2013). Production of microbial polysaccharides for use in food. pp. 413–468. *In*: McNeil, B., Archer, D., Giavasis, I. and Harvey, L. (eds.). Microbial Production of Food Ingredients, Enzymes and Nutraceuticals. Woodhead Publishing.

Giavasis, I. (2014). Bioactive fungal polysaccharides as potential functional ingredients in food and nutraceuticals. Curr. Opin. Biotechnol., 26: 162–173.

Gonzalez, L.F., Sarria, V. and Sánchez, O.F. (2010). Degradation of chlorophenols by sequential biological-advanced oxidative process using *Trametes pubescens* and TiO_2/UV. Bioresour. Technol., 101(10): 3493–3499.

Grenda, K., Arnold, J., Gamelas, J.A. and Rasteiro, M.G. (2017). Environmentally friendly cellulose-based polyelectrolytes in wastewater treatment. Water Sci. Technol., 76(6): 1490–1499.

Gunatillake, P.A. and Adhikari, R. (2003). Biodegradable synthetic polymers for tissue engineering. Eur. Cell Mater., 5(1): 1–16.

Guo, J., Yuan, C., Huang, M., Liu, Y., Chen, Y. et al. (2018). *Ganoderma lucidum*-derived polysaccharide enhances coix oil-based microemulsion on stability and lung cancer-targeted therapy. Drug Delivery., 25(1): 1802–1810.

Gutiérrez, A., Rencoret, J., Cadena, E.M., Rico, A., Barth, D. et al. (2012). Demonstration of laccase-based removal of lignin from wood and non-wood plant feedstocks. Bioresour. Technol., 119: 114–122.

Gutiérrez-Grijalva, E.P., Picos-Salas, M.A., Leyva-López, N., Criollo-Mendoza, M.S., Vazquez-Olivo, G. and Heredia, J.B. (2018). Flavonoids and phenolic acids from oregano: occurrence, biological activity and health benefits. Plants, 7(1): 2. 10.3390/plants7010002.

Guyton, K.Z., Kensler, T.W. and Posner, G.H. (2003). Vitamin D and vitamin D analogs as cancer chemopreventive agents. Nutr. Rev., 61(7): 227–238.

Hammel, K.E. and Cullen, D. (2008). Role of fungal peroxidases in biological ligninolysis. Curr. Opin. Plant Biol., 11(3): 349–355.

Han, C.H., Liu, Q.H., Ng, T.B. and Wang, H.X. (2005). A novel homodimeric lactose-binding lectin from the edible split gill medicinal mushroom *Schizophyllum commune*. Biochem. Bioph. Res. Com., 336(1): 252–257.

Hawksworth, D.L. (1995). Biodiversity: Measurement and Estimation. Springer Science & Business Media.

Heleno, S.A., Barros, L., Martins, A., Queiroz, M.J.R., Santos-Buelga, C. and Ferreira, I.C. (2012). Fruiting body, spores and *in vitro* produced mycelium of *Ganoderma lucidum* from Northeast Portugal: A comparative study of the antioxidant potential of phenolic and polysaccharidic extracts. Food Res. Int., 46(1): 135–140.

Heleno, S.A., Ferreira, I.C., Esteves, A.P., Ćirić, A., Glamočlija, J. et al. (2013). Antimicrobial and demelanising activity of *Ganoderma lucidum* extract, p-hydroxybenzoic and cinnamic acids and their synthetic acetylated glucuronide methyl esters. Food Chem. Toxicol., 58: 95–100.

Heleno, S.A., Martins, A., Queiroz, M.J.R. and Ferreira, I.C. (2015). Bioactivity of phenolic acids: Metabolites versus parent compounds: A review. Food Chem., 173: 501–513.

Hiraishi, T. and Taguchi, S. (2009). Enzyme-catalysed synthesis and degradation of biopolymers. Mini. Rev. Org. Chem., 6(1): 44–54.

Huang, H.Y., Korivi, M., Chaing, Y.Y., Chien, T.Y. and Tsai, Y.C. (2012). *Pleurotus tuber-regium* polysaccharides attenuate hyperglycemia and oxidative stress in experimental diabetic rats. Evid. Based Compl. Alter. Med.

Iqbal, H.M.N., Ahmed, I., Zia, M.A. and Irfan, M. (2011). Purification and characterisation of the kinetic parameters of cellulase produced from wheat straw by *Trichoderma viride* under SSF and its detergent compatibility. Adv. Biosci. Biotechnol., 2: 149–156.

Ikekawa, T. (2001). Beneficial effects of edible and medicinal mushrooms on health care. Int. J. Med. Mushrooms, 3(2-3). 10.1615/IntJMedMushr.v3.i2-3.30.

Ina, K., Kataoka, T. and Ando, T. (2013). The use of lentinan for treating gastric cancer. Anti-Canc. Ag. Med. Chem., 13(5): 681–688.

Ishibashi, K.I., Miura, N.N., Adachi, Y., Ohno, N. and Yadomae, T. (2001). Relationship between solubility of grifolan, a fungal 1, 3-β-D-glucan and production of tumor necrosis factor by macrophages *in vitro*. Biosci. Biotechnol. Biochem., 65(9): 1993–2000.

Ishola, M.M. and Taherzadeh, M.J. (2014). Effect of fungal and phosphoric acid pretreatment on ethanol production from oil palm empty fruit bunches (OPEFB). Bioresour. Technol., 165: 9–12.

Ismail, B. and Nampoothiri, K.M. (2010). Exopolysaccharide production and prevention of syneresis in starch using encapsulated probiotic *Lactobacillus plantarum*. Food Technol. Biotechnol., 48(4): 484–489.

Järvinen, J., Taskila, S., Isomäki, R. and Ojamo, H. (2012). Screening of white-rot fungi manganese peroxidases: a comparison between the specific activities of the enzyme from different native producers. AMB Express, 2(1): 1–9.

Jaszek, M., Osińska-Jaroszuk, M., Janusz, G., Matuszewska, A., Stefaniuk, D. et al. (2013). New bioactive fungal molecules with high antioxidant and antimicrobial capacity isolated from *Cerrena unicolor* idiophasic cultures. BioMed. Res. Int., 497492.

Kadokawa, J.I. and Kobayashi, S. (2010). Polymer synthesis by enzymatic catalysis. Curr. Opin. Chem. Biol., 14(2): 145–153.

Kakumu, S., Ishikawa, T., Wakita, T., Yoshioka, K., Ito, Y. and Shinagawa, T. (1991). Effect of sizofiran, a polysaccharide, on interferon gamma, antibody production and lymphocyte proliferation specific for hepatitis B virus antigen in patients with chronic hepatitis B. Int. Immunopharmacol., 13(7): 969–975.

Kamei, I., Hirota, Y., Mori, T., Hirai, H., Meguro, S. and Kondo, R. (2012). Direct ethanol production from cellulosic materials by the hypersaline-tolerant white-rot fungus *Phlebia* sp. MG-60. Bioresour. Technol., 112: 137–142.

Kamo, T., Imura, Y., Hagio, T., Makabe, H., Shibata, H. and Hirota, M. (2004). Anti-inflammatory cyathane diterpenoids from *Sarcodon scabrosus*. Biosci. Biotechnol. Biochem., 68(6): 1362–1365.

Kannan, M., Nesakumari, M., Rajarathinam, K. and Singh, A.J.A.R. (2010). Production and characterisation of mushroom chitosan under solid-state fermentation conditions. Adv. Biol. Res., 4(1): 10–13.

Kaur, M., Velmurugan, B., Rajamanickam, S., Agarwal, R. and Agarwal, C. (2009). Gallic acid, an active constituent of grape seed extract, exhibits anti-proliferative, pro-apoptotic and anti-tumorigenic effects against prostate carcinoma xenograft growth in nude mice. Pharm. Res., 26(9): 2133–2140.

Khammuang, S. and Sarnthima, R. (2013). Decolourisation of synthetic melanins by crude laccases of *Lentinus polychrous* Lév. Folia Microbiol., 58(1): 1–7.

Kim, J., Lee, S.M., Bae, I.Y., Park, H.-G., Lee, H.G. and Lee, S. (2011). (1–3)(1–6)-β-Glucan-enriched materials from *Lentinus edodes* mushroom as a high-fibre and low-calorie flour substitute for baked foods. J. Sci. Food Agric., 91(10): 1915–1919.

Kim, M.Y., Seguin, P., Ahn, J.K., Kim, J.J., Chun, S.C. et al. (2008). Phenolic compound concentration and antioxidant activities of edible and medicinal mushrooms from Korea. J. Agric. Food Chem., 56(16): 7265–7270.

Kim, S.W., Park, S.S., Min, T.J. and Yu, K.H. (1999). Antioxidant activity of ergosterol peroxide (5, 8-epidioxy-5α, 8α-ergosta-6, 22E-dien-3β-ol) in *Armillariella mellea*. Bull. Kor. Chem. Soc., 20(7): 819–823.

Kim, S.Y., Kang, M.Y. and Kim, M.H. (2008). Quality characteristics of noodle added with browned oak mushroom (*Lentinus edodes*). Korean J. Food Cook. Sci., 24(5): 665–671.

Kim, S.Y., Chung, S.I., Nam, S.H. and Kang, M.Y. (2009). Cholesterol-lowering action and antioxidant status improving efficacy of noodles made from unmarketable oak mushroom (*Lentinus edodes*) in high cholesterol fed rats. J. Kor. Soc. Appl. Biol. Chem., 52(3): 207–212.

Kim, Y.J. (2007). Antimelanogenic and antioxidant properties of gallic acid. Biol. Pharm. Bull., 30(6): 1052–1055.

Kim, Y.O., Park, H.W., Kim, J.H., Lee, J.Y., Moon, S.H. and Shin, C.S. (2006). Anti-cancer effect and structural characterisation of endo-polysaccharide from cultivated mycelia of *Inonotus obliquus*. Life Sci., 79(1): 72–80.

Kindred, C., Okereke, U. and Callender, V.D. (2013). Skin-lightening agents: An overview of prescription, office-dispensed, and over-the-counter products. Cutis., 18–26.

Kino, K., Yamashita, A., Yamaoka, K., Watanabe, J., Tanaka, S. et al. (1989). Isolation and characterisation of a new immunomodulatory protein, ling zhi-8 (LZ-8), from *Ganoderma lucidium*. J. Biol. Chem., 264(1): 472–478.

Kolya, H., Sasmal, D. and Tripathy, T. (2017). Novel biodegradable flocculating agents based on grafted starch family for the industrial effluent treatment. J. Polym. Environ., 25(2): 408–418.

Kondo, R., De Leon, R., Anh, T.K., Meguro, S., Shimizu, K. and Kamei, I. (2014). Effect of chemical factors on integrated fungal fermentation of sugarcane bagasse for ethanol production by a white-rot fungus, *Phlebia* sp. MG-60. Bioresour. Technol., 167: 33–40.

Kozarski, M., Klaus, A., Niksic, M., Jakovljevic, D., Helsper, J.P. and Van Griensven, L.J. (2011). Antioxidative and immunomodulating activities of polysaccharide extracts of the medicinal mushrooms *Agaricus bisporus, Agaricus brasiliensis, Ganoderma lucidum* and *Phellinus linteus*. Food Chem., 129(4): 1667–1675.

Kuo, Y.C., Weng, S.C., Chou, C.J., Chang, T.T. and Tsai, W.J. (2003). Activation and proliferation signals in primary human T lymphocytes inhibited by ergosterol peroxide isolated from *Cordyceps cicadae*. Br. J. Pharmacol., 140(5): 895–906.

Kurisawa, M., Chung, J.E., Uyama, H. and Kobayashi, S. (2003). Laccase-catalysed synthesis and antioxidant property of poly(catechin). Macromol. Biosci., 3(12): 758–764.

Lakhanpal, T.N. and Rana, M. (2005). Medicinal and nutraceutical genetic resources of mushrooms. Pl. Gen. Resour., 3(2): 288–303.

Lang, G., Cotteret, J. and Maubru, M. (2004). Oxidation dyeing composition for keratinous fibres containing a 3-aminopyridine azo derivative and dyeing method using said composition. U.S. Patent # 6,797,013.

Latha, K. and Baskar, R. (2014). Comparative study on the production, purification and characterisation of exopolysaccharides from oyster mushrooms, *Pleurotus florida* and *Hypsizygus ulmarius* and their applications. Proc. 8th Int. Conf. Mush. Biol. Mush. Prod., 192–198.

Lawal, T.O., Wicks, S.M., Calderon, A.I. and Mahady, G.B. (2019). Bioactive molecules, pharmacology and future research trends of *Ganoderma lucidium* as a cancer chemotherapeutic agent. pp. 159–178. *In*: Khan, M.S.A., Ahmad, I. and Chattopadhyay, D. (eds.). New Look to Phytomedicine. Academic Press.

Lazarova, V. and Manem, J. (1995). Biofilm characterisation and activity analysis in water and wastewater treatment. Water Res., 29(10): 2227–2245.

Lee, I., Ahn, B., Choi, J., Hattori, M., Min, B. and Bae, K. (2011). Selective cholinesterase inhibition by lanostane triterpenes from fruiting bodies of *Ganoderma lucidum*. Bioorg. Med. Chem. Lett., 21(21): 6603–6607.

Li, J.R., Cheng, C.L., Yang, W.J., Yang, C.R., Ou, Y.C. et al. (2014). FIP-gts potentiate autophagic cell death against cisplatin-resistant urothelial cancer cells. Anticanc. Res., 34(6): 2973–2983.

Li, S.Y., Shi, L.J., Ding, Y., Nie, Y. and Tang, X.M. (2015). Identification and functional characterisation of a novel fungal immunomodulatory protein from *Postia placenta*. Food Chem. Toxicol., 78: 64–70.

Li, Y.R., Liu, Q.H., Wang, H.X. and Ng, T.B. (2008). A novel lectin with potent antitumor, mitogenic and HIV-1 reverse transcriptase inhibitory activities from the edible mushroom *Pleurotus citrinopileatus*. Biochem. Biophys. Acta., 1780(1): 51–57.

Lin, A.N. and Nakatsui, T. (1998). Salicylic acid revisited. Int. J. Dermatol., 37(5): 335–342.

Lindequist, U., Niedermeyer, T.H. and Jülich, W.D. (2005). The pharmacological potential of mushrooms. Evid.-based Compl. Alt. Med., 2(3): 285–299.

Liu, Y. and Fang, H.H.P. (2003). Influence of extracellular polymeric substances (EPS) on flocculation, settling, and dewatering of activated sludge. Crit Rev. Environ. Sci. Technol., 33: 237–273.

Lo, T.C.T., Hsu, F.M., Chang, C.A. and Cheng, J.C.H. (2011). Branched α-(1, 4) glucans from *Lentinula edodes* (L10) in combination with radiation enhance cytotoxic effect on human lung adenocarcinoma through the toll-like receptor 4 mediated induction of THP-1 differentiation/activation. J. Agric. Food Chem., 59(22): 11997–12005.

López-Abelairas, M., Lu-Chau, T.A. and Lema, J.M. (2013). Enhanced saccharification of biologically pretreated wheat straw for ethanol production. Appl. Biochem. Biotechnol., 169(4): 1147–1159.

Lourenço, L.A., Magina, M.D.A., Tavares, L.B.B., de Souza, S.M.A.G.U., Román, M.G. and Vaz, D.A. (2018). Biosurfactant production by *Trametes versicolor* grown on two-phase olive mill waste in solid-state fermentation. Environ. Technol., 39(23): 3066–3076.

Ma, G., Yang, W., Fang, Y., Ma, N., Pei, F. et al. (2016). Antioxidant and cytotoxicites of *Pleurotus eryngii* residue polysaccharides obtained by ultrafiltration. LWT Food Sci. Technol., 73: 108–116.

Ma, K. and Ruan, Z. (2015). Production of a lignocellulolytic enzyme system for simultaneous biodelignification and saccharification of corn stover employing co-culture of fungi. Bioresour. Technol., 175: 586–593.

Mahajna, J., Dotan, N., Zaidman, B.Z., Petrova, R.D. and Wasser, S.P. (2008). Pharmacological values of medicinal mushrooms for prostate cancer therapy: The case of *Ganoderma lucidum*. Nutr. Canc., 61(1): 16–26.

Mahapatra, S. and Banerjee, D. (2013). Fungal exopolysaccharide: Production, composition and applications. Microbial Insights, 6: 1–16.

Majtan, J. (2013). Pleuran (β-Glucan from *Pleurotus ostreatus*): an effective nutritional supplement against upper respiratory tract infections? Med. Sport Sci., 59: 57–61.

Malik, D., Rahi, D.K. and Prabha, V. (2018). Exopolysaccharide producing potential of indigenous white rot fungi from foot hill forests of lower Shivalik ranges of Chandigarh. IOSR J. Pharm., 8(3): 1–7.

Malik, D., Rahi, D.K. and Prabha, V. (2019). Safety evaluation and *in vitro* antioxidant activity of exopolysaccharide produced by an indigenous species of *Pleurotus pulmonarius* RDM9. Toxicol. Int., 25(1): 40–47.

Mathiasen, T.E. (1995). Laccase and Beer Storage Patent # WO 9521240.

Mattila, P., Suonpää, K. and Piironen, V. (2000). Functional properties of edible mushrooms. Nutrition, 16(7-8): 694–696.

Mattila, P., Lampi, A.M., Ronkainen, R., Toivo, J. and Piironen, V. (2002). Sterol and vitamin D$_2$ contents in some wild and cultivated mushrooms. Food Chem., 76(3): 293–298.

Minato, K.I., Laan, L.C., van Die, I. and Mizuno, M. (2019). *Pleurotus citrinopileatus* polysaccharide stimulates anti-inflammatory properties during monocyte-to-macrophage differentiation. Int. J. Biol. Macromol., 122: 705–712.

Minussi, R.C., Pastore, G.M. and Durán, N. (2002). Potential applications of laccase in the food industry. Tr. Food Sci. Technol., 13(6-7): 205–216.

Mizuno, M. and Nishitani, Y. (2013). Immunomodulating compounds in Basidiomycetes. J. Clin. Biochem. Nutr., 52(3): 202–207.

Mothana, R.A.A., Ali, N.A., Jansen, R., Wegner, U., Mentel, R. and Lindequist, U. (2003). Antiviral lanostanoid triterpenes from the fungus *Ganoderma pfeifferi*. Fitoterapia, 74(1-2): 177–180.

Mukhopadhyay, M., Kuila, A., Tuli, D.K. and Banerjee, R. (2011). Enzymatic depolymerisation of *Ricinus communis*, a potential lignocellulosic for improved saccharification. Biomass Bioenerg., 35(8): 3584–3591.

Ng, T.B., Lam, Y.W. and Wang, H. (2003). Calcaelin, a new protein with translation-inhibiting, antiproliferative and antimitogenic activities from the mosaic puffball mushroom *Calvatia caelata*. Pl. Med., 69(03): 212–217.

Ngai, P.H. and Ng, T.B. (2003). Lentin, a novel and potent antifungal protein from shitake mushroom with inhibitory effects on activity of human immunodeficiency virus-1 reverse transcriptase and proliferation of leukemia cells. Life Sci., 73(26): 3363–3374.

Nicolis, E., Lampronti, I., Dechecchi, M.C., Borgatti, M., Tamanini, A. et al. (2008). Pyrogallol, an active compound from the medicinal plant *Emblica officinalis*, regulates expression of pro-inflammatory genes in bronchial epithelial cells. Int. Immunopharmacol., 8(12): 1672–1680.

Nielsen, P.H. and Jahn, A. (1999). Extraction of EPS. pp. 43–72. *In*: Wingender, J., Neu, T.R. and Flemming, H.C. (eds.). Microbial Extracellular Polymeric Substances: Characterisation, Structure and Function. Springer-Verlag, Heidelberg.

Nirwan, B., Choudhary, S., Sharma, K. and Singh, S. (2016). *In vitro* studies on management of root rot disease caused by *Ganoderma lucidum* in *Prosopis cineraria*. Curr. Life Sci., 2(4): 118–126.

Nitschke, M. and Costa, S.G.V.A.O. (2007). Biosurfactants in food industry. Tr. Food Sci. Technol., 18(5): 252–259.

Okaiyeto, K., Nwodo, U.U., Okoli, S.A., Mabinya, L.V. and Okoh, A.I. (2016). Implications for public health demands alternatives to inorganic and synthetic flocculants: Bioflocculants as important candidates. Microbiology Open, 5(2): 177–211.

Okamura-Matsui, T., Takemura, K., Sera, M., Takeno, T., Noda, H. et al. (2001). Characteristics of a cheese-like food produced by fermentation of the mushroom *Schizophyllum commune*. J. Biosci. Bioeng., 92(1): 30–32.

Olempska-Beer, Z. (2004). Laccase from *Myceliophthora thermophila* expressed in *Aspergillus Oryzae*, 61st JECFA-Chemical and Technical Assessment (CTA), Geneva, FAO, 4.

Osińska-Jaroszuk, M., Jaszek, M., Mizerska-Dudka, M., Błachowicz, A., Rejczak, T.P. et al. (2014). Exopolysaccharide from *Ganoderma applanatum* as a promising bioactive compound with cytostatic and antibacterial properties. BioMed Res. Int., 743812.

Piazzon, A., Vrhovsek, U., Masuero, D., Mattivi, F., Mandoj, F. and Nardini, M. (2012). Antioxidant activity of phenolic acids and their metabolites: Synthesis and antioxidant properties of the sulfate derivatives of ferulic and caffeic acids and of the acyl glucuronide of ferulic acid. J. Agric. Food Chem., 60(50): 12312–12323.

Pilz, R., Hammer, E., Schauer, F. and Kragl, U. (2003). Laccase-catalysed synthesis of coupling products of phenolic substrates in different reactors. Appl. Microbiol. Biotechnol., 60(6): 708–712.

Placido, J. and Capareda, S. (2015). Ligninolytic enzymes: A biotechnological alternative for bioethanol production. Bioresour. Bioproc., 2(1): 23. 10.1186/s40643-015-0049-5.

Puttaraju, N.G., Venkateshaiah, S.U., Dharmesh, S.M., Urs, S.M.N. and Somasundaram, R. (2006). Antioxidant activity of indigenous edible mushrooms. J. Agric. Food Chem., 54(26): 9764–9772.

Rahi, D.K., Shukla, K.K., Rajak, R.C. and Pandey, A.K. (2004). Mushrooms and their sustainable utilisation. Everyman Sci., 38: 357–365.

Rahi, D.K., Rajak, R.C. and Pandey, A.K. (2005). Mushroom nutriceuticals: An emerging health care aid. pp. 481–496. *In*: Mukherjee, K.G., Tilak, K.V.B.R., Reddy, S.M., Ganwane, L.V., Prakash, P. and Kunwar, I.K. (eds.). Frontiers in Plant Sciences. I.K. International Pvt. Ltd., New Delhi, India.

Rahi, D.K., Rajak, R.C., Shukla, K.K. and Pandey, A.K. (2005). Diversity and nutriceutical potential of wild edible mushrooms of Central India. pp. 967–980. *In*: Sattyanarayana, T. and Johri, B.N. (eds.). Microbial Diversity: Current Perspectives and Potential Applications. I.K. International Pvt. Ltd., New Delhi, India.

Rahi, D.K. and Soni, S.K. (2007). Application and commercial uses of microorganisms. pp. 69–126. *In*: Soni, S.K. (ed.). Microbes: A Source of Energy for 21st Century. New India Publishing House, New Delhi.

Rahi, D.K. and Rahi, S. (2010). Microbial polysaccharides and potential applications. pp. 825–849. *In*: Chauhan, A.K. and Varma, A. (eds.). Text Book of Molecular Biotechnology. I.K. International Pvt. Ltd., New Delhi, India.

Rahi, D.K. (2015). Fungal production of exopolysaccharides: applications & perspectives. pp. 195–221. *In*: Sobti, R.C., Sharma, P. and Puri, S. (eds.). Emerging Trends in Microbial Biotechnology Energy and Environment. Narendra Publishing House, Delhi, India.

Rahi, D.K. and Urvashi. (2015). Production, characterisation and evaluation of potential applications of bioemulsifier produced by an endophytic actinomycete EARN5 isolated from *Andrographis paniculata*. Res. Biotechnol., 6(6): 1–16.

Rahi, D.K., Bhrigu and Malik, D. (2015). Production of bioemulsifier by an Indigenous species of *Streptomyces rubiginosus* AOBR5 isolated from rhizosphere of Medicinal Plant '*Ocimum sanctum*'. World J. Pharm. Pharmceut. Sci., 4(10): 2174–2198.

Rahi, D.K., Bibra, M. and Rahi, S. (2015). Potential of an indigenous isolate of *Penicillium purpurogenum* RBP04 to produce exopolysaccharides under submerged fermentation. Int. J. Res. Pure Appl. Microbiol., 5(2): 11–17.

Rahi, D.K. and Barwal, M. (2015a). Biosynthesis of silver nano particles by *Ganoderma applanatum*, evaluation of their antibacterial and antibiotic activity enhancing potential. World J. Pharm. Pharmceut. Sci., 4(10): 1234–1247.

Rahi, D.K. and Barwal, M. (2015b). Potential of *Pleurotus sajor-caju* to synthesise silver nanoparticles: Evaluation of antibacterial activity and their role as antibiotic activity enhancer. KAVAKA, 43: 74–78.

Rahi, D.K. and Malik, D. (2016). Diversity of mushrooms and their metabolites of nutraceutical and therapeutic significance. J. Mycol., 2016. 7654123. 10.1155/2016/761-18.7654123.

Rahi, D.K., Chaudhary, E. and Malik, D. (2018). Production of Exopolysaccharide by Enterobacter cloacae, a root nodule isolate from *Phaseolus vulgaris*: Parameter optimisation for yield enhancement under submerged fermentation. Int. J. Basic Appl. Res., 8(9): 936–948.

Rahi, D.K., Richa and Kaur, M. (2018). Production of xylooligosaccharide from corn cob xylan by xylanase obtained from *Aureobasidium pullulans*. Int. J. Sci. Res. Sci. Eng. Technol., 4(4): 1142–1448.

Rehman, S., Farooq, R., Jermy, R., Asiri, S.M., Ravinayagam, V., Jindan, R.A. and Khan, F.A. (2020). A wild *Fomes fomentarius* for biomediation of one pot synthesis of titanium oxide and silver nanoparticles for antibacterial and anticancer application. Biomolecules, 10(4).

Ribeiro, D.S., Henrique, S.M., Oliveira, L.S., Macedo, G.A. and Fleuri, L.F. (2010). Enzymes in juice processing: A review. Int. J. Food Sci. Technol., 45(4): 635–641.

Saeed, A., Fatehi, P. and Ni, Y. (2011). Chitosan as a flocculant for pre-hydrolysis liquor of kraft-based dissolving pulp production process. Carbohyd. Polym., 86(4): 1630–1636.

Salvachúa, D., Prieto, A., López-Abelairas, M., Lu-Chau, T., Martínez, Á.T. and Martínez, M.J. (2011). Fungal pretreatment: An alternative in second-generation ethanol from wheat straw. Bioresour. Technol., 102(16): 7500–7506.

Sánchez, O., Sierra, R. and Alméciga-Díaz, C.J. (2011). Delignification process of agro-industrial wastes an alternative to obtain fermentable carbohydrates for producing fuel. Altern. Fuel, 7.

Santori, G., Di Nicola, G., Moglie, M. and Polonara, F. (2012). A review analysing the industrial biodiesel production practice starting from vegetable oil refining. Appl. Energy., 92: 109–132.

Sasidhara, R. and Thirunalasundari, T. (2012). Antimicrobial activity of mushrooms. Acta Pharmacol. Sin., 23(9): 787–791.

Seo, H.W., Hung, T.M., Na, M., Jung, H.J., Kim, J.C. et al. (2009). Steroids and triterpenes from the fruit bodies of *Ganoderma lucidum* and their anti-complement activity. Arch. Pharmacal. Res., 32(11): 1573–1579.

Sermwittayawong, D., Patninan, K., Phothiphiphit, S., Boonyarattanakalin, S., Sermwittayawong, N. and Hutadilok-Towatana, N. (2018). Purification, characterisation, and biological activities of purified polysaccharides extracted from the gray oyster mushroom (*Pleurotus sajor-caju* (Fr.) Sing.). J. Food Biochem., 42(5): e12606.

Stachowiak, B. and Reguła, J. (2012). Health-promoting potential of edible macromycetes under special consideration of polysaccharides: A review. Eur. Food Res. Technol., 234(3): 369–380.

Su, H.J., Fann, Y.F., Chung, M., Won, S.J. and Lin, C.N. (2000). New Lanostanoids of *Ganoderma tsugae*. J. Nat. Prod., 63(4): 514–516.

Su, P., Cheng, Q., Wang, X., Cheng, X., Zhang, M., Tong, Y. and Huang, L. (2014). Characterisation of eight terpenoids from tissue cultures of the Chinese herbal plant, *Tripterygium wilfordii*, by high-performance liquid chromatography coupled with electrospray ionisation tandem mass spectrometry. Biomed. Chromatogr., 28(9): 1183–1192.

Sugiyama, Y. (2016). Polysaccharides. pp. 37–50. *In*: Yamaguchi, Y. (ed.). Immunotherapy of Cancer. Springer, Berlin.

Suwannarach, N., Kumla, J., Sujarit, K., Pattananandecha, T., Saenjum, C. and Lumyong, S. (2020). Natural bioactive compounds from fungi as potential candidates for protease inhibitors and immunomodulators to apply for corona viruses. Molecules, 25(8): 1800. 10.3390/molecules25081800.

Takei, T., Yoshida, M., Ohnishi-Kameyama, M. and Kobori, M. (2005). Ergosterol peroxide, an apoptosis-inducing component isolated from *Sarcodon aspratus* (Berk.). Biosci. Biotechnol. Biochem., 69(1): 212–215.

Tanrıöven, D. and Ekşi, A. (2005). Phenolic compounds in pear juice from different cultivars. Food Chem., 93(1): 89–93.

Teichmann, A., Dutta, P.C., Staffas, A. and Jägerstad, M. (2007). Sterol and vitamin D_2 concentrations in cultivated and wild grown mushrooms: Effects of UV irradiation. LWT- Food Sci. Technol., 40(5): 815–822.

Thompson, W. and Meyer, S. (2013). Second generation biofuels and food crops: Co-products or competitors? Global Food Sec., 2(2): 89–96.

Tsai, S.W., Liu, R.L., Hsu, F.Y. and Chen, C.C. (2006). A study of the influence of polysaccharides on collagen self-assembly: Nanostructure and kinetics. Biopolymers, 83(4): 381–388.

Tsukamoto, S., Macabalang, A.D., Nakatani, K., Obara, Y., Nakahata, N. and Ohta, T. (2003). Tricholomalides A–C, new neurotrophic Diterpenes from the mushroom *Tricholoma* sp. J. Nat. Prod., 66(12): 1578–1581.

Ulery, B.D., Nair, L.S. and Laurencin, C.T. (2011). Biomedical applications of biodegradable polymers. J. Polym. Sci. Pol. Phys., 49(12): 832–864.

Van Nevel, C.J., Decuypere, J.A., Dierick, N. and Molly, K. (2003). The influence of *Lentinus edodes* (Shiitake mushroom) preparations on bacteriological and morphological aspects of the small intestine in piglets. Arch. Anim. Nutr., 57(6): 399–412.

Vannucci, L., Krizan, J., Sima, P., Stakheev, D., Caja, F., Rajsiglova, L. and Saieh, M. (2013). Immunostimulatory properties and antitumor activities of glucans. Int. J. Oncol., 43(2): 357–364.

Veverka, M., Dubaj, T., Veverková, E. and Šimon, P. (2018). Natural oil emulsions stabilised by β-glucan gel. Coll. Surf. A: Physicochem. Eng. Asp., 537: 390–398.

Vroman, I. and Tighzert, L. (2009). Biodegradable polymers. Materials, 2(2): 307–344.

Wang, F.Q., Xie, H., Chen, W., Wang, E.T., Du, F.G. and Song, A.D. (2013). Biological pretreatment of corn stover with ligninolytic enzyme for high efficient enzymatic hydrolysis. Bioresour. Technol., 144: 572–578.

Wang, H., Ng, T.B. and Ooi, V.E. (1998). Lectins from mushrooms. Mycol. Res., 102(8): 897–906.

Wang, H. and Ng, T.B. (2001). Isolation and characterisation of velutin, a novel low-molecular-weight ribosome-inactivating protein from winter mushroom (*Flammulina velutipes*) fruiting bodies. Life Sci., 68(18): 2151–2158.

Wang, H. and Ng, T.B. (2006). Ganodermin, an antifungal protein from fruiting bodies of the medicinal mushroom *Ganoderma lucidum*. Peptides, 27(1): 27–30.

Wang, H.X. and Ng, T.B. (2000). Flammulin: A novel ribosome-inactivating protein from fruiting bodies of the winter mushroom *Flammulina velutipes*. Biochem. Cell Biol., 78(6): 699–702.

Wang, S.J., Li, Y.X., Bao, L., Han, J.J., Yang, X.L. et al. (2012). Eryngiolide A, a cytotoxic macrocyclic diterpenoid with an unusual cyclododecane core skeleton produced by the edible mushroom *Pleurotus eryngii*. Org. Lett., 14(14): 3672–3675.

Wang, S., Bao, L., Zhao, F., Wang, Q., Li, S. et al. (2013). Isolation, identification, and bioactivity of monoterpenoids and sesquiterpenoids from the mycelia of edible mushroom *Pleurotus cornucopiae*. J. Agric. Food Chem., 61(21): 5122–5129.

Wang, S., Li, J., Sun, J., Zeng, K.W., Cui, J.R. et al. (2013). NOo inhibitory guaianolide-derived terpenoids from *Artemisia argyi*. Fitoterapia, 85: 169–175.

Wang, W., Yuan, T., Cui, B. and Dai, Y. (2012). Pretreatment of *Populus tomentosa* with *Trametes velutina* supplemented with inorganic salts enhances enzymatic hydrolysis for ethanol production. Biotechnol. Lett., 34(12): 2241–2246.

Wang, Y., Liu, Y., Yu, H., Zhou, S., Zhang, Z. et al. (2017). Structural characterisation and immuno-enhancing activity of a highly branched water-soluble β-glucan from the spores of *Ganoderma lucidum*. Carbohydr. Polym., 167: 337–344.

Weete, J.D. (1989). Structure and function of sterols in fungi. pp. 115–167. *In*: Paoletti, R. and Kritchevsky, D. (eds.). Advances in Lipid Research. Elsevier.

Weete, J.D. and Gandhi, S.R. (1996). Biochemistry and molecular biology of fungal sterols. pp. 421–438. *In*: Brambl, R. and Marzluf, G.A. (eds.). Biochemistry and Molecular Biology. Springer, Berlin, Heidelberg.

Weng, C.J., Chau, C.F., Chen, K.D., Chen, D.H. and Yen, G.C. (2007). The anti-invasive effect of lucidenic acids isolated from a new *Ganoderma lucidum* strain. Mol. Nutr. Food. Res., 51(12): 1472–1477.

Wu, G.S., Lu, J.J., Guo, J.J., Li, Y.B., Tan, W. et al. (2012). Ganoderic acid DM, a natural triterpenoid, induces DNA damage, G1 cell cycle arrest and apoptosis in human breast cancer cells. Fitoterapia, 83(2): 408–414.

Wu, M., Luo, X., Xu, X., Wei, W., Yu, M. et al. (2014). Antioxidant and immunomodulatory activities of a polysaccharide from *Flammulina velutipes*. J. Tradit. Chin. Med., 34(6): 733–740.

XiaoPing, C., Yan, C., ShuiBing, L., YouGuo, C., JianYun, L. and LanPing, L. (2009). Free radical scavenging of *Ganoderma lucidum* polysaccharides and its effect on antioxidant enzymes and immunity activities in cervical carcinoma rats. Carbohydr. Res., 77(2): 389–393.

Xu, H., Kong, Y.Y., Chen, X., Guo, M.Y., Bai, X. et al. (2016). Recombinant FIP-gat, a fungal immunomodulatory protein from *Ganoderma atrum*, induces growth inhibition and cell death in breast cancer cells. J. Agric. Food Chem., 64(13): 2690–2698.

Xu, T. and Beelman, R.B. (2015). The bioactive com-pounds in medicinal mushrooms have potential protective effects against neu-rodegenerative diseases. Adv. Food Technol. Nutr. Sci. Open J., 1(2): 62–66.

Yadav, K.L., Rahi, D.K. and Soni, S.K. (2014). An indigenous hyperproductive species of *Aureobasidium pullulans* RYLF-10: Influence of fermentation conditions on exopolysaccharide (EPS) production. Appl. Biochem. Biotechnol., 172(4): 1898–1908.

Yadav, K.L., Rahi, D.K. and Soni, S.K. (2014). Bioemulsifying potential of exopolysaccharide produced by an indigenous species of *Aureobasidium pullulans* RYLF10. Peer J., 10.7287/peerj.preprints.726v1.

Yagüe, S., Terrón, M.C., González, T., Zapico, E., Bocchini, P. et al. (2000). Biotreatment of tannin-rich beer-factory wastewater with white-rot basidiomycete *Coriolopsis gallica* monitored by pyrolysis/gas chromatography/mass spectrometry. Rapid Commun. Mass Spectrom., 14(10): 905–910.

Yang, R., Li, H., Huang, M., Yang, H. and Li, A. (2016). A review on chitosan-based flocculants and their applications in water treatment. Water Res., 95: 59–89.

Yang, S.X., Yu, Z.C., Lu, Q.Q., Shi, W.Q., Laatsch, H. and Gao, J.M. (2012). Toxic lanostane triterpenes from the basidiomycete *Ganoderma amboinense*. Phytochem. Lett., 5(3): 576–580.

Yaoita, Y., Yoshihara, Y., Kakuda, R., Machida, K. and Kikuchi, M. (2002). New sterols from two edible mushrooms, *Pleurotus eryngii* and *Panellus serotinus*. Chem. Pharm. Bull., 50(4): 551–553.

Yaropolov, A.I., Skorobogat'Ko, O.V., Vartanov, S.S. and Varfolomeyev, S.D. (1994). Laccase. Appl. Biochem. Biotechnol., 49(3): 257–280.

Yasukawa, K., Aoki, T., Takido, M., Ikekawa, T., Saito, H. and Matsuzawa, T. (1994). Inhibitory effects of ergosterol isolated from the edible mushroom *Hypsizigus marmoreus* on TPA-induced inflammatory ear oedema and tumour promotion in mice. Phytother. Res., 8(1): 10–13.

Yegenoglu, H., Aslim, B. and Oke, F. (2011). Comparison of antioxidant capacities of *Ganoderma lucidum* (Curtis) P. Karst and *Funalia trogii* (Berk.) Bondartsev and Singer by using different *in vitro* methods. J. Med. Food, 14(5): 512–516.

Yin, M.C., Lin, C.C., Wu, H.C., Tsao, S.M. and Hsu, C.K. (2009). Apoptotic effects of protocatechuic acid in human breast, lung, liver, cervix, and prostate cancer cells: potential mechanisms of action. J. Agric. Food Chem., 57(14): 6468–6473.

Yip, E.C.H., Chan, A.S.L., Pang, H., Tam, Y.K. and Wong, Y.H. (2006). Protocatechuic acid induces cell death in HepG2 hepatocellular carcinoma cells through a c-Jun N-terminal kinase-dependent mechanism. Cell Biol. Toxicol., 22(4): 293–302.

Yogita, R., Simanta, S., Aparna, S., Shraddha, G. and Kamlesh, S. (2011). Screening for exopolysaccharide production from Basidiomycetes of Chhattisgarh. Curr. Bot., 2(10): 11–14.

You, Y.H. and Lin, Z.B. (2002). Protective effects of *Ganoderma lucidum* polysaccharides peptide on injury of macrophages induced by reactive oxygen species. Acta Pharmacol. Sin., 23(9): 787–791.

Yuan, J.P., Wang, J.H., Liu, X., Kuang, H.C. and Zhao, S.Y. (2007). Simultaneous determination of free ergosterol and ergosteryl esters in *Cordyceps sinensis* by HPLC. Food Chem., 105(4): 1755–1759.

Yuen, J.W. and Gohel, M.D.I. (2005). Anticancer effects of *Ganoderma lucidum*: A review of scientific evidence. Nutr. Canc., 53(1): 11–17.

Zhang, D., Hou, Z., Liu, Z. and Wang, T. (2013). Experimental research on *Phanerochaete chrysosporium* as coal microbial flocculant. Int. J. Min. Sci. Technol., 23(4): 521–524.

Zhang, L., Fan, C., Liu, S., Zang, Z., Jiao, L. and Zhang, L. (2011). Chemical composition and antitumor activity of polysaccharide from *Inonotus obliquus*. J. Med. Plant Res., 5(7): 1251–1256.

Zhang, L., Khoo, C., Koyyalamudi, S.R., Pedro, N.D. and Reddy, N. (2017). Antioxidant, anti-inflammatory and anticancer activities of ethanol soluble organics from water extracts of selected medicinal herbs and their relation with flavonoid and phenolic contents. Pharmacologia, 8(2): 59–72.

Zhang, M., Cui, S.W., Cheung, P.C.K. and Wang, Q. (2007). Antitumor polysaccharides from mushrooms: A review on their isolation process, structural characteristics and antitumor activity. Tr. Food Sci. Technol., 18(1): 4–19.

Zhang, Y., Kong, H., Fang, Y., Nishinari, K. and Phillips, G.O. (2013). Schizophyllan: A review on its structure, properties, bioactivities and recent developments. Bioact. Carbohyd. Diet. Fibre, 1(1): 53–71.

Zheng, S., Liu, Q., Zhang, G., Wang, H. and Ng, T.B. (2010). Purification and characterisation of an antibacterial protein from dried fruiting bodies of the wild mushroom *Clitocybe sinopica*. Acta Biochem. Pol., 57(1): 43–48.

Zheng, W., Zhao, Y., Zheng, X., Liu, Y., Pan, S. et al. (2011). Production of antioxidant and antitumor metabolites by submerged cultures of *Inonotus obliquus* cocultured with *Phellinus punctatus*. Appl. Microbial. Biotechnol., 89(1): 157–167.

Zong, A., Cao, H. and Wang, F. (2012). Anticancer polysaccharides from natural resources: A review of recent research. Carbohyd. Polym., 90(4): 1395–1410.

Cosmeceuticals

13

Mushroom Bioactive Ingredients in Cosmetic Industries

Nur Izyan Wan Azelee,[1,2] *Nor Hasmaliana Abdul Manas,*[1,2] *Daniel Joe Dailin,*[1,2] *Roslinda Malek,*[1] *Neo Moloi,*[3] *Joe Gallagher,*[4] *Ana Winters,*[4] *Ong Mei Leng*[5] *and Hesham Ali El Enshasy*[1,2,6,*]

1. INTRODUCTION

Almost 2,400 years ago, Hippocrates stated, "Let your food be your medicine and your medicine be your food" (Khan et al., 2018). Mushrooms are well documented in ancient texts on traditional medicines, with accounts describing their phenomenal healing powers (El Enshasy et al., 2013). The Pharaohs cherished mushrooms as a delicacy; the Greeks believed that mushrooms offer power to fighters in combat; while the Romans considered mushrooms as the 'food of the gods' and had mushrooms on the menu during celebratory events (Rahi and Malik, 2016). Presently, about 2,000 edible mushrooms species with momentous nutritional value are found around the world (Rathore et al., 2019). A large number of mushroom species are well recognised

[1] School of Chemical and Energy Engineering, Faculty of Engineering, Universiti Teknologi Malaysia (UTM), 81310 Skudai, Johor, Malaysia.
[2] Institute of Bioproduct Development (IBD), Universiti Teknologi Malaysia (UTM), 81310 Skudai, Johor, Malaysia.
[3] Sawubone Mycelium Co., Centurion, Gauteng, South Africa.
[4] Institute of Biological, Environmental and Rural Sciences, Aberystwyth University, Aberystwyth, UK.
[5] Harita Go Green Sdn. Bhd., Johor Bahru, Johor, Malayisa.
[6] City of Scientific Research and Technology Applications (SRTA), New Burg Al Arab, Alexandria, Egypt.
* Corresponding author: henshasy@ibd.utm.my

for their therapeutic properties. Mushrooms are high in numerous plant vitamins, protein and minerals and are viewed as a low-calorie food.

A small number of edible and appetising species of mushroom are appropriate for human intake due to their low toxicity, texture and flavour (Govorushko et al., 2019). There are some highly cultivated species, such as oyster mushrooms (*Pleurotus* spp.), shiitake mushrooms (*Lentinula edodes*) and button mushrooms (*Agaricus bisporus*); among these are found a small number of edible mushrooms (Wan Mahari et al., 2020). These three mushroom types have a high market demand and are commercially cultivated. Oyster mushrooms are often cultivated in lowlands, while shiitake and button mushrooms are cultivated in the uplands and cold environments (Haimid et al., 2013). The gilled mushrooms, which belong to the genus *Pleurotus*, such as red oyster (*Pleurotus flabellatus*), king oyster (*Pleurotus eryngii*) and white oyster (*Pleurotus florida*) account for more than 16 per cent of global mushroom production (Phan and Sabaratnam, 2012). Besides that, *Ganoderma lucidum* (known as Reishi or Ling Zhi in Japan and China, respectively) is one of the most prevalent traditional herbs used in Asia. This mushroom has high commercial value due to its medicinal properties. Additionally, the mushrooms of the genus *Cordyceps* have long been consumed traditionally in Asian countries and is believed to provide a long and healthy life. Many studies, both *in vitro* and *in vivo*, have been performed on the metabolic activity of *Cordyceps*. The *Cordyceps* genus contains some of the most highly prized and valued of all medicinal fungi. The most famous and widely used species of *Cordyceps* is *Cordyceps sinensis* (Berk.) Sacc. (Elkhateeb et al., 2019).

Mushrooms have a long history of usage in traditional medicine to avert and treat various illnesses (Masri et al., 2017). Research has proven the existence of precious bioactive compounds in mushrooms that contribute to their healing properties. Today, mushrooms are commercialised as nutritional supplements for their properties. However, little is known about the potential use of mushrooms for cosmeceutical purposes. As much as it is beneficial to the body, it is believed that the mushroom's active ingredients are also beneficial to the largest human organ (e.g., skin). Therefore, this chapter gives an insight on the mushroom characteristics, their bioactive compounds and functional properties, potential application in cosmeceuticals, cosmeceutical formulation and commercial potential for use in cosmeceutical industries.

2. Characteristics of Mushrooms and Their Healing Properties

Mushrooms belong to the group of macrofungi, Ascomycetes and Basidiomycetes. They obtain their nutrition through being parasites, saprotrophs, or symbiotic as mycorrhiza. Mushrooms' growth stages consist of a vegetative phase (mycelia) and a reproductive phase (fruit bodies). A mushroom grows from a pinhead or nodule of less than two millimetres in diameter (Singh and Prasad, 2019). This is known as the primordium and is typically situated on or close to the ground surface. Numerous species of mushrooms grow rapidly, often overnight. This rapid development and increase in size is termed mushrooming. In nature, mushrooms, such as *Pleurotus* spp., generally grow on discarded resources and inhabit dead biological materials, namely dead maple, oak, or cottonwood (Wan Mahari et al., 2020). They function as key decomposers, changing dead or living materials into the nutrients they require for growth.

The diversity of mushrooms has attracted attention as a source for new compounds with new mode of action against severe diseases. Biologically active compounds are normally present as cell wall components, such as polysaccharides, proteins, or as organic secondary metabolites like phenolic compounds, terpenes and steroids (Elkhateeb et al., 2019). The production of these compounds is heavily reliant on many factors, such as the forms of mushroom, growth stage and cultivation conditions (Guillamón et al., 2010). Initial indications show that mushrooms, such as *Amauroderma, Agaricus, Coprinus, Ganoderma, Grifola, Lentinula, Phellinus, Pleurotus* and *Polyozellus* genera can improve the immune and inflammatory system (Figueiredo and Régis, 2017). Mushrooms have long been recognised for their significant role in conserving good health and treating many health complications, such as inflammation, obesity, hypercholesterolemia, hypertension, immunodeficiency, cancer and hyperlipidemia (Ma et al., 2018). Their antimicrobial, antioxidant, and anti-inflammatory activity have been highlighted and intensively studied by researchers, making them flexible and multipurpose for use as cosmetic ingredients. Table 1 shows the medicinal properties of different types of mushrooms.

Table 1. Medicinal properties of different varieties of mushroom species.

Mushroom species	Medicinal properties	References
Agaricus blazei	Antiallergic	Hetland et al., 2020
Agaricus brasiliensis	Antitumour	Sovrani et al., 2017; Rubel et al., 2018
	Antifibrotic	Nakamura et al., 2019
Coprinus comatus	Anticancer (ovarian cancer)	Rouhana-Toubi et al., 2015
Ganoderma lucidum	Anticancer (lung cancer)	Lin et al., 2017
	Anticancer (melanoma)	Barbieri et al., 2017
	Anticancer (liver cancer)	Li et al., 2015
	Anticancer (breast cancer)	Rossi et al., 2018
	Immunomodulatory	Rubel et al., 2018
Grifola frondosa	Antitumour	Masuda et al., 2017; Mao et al., 2018
	Regulation of gut microbiota	Li et al., 2019
Ganoderma atrum	Anticancer (colon cancer)	Yu et al., 2015
	Anticancer (leukemia)	Zhang et al., 2017
Hericium erinaceus	Anti-inflammatory	Diling et al., 2017
Imleria badia	Anti-inflammatory	Dogan et al., 2016; Grzywacz et al., 2016

2.1 Bioactive Metabolites

Large quantities of bioactive compounds have been found to be responsible for the medicinal properties of mushrooms. In edible mushrooms, the content and form of biologically active substances can vary considerably. Some of the important bioactive components that can be obtained from mushrooms are polysaccharides, glucans, proteins, fats, phenolic compounds, terpenoids and tocopherols among others. Concentrations of these substances are influenced by variations in stage of growth, storage, strain, substratum, age, processing conditions and production (Valverde et al., 2015).

2.2 *Polysaccharides and β-Glucans*

Polysaccharides are one of the major components of mushroom ingredients with high nutraceutical/medicinal value and have been studied extensively during the past two decades. As a major constituent of edible mushrooms and due to their varied biological activities and complex structure, polysaccharides have received considerable attention (Gong et al., 2020). Commercial mushroom polysaccharides are mainly mined from the fruit bodies, which contribute 80–85 per cent of mushroom products with the remaining 15 per cent obtained from fungal mycelia (Wu, 2015). This polymer has diverse medical activities including antitumour, anti-inflammatory, antibacterial, and immunomodulation properties (Elsayed et al., 2014; Valverde et al., 2015; Friedman, 2016; Yang et al., 2019; Gong et al., 2020). Recent studies show that mushroom polysaccharides, such as galactans, mannans, α- and β-glucans have the potential for use as prebiotics as they are able to reach the colon and stimulate the growth of beneficial bacteria (Singdevsachan et al., 2016; Nowak et al., 2018). In addition, in other studies, mushroom polysaccharides exhibited anti-diabetic properties (Jiang et al., 2020). For example, the mycelium zinc-polysaccharides from *Pleurotus djamor* display renoprotective and hepatoprotective effects via oxidative stress regulation and can treat diabetic complications (Zhang et al., 2015). *Gomphidiaceae rutilus* polysaccharides showed enhanced insulin-stimulated glucose uptake in high glucose and fatty acid-treated hepatic cells (Yang et al., 2020).

The β-glucans are polysaccharide compounds that can reduce coronary heart disease, high blood glucose levels, coloreactal cancer, high cholesterol levels and insulin resistance (Valverde et al., 2015; Ayeka, 2018; Afiati et al., 2019). β-glucans are lentinan, pleuran, schizophyllan, ganoderan and fucomanogalactan (Chaturvedi et al., 2018). They are the components of the cell, known as biological response modifiers (BRM) and they consist of glucopyranose molecules linked through β (1-3), β (1-4), or β (1-6) linkages. A study reported that lentinan from *Lentinus edodes* reduced pro-inflammatory cytokines and oxidative stress (Murphy et al., 2020). Findings from *in vitro* analysis have also shown the potential of *L. edodes* β-glucan to treat lung injury. Recently, the β-glucans extracted from wild mushrooms were observed to have the potential to significantly reduce the pathological immune cascade patients with COVID-19 disorders, such as acute respiratory distress syndrome (ARDS) (Murphy et al., 2020). Urbancikova et al. (2020) also reported the function of pleuran in respiratory symptoms and treatment of acute Herpes Simplex Virus Type 1 infection (HSV-1). Furthermore, there is evidence that demonstrates the use of pleuran in children's clinical care and prevention of recurrent respiratory tract infections (RRTIs), indicating a substantial decrease in the number of various forms of respiratory tract infections, such as otitis, laryngitis, bronchitis and flu (Jesenak et al., 2013).

2.3 Proteins and Fats

Mushrooms are important sources of natural bioactive proteins, with their protein contents exceeding those found in most vegetables. Leucine, valine, glutamine, glutamic and aspartic acids are the most abundant amino acids in 200–250 g protein/ kg of dry matter (Valverde et al., 2015). A study conducted by Oluwafemi et al. (2016) on *Pleurotus ostreatus* protein fractionation showed that globulin was most abundant with 47.31 per cent in the cap, 23.31 per cent% in the stalk, and 44.65 per cent present in cap with a stalk. For albumin, the percentage varied between 3.29 and 4.18 per cent in stalk, cap and cap with a stalk. Novel proteins with biological activities have also been discovered and they can be applied in biotechnological processes, including lignocellulose-degrading enzymes (Tovar-Herrera et al., 2018), hydrophobins (Goyal et al., 2018) and lectins (Singh et al., 2020). Lipids exist in the mushroom's cell wall, which is vital for the storage of vitamin D (Reid et al., 2017). The studies on edible mushrooms showed that they are low in lipids; however, this fraction is rich in polyunsaturated fatty acids (Sande et al., 2019; Ho et al., 2020). Mushroom lipids essentially comprise of healthy unsaturated fatty acids, such as linoleic and oleic acids in low amounts, but are beneficial when consumed. Mushrooms with a lipid content of 20–30 g/kg of dry matter contain oleic (C18:1), linoleic (C18:2) and palmitic (C16:0) acids, which are key fatty acids (Valverde et al., 2015).

2.4 Enzymes

Mushrooms produce a variety of enzymes, such as laccases, lignin peroxidases, manganese peroxidases, oxidase of aryl alcohol, dehydrogenases of alcohol, or quinone reductase, xylanase, and cellulase for degradation of lignocellulose substrates. Cellulose and cellobiosis degradants include xylanase, cellulase and cellobiosis reductase (Sánchez, 2010; Kabel, 2017; Vos et al., 2017). *Ganoderma lucidum* shows the capacity to produce β-N acetylhexosaminidase, α-1,2-mannasidase and endo-β-1,3-glucanase with glutamic protease as the main protein in the extract (Kumakura et al., 2019; Yang et al., 2019). Protease isolated from a new basidiomycete fungus, *Pleurotus sajor-caju* (oyster mushroom), showed high tolerance towards organic solvent together with considerable detergent stability as compared to commercial proteases. This provides new and exciting bioprocess possibilities, primarily in peptide synthesis and detergent formulation for industrial and biotechnological perspectives (Benmrad et al., 2019). Studies on edible mushroom, *Pleurotus albidus*, reveal a new source of milk-clotting proteases with a high coagulant ratio (Martim et al., 2017). The mushrooms, *Lentinus crinitus*, *Lentinus citrinus*, *Pleurotus ostreatoroseus*, *P. florida* and *P. albidus* were identified as major sources of enzymes, including proteases (Souza et al., 2016; Machado et al., 2017; Martim et al., 2017).

2.5 Phenolic Compounds

Mushrooms produce a range of secondary metabolites that include phenolic compounds, molecules with one or more hydroxyl groups bonded directly to an aromatic ring. Structures may include simple phenolic or complex polymer molecules.

Phenolic compounds contribute to growth; anticancer, antimutagenic, antimicrobial activities; have antioxidant properties; act as oxygen scavengers and inhibitors of free radicals (Abugri and McElhenney, 2013; Valverde et al., 2015; Panche et al., 2016; Elkhateeb et al., 2019; Gąsecka et al., 2020). Examples of these compounds include pyrogallol, which has been identified in mushrooms and is now included in various human diet-foods. Pyrogallol has been extracted from *A. bisporus* (Elsayed et al., 2014; Elkhateeb et al., 2019). High-performance liquid chromatography analysis revealed the phenolic acid profile of 26 types of mushrooms and the foremost phenolic acid substances identified asgallic acid, fumaric acid and catechin hydrate (Çayan et al., 2020). Gallic acid was detected as the principal phenolic compound in *Russula aurora* (3 μg/g), and in *Armillaria tabescens* (2.1 μg/g) and *L-leucothites* (9 μg/g); 6,7-dihydroxy-coumarin were known as the main phenolic compound (Çayan et al., 2020). Other bioactive phenolic compounds were isolated from mushrooms, such as grifolin obtained from *Albatrellus ovinus*, myricetion isolated from *Craterellus cornucopioides,* gallic acid, catechin and quercetin isolated from *L. subsericatus,* gallic acid and benzoic acid from *Amanita rubescens* and hericenones extracted from *Hericium erinaceus* (Elkhateeb et al., 2019). Lee et al. (2006) reported that hispidin, methylinoscavin A, inoscavin B, methylinoscavin B are the major metabolites from fruiting bodies of *Inonotus xeranticus.*

2.6 L-Ergothioneine

Mushrooms are the primary source of water-soluble L-ergothioneine, an unusual thio-histidine betaine amino acid with a high antioxidant potential (Borodina et al., 2020). In general, this type of amino acid can be synthesised by microbes, including fungal groups and is found in mushroom fruiting bodies and actinobacteria. It is often referred to as 2-mercaptohistidine trimethylbetaine, with IUPAC name (2S)-3-(2-thioxo 2,3-dihydro-1H-imidazol-4-yl)-2-(trimethylamonio) propanoate. The best sources of L-ergothioneine are edible mushrooms, like *A. bisporus* (Kalaras et al., 2017), *L. edodes* and *Grifola frondosa.* Studies have shown delayed astaxanthin degradation and regulation of lipid oxidation in the presence of ergothioneine-rich mushroom extract in a liposomal setting (Pahila et al., 2019).

2.7 Ergosterol/Vitamin D

Ergosterol ($C_{28}H_{43}OH$), IUPAC name; Ergosta-5,7,22-trien-3B-01, is a typical fungal sterol and an important building block of membrane structures of fungal cells. Phillips et al. (2011) demonstrated that ergosterol concentration in all types of mushrooms was in the range of 26–85 mg/100 g. Mushrooms are the only non-animal foodstuff with large quantities of bioavailable vitamin D and have the potential to be a key food supply of dietary vitamin D for vegetarians. Ergosterol is a precursor of various pharmaceutical products, such as vitamin D_2, progesterone, hydrocortisone, and brassinolide. Biosynthesis of ergosterol produces byproducts that could act as drug precursors. *Flammulina velutipes* ergosterol has been shown to be an important precursor of new anticancer and anti-HIV drugs (Wang et al., 2019). *Pleurotus* mushrooms were shown to exhibit relatively higher levels and better kinetics of conversion of ergosterol into vitamin D_2 than those of other cultivated species (Taofiq et al., 2020). It was noted that 14 ergosterol derivatives of *G. lucidum* were isolated, including a new substance, that have been studied at different levels *in vitro* in anti-proliferative and anti-angiogenic activities (Chen et al., 2017).

2.8 Flavonoids

The flavonoids are important components of the polyphenolic group and are prevalent in many fruits, vegetables and mushrooms. According to their chemical characteristics, flavonoids are classified as flavones, flavanones, catechins, anthocyanins, isoflavones and chalcones. Natural flavonoids and phenolics are bioactives, often produced by plants and contain at least two aromatic rings with at least one hydroxyl group bound to an aromatic ring (Panche et al., 2016). Antioxidant, anticancer, antimutagenic, antimicrobial and antiradical properties have been identified in flavonoids (Panche et al., 2016). The presence of flavonoids in mushrooms can be attributed to the fact that these species can absorb several nutrients and compounds from their substrates or from the adjacent plants (Gil-Ramírez et al., 2016).

2.9 Terpenes and Terpenoids

Terpenes are known to have anticancer, antimalarial, antiviral activity, antiaging, immunoregulation, insect resistance and anti-inflammatory activities (Dudekula et al., 2020; Enshasy and Hatti-kaul, 2013). Basidiomycetes terpenes are terpenoids or isoprenoids. Terpenoids are categorised by the number of carbon units found in their building blocks, such as C10 and C15 units (sesquiterpenoids), C20 (diterpenes), C40 (carotenoid pigments), and C30 (triterpenoids and sterols) (Dudekula et al., 2020). More than 130 triterpenoids (type Lanostane) were isolated from the edible mushroom *G. lucidum* with a molecular weight ranging between 400–600 kDa. Triterpenoids have been isolated from *G. lucidum* spore and exhibit potent activities in alleviating vinorelbine-induced neutropenia or macrophage deficiency and increased immunomodulatory activity (Li et al., 2020). Ye et al. (2018) identified important triterpenoid biosynthetic genes in mycelia, primordia and fruiting bodies of *G. lucidum*. Lanostane-type triterpenoid (leucocontexins S-X (1-6) with a molecular formula $C_{30}H_{48}O_6$ have been isolated from the fruiting bodies of *Ganoderma leucocontextum* (Zhao et al., 2016). Ergosterol peroxide exhibits anti-proliferative and apoptosis-induction properties (Martínez-montemayor et al., 2019).

2.10 Tocopherols/Vitamin E

Tocopherol and tocotrienol, also known as vitamin E, play an important role as antioxidant and is composed of eight different bioactive compounds, including alpha and beta-tocopherols and four other corresponding tocotrienols (Selvamani et al., 2018). Bouzgarrou et al. (2018) reported that the mycelia from *G. lucidum*, *P. ostreatus* and *P. eryngii* have been used as alternatives for tocopherols as the main ingredients added to yoghurt, especially in the case *G. lucidum* mycelium, which displayed the greatest potential for antioxidants, mostly based on its tocopherol profile. Tocopherol content was reported and is well known in edible mushrooms of *P. ostreatus* type (Selvamani et al., 2018). A study on antioxidants and properties of 18 wild Portuguese edible mushrooms showed beta-tocopherol was the vitamin detected in highest amounts, while alpha-tocopherol was not detected in the majority of these mushroom species (Sandrina et al., 2010).

3. Potential Application of Mushroom in Cosmetics

Cosmetic industries have existed for centuries and the market undergoes continuous growth. The cosmetic industry has shifted its paradigm to using natural-based ingredients due to several issues on the adverse effects of chemical-based ingredients. Consumers are searching for natural sources of ingredients that have skin benefits, with environment-friendly implications and with a less toxic effect on the skin. As people are becoming more aware of their health and wellness, the demand for nature-based cosmetics is rapidly increasing. Having various healing properties, researchers have found that bioactive metabolites of mushroom are not only beneficial for body consumption, but can also exert their effects on the skin. The mushroom industry has been growing rapidly over the years as the bioactive molecules of mushrooms have an excellent potential for inclusion in skincare products (Hyde et al., 2010). Skincare products with excellent skin protection and healing properties as well as in improving the skin texture are in high demand. Hence, the cosmetic market is growing worldwide, strengthened by various claims and functional properties that attract more consumers to purchase their products. The most bioactive compounds in high demand are those that have multifunctional properties with multifunctional benefits on the skin. Table 2 summarises the wide applications of mushroom bioactive ingredients in cosmeceuticals.

3.1 UV Protection

Generally, sunlight emits three types of ultraviolet (UV) radiation, namely ultraviolet A (UVA), ultraviolet B (UVB) and ultraviolet C (UVC). All these types of radiation have the potential to damage the skin differently, with UBC giving the most adverse effects. The high energy from these UVs can be mutagenic or carcinogenic (Savoye et al., 2018). The UVB can result in DNA damage, promoting mutations that cause skin cancer (Grether-Beck et al., 2014). Additionally, there is increasing evidence that visible light and infrared radiations are able to penetrate deep into the skin and generate reactive oxygen species (ROS). These ROS include the skin oxidative stress factor which is responsible in stimulating skin aging, carcinogenesis, as well as disturbing the mitochondrial integrity (Dupont et al., 2013). *Coprinus comatus* mushroom extract was demonstrated to contain high levels of L-Ergothioneine (LE). From some *in vitro* studies, LE shows the ability to inhibit myeloperoxidase (MPO). The MPO is a molecule which is responsible for the production of ROS molecules and contributes towards skin inflammation. Extreme exposure for an extended period of time to sunlight activates the production of ROS inside the skin. The activation of ROS is primarily caused by UVB radiation, which comprises approximately 4 per cent of sunlight radiation (Asahi et al., 2016). The LE is readily absorbed on the skin. Hence, LE is one of the essential ingredients that has been used in sunscreen lotions and many other cosmetic products for protecting against UV radiation (Bazela et al.,

Table 2. Wide application of various mushroom extracts in cosmeceutical industries.

Function	Bioactive compounds	Type of mushroom	Applications	References
UV protection	L-Ergothioneine and beta-glucan	*G. lucidum*	Sunscreen protection lotions, compact and loose powders, foundations	Pillai and Uma Devi, 2013
Skin whitening and anti-pigmentation	L-Ergothioneine and Clitocybin A	*Clitocybe aurantiaca*	Creams and lotions	Lee et al., 2017
Anti-ageing and anti-wrinkle	Protocatechuic, syringic acids and ergosterols	*A. bisporus, P. ostreatus, L. edodes* and *G. lucidum*	Serums, creams and lotions	Weng et al., 2011; Taofiq et al., 2019
Wound healing and anti-inflammatory	L-Ergothioneine, lentinan, phenolic acids (cinnamic acids, p-hydroxybenzoic acids, p-coumaric acid, protocatechuic acid) and ergosterol	*A. bisporus, L. edodes* and *P. ostreatus*	Healing creams, nappy rash creams and base creams	Zembron-Lacny et al., 2013; Taofiq et al., 2016a
Antioxidant	L-ergothioneine, lentinan, phenolic acids (cinnamic acids, p-hydroxybenzoic acids, p-coumaric acid, protocatechuic acid) and ergosterol	*A. bisporus, L. edodes* and *P. ostreatus*	Creams, moisturiser, lotions and cosmetics base creams	Taofiq et al., 2016b
Anti-microbial	Phenolic acids (cinnamic and p-hydroxybenzoic acids, p-coumaric acid) and ergosterol	*A. bisporus* and *P. ostreatus*	Antiseptic creams and lotions, ointments, talcum powder, facial cleanser and base creams	Taofiq et al., 2016a

2014). Pillai and Uma Devi (2013) performed a study on the potential of beta-glucan from *G. lucidum* mushroom against radiation-induced damage. The study assessed mouse survival, hematology, liver reduced glutathione, liver malondialdehyde and bone marrow chromosomal aberrations as endpoints. The result obtained showed 83 per cent survival of Swiss albino mice with beta-glucan protection compared to 100 per cent mortality after 30 days without the β-glucan protection. Hence, β-glucan is seen as one of the potential mushroom active ingredients to be used in cosmetics for radiation protection. The application for high-UV protection cosmetics is more demanding in Western countries as they have less melanin inside the skin and are usually prone to skin inflammation (Obayashi et al., 2005).

3.2 *Skin Whitening* and *Anti-pigmentation*

Pigmentation refers to skin colouring, which is generally caused by melanin. The higher the melanin content inside the skin, the darker the skin colour. Hyperpigmentation is a condition that causes excessive darkening of the skin. People living primarily in ASEAN countries tend to search for cosmetics with anti-pigmentation formulas. Tyrosinase is a rate-limiting enzyme in melanin biosynthesis pathway. Tyrosinase's mode of action is to convert tyrosine to dihydroxyphenylalanine (DOPA) and further oxidise DOPA to dopaquinone (Meng et al., 2012). Hence, the tyrosine inhibition potential from the mushrooms' bioactive ingredients can suppress the melanin, which dramatically leads to skin hyperpigmentation and photo-ageing (Thring et al., 2009; Taofiq et al., 2016a). The LE contained in mushroom extracts is also attributed to the whitening effect of the skin. The LE has the capability to inhibit tyrosine at a specific dosage and thus to stop melanin synthesis in the skin (Nachimuthu et al., 2019). Skin-whitening cosmeceutical products have a high demand in ASEAN countries and is added as a bioactive ingredient, especially in skin lotions, face moisturisers and hand lotions.

3.3 *Anti-ageing and Anti-wrinkle*

Skin ageing is a natural mechanism which affects the skin and also the body organs due to the hormonal changes. The natural changes in hormones are due to the factor of age and intensive exposure to ultra-violet radiation which cause the generation of free radical species inside the skin layer (Papakonstantinou et al., 2012). The free radicals, such as ROS, induces a transcription factor called activator protein-1 (AP-1) which regulates the matrix metalloproteinases (MMPs) (Taofiq et al., 2016a). The regulation of this matrix eventually promotes skin elastin and collagen breakdown, causing a reduction in the elasticity and tensile strength of the skin as well as skin inflammation (Yadav et al., 2015). This reduces skin firmness and eventually collapses the skin's inner structure, resulting in the formation of wrinkles (Thring et al., 2009). The bioactive compounds in mushrooms are believed to be able to suppress the production of collagenase and elastase, the enzymes responsible in the degradation of skin extracellular matrix. Mushroom extracts containing LE have also been tested for inhibition of hyaluronic acid (HA)-degrading enzyme in the skin. The HA is a vital matrix-binding biomolecule that helps to maintain skin integrity and firmness. Hence, the extract is suitable to be added in night creams and firming body lotions. Under physiological conditions, the LE is a stable and unique type of antioxidant which is prone to accumulate in the cells that are highly exposed to oxidative stress. Interestingly, LE found in mushroom extracts is much higher than other dietary sources (Nachimuthu et al., 2019). The protocatechuic and syringic acids from the ethanolic extract of *P. ostreatus* and *G. lucidum* also show beneficial properties for topical application and offer several advantages, such as protecting against photo-ageing of the skin, giving antioxidative support to the skin and suppressing oxidative stress (Taofiq et al., 2019). Lee and co-workers (2017) had investigated the anti-wrinkle effects of clitocybin A, an isoindolinone isolated from *Clitocybe aurantiaca* mushroom mycelium. The compound strongly scavenged radical activities on human primary dermal fibroblast-neonatal (HDF-N) cells, depending on the concentration following irradiation with UVB. Their positive findings suggest that clitocybin A from mushrooms may be an effective ingredient for use in future anti-wrinkle cosmetic products.

3.4 Wound-healing and Anti-inflammatory

The anti-inflammatory properties of different mushroom extracts is related to its ability to inhibit specific steps in the pathway leading to nuclear factor kappa B (NF-κB) release (Taofiq et al., 2016c). The existence of triterpenes and phenolic acids bioactive ingredients in mushrooms show excellent anti-inflammatory properties (Taofiq et al., 2015). The presence of these bioactive compounds can reduce the production of inducible nitric oxide synthase (iNOS), one of the oxidative enzymes associated with skin inflammatory diseases (Thring et al., 2009). *A. bisporus* and *L. edodes* are among the widely cultivated mushrooms worldwide and have been demonstrated to have anti-inflammatory properties (Zembron-Lacny et al., 2013; Stojković et al., 2014). Anti-inflammatory studies conducted by Taofiq et al. (2016c) on ethanolic extract of three types of mushrooms (*A. bisporus* and *L. edodes, P. ostreatus*) was assessed upon stimulation of RAW 264.7 macrophages with lipopolysaccharide for the production of inflammatory mediator (NO). Among the three types of mushroom extracts tested, the *L. edodes* exhibited the highest anti-inflammatory activity. This may be due to its high content of protocatechuic acid and the synergistic effect with other metabolites, such as terpenes, polysaccharides and lipid metabolites. However, different methods of extraction may also contribute to the different anti-inflammatory activities of the extracts.

3.5 Antioxidant

Nowadays, in designing the cosmeceutical formulations from mushrooms, extracting the best bioactive metabolites with radical scavenging activity becomes one of the most important parameters. Stanikunaite et al. (2009) successfully extracted two types of bioactive compounds from *Elaphomyces granulatus*. From the *in vitro* studies performed, they have shown that the presence of syringic acid and syringaldehyde in the truffle-like mushroom displayed an excellent antioxidant activity with IC_{50} of 41 μg/ml. The *E. granulatus* extract can inhibit up to 68 per cent of COX-2 activity at 50 μg/ml. Syringic acid had shown the most potent antioxidant activity with an IC_{50} of 0.7 μg/ml, while interestingly, syringaldehyde showed negligible activity when it was tested separately. Hence, the synergistic effect between the two bioactive compounds resulted in high antioxidant activity. The famous 'shiitake' mushroom, with its scientific name *L. edodes*, has been reported to have several bioactive compounds, namely L-ergothioneine and lentinan which are known to exert strong antioxidant activity (Zembron-Lacny et al., 2013). Two different studies on *A. Bisporus* extracts showed the highest DPPH (2,2-diphenyl-1-picrylhydrazyl) radical-scavenging activity and reducing power (Reis et al., 2012; Taofiq et al., 2016b).

3.6 Anti-Microbial

Taofiq et al. (2018) performed anti-microbial tests by using different ethanolic extracts for cosmeceutical formulations from *A. bisporus* and *P. ostreatus*. Interestingly, all formulations with the mushroom extracts showed a positive result for Gram-positive bacteria (*Enterococcus faecalis*, Methicillin-sensitive *Staphylococcus aureus* and Methicillin resistant *Staphylococcus aureus*) and Gram-negative (*Escherichia coli* and *Pseudomonas aeruginosa*) as well as yeast strains of *Candida albicans* (Taofiq et al., 2018). Previously, Taofiq et al. (2016a) had shown that mushroom extracts (*A. bisporus, L. edodes* and *P. ostreatus*) showed anti-microbial potential against MRSA and MSSA. However, most of the mushroom extracts failed to inhibit the growth of clinical isolates with antibiotic resistance Gram-negative bacteria strains (Barros et al., 2008; Taofiq et al., 2016a). Regardless of its good potential as an anti-microbial agent, the anti-microbial activity is usually totally reduced after only six months of storage. Hence, Taofiq and his co-workers developed an atomisation/ coagulation technique to microencapsulate the extracts. The slow and gradual release of the extract ensures continual effectiveness against microbial growth and makes the process suitable for use in designing various cosmeceutical formulations. A report from Ali and Yosipovitch (2013) suggested the pH of any topical formulations to be critically maintained around pH 3.5–6.5, as it is close to the pH of the skin which is at pH 5.5. At this pH range and condition, the microbes of the skin could be maintained. Most of the pathogenic bacteria typically started to thrive best at neutral pH, and thus all cosmeceutical formulations must avoid use of this pH for any of the topical formulations.

Figure 1 summarises the wide range of mushroom bioactive compounds, the functional properties and potential application in cosmeceutical industry. In general, most of the bioactive compounds are multi-functional, which means that they can exert several healing or protective functions when incorporated into skincare or cosmetics. This property makes the mushroom-based skincare cosmetics even more valuable. Nevertheless, most of the activities of the bioactive compounds were affected adversely when they were incorporated into the cream formulations (Taofiq et al., 2016a). This may be due to some interference between the original formulation

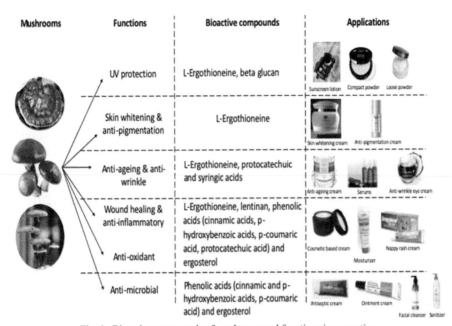

Fig. 1. Bioactive compounds of mushroom and functions in cosmetics.

in the base cream that caused activity reduction of the bioactive compound. Hence, optimisation is required in order to obtain a cosmeceutical formulation with a maintained bioactive compound.

4. Mushroom Bioactive Formulation in Cosmetic Industry

Mushroom-based cosmetics are regarded as natural cosmetics and have attracted extensive attention from consumers. Bioactive metabolites-rich mushroom extracts have been incorporated in various cosmetic products that serve as cosmeceuticals. Table 3 lists some of the commercial cosmetic products that contain mushroom

Table 3. Current mushroom-based cosmetic product in the market.

Product name	Mushroom extract	Function	References
Aveeno® Positively Ageless® range, US	*G. lucidum* extract	Anti-aging, brightening and moisturising effect	https://www.aveeno.com/
One Love Organics® Botanical D moisture mist, UK	*L. edodes* extract	Moisturising	https://shop. oneloveorganics.com/
Osmia Organics® Brighten Facial Serum, US	*L. edodes* extract	Brightening and moisturising effect	https://osmiaorganics. com/
CV Skinlabs® Body Repair Lotion, US	*G. lucidum* extract	Wound-healing and anti-inflammatory	https://cvskinlabs.com/
Dr. Andrew Weil for Origins™ Mega-Mushroom Skin Relief Face Mask, US	*Hypsizygus ulmarius* mycelium, *G. lucidum* and *C. sinensis* (Berk.) Sacc. extracts	Anti-inflammatory properties and anti-aging	https://www.origins.com/
Volition Snow Mushroom Water Serum, US	*Tremella fuciformis* extract	Moisturising and anti-aging	https://volitionbeauty. com/
Blithe Tundra Chaga Pressed Serum, Korea	*Inonotus obliquus* extract	Anti-aging	https://blithecosmetic. com/
REN Evercalm™ Ultra Comforting Rescue Mask, UK	*Albarellus ovinus* extract	Anti-inflammatory properties and moisturising	https://www.renskincare. com/
Menard Embellir range, Japan	*G. lucidum* extract	Eliminate toxins, repair skin damage associated UV radiation and free radical	www.menard-cosmetic. com
Estée Lauder Re-Nutriv sun care product, U.S.	*G. lucidum* extract	Brightening and anti-aging effects of UV radiation	www.esteelauder.com
Wexler DermatologyInstant De-Puff Eye Gel, US	*A. bisporus* L. extract	Antiaging and anti-wrinkles	www.wexlerdermatology. com
Sulwhasoo Timetreasure Honorstige Serum, Korea	*G. lucidum* extract	Antiaging and tone skin texture	www.sulwhasoo.com

extracts and the functions on the skin and produced by cosmetic companies worldwide. The most claimed functions of the mushroom extracts in cosmetics are moisturising function, antiaging, brightening, wound-healing, anti-inflammatory and anti-UV damage. These functions are corroborated with the active ingredients found in the mushroom extract that imparts these properties.

Mushroom bioactive compounds are normally incorporated in the skincare and cosmetics as mushroom extracts. Mushroom extracts contain different ranges of bioactive compounds at various concentrations, depending on the mushroom source and extraction technique. Therefore, prior to formulation for skincare and cosmetics, the mushroom extracts are normally characterised for their bioactive molecule composition and functional properties. Effective concentrations of mushrooms' active ingredients to exert their respective functions have been widely studied and some are shown in Table 4. A study reported that *Tremella* polysaccharides function at 0.05–0.10 per cent concentration to display its moisturising effect (Liu and He, 2012). The carboxymethylated polysaccharide (CATP) from *Tremella fuciformis* functions best for moisture retention and antioxidant at 1.64 mg/ml concentration (Wang et al., 2015). In another study, polysaccharide-rich aqueous extract of *Auricularia fuscosuccinea* exhibited nearly equal moisture retention performance to sodium hyaluronic acid at 2 mg/ml of concentration (Liao et al., 2014). Phenolic acids and ergosterol function are seen at various concentrations for anti-inflammatory, anti-tyrosinase activity, antioxidant, radical scavenging and anti-bacterial activity, depending on the source of the extract.

The cosmetic formulations of mushroom extracts vary, depending on the product and the variety of mushroom used. For example, mushroom extract powder from *Fomes officinalis* or *Polyporus officinalis* is cosmetically effective in reducing the shiny appearance of skin and improving the appearance of skin imperfections at a concentration range of 0.01–10 per cent (w/w). The more preferable concentration was 0.05–8 per cent (w/w) with 0.1–7 per cent (w/w) being reported to be the most effective range (Sandewicz et al., 2003). Taofiq et al. (2016) incorporated 100 mg of ethanolic extracts of *A. bisporus, P. ostreatus* and *L. edodes* per gram of base cream. The mushroom extracts incorporated into base-cream showed strong anti-inflammatory, antityrosinase, antioxidant, and antibacterial properties. The concentrations of phenolic acids and ergosterols in the final formulations were maintained at 85–100 per cent and all bioactivities were preserved. Interestingly, the significant antioxidant and antibacterial bioactivities make the mushroom extracts a good preservative for cosmetics. This formulation can be used in cosmetics to combat skin aging, inflammation and hyperpigmentation.

Incorporation of mushrooms' active compounds in cosmetic ingredients often faced instability limitation due to many environmental factors. Reduction of bioactivities was reported when mushroom extracts were incorporated into cosmetics compared to pure extracts (Taofiq et al., 2016a). Degradation of active compound reduces its effectiveness with time. Microencapsulation with polymer material can protect the active compounds from exposure to harsh environmental conditions and prevent interference from other ingredients. Micro encapsulation of *A. bisporus* and *P. ostreatus* extracts in sodium alginate showed slow release of anti-tyrosinase and

Table 4. Formulation of different mushroom active compounds from mushroom in cosmetics (*GAE, gallic acid equivalent).

Tested compounds	Source	Effects	Concentration in extract	Effective concentration of extract/ compound	References
Polysaccharide	*Tremella* sp.	Lightening effect (inhibits melanin formation) and moisturizing effects	Pure extract	0.05–0.10 per cent polysaccharide extract	Liu and He, 2012
Carboxymethylated polysaccharide (CATP)	*T. fuciformis*	Moisturizing and anti-oxidative effects	Pure compound	1.64 mg/ml of CATP4 compound	Wang et al., 2015
Phenolic acid-rich extract	*Auricularia fuscosuccinea*	Anti-oxidative effects and moisture retention capacity	35.9 mg *GAE/g sample	EC_{50} of DPPH free radical scavenging activity: 1.31 mg/ml of aqueous extract	Liao et al., 2014
			195.5 mg *GAE/g sample	EC_{50} of DPPH free radical scavenging activity: 1.15 mg/ml of ethanol extract	
Phenolic acid-rich methanol extract	*Volvariella volvacea*	Anti-aging effects (antioxidant activity and free radical scavenging abilities)	0.25 μmoles of *GAE/mg of extract	IC_{50} of lipid peroxidation: 0.109 mg/ml dichloromethane subfractions	Cheung and Cheung, 2005
Phenolic acid-rich aqueousextract	*L. edodes*		0.118 μmoles of *GAE/mg of extract	IC_{50} of lipid peroxidation: 1.05 mg/ml low molecular weight subfractions	
Phenolic acid-rich extract	*Termitomyces heimii*	Antioxidant activity	37 mg/g of aqueous extract	IC_{50} of free radical scavenging activity: 1.1 mg (dry weight)/ml	Puttaraju et al., 2006
			11.2 mg/g of methanolic extract	IC_{50} of free radical scavenging activity 2.7 mg (dry weight)/ml	
Phenolic acids (cinnamic acid, *p*-hydroxybenzoic acid, *p*-coumaric acid and protocatechuic acid) and ergosterol-rich ethanolic extract	*A. bisporus*	Anti-inflammatory, anti-tyrosinase activity, antioxidant, radical scavenging, anti-bacterial activity	90.06 μg/g (phenolic acid), 44.79 mg/g (ergosterol)	EC_{50} of anti-inflammatory: 0.18 mg/ml; EC_{50} of anti-tyrosinase: 0.16 mg/ml; EC_{50} of antioxidant: 7.04 mg/ml; Reducing power: 2.34	Taofiq et al., 2016b
	P. ostreatus		584.24 μg/g (phenolic acid), 78.20 mg/g (ergosterol)	EC_{50} of anti-inflammatory: 0.29 mg/ml; EC_{50} of anti-tyrosinase: 0.86 mg/ml; EC_{50} of antioxidant: 7.69 mg/ml; Reducing power: 2.36	
	L. edodes		142.81 μg/g (phenolic acid), 8.94 mg/g (ergosterol)	EC_{50} of anti-inflammatory: 0.16 mg/ml; EC_{50} of anti-tyrosinase: 0.82 mg/ml; EC_{50} of antioxidant: 23.36 mg/ml; Reducing power: 3.03	

anti-microbial activities. This ensures controlled bioactivity and gradual release of active compounds over time (Taofiq et al., 2018).

5. Safety of Mushroom Bioactive Ingredients

The increase in consumer awareness and consciousness about the origin of the ingredients used, the adverse effects on the skin, safety and environmental issues has encouraged interest in exploring greener and more sustainable raw materials for cosmetic formulations (Wang et al., 2015). However, the diverse chemical compounds in mushroom bioactive extracts may induce unwanted skin problems, such as skin irritation, skin photosensitive, dermatitis, hair and nail impairment, hyper- and hypo-pigmentation and systemic effects (Gao et al., 2008). Throughout the years, scientists have developed *in vitro* model systems (mimic skin cells) to help in assessing the safety and potential toxicity against skin cells (Rodrigues et al., 2013). Table 5 shows some of the studies that have been conducted to investigate the safety of mushroom bioactive compounds. LE extract has been tested on animal models and human volunteers, and currently, no adverse effect has been reported in the behavioural and histopathological changes (Schauss et al., 2011; Forster et al., 2015; Asahi et al., 2016). Additionally, the LE has also been accepted with significant health benefits in food supplements for inner health and beauty. Thus, LE extract from mushrooms can be used as one of the healthy, beneficial bioactive compounds for a broad application in cosmeceutical products. Based on scientific studies, the European Food Safety Authority (EFSA) revealed that there is no precise relationship between the intake of LE and the development of diabetes mellitus or chronic inflammatory disease (Nachimuthu et al., 2019).

Taofiq et al. (2019) evaluated the safety of their final cosmetic formulation which was supplemented with phenolic and triterpenoid rich extracts obtained from *G. lucidum* and *P. ostreatus* mushrooms. The effect of their formulation on the cell viability (keratinocytes and fibroblasts cells) shows approximately 60 per cent of cell survival rate for all concentrations below 100 µg/ml demonstrating its safety. However, exposure at concentration above 100 µg/ml affected the viability of fibroblast cells. An *ex vivo* skin permeation analysis using a pig skin model also showed a significant percentage of some earlier detected compounds (protocatechuic and syringic acids) being retained in the skin layer after 8 hours. This finding can be significant if the retained compound can cause a local cosmeceutical effect. Cinnamic, p-hydroxybenzoic and p-coumaric acids from *P. ostreatus* showed no skin penetration after 8 hours (Taofiq et al., 2019).

Several compounds, such as triterpenes, terpenes and sesquiterpenes were categorised as 'generally regarded as safe' (GRASS) by the Food and Drug Administration (Aqil et al., 2007). The ultimate target is to enable the penetration of active compounds in the *stratum cornuem*, a rate-limiting barrier before the compound can successfully permeate through the epidermis (Haq et al., 2018). The β-glucan administered at 500 g/kg body wt. of animal model produced no toxicity and showed 100 per cent survival (Pillai and Uma Devi, 2013).

Table 5. Safety of mushroom extract as cosmetic and skin care ingredient.

Compounds	Risk	Recommendation	References
L-Ergothioneine (LE)	No toxicity	In human subjects - Up to 25 mg/day for 1 week. In animal models - Up to 1600 mg per kg of body weight	Schauss et al., 2011; Forster et al., 2015
	No mutagenic activity	5000 µg/ml	
	No genotoxicity	30 mg/day for adults and 20 mg per day for children	
	No toxicity	1500 mg/kg in the *in vivo* mouse mammalian erythrocyte micronucleus test	Asahi et al., 2016
Beta-glucan	No toxicity	500 g/kg body weight of animal model	Pillai and Uma Devi, 2013
Phenolic acid	60% fibroblast cells viability maintained	100 µg/ml *G. lucidum* extract	Taofiq et al., 2019
	90% fibroblast cells viability maintained	100 µg/ml *P. ostreatus* extract	
	90% keratinocytes cell viability maintained	100 µg/ml of *P. ostreatus* and *G. lucidum*, respectively	

6. Conclusion and Outlook

Mushroom has emerged as a powerful ingredient, which not only exerts its powerful healing effect on the body but also contributes to skin beauty. Research has discovered the important mushroom bioactive compounds that confer moisturising, anti-inflammatory, anti-tyrosinase, antioxidant, and antibacterial properties. Moreover, each bioactive compound exhibits multi-functional properties which make it versatile for a wide range of applications in cosmetics. Mushroom-based cosmetics are claimed and proven to have UV protection, whitening, antipigmentation, antiageing, antiwrinkle, wound healing and anti-inflammatory effects. However, research in also needed to investigate the bioactive molecules of rare mushrooms. In addition, further research has to investigate in depth the mechanism of action of the mushroom bioactive molecules (structure/function relationship) with some clinical trials, especially for claims related to antiaging, antipigmentation and wound-healing properties. Therefore, mushroom 'super ingredients' are predicted to evolve and their demand in cosmeceutical industry will continue accordingly.

References

Abugri, D.A. and McElhenney, W.H. (2013). Extraction of total phenolic and flavonoids from edible wild and cultivated medicinal mushrooms as affected by different solvents. J. Nat. Prod. Pl. Resour., 3(3): 37–42.

Afiati, F., Firza, Kusmiati, S.F. and Aliya, L.S. (2019). The effectiveness β-glucan of shiitake mushrooms and *Saccharomyces cerevisiae* as antidiabetic and antioxidant in mice Sprague Dawley induced alloxan. AIP Conf. Proc., 2120(1): 070006.10.1063/1.5115723.

Ali, S.M. and Yosipovitch G. (2013). Skin pH: From basic science to basic skin care. Acta Derm. Venereol., 93(3): 261–267.

Aqil, M.A., Ahad, Y., Sultana and Ali, A. (2007). Status of terpenes as skin penetration enhancers. Drug Discov. Today, 12(23-24): 1601–1607.

Asahi, T., Wu, X., Shimoda, H., Hisaka, S., Harada, E. et al. (2016). A mushroom-derived amino acid, ergothioneine, is a potential inhibitor of inflammation-related DNA halogenation. Biosci. Biotechnol. Biochem., 80(2): 313–317.

Ayeka, P.A. (2018). Potential of mushroom compounds as immunomodulators in cancer immunotherapy: A review. Evid. Based Compl. Alt., 2018. 7271509.10.1155/2018/7271509.

Barbieri, A., Quagliariello, V., Del Vecchio, V., Falco, M., Luciano, A. et al. (2017). Anticancer and anti-inflammatory properties of *Ganoderma lucidum* extract effects on melanoma and triple-negative breast cancer treatment. Nutrients, 9(3): 210.10.3390/nu9030210.

Barros, L., Cruz, T., Baptista, P., Estevinho, L.M. and Ferreira, I.C.F.R. (2008). Wild and commercial mushrooms as source of nutrients and nutraceuticals. Food. Chem. Toxicol., 46(8): 2742–2747.

Bazela, K., Solyga-Zurek, A., Debowska, R., Rogiewicz, K., Bartnik, E. and Eris, I. (2014). L-ergothioneine protects skin cells against UV-induced damage—A preliminary study. Cosmetics, 1(1): 51–60.

Benmrad, M.O., Mechri, S., Jaouadi, N.Z., Elhoul, M., Rekik, H. et al. (2019). Purification and biochemical characterisation of a novel thermostable protease from the oyster mushroom. BMC Biotechnol., 19(43): 1–18.

Borodina, I., Kenny, L.C.C.M., McCarthy, K., Paramasivan, K., Pretorius, E. et al. (2020). The biology of ergothioneine, an antioxidant nutraceutical. Nutr. Res. Rev., 2020: 1–28.10.1017/S0954422419000301.

Bouzgarrou, C., Amara, K., Reis, F.S., Barreira, J.C.M., Skhiri, F. et al. (2018). Incorporation of tocopherol-rich extracts from mushroom mycelia into yogurt. Food Funct., 9(6): 3166–3172.

Çayan, F., Deveci, E., Tel-Çayan, G. and Duru, M.E. (2020). Identification and quantification of phenolic acid compounds of twenty-six mushrooms by HPLC-DAD. J. Food Meas. Charact., 14(4): 1690–1698.

Chaturvedi, V.K., Agarwal, S., Gupta, K.K., Ramteke, P.W. and Singh, M.P. (2018). Medicinal mushroom: Boon for therapeutic applications. 3 Biotech., 8(8): 334.10.1007/s13205-018-1358-0.

Chen, S., Yong, T., Zhang, Y., Su, J., Jiao, C. and Xie, Y. (2017). Anti-tumor and anti-angiogenic ergosterols from *Ganoderma lucidum*. Front. Chem., 5(85): PMC5670154.

Cheung, L.M. and Cheung, P.C.K. (2005). Mushroom extracts with antioxidant activity against lipid peroxidation. Food Chem., 89(3): 403–409.

Diling, C., Xin, Y., Chaoqun, Z., Jian, Y., Xiaocui, T. et al. (2017). Extracts from *Hericium erinaceus* relieve inflammatory bowel disease by regulating immunity and gut microbiota. Oncotarget., 8(49): 85838–85857.

Dogan, H., Coteli, E. and Karatas, F. (2016). Determination of glutathione, selenium and malondialdehyde in different edible mushroom species. Biol. Trace Elem. Res., 174(2): 459–463.

Dudekula, U.T., Doriya, K. and Devarai, S.K. (2020). A critical review on submerged production of mushroom and their bioactive metabolites. 3 Biotech., 10(337): 1–12.

Dupont, E., Gomez, J. and Bilodeau, D. (2013). Beyond UV radiation: A skin under challenge. Int. J. Cosmet. Sci., 35(3): 224–232.

Elkhateeb, W.A., Daba, G.M., Thomas, P.W. and Wen, T. (2019). Medicinal mushrooms as a new source of natural therapeutic bioactive compounds. Egypt. Pharm. J., 18(2): 88–101.

Elsayed, E.A., El Enshasy, H., Wadaan, M.A.M. and Aziz, R. (2014). Mushrooms: A potential natural source of anti-inflammatory compounds for medical applications. Mediators of Inflammation, 2014: 805841.10.1155/2014/805841.

El Enshasy, H., Elsayed, E.A., Aziz, R. and Wadaan, M.A. (2013). Mushrooms and truffles: Historical biofactories for complementary medicine in Africa and in the Middle East. eCAM.: 2013. 620451. 10.1155/2013/620451.

El Enshasy, H.A. and Hatti-kaul, R. (2013). Mushroom immunomodulators: Unique molecules with unlimited applications. Tr. Biotechnol., 31(12): 668–677.

El Enshasy, H.A., Masri, H.M., Maftoun, P., Abd Malek, R., Boumehira, A.Z. et al. (2017). The edible mushroom *Pleurotus* spp. - II. Medicinal values. Int. J. Biotechnol. Wellness Industries, 6(1): 1–11.

Figueiredo, L. and Régis, W.C.B. (2017). Medicinal mushrooms in adjuvant cancer therapies: An approach to anticancer effects and presumed mechanisms of action. Nutrire, 42(1): 28. 10.1186/s41110-017-0050-1.

Forster, R., Spézia, F., Papineau, D., Sabadie, C., Erdelmeier, I. et al. (2015). Reproductive safety evaluation of L-Ergothioneine. Food Chem. Toxicol., 80: 85–91.

Friedman, M. (2016). Mushroom polysaccharides: Chemistry and antiobesity, antidiabetes, anticancer, and antibiotic properties in cells, rodents and humans. Foods, 5(4): 80. 10.3390/foods5040080.

Gao, X.-H., Zhang, L., Wei, H. and Chen, H.D. (2008). Efficacy and safety of innovative cosmeceuticals. Clin. Dermatol., 26(4): 367–374.

Gąsecka, M., Siwulski, M., Magdziak, Z., Budzyńska, S., Stuper-Szablewska, K. et al. (2020). The effect of drying temperature on bioactive compounds and antioxidant activity of *Leccinum scabrum* (Bull.) Gray and *Hericium erinaceus* (Bull.) Pers. J. Food Sci. Technol., 57(2): 513–525.

Gil-Ramírez, A., Pavo-Caballero, C., Baeza, E., Baenas, N., Garcia-Viguera, C. et al. (2016). Mushrooms do not contain flavonoids. J. Funct. Foods, 25: 1–13.

Gong, P., Wang, S., Liu, M., Chen, F., Yang, W. et al. (2020). Extraction methods, chemical characterizations and biological activities of mushroom polysaccharides: A mini-review. Carbohydr. Res., 494: 108037. 10.1016/j.carres.2020.108037.

Govorushko, S., Rezaee, R., Dumanov, J. and Tsatsakis, A. (2019). Poisoning associated with the use of mushrooms: A review of the global pattern and main characteristics. Food Chem. Toxicol., 128: 267–279.

Goyal, A., Kalia, A. and Sodhi, H.S. (2018). Profiling of intra-and extracellular enzymes involved in fructification of the lingzhi or reishi medicinal mushroom, *Ganoderma lucidum* (Agaricomycetes). Int. J. Med. Mushrooms, 20(12): 1209–1221.

Grether-Beck, S., Marini, A., Jaenicke, T. and Krutmann, J. (2014). Photoprotection of human skin beyond ultraviolet radiation. Photodermatol. Photoimmunol. Photomed., 30(2-3): 167–174.

Grzywacz, A., Argasińska, J.G., Kała Opoka, K. and Muszyńska, B. (2016). Anti-inflammatory activity of biomass extracts of the bay mushroom, *Imleria badia* (Agaricomycetes), in RAW 264.7 cells. Int. J. Med. Mushrooms, 18(9): 769–779.

Guillamón, E., García-Lafuente, A., Lozano, M., D́arrigo, M., Rostagno, M.A. et al. (2010). Edible mushrooms: Role in the prevention of cardiovascular diseases. Fitoterapia, 81(7): 715–723.

Haimid, M.T., Rahim, H. and Dardak, R.A. (2013). Understanding the mushroom industry and its marketing strategies for fresh produce in Malaysia. e-ETMR, 8: 27–37.

Haq, A., Goodyear, B., Ameen, D., Joshi, V. and Michniak-Kohn, B. (2018). Strat-M® synthetic membrane: Permeability comparison to human cadaver skin. Int. J. Pharm., 547(1): 432–437.

Hetland, G., Tangen, J., Mahmood, F., Mirlashari, M.R., Tjønnfjord, G.E. and Johnson, E. (2020). Related Medicinal basidiomycetes mushrooms, preclinical and clinical studies. Nutrients, 12(1339): 1–19.

Ho, L.-H, Zulkifli, N.A. and Tan, T.-C. (2020). Edible mushroom: Nutritional properties, potential nutraceutical values, and its utilisation in food product development. pp. 1–19. *In*: Passari, A.K. and Sánchez, S. (eds.). Introduction to Mushroom. IntechOpen.10.5772/intechopen.91827.

Hyde, K.D., Bahkali, A.H. and Moslem, M.A. (2010). Fungi—An unusual source for cosmetics. Fungal Divers., 43(1): 1–9.

Jesenak, M., Majtan, J., Rennerova, Z., Kyselovic, J., Banovcin, P. and Hrubisko, M. (2013). Immunomodulatory effect of pleuran (β-glucan from *Pleurotus ostreatus*) in children with recurrent respiratory tract infections. Int. Immunopharmacol., 15(2): 395–399.

Jiang, X., Meng, W., Li, L., Meng, Z. and Wang, D. (2020). Adjuvant therapy with mushroom polysaccharides for diabetic complications. Front. Pharmacol., 11(168). 10.3389/fphar.2020.00168.

Kabel, M.A. (2017). Occurrence and function of enzymes for lignocellulose degradation in commercial *Agaricus bisporus* cultivation. Appl. Microbiol. Biotechnol., 101(11): 4363–4369.

Kalaras, M.D., Richie, J.P., Calcagnotto, A. and Beelman, R.B. (2017). Mushrooms: A rich source of the antioxidants ergothioneine and glutathione. Food Chem., 233: 429–433.

Khan, A.A., Gani, A., Khanday, F.A. and Masoodi, F.A. (2018). Biological and pharmaceutical activities of mushroom β-glucan discussed as a potential functional food ingredient. Bioact. Carbohyd. Diet. Fibre, 16: 1–13.

Kumakura, K., Hori, C., Matsuoka, H., Igarashi, K. and Samejima, M. (2019). Protein components of water extracts from fruiting bodies of the reishi mushroom *Ganoderma lucidum* contribute to the production of functional molecules. J. Sci. Food Agric., 99(2): 529–535.

Lee, I.-K., Seok, S.-J., Kim, W.-K. and Yun, B.S. (2006). Hispidin derivatives from the mushroom inonotus xeranticus and their antioxidant activity. J. Nat. Prod., 69(2): 299–301.

Lee, J.E., Lee, I.S., Kim, K.C., Yoo, I.D. and Yang, H.M. (2017). Ros scavenging and anti-wrinkle effects of clitocybin a isolated from the mycelium of the mushroom *Clitocybe aurantiaca*. J. Microbiol. Biotechnol., 27(5): 933–938.

Li, A., Shuai, X., Jia, Z., Li, H., Liang, X. et al. (2015). *Ganoderma lucidum* polysaccharide extract inhibits hepatocellular carcinoma growth by downregulating regulatory T cells accumulation and function by inducing microRNA-125b. J. Transl. Med., 13(1): 100.10.1186/s12967-015-0465-5.

Li, X., Zeng, F., Huang, Y. and Liu, B. (2019). The positive effects of *Grifola frondosa* heteropolysaccharide on NAFLD and regulation of the gut microbiota. Int. J. Mol. Sci., 20(21): 5302.

Li, Z., Shi, Y., Xu, X., Zhang, J., Wang, H. et al. (2020). Screening immunoactive compounds of *Ganoderma lucidum* spores by mass spectrometry molecular networking combined with *in vivo* zebrafish assays. Front. Pharmacol., 11: 1–19.

Liao, W.C., Hsueh, C.Y. and Chan, C.F. (2014). Antioxidative activity, moisture retention, film formation, and viscosity stability of *Auricularia fuscosuccinea*, white strain water extract. Biosci. Biotechnol. Biochem., 78(6): 1029–1036.

Lin, T.Y., Hsu, H.Y., Sun, W.H., Wu, T.H. and Tsao, S.M. (2017). Induction of Cbl-dependent epidermal growth factor receptor degradation in Ling Zhi-8 suppressed lung cancer. Canc. J. Int., 140(11): 2596–2607.

Liu, H. and He, L. (2012). Comparison of the moisture retention capacity of *Tremella* polysaccharides and hyaluronic acid. J. Anhui Agric. Sci., 40: 13093–13094.

Ma, G., Yang, W., Zhao, L., Pei, F., Fang, D. and Hu, Q. (2018). A critical review on the health promoting effects of mushrooms nutraceuticals. Food Sci. Hum. Well., 7(2): 125–133.

Machado, A.R.G., Martim, S.R., Alecrim, M.M. and Teixeira, M.F.S. (2017). Production and characterisation of proteases from edible mushrooms cultivated on amazonic tubers. Afr. J. Biotechnol., 16(46): 2160–2166.

Mao, G., Li, Q., Deng, C., Wang, Y., Ding, Y. et al. (2018). The synergism and attenuation effect of Selenium (Se)-enriched *Grifola frondosa* (Se)-polysaccharide on 5-Fluorouracil (5-Fu) in Heps-bearing mice. Int. J. Biol. Macromol., 107: 2211–2216.

Martim, S.R., Silva, L.S.C., de Souza, L.B., do Carmo, E.J., Alecrim, M.M. et al. (2017). *Pleurotus albidus*: A new source of milk-clotting proteases. Afr. J. Microbiol. Res., 11(7): 660–667.

Martínez-montemayor, M.M., Ling, T., Suárez-Arroyo, I.J., Oritz-Soto, G., Santiago-Negrón, C.L. et al. (2019). Identification of biologically active *Ganoderma lucidum* compounds and synthesis of improved derivatives that confer anti-cancer activities *in vitro*. Front. Pharmacol., 10: 1–17.

Masuda, Y., Nakayama, Y., Tanaka, A., Naito, K. and Konishi, M. (2017). Antitumor activity of orally administered maitake α-glucan by stimulating antitumor immune response in murine tumor. PLoS ONE, 12(3): e0173621.10.1371/journal.pone.0173621.

Meng, T.X., Zhang, C.F., Miyamoto,T., Ishikawa, H., Shimizu, K. et al. (2012). The melanin biosynthesis stimulating compounds isolated from the fruiting bodies of *Pleurotus citrinopileatus*. J. Cosmet. Dermatol., 2(3): 151–157.

Murphy, E.J., Masterson, C., Rezoagli, E., O'Toole D., Major, I. et al. (2020). β-glucan extracts from the same edible shiitake mushroom *Lentinus edodes* produce differential *in-vitro* immunomodulatory and pulmonary cytoprotective effects – Implications for coronavirus disease (Covid-19) immunotherapies. Sci. Total Environ., 732: 139330.10.1016/j.scitotenv.2020.139330.

Nachimuthu, S., Kandasamy, R., Ponnusamy, R., Deruiter, J., Dhanasekaran, M. and Thilagar, S. (2019). L-Ergothioneine: A potential bioactive compound from edible mushrooms. Med. Mushrooms, 391–407.

Nakamura, A., Zhu, Q., Yokoyama, Y., Kitamura, N., Uchida, S. et al. (2019). *Agaricus brasiliensis* KA21 May prevent diet-induced nash through its antioxidant, anti-inflammatory, and anti-fibrotic activities in the liver. Foods, 8(11): 546.10.3390/foods8110546w.

Nowak, R., Nowacka-Jechalke, N., Juda, M. and Malm, A. (2018). The preliminary study of prebiotic potential of polish wild mushroom polysaccharides: The stimulation effect on *Lactobacillus* strains growth. Eur. J. Nutr., 57(4): 1511–1521.

Obayashi, K., Kurihara, K., Okano, Y., Masaki, H. and Yarosh, D.B. (2005). L-ergothioneine scavenges superoxide and singlet oxygen and suppresses TNF-α and MMP-1 expression in UV-irradiated human dermal fibroblasts. J. Cosmet. Sci., 56(1): 17–27.

Oluwafemi, G.I., Seidu, K.T. and Fagbemi, T.N. (2016). Chemical composition, functional properties and protein fractionation of edible oyster mushroom (*Pleurotus ostreatus*). Food Sci. Technol., 17(1): 218–223.

Pahila, J., Ishikawa, Y. and Ohshima, T. (2019). Effects of ergothioneine-rich mushroom extract on the oxidative stability of astaxanthin in liposomes. J. Agric. Food Chem., 67(12): 3491–3501.

Panche, A.N., Diwan, A.D. and Chandra, S.R. (2016). Flavonoids: An overview. J. Nutr. Sci., 5(47): 1–15.

Papakonstantinou, E., Roth, M. and Karakiulakis, G. (2012). Hyaluronic acid: A key molecule in skin aging. Dermatoendocrinol., 4(3): 253–258.

Phan, C.W. and Sabaratna, V. (2012). Potential uses of spent mushroom substrate and its associated lignocellulosic enzymes. Appl. Microbiol. Biotechnol., 96(4): 863–873.

Phillips, K.M., Ruggio, D.M., Horst, R.L., Minor, B., Simon, R.R. et al. (2011). Vitamin D and sterol composition of 10 types of mushrooms from retail suppliers in the United States. J. Agric. Food Chem., 59(14): 7841–7853.

Pillai, T.G. and Uma Devi, P. (2013). Mushroom beta glucan: Potential candidate for post irradiation protection. Mutat. Res., 751(2): 109–115.

Puttaraju, N.G., Venkateshaiah, S.U., Dharmesh, S.M., Urs S.M.N. and Somasundaram, R. (2006). Antioxidant activity of indigenous edible mushrooms. J. Agric. Food Chem., 54(26): 9764–9772.

Rahi, D.K. and Malik, D. (2016). Diversity of mushrooms and their metabolites of nutraceutical and therapeutic significance. J. Mycol., 1–18.

Rathore, H., Prasad, S., Kapri, M., Tiwari, A. and Sharma, S. (2019). Medicinal importance of mushroom mycelium: Mechanisms and applications. J. Funct. Foods, 56: 182–193.

Reid, T., Munyanyi, M. and Mduluza, T. (2017). Effect of cooking and preservation on nutritional and phytochemical composition of the mushroom *Amanita zambiana*. Food Sci. Nutr., 5(3): 538–544.

Reis, F.S., Martins, A., Barros, L. and Ferreira, I.C.F.R. (2012). Antioxidant properties and phenolic profile of the most widely appreciated cultivated mushrooms: A comparative study between *in vivo* and *in vitro* samples. Food Chem. Toxicol., 50(5): 1201–1207.

Rodrigues, F., Palmeira-de-Oliveira, A., das Neves, J., Sarmento, B., Amaral, M.H. and Oliveira, M.B. (2013). *Medicago* spp. extracts as promising ingredients for skin care products. Ind. Crops Prod., 49: 634–644.

Rossi, P., Difrancia, R., Quagliariello, V., Savino, E., Tralongo, P., Randazzo, C.L. and Berretta, M. (2018). B-glucans from *Grifola frondosa* and *Ganoderma lucidum* in breast cancer: An example of complementary and integrative medicine. Oncotarget, 9: 24837–24856.

Rouhana-Toubi, A., Wasser, S.P. and Fares, F. (2015). The shaggy ink cap medicinal mushroom, *Coprinus comatus* (higher basidiomycetes) extract induces apoptosis in ovarian cancer cells via extrinsic and intrinsic apoptotic pathways. Int. J. Med. Mushrooms, 17(12): 1127–1136.

Rubel, R., Dalla Santa, H.S., Dos Santos, L.F., Fernandes, L.C., Figueiredo, B.C. and Soccol, C.R. (2018). Immunomodulatory and antitumoral properties of *Ganoderma lucidum* and *Agaricus brasiliensis* (Agaricomycetes) medicinal mushrooms. Int. J. Med. Mushrooms, 20(4): 393–403.

Sánchez, C. (2010). Cultivation of *Pleurotus ostreatus* and other edible mushrooms. Appl. Microbiol. Biotechnol., 85(5): 1321–1337.

Sande, D., de Oliveira, G.P., Moura, M.A.F., Martins, B.A., Lima, M.T.N.S. and Takahashi, J.A. (2019). Edible mushrooms as a ubiquitous source of essential fatty acids. Food Res. Int., 125: 108524. 10.1016/j.foodres.2019.108524.

Sandewicz, I.M., Russ, J.G. and Zhu, V.X. (2003). Anhydrous Cosmetic Compositions Containing Mushroom Extract. United States Patent # US 6,645,502.

Sandrina, H., Lillian, B., Maria João, S., Anabela, M. and Isabel, F. (2010). Tocopherols composition of Portuguese wild mushrooms with antioxidant capacity. Food Chem., 119(4): 1443–1450.

Savoye, I., Olsen, C.M., Whiteman, D.C., Bijon, A., Wald, L. et al. (2018). Patterns of ultraviolet radiation exposure and skin cancer risk: The E3N-Sun exp. study. J. Epidemiol., 28(1): 27–33.

Schauss, A.G., Béres, E., Vértesi, A., Frank, Z., Pasics, I. et al. (2011). The effect of ergothioneine on clastogenic potential and mutagenic activity: Genotoxicity evaluation. Int. J. Toxicol., 30(4): 405–409.

Selvamani, S., El-Enshasy, H.A., Dailin, D.J., Malek, R.A., Hanapi, S.Z. et al. (2018). Antioxidant compounds of the edible mushroom *Pleurotus ostreatus*. Int. J. Biotechnol., Well. Ind., 7(7): 1–14.

Singdevsachan, S.K., Auroshree, P., Mishra, J., Baliyarsingh, B., Tayung, K. and Thatoi, H. (2016). Mushroom polysaccharides as potential prebiotics with their antitumor and immunomodulating properties: A review. Bioact. Carbohydr. Diet. Fibre, 7(1): 1–14.

Singh, K. and Prasad, G. (2019). Biology and growth characteristics of edible mushroom: *Agaricus compestris, Agaricus bisporous, Coprinus comatus*. IOSR J. Biotechnol. Biochem., 5(1): 46–59.

Singh, R.S., Walia, A.K. and Kennedy, J.F. (2020). Mushroom lectins in biomedical research and development. Int. J. Biol. Macromol., 151: 1340–1350.

Souza, R.Á.T., da Fonseca, T.R.B., Kirsch, L.S., Silva, L.S.C., Alecrim, M.M. et al. (2016). Nutritional composition of bioproducts generated from semi-solid fermentation of pineapple peel by edible mushrooms. Afr. J. Biotechnol., 15(12): 451–457.

Sovrani, V., Da Rosa, J., Drewinski, M.D.P., Colodi, F.G., Tominaga, T.T. et al. (2017). *In vitro* and *in vivo* antitumoral activity of exobiopolymers from the royal sun culinary-medicinal mushroom *Agaricus brasiliensis* (agaricomycetes). Int. J. Med. Mushrooms, 19(9): 767–775.

Stanikunaite, R., Khan, S.I., Trappe, J.M. and Ross, S.A. (2009). Cyclooxygenase-2 inhibitory and antioxidant compounds from the truffle *Elaphomyces granulatus*. Phytother. Res., 23(4): 575–578.

Stojković, D., Reis, F.S., Glamočlija, J., Ćirić, A., Barros, L. et al. (2014). Cultivated strains of *Agaricus bisporus* and *A. brasiliensis*: Chemical characterisation and evaluation of antioxidant and antimicrobial properties for the final healthy product-natural preservatives in yoghurt. Food Funct., 5(7): 1602–1612.

Taofiq, O., Calhelha, R.C., Heleno, S., Barros, L., Martins, A. et al. (2015). The contribution of phenolic acids to the anti-inflammatory activity of mushrooms: Screening in phenolic extracts, individual parent molecules and synthesised glucuronated and methylated derivatives. Food Res. Int., 76: 821–827.

Taofiq, O., González-Paramás, A.M., Martins, A., Barreiro, M.F. and Ferreira, I.C.F.R. (2016a). Mushrooms extracts and compounds in cosmetics, cosmeceuticals and nutricosmetics—A review. Ind. Crops Prod., 90: 38–48.

Taofiq, O., Heleno, S.A., Calhelha, R.C., Alves, M.J., Barros, L. et al. (2016b). Development of mushroom-based cosmeceutical formulations with anti-inflammatory, anti-tyrosinase, antioxidant, and antibacterial properties. Molecules, 21(10): 1–12.

Taofiq, O., Martins, A., Barreiro, M.F. and Ferreira, I.C.F.R. (2016c). Anti-inflammatory potential of mushroom extracts and isolated metabolites. Tr. Food Sci. Technol., 50: 193–210.

Taofiq, O., Heleno, S.A., Calhelha, R.C., Fernandes, I.P., Alves, M.J. et al. (2018). Mushroom-based cosmeceutical ingredients: Microencapsulation and *in vitro* release profile. Ind. Crops Prod., 124: 44–52.

Taofiq, O., Rodrigues, F., Barros, L., Barreiro, M.F., Ferreira, I.C.F.R. and Oliveira, M.B.P.P. (2019). Mushroom ethanolic extracts as cosmeceuticals ingredients: Safety and *ex vivo* skin permeation studies. Food Chem. Toxicol., 127: 228–236.

Taofiq, O., Silva, A.R., Costa, C., Ferreira, I., Nunes, J. et al. (2020). Optimisation of ergosterol extraction from *Pleurotus* mushrooms using response surface methodology. Food Funct., 11(7): 5887–5897.

Thring, T.S.A., Hili, P. and Naughton, D.P. (2009). Anti-collagenase, anti-elastase and anti-oxidant activities of extracts from 21 plants. BMC Complem. Med., 9(1): 27.10.1186/1472-6882-9-27.

Tovar-Herrera, O.E., Martha-Paz, A.M., Pérez-LLano, Y., Aranda, E., Tacoronte-Morales, J.E. et al. (2018). Schizophyllum commune: An unexploited source for lignocellulose degrading enzymes. Microbiology Open, 7(3): e00637.

Urbancikova, I., Hudackova, D., Majtan, J., Rennerova, Z., Banovcin, P. and Jesenak, M. (2020). Efficacy of Pleuran (β-Glucan from *Pleurotus ostreatus*) in the management of Herpes simplex virus type 1 infection. Evid. Based Compl. Alt. Med., 1–8. 856230.10.1155/2020/8562309.

Valverde, M.E., Hernández-Pérez, T. and Paredes-López, O. (2015). Edible mushrooms: Improving human health and promoting quality life. Int. J. Microbiol., 2015: 376387.10.1155/2015/376387.

Vos, A.M., Jurak, E., Pelkmans, J.F., Herman, K., Pels, G. et al. (2017). H_2O_2 as a candidate bottleneck for MnP activity during cultivation of *Agaricus bisporus* in compost. AMB Express, 7(1): 124.10.1186/s13568-017-0424-z.

Wan Mahari, W.A., Peng, W., Nam, W.L., Yang, H., Lee, X.Y. et al. (2020). A review on valorization of oyster mushroom and waste generated in the mushroom cultivation industry. J. Hazard. Mater., 400: 123156.

Wang, R., Ma, P., Li, C., Xiao, L., Liang, Z. and Dong, J. (2019). Combining transcriptomics and metabolomics to reveal the underlying molecular mechanism of ergosterol biosynthesis during the fruiting process of *Flammulina velutipes*. BMC Genom., 20(1): 1–12.

Wang, X., Zhang, Z. and Zhao, M. (2015). Carboxymethylation of polysaccharides from *Tremella fuciformis* for antioxidant and moisture-preserving activities. Int. J. Biol. Macromol., 72: 526–530.

Weng, Y., Lu, J., Xiang, L., Matsuura, A., Zhang, Y. et al. (2011). Ganodermasides C and D, two new anti-aging ergosterols from spores of the medicinal mushroom *Ganoderma lucidum*. Biosci. Biotechnol. Biochem., 75(4): 800–803.

Wu, J.Y. (2015). Polysaccharide-protein complexes from edible fungi and applications. pp. 927–937. *In*: Ramawat, K. and Mérillon, J.M. (eds.). Polysaccharides. Springer Cham. 10.1007/978-3-319-16298-0_38.

Yadav, T., Mishra, S., Das, S., Aggarwal, S. and Rani, V. (2015). Anticedants and natural prevention of environmental toxicants induced accelerated aging of skin. Environ. Toxicol. Pharmacol., 39(1): 384–391.

Yang, C.H., Su, C.H., Liu, S.C. and Ng, L.T. (2019). Isolation, anti-inflammatory activity and physicochemical properties of bioactive polysaccharides from fruiting bodies of cultivated cordyceps cicadae (Ascomycetes). Int. J. Med. Mushrooms, 21(10): 995–1006.

Yang, S., Qu, Y., Zhang, H., Xue, Z., Liu, T. et al. (2020). Hypoglycemic effects of polysaccharides from: *Gomphidiaceae rutilus* fruiting bodies and their mechanisms. Food Funct., 11(1): 424–434.

Yang, Y., Zhang, H., Zuo, J., Gong, X., Yi, F. et al. (2019). Advances in research on the active constituents and physiological effects of *Ganoderma lucidum*. Biomed. Dermatol., 3: 6.10.1186/s41702-019-0044-0.

Ye, L., Liu, S., Xie, F., Zhao, L. and Wu, X. (2018). Enhanced production of polysaccharides and triterpenoids in *Ganoderma lucidum* fruit bodies on induction with signal transduction during the fruiting stage. PLoS ONE, 13(4): e0196287.10.1371/journal.pone.0196287.

Yu, Q., Nie, S.P., Wang, J.Q., Huang, D.F., Li, W.J. and Xie, M.Y. (2015). Toll-like receptor 4 mediates the antitumor host response induced by *Ganoderma atrum* polysaccharide. J. Agr. Food Chem., 63(2): 517–525.

Zembron-Lacny, A., Gajewski, M., Naczk, M. and Siatkowski, I. (2013). Effect of shiitake (*Lentinus edodes*) extract on antioxidant and inflammatory response to prolonged eccentric exercise. J. Physiol. Pharmacol., 64(2): 249–254.

Zhang, J., Meng, G., Zhang, C., Lin, L., Xu, N. et al. (2015). The antioxidative effects of acidic-, alkalic-, and enzymatic-extractable mycelium zinc polysaccharides by *Pleurotus djamor* on liver and kidney of streptozocin-induced diabetic mice. BMC Compl. Altern. Med., 15(1): 440.10.1186/s12906-015-0964-1.

Zhang, P., Li, K., Yang, G., Xia, C., Polston, J.E. et al. (2017). Cytotoxic protein from the mushroom *Coprinus comatus* possesses a unique mode for glycan binding and specificity. Proceed. Nat. Aca. Sci. USA, 114(34): 8980–8985.

Zhao, Z.-Z., Chen, H., Huang, Y., Li, Z.-H., Zhang, L. et al. (2016). Lanostane triterpenoids from fruiting bodies of *Ganoderma leucocontextum*. Nat. Prod. Bioprospect., 6(2): 103–109.

14

Cosmeceuticals from Mushrooms

Muhammad Fazril Razif* and Shin-Yee Fung*

1. INTRODUCTION

Over the past decade, consumers have been spending a major share of their income on cosmetics, cosmeceuticals and nutricosmetics. The cosmetics industry, currently valued at USD\$532 billion (Edited, 2019) is generally divided into five major categories: skin care, hair care, colour cosmetics, fragrances and personal care products. Of these, skin care occupies a major position (accounting for 36.4 per cent of global market sales in 2016) with the Asia-Pacific region emerging as the biggest cosmetic market in the world (Statista, 2020). According to the 1938 Federal Food, Drug and Cosmetics Act, cosmetics are defined as '*articles intended to be applied to the human body for cleansing, beautifying, promoting attractiveness, or altering the appearance without affecting the body's structure or functions*' (US Food and Drug Administration, 2020). Consumer behaviour remains the greatest influence behind the stable growth of cosmetics with an increasing focus on product efficacy, advancements in new formulations and delivery technology. In recent times, consumers have become more educated with regards to the ingredients that are included in their products, with increasingly more of them opting for products with minimal ingredients of high potency, non-toxic and cruelty-free. In addition, there remains to be a sustained interest in natural skin care products despite the increase of more specialised products in the market.

Cosmetics have traditionally been focused towards women, but men are now progressively using more cosmetics in their daily routine, with an overall rise in skin care and personal care products. A combination of cosmetic and cosmeceutical

Medicinal Mushroom Research Group (MMRG), Department of Molecular Medicine, Faculty of Medicine, University of Malaya, 50603 Kuala Lumpur, Malaysia.
* Corresponding authors: fazril.razif@um.edu.my; syfung@ummc.edu.my

products used in routine can play a significant role in assisting consumers to maintain or restore their skin health. This rising demand for cosmetic products has in turn led to the growth of the cosmetics market around the world. There has been a growing trend in the use of natural ingredients in cosmetic products among consumers, with the opinion being that these products have minimal possible side effects. Several botanicals are well known for their protective action against free radicals and both oxidative stress and damage. Hence, cosmetic laboratories and manufacturers are now focused on developing new sustainable products, using naturally derived ingredients to cater to the needs of consumers. Skin care has become the most technically advanced and diverse categories of cosmetics as these products are subjected to rigorous scientific testing and dermatological evaluations before they are introduced into the market. At present, manufacturers are developing new formulations for skin care products to ensure that their products are stable, possess high efficacy and are suitable for most skin types.

2. Cosmeceuticals and Nutraceuticals

Cosmeceutical can be defined as a cosmetic product with bioactive ingredients that claim to have pharmaceutical action on the skin (Reed, 1964), while nutraceutical, a term coined in 1989 by Stephen L. Defelice, is defined as '*any substance that is a food or part of a food and provides medical or health benefits, including the prevention and treatment of disease*' (Aronson, 2017). Both products are marketed to contain bioactive ingredients from either natural sources or synthetically derived, with the principle that they are antiaging, anti-inflammatory, treat pigmentation and improve skin complexion. It is important to note that most governmental bodies, like the USFDA, currently do not recognise the term '*cosmeceutical*'; hence most products are only self-regulated by the cosmetics industry (Draelos, 2009; US Food and Drug Administration, 2020). According to London-based analysts Future Market Insights (2020), the global natural cosmetics market is expected to be worth US$54 billion by 2027, with a 5.2 per cent compound annual growth rate (CAGR). On the other hand, Technavio (2019) reported that the global nutricosmetics market is expected to post a year-over-year growth rate of 5.2 per cent between 2020–2024, amounting to an increase of US$1.51 billion. Hyderabad-based analysts, Mordor Intelligence, reported that the Asia-Pacific nutricosmetics market is forecasted to reach US$2.4 billion by 2024, with a CAGR of 9.24 per cent (Mordor Intelligence, 2019). For nutraceuticals, a fundamental understanding of nutrition, including vitamins, antioxidants and other active ingredients that provide nutritional support and protection against ageing and stress, has become the holy grail. The growing popularity of this notion is rapidly changing the perception of consumers, contributing to the adoption of functional foods and dietary supplements. There has been a steady increase in research on nutrients that improve the skin health, with emphasis that the skin is a key barrier to external factors. Akin to traditional medicine, the benefits of nutricosmetic products and their ingredients are mainly based on the historical use and by word of mouth as many of which have not been studied extensively. Similarly, cosmeceutical products are often marketed by associating an active ingredient with unsupported claims and benefits of a formulation.

The recent linkage between skin health and nutrition has resulted in a growth of peer-reviewed scientific articles that examine the biochemical and molecular mechanisms of the action of natural ingredients and extracts when applied topically and/or consumed orally. The demand for these natural products has surged in the past few years due to an increasing awareness about the potential benefits they offer. Nowadays, it is common to find cosmetics containing organic ingredients, such as oils from barks, roots and seeds, fruit and leaf extracts, natural fruit acids and flavonoids. These are often marketed by being declared as rich in antioxidants and co-enzymes that help nourish and maintain skin hydration, reduce pigmentation and the appearance of dark circles and wrinkles.

Cosmeceuticals are applied topically given that the epidermal layers of the skin lack blood vessels. Topical applications are considered as a more efficient method for supplying nutrients to the skin, especially to the epidermis. However, the chemical structure of the stratum corneum can inhibit the movement of various molecules. For example, lipid-soluble molecules are more likely to pass through the epidermis to potentially reach the dermis. Highly-charged proteins and peptides with molecular weights > 1000 kDa often encounter difficulty in penetrating an intact stratum corneum. The benefits of enhancing skin health are no longer found only in topical creams and serums, but numerous studies have found that oral consumption of natural ingredients and supplements resulted in significant improvements in dermal parameters. There are studies that reported improvements in overall skin health, including enhanced protection against photoaging and improving skin hydration and elasticity (Lassus et al., 1991; Kieffer and Ehsen, 1998; Hong et al., 2017; Kim et al., 2018; Czajka et al., 2018). The most commonly used compounds in cosmeceuticals include vitamin C and E, peptides, ceramides, hyaluronic acid, resveratrol and niacinamide. On the other hand, nutraceuticals commonly contain ingredients that are high in alpha-lipoic acid, coenzyme Q10, vitamin B, vitamin C, retinoids, zinc and glucosamine, to name a few. For most of the ingredients used, the quality of the final product is highly dependent on its source of the flora and growth conditions. Also, the physicochemical properties of the ingredient as well as the extraction method used to prepare or extract these active compounds are critical.

3. Natural vs Synthetic Compounds

Due to consumers' demand and sustainability risk, the industry is in a continuous search for new natural ingredients that can be obtained from sustainable and ecological sources. Natural ingredients in cosmetics are popular due to their scientifically proven protective and defensive roles against environmental stress generated from free radicals, reactive oxygen species (ROS), reactive nitrogen species (RNS) and oxidative enzymes. All these factors are associated with increased production of inflammatory enzymes, collagenase and elastase, that ultimately lead to degradation of the extracellular matrix of the skin (Tamsyn et al., 2009; Yadav et al., 2015). Most often, these ingredients face less scrutiny than their synthetic equivalents, as the latter often encounter restrictions according to each country's regulations and their perceived effects on human health and the environment. Natural ingredients are being increasingly consumed orally or included in various cosmeceutical formulations.

These include various types of ferments, phytonutrients, microbial metabolites and animal protein components. Recently, both pre-clinical and clinical studies have provided scientific validation for the use of certain botanical ingredients (Zillich et al., 2015). Botanical compounds for which cosmetic applications are in use, include olive oil, rose hip oil, chamomile, acai berry, green tea, pomegranate and soya.

There has also been a growing interest in studying the compounds and residues found in the waste produced after processing these ingredients as phenolic compounds, minerals and fatty acids that have been shown to possess biological properties have been discovered. Furthermore, alpha and beta hydroxy acids, isolated from various ingredients, are currently heavily utilised in many new skin care formulations for their exfoliant, emollient, anti-wrinkle and anti-pigmentation properties. Lactic, salicylic and glycolic acids are the three most popular of them (Fig. 1).

The appeal of using natural compounds in nutricosmetics stems from consumers' ever-growing pursuit of wellness, steering them towards natural alternatives when it comes to treating and/or preventing health ailments. In addition, naturally-sourced or natural-positioned cosmeceutical products increasingly appeal to consumers as evidenced by their rapid growth across most consumer products markets.

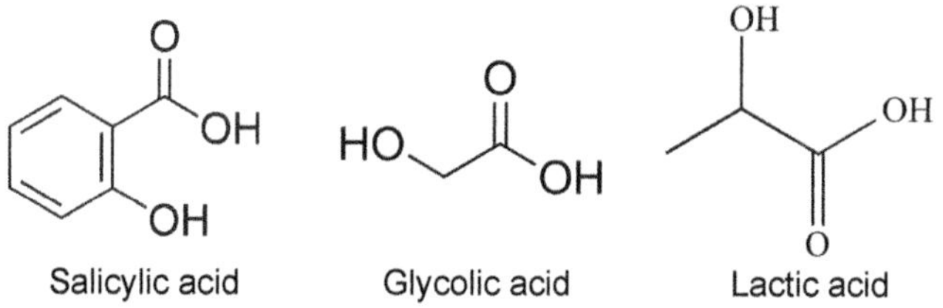

Fig. 1. Structures of salicylic, glycolic and lactic acids.

4. Mushrooms in Cosmeceuticals

Mushrooms are a common staple in most diets and have been consumed for their nutritional and medicinal properties for thousands of years. Studies show that they are high in glycoproteins, potent antioxidants, enzymes and phenolic acids. Fundamental research has shown that certain mushrooms exhibit antioxidant, antimicrobial, anti-inflammatory, antitumor and immunomodulatory activities. Data from *in vitro* studies indicate that mushrooms possess bioactive metabolites that are capable of anti-elastase, anti-hyaluronidase, anti-collagenase and anti-tyrosinase activities. However, understanding on the mechanism of action of these bioactive metabolites are still lacking. Mushrooms differ in their nutritional and bioactive composition—their individual efficacies being typically affected by variations in cultivation methods and extract preparation (Fig. 2).

Manufacturers are constantly searching for natural compounds or extracts with bioactive properties to include in their cosmeceutical formulations. Cleansing, moisturising, supplementing and protecting are the key components of an effective skincare routine. Extrinsic or photoaging accounts for ~ 80 per cent of the visible signs of aging attributed to pollution, ultraviolet radiation and lifestyle choices (Cevvini et al., 2008). Signs of extrinsic aging include pigmentation, dullness, dryness and

Fig. 2. Various types of mushrooms used in cosmeceuticals (*from left*: a. *Hericium erinaceum*, b. *Ganoderma lucidum* and c. *Phellinus linteus*) (pictures: courtesy, Ligno Biotech Sdn. Bhd., Malaysia).

wrinkles. On the other hand, intrinsic or age-dependent aging is attributed to genetics, including hormonal changes and free radical imbalance (Jenkins, 2002; Buckingham and Klingelhutz, 2011; Murina et al., 2012). Although certain species of mushrooms have been well studied, a considerable amount of progress has yet to be made in the biotechnological application of mushroom extracts in cosmetic formulations (Table 1).

4.1 Bioactivities of Mushrooms and Their Associated Compounds

There are approximately 14,000 known species of mushrooms though there are an estimated 140,000 species on Earth. Of those that have been characterised, scientists have focused their research on a selected few that include *Lentinula edodes* (Shiitake), *Ganoderma lucidum* (Lingzhi), *Cordyceps sinensis*, *Sparassis crispa* (cauliflower), *Agaricus bisporus* (Portobello), *Pleurotus ostreatus* (Oyster) and *Phellinus linteus* (Chaga). There are numerous mycochemicals from mushrooms that can potentially be exploited for cosmeceutical or nutraceutical applications. The mycochemicals that have been characterised in different types of mushrooms include alkaloids, chitosan, polysaccharides, ergosterols, flavanoids, glycosides, lectins, phenols/polyphenols, proteins/amino acids, saponins, steroids, tannins and triterpenoids. These mycochemicals are generally derived from the fruiting bodies and mycelia of the mushrooms. They also contain a complex mixture of bioactive metabolites. Studies show that these mycochemicals/bioactive metabolites possess anti-inflammatory, anti-tyrosinase, anti-collagenase, anti-elastase and antioxidant activities.

4.1.1 Anti-Inflammatory

Chronic inflammation is a major concern in skin aging. The primary source of inflammation arises from oxidative stress induced by exposure to ultraviolet (UV) radiation. When epidermal cells get damaged, a chain of downstream reactions occur, leading to inflammation, including: (i) release of inflammatory cytokines [e.g., interleukin-1 (IL-1), tumor necrosis factor-alpha (TNF-α), etc.] by epidermal keratinocytes; (ii) peroxidation of membrane lipids; (iii) generation of prostaglandins and leucotrienes by mast cells; (iv) infiltration and activation of macrophages that release matrix metalloproteinases (MMPs) and pro-inflammatory cytokines; and (v) activation of neutrophils that release elastase and MMPs (Kawaguchi et al., 1996; Gonzalez and Pathak, 1996; Yoshida et al., 1998; Takahara et al., 2003; Rijken

Table 1. Listed below are examples of mushroom cosmetics that are currently available in the market (List was compiled based on product ingredients listed on: INCI Decoder, 2020; Ulta Beauty, 2020; Sephora, 2020).

Mushroom	Product name
Sparassis crispa (Cauliflower fungus)	Acwell Betaglution Ultra Moisture Toner Blithe Tundra Chaga Pressed Serum Nature Republic Collagen Dream 70 Essence Skin & Co Premium Grade Moisturising Balm for Men Some by Mi Galactomyces Pure Vitamin C Glow Serum & Toner
Agaricus bisporus (Common)	Umoua Barrier Repair Essence Advanced Lush Full of Grace Zelens Provitamin D Treatment Drops Cosmetics Bye Bye Redness Sensitive Skin Moisturiser
Inonotus obliquus (Chaga)	Blithe Tundra Chaga Pressed Serum Dr Dennis Gross B$_3$ Adaptive Super Foods Stress Rescue Super Serum Face Republic Perfect Cover BB Cream Kypris Cerulean Soothing Hydration Treatment Mask Lumene Harmonia [Balance] Nutri-recharging Intense Moisturiser Mad Hippie Triple C Night Cream Make Prem Chaga Concentrate Essence Mizon Black Snail All-in-One Cream Nature Republic Collagen Dream 70 Essence Neogen Code 9 Black Volume Cream One-o-seven Low pH Chaga Cleanser Plantioxidants Chaga & Ginseng Serum So Natural Cera Plus Peptide Toner Essence SU:M37 Water-full Aqua Sleeping Pack Supermood Ego Boost One-minute Facelift Serum
Tricholoma matsutake (Matsutake)	Amore Pacific Moisture Bound Sleeping Recovery Mask Coxir Ceramide Milk Drop Cream Cremorlab TE.N. Cremor Skin Renewal Cream Dr. Jart+ V7 Serum Goodal Double Bright Intense Serum Missha Misa Geum Sul Wild Ginseng Exfoliating Gel Neogen Bio-Peel Gauze Peeling Lemon Osea Malibu Brightening Serum Palmer's Eventone Dark Spot Corrector Shea Moisture Coconut & Hibiscus Radiance Mud Mask SU:M37 Water-full Aqua Sleeping Pack Sulwhasoo Time Treasure Invigorating Collection The Face Shop Yehwadam Heaven Grade Ginseng Collection The History of Whoo Hwan Yu Collection Tosowoong Spot Whitening Vita Clinic Vitamin Tree Fruits Extract

Table 1 Contd. ...

...Table 1 Contd.

Mushroom	Product name
Ganoderma lucidum (Lingzhi)	Alterna Caviar Anti-ageing Blonde Shampoo Aveeno Positively Ageless® Collection CV Skin Labs Calming Moisture Eminence Organics Snow Mushroom Moisture Cloud Eye Cream Four Sigmatic Superfood Serum, Hydrate with Oils & Reishi Honest Beauty Soothing Daily Moisturiser Innis free Perfect 9 Collection Holy Snails First Snow Essence Mirai Clinical 3-In-1 Multi-tasking Face Serum with Astaxanthin Moon Juice Beauty Shroom Exfoliating Acid Potion Nature Republic Collagen Dream 70 Essence Origins Dr. Andrew Weil for Origins Mega-Mushroom Collection Shiseido Ultimune Eye Power Infusing Concentrate Youth to the People Adaptogen Deep Moisture Cream YUESAI Rejuvenate Cordyceps Personalised Serum
Phellinus linteus (Sanghwang)	9wishes Miracle White Perfect Ampoule Serum Be the Skin Purifying White Waterful Collection Botanic Farm Rice Ferment First Essence Clinelle Purifying Gel Moisturiser Dr. G. Actifirm Real Lifting Collection May Coop Raw Eye Contour Missha Misa Cho Gong Jin Collection Nature Republic Collagen Dream 70 Collection One by Kose Melanoshot White Brightening Serum Pixi Plump Collagen Boost Sheet Mask Rose by Dr. Dream Dream Age Rejuvenating Cream, Serum Shinetree Squeeze & Go Soothing & Brightening Aloe & Green Tea Gel Some by Mi Snail Truecica Miracle Repair Toner Thank You Farmer Miracle Age Repair Collection The Saem Chaga Anti-wrinkle Eye Cream, Serum
Tremella fuciformis (Snow fungus)	2Sol Elixir Propolis Serum Aisuanis Empowerment Azelaic Acid Cosme Decorte Aq Meliority Intensive Revitalising Lotion Edible Beauty Turmeric Beauty Latte - Clear & Brighten Etude House 99 per cent Aloe Soothing Gel Hada Labo Air Aqua UV - Fresh SPF 50 Pa+++ Herbivore Bakuchiol Retinol Alternative Serum Nutox Astringent Toner, Moisture Emulsion SPF 25 Pa++ Sephora Sleeping Mask - Toning and Firming, Sleeping Mask Green Tea Tarte Face Tape Foundation The Inkey List Snow Mushroom Volition Snow Mushroom Water Serum Wander Beauty Dive In™ Moisturiser Yello Skincare Snowmeric Hydra-Boost Brightening Serum Banila Co VV Vitalising Bio Cellulose Mask
Cordyceps militaris	Touch in Sol Pretty Filter Waterful Glow Cream

Table 1 Contd. ...

...Table 1 Contd.

Mushroom	Product name
Cordyceps sinensis	Acure Ultra Hydrating Green Juice Cleanser BEYOND Phyto Aqua Cream Bobbi Brown Intensive Skin Serum Foundation SPF40 CNP Laboratory Invisible Peeling Booster Dermalogica Invisible Physical Defense Sunscreen Spf 30 Estee Lauder Re-nutriv Ultimate Diamond Transformative Energy Creme, Mizon Mark-X Blemish after Cream Neogen Bio-Peel Gauze Peeling Lemon Origins VitaZing™ SPF15 Energy-boosting Tinted Moisturiser with Mangosteen Origins Ginzing Refreshing Eye Cream, Energy-Boosting Tinted Moisturiser SPF40 Peter Thomas Roth Acne-clear Oil-free Matte Moisturiser Shangpree Phyto Essence UV Sunscreen SPF 50 Pa++++ The History of Whoo Gongjinhyang Intensive Nutritive Eye Cream YUESAI Quintessence Jade Illuminating Serum Yuripibu Cucu Black Truffle Serum
Lentinus edodes (Shiitake)	100% Pure Fermented Rice Water Cleanser, Toner Armani Prima Eye & Lip Contour Perfector Chanel Sublimage Masque La Prairie Anti-Aging Stress Cream Epionce Intense Defence Serum Herbivore Emerald Cannabis Sativa Hemp Seed Deep Moisture Glow Oil Manyo Factory Lactobacillus Bifida Mist Mirai Clinical 3-In-1 Multi-Tasking Face Serum with Astaxanthin Missha Super Aqua Cell Renew Snail B.B. Cream SPF30 Pa++ Murad Invisiblur Perfecting Shield Broad Spectrum SPF 30 Rael Good Chemistry Advanced Antioxidant Serum Skinfood Gold Caviar Collagen Serum STARSKIN Close-Up™ Firming Bio-Cellulose Second Skin Face Mask SU:M37 Losec Summa Elixir Collection Sulwhasoo Everefine Lifting Ampoule Serum, Snowise Brightening Spot Serum Yves Saint Laurent Blanc Pur Couture Brightening Cream
Grifola frondosa (Hen-of-the-wood)	Acwell Betaglution Ultra Moisture Toner DHC Super Collagen Supreme Helena Rubinstein Collagenist Re-Plump Day Cream Dry Skin SPF 15 Lancôme Blanc Expert Skin Tone Brightening Emulsion Missha Time Revolution Night Repair New Science Activator Ampoule Mask Skin79 Phyto-Hyaluron Foam Cleanser SU:M37 Losec Summa Elixir Day Collection Swanicoco Golden Time Leap Mask Pack
Trametes versicolor (Turkey tail)	BEYOND Phyto Aqua Cream Briogeo Mushroom + Bamboo Color Protect Shampoo Clinique Even Better Clinical™ Radical Dark Spot Corrector + Interrupter CNP Laboratory 2-Step Quick Soothing S.O.S Mask, Invisible Peeling Booster Dermalogica Invisible Physical Defence Sunscreen Spf 30 Peter Thomas Roth Acne-Clear Oil-Free Matte Moisturiser

Table 1 Contd. ...

...Table 1 Contd.

Mushroom	Product name
Fomes officinalis	Allies of Skin Fresh Slate Purifying Cleanser + Masque
	Annique Hydrafine Skin Refining Freshener
	Belei Vitamin C + Hyaluronic Acid Serum
	Benefit Professional Matte Rescue Gel
	Bioderma Sebium Lotion
	Chantecaille Detox Clay Face Mask
	Charlotte Tilbury Magic Foundation
	Dr. Dennis Gross Skincare Alpha Beta® Pore Perfecting & Refining Serum
	Ebanel Ultimate Brightening Peeling Gel
	Elizabeth Arden Visible Difference Replenishing Hydragel Complex
	Hylamide Booster Pore Control
	Murad Intensive-C Radiance Peel
	Papa Recipe Honey Moist Sun Essence Spf 50+ Pa+++
	Simple Regeneration Night Cream Age Resisting
	Youth Lab Candy Scrub & Mask for all Skin Types
Pleurotus ferulae (White Elf)	CLIV Ginseng Berry Premium Ampoule
	PureHeal's Ginseng Berry 80 Eye Essence
Agaricus blazei (Almond)	May Coop Raw Activator
	May Coop Raw Eye Contour
	Neogen Blueberry Real Fresh Foam Cleanser
	Neogen Code 9 Black Volume Cream
Hericium erinaceum (Lion's Mane)	Liz K First C Serum
	The Plant Base Time Stop Collagen Ampoule
	Yuripibu Cucu Black Truffle Serum
Schizophyllum commune (Split Gills)	Cosmetics Bye Bye Foundation Full Coverage Moisturiser with Spf 50+
	Cosmetics Secret Sauce Clinically Advanced Miraculous Anti-aging Moisturiser
	OSSOLA Skincare the Turmeric Emulsion
	Swanicoco Fermentation Snail Care Skin Toner

Table 2. Compounds found in mushrooms that are used in cosmetics.

Compounds	Mushroom	References
Lactic acid	*Rhizopus* strains	Zhang et al., 2007
Ceramide	*Phellinus pini* *Tuber indicum*	Lourenço et al., 1996 Gao et al., 2004
Chitin-glucan	*Aspergillus niger*	Gautier et al., 2008
Superoxide dismutase	*Pleurotus citrinopileatus* *Agaricus bisporus,* *Pleurotus ostreatus,* *Volvariella volvacea* *Agrocybe aegerita*	Cheng et al., 2012
Polysaccharides: lentinan, schizophyllan and exopolysaccharide	*Pleurotus* species *Lentinula edodes* *Schizophyllum commune* *Grifola frondosa*	Synytsya et al., 2009 Bae et al., 2005
L-ergothioneine	*Agaricus bisporus*	Dubost et al., 2006
2-amino-3H-phenoxazin-3-one	*Agaricus bisporus*	Miyake et al., 2010
Betulin and trametenolic acid	*Inonotus obliquus*	Yan et al., 2014
2-Hydroxytyrosol	*Metarhizium anisopliae*	Uchida et al., 2014

et al., 2005; Handoko et al., 2013). Several gene expression studies have uncovered an unequivocal link between genes encoding inflammatory cytokines and the immune system with aging (Urschitz et al., 2002; Robinson et al., 2009; Yan et al., 2013). Therefore, repeated cycles of inflammation in skin eventually cause damage to the extracellular matrix (ECM) of the dermis. Over time, dermal fibroblasts responsible for regenerating connective tissue for healing can no longer repair the ECM effectively.

The numerous anti-inflammatory compounds discovered in mushrooms have been attributed to the presence of polysaccharides, proteoglucans, terpenoid, phenolic, steroids and lectin (Elsayed et al., 2014). The crude extracts with anti-inflammatory properties that are known and recognised are from the following species: *Amanita muscaria* (Michelot and Melendez-Howell, 2003), *Cordyceps militaris* (Won and Park, 2005), *Cordyceps pruinosa* (Kim et al., 2003), *Geastrum saccatum* (Van et al., 2009), *Inonotus obliquus* (Park et al., 2005; Debnath et al., 2012), *Lentinus polychrous* (Fangkrathok et al., 2013), *Phellinus linteus* (Kim et al., 2007), *Antrodia camphorate* (Geethangili and Tzeng, 2011), *Ganoderma lucidum* (Chu et al., 2015), *Lignosus rhinocerus* (Lee et al., 2014) and *Termitomyces albuminosus* (Lu et al., 2008). In these extracts, polysaccharides, proteoglucans, terpenoids and phenolics are commonly recognised as the active compounds that give rise to anti-inflammatory properties. These compounds are purified by using various methods of extraction (methanol, ethanol, hexane, ethyl acetate, aqueous, dichloromethane and so on). These extracts have been tested for their anti-inflammatory properties using various *in vitro* and *in vivo* biological techniques (Lu et al., 2008; Jedinak et al., 2011; Lee et al., 2014). The anti-inflammatory potential of mushroom extracts may be exerted through a few biochemical pathways, which involves compounds known as inflammatory markers: interleukins (IL-1β, IL-6; IL-8); tumor necrosis factor (TNF-α); nuclear factor-κB (NF-κB), intercellular adhesion molecule-1 (ICAM-1); inducible type cyclooxygenase- (COX-) 2, prostaglandin E2 (PGE2); 5-lipooxygenase (5-LOX) and inducible nitric oxide synthase (iNOS), which lead to the production of reactive nitrogen species, such as nitric oxide (NO) and their action in inhibiting their release and/or affecting their activities (Jedinak et al., 2011; Taofiq et al., 2016). Skin inflammation may also stem from a systemic disorder, such as scleroderma and psoriasis. Hence anti-inflammatory compounds can also be consumed orally (based on thorough safety studies) in addition to topical applications in the form of cream, gels or lotions.

Relevant to UV-induced damage, mushrooms are also a rich sustainable source of chitin, an N-acetylglucosamine polymer typically found in the exoskeleton of crustaceans and insects. Chitosan, a derivative biopolymer generated via the deacetylation of chitin (Ospina et al., 2015), is utilised in many industries including textiles, agriculture and pharmaceuticals. The UV absorption range of chitosan films was revealed to be below 400 nm; hence, they may potentially be used in sunscreen formulations (Gomaa et al., 2010). Interestingly, a study by Chou and Wang (2015) found that the transmittance for UVA-UVB of mushroom chitosan film was lower when compared with a shrimp-derived chitosan film, implying that the former had better UV resistance. It is important to note that chitin and chitosan from different mushroom species exhibit dissimilar physicochemical properties (Kumirska et al., 2011; Erdogan et al., 2017).

4.1.2 *Anti-tyrosinase and Anti-melanogenic*

The melanin biosynthesis pathway has tyrosinase as the rate-limiting enzyme, where it converts tyrosine to dihydroxyphenylalanine (DOPA) and further oxidises DOPA into dopaquinone (Meng et al., 2012) (Fig. 3). Hyperpigmentation in skin is caused by exposure to UV radiation and the release of melanocyte-stimulating hormone, which triggers over-secretion of melanin from melanocytes (Ali et al., 2015).

Some mushrooms contain inhibitors of tyrosinase in varying degrees (Sharma et al., 2004; Alam et al., 2010; Miyake et al., 2010; Yoon et al., 2011; Yan et al., 2014; Nagasaka et al., 2015; Kaewnarin et al., 2016; Kim et al., 2016; Chien et al., 2008). These mushrooms have been used in cosmetic products as components for skin lightening and/or brightening effects. Compounds with anti-tyrosinase or anti-melanogenic activity require rigorous clinical testing to confirm that they are not cytotoxic and have no mutagenic effects on melanocytes and mammalian cells. Numerous studies have examined the inhibitory effects of these mushroom-derived compounds on melanogenesis, using various cells lines. For instance, ganodermanondiol from *Ganoderma lucidum* was found to inhibit the activity and expression of cellular tyrosinase, the expression of tyrosinase-related protein-1 (TRP-1), TRP-2, and microphthalmia-associated transcription factor (MITF), thereby decreasing melanin production.

Ganodermanondiol was also found to affect the mitogen-activated protein kinase (MAPK) cascade and cyclic adenosine monophosphate (cAMP)-dependent signalling

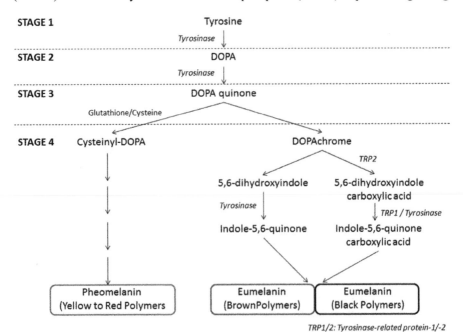

TRP1/2: Tyrosinase-related protein-1/-2

Fig. 3. Melanin biosynthesis pathway in human skin.

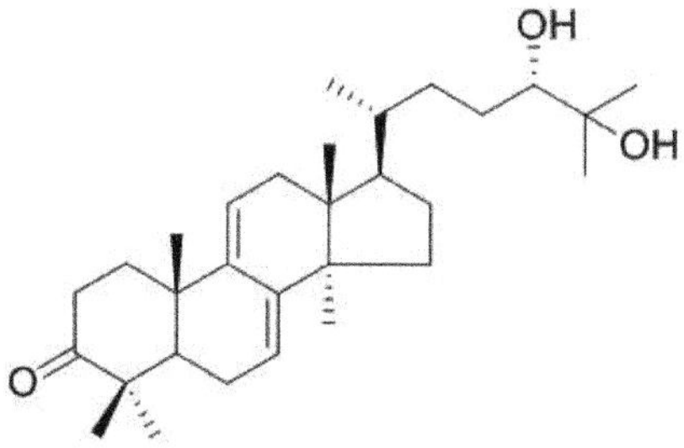

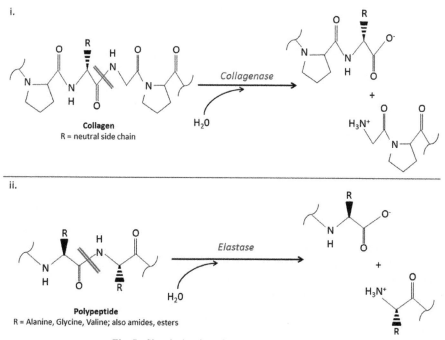

Fig. 4. Chemical structure of ganodermanondiol.

pathway, which are involved in the melanogenesis of B16F10 melanoma cells (Kim et al., 2016) (Fig. 4). Interestingly, the name *Ganoderma* means 'bright skin' in Greek and hence, has been used in cosmeceuticals, such as face masks. Im et al. (2019) also showed that the methanolic and hot water extracts of *Phellinus vaninii* (sanghuang) possess *in vitro* tyrosinase-inhibitory activities comparable to kojic acid.

4.1.3 Anti-collagenase and Anti-elastase

Dermal fibroblasts are present in the ECM and they generate collagen and elastin fibres in tissues. They also represent the extracellular part of the skin's multicellular structure. Collagen and elastin are synthesised and decomposed repeatedly in order to maintain the dermal structure (Fig. 5). Exposure to extrinsic factors, such as UV, can decrease collagen and elastin fibres in the connective tissue of the skin. These fibres are degraded by matrix metalloproteinases (MMP), collagenase or serine proteinase.

Fig. 5. Chemical action of elastase and collegenase.

There are four collagenases in humans: interstitial collagenase (MMP-1), neutrophil collagenase (MMP-8), collagenase 3 (MMP-13) and collagenase 4 (MMP-18). There are also two gelatinases, such as A (MMP-2) and B (MMP-9) which degrade native and denaturated collagens, gelatin and laminin (Aimes and Quigley, 1995; Patterson et al., 2001), and metalloelastase (MMP-12), mainly produced in macrophages that degrade elastin and other ECM proteins (Shapiro et al., 1993; Shipley et al., 1996). Takeuchi et al. (2010) reported that neutrophil elastase is indirectly involved in collagen and elastin degradation through MMP-1 and MMP-2 activation as well as by direct degradation. It has also been shown that excessive generation of reactive oxygen species also leads to deoxyribonucleic acid (DNA) damage in normal cells (*see* Antioxidant Section), increasing MMP (collagenase) levels and promoting skin aging (Dupont et al., 2013). During aging, fragmentations of collagen fibrils lead to the disruption of fibroblast-ECM interactions, which cause a loss of mechanical force in the skin. This problem is attenuated by the inability of fibroblasts to synthesise more ECM proteins and less MMPs.

Mushrooms are a potential source of inhibitors for elastase and collagenase and are beneficial as ingredients in cosmetics that aim to reduce the rate of skin aging. Kim et al. (2014) reported on the anti-elastase activity of the extract from *T. matsutake* (*Tricholomataceae*) mycelium, showing that it also has the capacity to reduce the levels of MMPs. It is known that the degradation of collagen and elastin fibres in the skin is mainly caused by the expression of MMP-1 and elastase. There are 24 genes-encoding MMPs in humans, including a duplicated MMP-23 gene (Murphy and Nagase, 2008). Another mushroom, *Phellinus vaninii*, has also been shown to possess anti-collagenase and moderate anti-elastase activities, which can be used for the development of novel anti-wrinkle cosmeceuticals. A study by Kawagishi et al. (2002) discovered a novel hydroquinone, (E)-2-(4-hydroxy-3-methyl-2-butenyl)-hydroquinone from the mushroom *Piptoporus betulinus* that acts as an MMP inhibitor. Im et al. (2019) showed that the methanol extract has better anti-collagenase activity than the hot water extract of the same species. However, since both types of extracts (methanol and hot water) exhibit notable inhibition of collagenase, the extracts may protect the ECM and collagen from degradation caused by photoaging. These studies provide evidence that mushrooms have many uncharacterised compounds that are not only beneficial for cosmetic formulations, but can also be developed for use in other therapeutics.

4.1.4 Antioxidant

Antioxidant activity of mushrooms can be demonstrated with a variety of assays, such as ferric reducing ability of plasma (FRAP) assay, looking at 2,2-diphenyl-1-picrylhydrazyl (DPPH) radical scavenging activity, ABTS [2,2'-azino-bis(3-ethylbenzothiazoline-6-sulfonic acid)] radical scavenging activity, superoxide anion radical scavenging activity, investigating the inhibition of superoxide anion radical in cell-based xanthine/xanthine oxidase assay and so on. A variety of antioxidant compounds (phenolics, flavonoids, glycosides, polysaccharides, copherols, ergothioneine, carotenoids, and ascorbic acid) are found in fruit bodies, mycelium and broth of mushrooms (Chen et al., 2012; Kozarski et al., 2015; *see* for a comprehensive list in References 72–148 of Kozarski et al., 2015). A Malaysian medicinal mushroom, *Lignosus rhinocerus* (tiger milk mushroom) was reported to have antioxidative activity which is correlated with its phenolic content. These are secondary metabolites ubiquitously found in plants, including mushrooms and usually exhibit high antioxidant potential. Phenolic content of hot water, cold water and methanol extracts of the sclerotial powder of *L. rhinocerus* ranged between 19.32–29.42 mg gallic acid equivalents/g extract, while the ferric reducing antioxidant power values ranged between 0.006–0.016 mmol/min/g extract. The DPPH•, ABTS•+, and superoxide anion radical-scavenging activities of the extracts ranged between 0.52–1.12, 0.05–0.20, and –0.98–11.23 mmol trolox equivalents/g extract, respectively. Both strains exhibited strong superoxide anion radical-scavenging activity comparable to rutin (Yap et al., 2013). Other mushrooms with reported antioxidant activities that are related to compounds, such as superoxidase dismutase (SOD), glutathione peroxidase (GPx) and L-ergothionine, include *Lentinula edodes, Volvariella volvacea, Pleurotus ostreatus, Pleurotus eryngii, Grifola frondosa* and *Agaricus bisporus* (Cheung et al., 2003; Cheung and Cheung, 2005; Puttaraju et al., 2006; Dubost et al., 2006). Glucan has also been implicated to be responsible for antioxidant properties with immune activation of macrophages, splenocytes and thymocytes which enable the activities observed *in vitro* to be applied *in vivo* after mushroom consumption as food or nutraceutical (Sun et al., 2004; Maity et al., 2011). Endogenous antioxidant defence mechanisms and dietary intake of antioxidants from mushrooms potentially regulate the oxidative homeostasis and can be a good source for nutraceutical applications.

4.1.5 Moisturising

Water is a major component of the body (~ 60 per cent). The association between water and healthy skin is widely accepted but not clearly established (Palma et al., 2012, 2013). Cutaneous water content plays an important role in maintaining the barrier and envelope functions of the skin with numerous dermatological dysfunctions linked to dehydration (Rosado et al., 2005; Wolf et al., 2010). Hence, moisturising products are commonly used to improve skin hydration and minimise transepidermal water loss (TEWL). There are three major types of moisturisers commonly found in cosmetic formulations: occlusives, humectants and emollients (Fig. 6). Occlusives are generally oils, waxes or silicone-based formulations that provide a physical barrier over the surface of the epidermis to prevent the TEWL. On the other hand, humectants, like hyaluronic acid (present in skin), glycerin and sorbitol, are water-binding substances that exert an osmotic effect to attract water from the atmosphere as well as from the dermis towards the stratum corneum. Finally, emollients (e.g., lanolin, ceramide, silicone and squalene) are characteristically lipid-based compounds that seal the gaps between corneocytes in the stratum corneum and aid in skin barrier function and overall skin texture.

A study by Kim et al. (2014) examining the exopolymer produced by the fungus, *Aureobasidium pullulans*, found that it possessed not only antioxidant, anti-melanogenic and anti-wrinkle effects, but the exopolymer, when applied on mouse skin, resulted in a significant increase in skin's water content. The extract from *Auricularia auricula-judae* (also known as wood ear) was shown to promote biosynthesis of procollagen and the expression of HAS-3 (hyaluronic acid synthase)

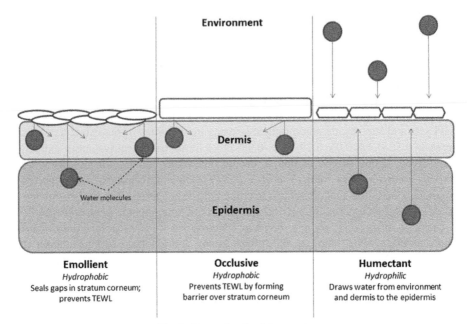

Fig. 6. Types of moisturising agents.

in HaCaT cells. Liao et al. (2014) revealed that the water extracts of *Auricularia fuscosuccinea* possessed moisture-retention capacity equally as potent as sodium hyaluronate, albeit less than *Tremella fuciformis* sporocarp extracts, exhibiting an excellent potential as a topical compound for moisturising formulations. Yang et al. (2006) discovered glucuronoxylomannan, an acid polysaccharide isolated from the hot-water extract of *Tremella*, was able to inhibit melanin formation and demonstrated excellent moisturising properties. A distinctive feature of *Tremella* is that 60–70 per cent of the dry fruit body is composed of pharmacologically-active, water-absorbing polysaccharides. Another study showed that 0.05 per cent *Tremella* polysaccharides had better moisture-retention capacity (0.02 per cent) than hyaluronic acid on incorporation into a cosmetic formulation (Liu et al., 2012). These studies demonstrate the potential use of mushroom polysaccharides, biofilms and/or extracts as effective moisturising agents in cosmetic formulations.

5. Clinical Trials

The application of topical products, such as serums, essences and creams to the skin can result in subclinical alterations in certain epidermal functions, including hydration, transepidermal water loss (TEWL) and pH (Loden, 2005; Buraczewska et al., 2007). Topically-applied cosmeceuticals are formulated to address a range of skin problems, including deceleration of the aging process, reduction of wrinkles and protection of the skin barrier against sun damage and free radicals. It is important to note that topically-applied cosmeceuticals do not cause changes in the skin barrier. Thus far, there have been numerous clinical studies that examine the consumption of various types of mushrooms and their effects on gut health, metabolic syndrome, immunity and oxygen kinetics (Clinical Trials.gov., 2020). Of interest, a few studies have examined the effects of mushroom consumption on vitamin D bioavailability. Results of these studies may have numerous downstream implications on skin health if additional nutraceutical research is pursued. However, there have not been any clinical trials examining the topical application of mushroom formulations and their potential effects on the skin. The only claims made thus far are the results of internal testing by manufacturers producing these products. This issue highlights the immense influence that manufacturers have over consumers as new, untested formulations with potentially little to no scientific results are heavily promoted.

Assessment of skin parameters after the application of a topical product will enable better understanding of the safety and efficacy of a product. A study by Taofiq et al. (2016) revealed that incorporation of ethanolic extracts of *A. bisporus, P. ostreatus*, and *L. edodes* into a base cosmetic cream did not affect their antioxidant and anti-inflammatory activities, revealing that mushrooms contain sustainable bioactive compounds which are valuable for testing further through clinical trials. One of the challenges for preclinical development of cosmeceuticals is that epidermal properties and changes are difficult to study as proper evaluation require invasive assessments. Non-invasive methods for determining pre- and post-application changes of a topical cream will be beneficial as a screening tool in future product development. Clinical trials will most likely require the use of non-invasive equipment to measure skin characteristics, such as elasticity, texture, water content and associated TEWL to verify claims made by the manufacturers. In addition, the use of pre- and post-treatment imaging and dermoscopy are useful tools to aid in assessing these skin changes. The general criteria of any clinical trial should be applied, including randomisation of a large study cohort, the use of placebo and clear research objectives. It is important that these products are scrutinised by regulatory bodies to ensure public safety.

6. Future Prospects

Cosmetic laboratories and manufacturers are in a never-ending search for new, multifunctional ingredients from natural sources. In order to fully utilise these new compounds in cosmeceutical formulations, optimisation of extraction techniques and effective delivery methods need to be refined in order to substantiate their effectiveness. In addition, skin permeation studies are critical in evaluating the topical availability of the bioactive compounds, followed by proper clinical testing to ensure the safety of the products. Tremendous progress is being made with regards to the innovation of novel technologies and improvements in formulations with the aim of producing safer and more effective cosmeceuticals (Fig. 7). Of these, the use of nanotechnology, liposomes, microencapsulation and penetration enhancers is increasing. These could potentially transform how active ingredients are delivered to their target site as well as improve the long-term stability of the products. The choice of delivery method will be dependent on the particle size of choice, stability (physical and chemical) of the ingredient, concentration, desired method of release (pH, heat and biodegradation) and cost of production.

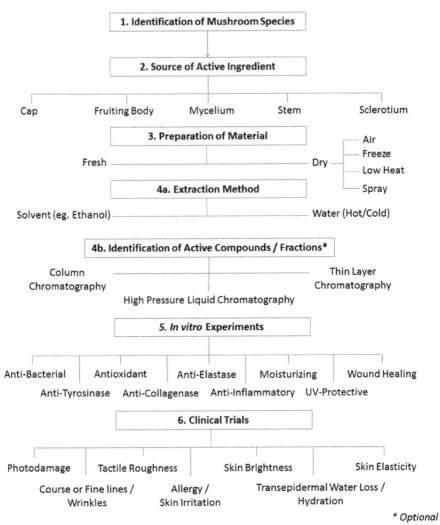

Fig. 7. Various stages in the discovery of new bioactive compounds for the development of cosmeceuticals using mushrooms.

The biggest challenge in using compounds derived from natural ingredients is their propensity for oxidation and degradation, along with the potentially poor permeability across the stratum corneum. Physicochemical parameters (e.g., lipophilic/hydrophilic balance) have been reported to affect the permeability of compounds in cosmetic formulations through the skin. Consumers are looking for scientifically-proven products with recognisable ingredients that are natural, effective and safe. In the past, the industry offered consumers minimal scientific evidence on the efficacy and quality of their products. Consumers also need to understand how the various ingredients within a formulation and/or different products can work synergistically. Further studies are required to determine the mechanism of action underlying the bioactive properties found in mushrooms and their extracts/

compounds. Ultimately, this will provide a solid foundation for the successful development of cosmetic formulations that will benefit and protect the skin overall.

References

Aimes, R.T. and Quigley, J.P. (1995). Matrix metalloproteinase-2 is an interstitial collagenase. Inhibitor-free enzyme catalyses the cleavage of collagen fibrils and soluble native type I collagen generating the specific 3/4- and 1/4-length fragments. J. Biol. Chem., 270: 5872–5876.

Alam, N., Yoon, K.N., Lee, K.R., Shin, P.G., Cheong, J.C., Yoo, Y.B., Shim, M.J., Lee, M.W., Lee, U.Y. and Lee, T.S. (2010). Antioxidant activities and tyrosinase inhibitory effects of different extracts from *Pleurotus ostreatus* fruiting bodies. Mycobiology, 38: 295–301.

Ali, S.A., Choudhary, R.K., Naaz, I. and Ali, A.S. (2015). Understanding the challenges of Melanogenesis: Key role of bioactive compounds in the treatment of hyperpigmentory disorders. J. Pigment. Dis., 2: 1–9.

Aronson, J. (2017). Defining 'nutraceuticals': Neither nutritious nor pharmaceutical. Br. J. Clin. Pharmacol., 83(1): 8–19.

Bae, J.T., Sim, G.S., Lee, D.H., Lee, B.C., Pyo, H.B., Choe, T.B. and Yun, J.W. (2005). Production of exopolysaccharide from mycelial culture of *Grifola frondosa* and its inhibitory effect on matrix metalloproteinase-1 expression in UV-irradiated human dermal fibroblasts. FEMS Microbiol. Lett., 251: 347–354.

Buckingham, E.M. and Klingelhutz, A.J. (2011).The role of telomeres in the ageing of human skin. Exp. Dermatol., 20: 297–302.

Buraczewska, I., Berne, B., Lindberg, M., Torma, H. and Loden, M. (2007). Changes in skin barrier function following long-term treatment with moisturisers, a randomised controlled trial. Br. J. Dermatol., 156(3): 492–498.

Cevenini, E., Invidia, L., Lescai, F., Salvioli, S., Tieri, P., Castellani, G. and Franceschi, C. (2008). Human models of aging and longevity. Exp. Opin. Biol. Ther., 8(9): 1393–405.

Chen, S.Y., Ho, K.J., Hsieh, Y.J., Wang, L.T. and Mau, J.L. (2012). Contents of lovastatin, γ-aminobutyric acid and ergothioneine in mushroom fruiting bodies and mycelia. LWT Food Sci. Technol., 47: 274–278.

Cheng, G.Y., Liu, J., Tao, M.X., Lu, C.M. and Wu, G.R. (2012). Activity, thermostability and isozymes of superoxide dismutase in 17 edible mushrooms. J. Food Compost. Anal., 26(1-2): 136–143.

Cheung, L.M., Cheung, P.C. and Ooi, V.E. (2003). Antioxidant activity and total phenolics of edible mushroom extracts. Food Chem., 81: 249–255.

Cheung, L.M. and Cheung, P.C. (2005). Mushroom extracts with antioxidant activity against lipid peroxidation. Food Chem., 89: 403–409.

Chien, C.C., Tsai, M.L., Chen, C.C., Chang, S.J. and Tseng, C.H. (2008). Effects on tyrosinase activity by the extracts of *Ganoderma lucidum* and related mushrooms. Mycopathol., 166(2): 117–120.

Chou, K.F. and Wang, L.F. (2015). Addition of titanium dioxide and sources effects on UV transmittance and hydrophilicity of chitosan film. *In*: Proc. 5th Int. Conf. Biomed. Eng. Technol., IACSIT Press, Seoul, Korea, 81: 20–24.

Chu, C.H., Ye, S.P., Wu, H.C. and Chiu, C.M. (2015). The *Ganoderma* extracts from Tein-Shan *Ganoderma* capsule suppressed LPS-induced nitric oxide production in RAW264.7 cells. MC-Trans. Biotechnol., 7: 1–11.

Clinical Trials. (2020). https://clinicaltrials.gov/ (Accessed 16 April 2020).

Czajka, A., Kania, E.M., Genovese, L., Corbo, A., Merone, G., Luci, C. and Sibilla, S. (2018). Daily oral supplementation with collagen peptides combined with vitamins and other bioactive compounds improves skin elasticity and has a beneficial effect on joint and general wellbeing. Nutr. Res., 57: 97–108.

Debnath, T., Hasnat, M.A., Pervin, M., Lee, S.Y., Park, S.R., Kim, D.H., Kweon, H.J., Kim, J.M. and Lim, B.O. (2012). Chaga mushroom (*Inonotus obliquus*) grown on germinated brown rice suppresses inflammation associated with colitis in mice. Food Sci. Biotechnol., 21(5): 1235–1241.

Draelos, Z.D. (2009). Cosmeceuticals: Undefined, unclassified, and unregulated. Clin. Dermatol., 27: 431–434.

Dubost, N.J., Beelman, R.B., Peterson, D. and Royse, D.J. (2006). Identification and quantification of ergothioneine in cultivated mushrooms by liquid chromatography-mass spectroscopy. Int. J. Med. Mush., 8: 215–222.

Dupont, E., Gomez, J. and Bilodeau, D. (2013). Beyond UV radiation: A skin under challenge. Int. J. Cosmet. Sci., 35: 224–232.

Edited. (2019). https://edited.com/resources/what-the-beauty-industry-looks-like-in-the-future/ (Accessed on March 01, 2020).

Erdogan, S., Kaya, M. and Akata, I. (2017). Chitin extraction and chitosan production from cell wall of two mushroom species (*Lactariusvellereus* and *Phyllophoraribis*). AIP Conf. Proc., 1809: 020012. https://doi.org/10.1063/1.4975427.

Elsayed, E.A., Enshasy, H.E., Wadaan, M.A.M. and Aziz, R. (2014). A potential natural source of anti-inflammatory compounds for medical applications. Mediators of Inflammation, Article ID 805841. 10.1155/2014/805841.

Fangkrathok, N., Junlatat, J. and Sripanidkulchai, B. (2013). *In vivo* and *in vitro* anti-inflammatory activity of *Lentinuspolychrous* extract. J. Ethnopharmacol., 147(3): 631–637.

Future Market Insights. (2020). 2019 analysis and review of natural cosmetics market by product type – Skin care, hair care, color cosmetics, fragrance, oral care, toiletries for 2019–2027. REP-GB-8588, published, 5 March, 2020.

Gao J.-M., Zhang A.-L., Chen, H. and Liu, J.-K. (2004). Molecular species of ceramides from the ascomycete truffle *Tuber indicum*. Chem. Phys. Lipids, 131: 205–213

Gautier, S., Xhauflaire-Uhoda, E., Gonry, P. and Piérard, G.E. (2008) Chitinglucan, a natural cell scaffold for skin moisturisation and rejuvenation. Int. J. Cosmet. Sci. 30: 459–469.

Geethangili, M. and Tzeng, Y.M. (2011). Review of pharmacological effects of *Antrodiacamphorate* and its bioactive compounds. Evid. Based Compl. Alt., Med., Article ID 212641: 1–17. https://doi.org/10.1093/ecam/nep108.

González, S. and Pathak, M.A. (1996). Inhibition of ultraviolet-induced formation of reactive oxygen species, lipid peroxidation, erythema and skin photosensitisation by polypodium leucotomos. Photodermatol. Photoimmunol. Photomed., 12(2): 45–56.

Gomaa, Y.A., El-Khordagui, L.K., Boraei, N.A. and Darwish, I.A. (2010). Chitosan microparticles incorporating a hydrophilic sunscreen agent. Carbohydr. Polym., 81: 234–242.

Handoko, H.Y., Rodero, M.P., Boyle, G.M., Ferguson, B., Engwerda, C., Hill, G., Muller, H.K., Khosrotehrani, K. and Walker, G.J. (2013). UVB-induced melanocyte proliferation in neonatal mice driven by CCR2-independent recruitment of Ly6c(low)MHCII(hi) macrophages. J. Invest. Dermatol., 133(7): 1803–1812.

Hong, Y.H., Chang, U.J., Kim, Y.S., Jung, E.Y. and Suh, H.J. (2017). Dietary galacto-oligosaccharides improve skin health: A randomised double-blind clinical trial. Asia Pac. J. Clin. Nutr., 26(4): 613–618.

Im, K.H., Baek, S.A., Choi, J. and Lee, T.S. (2019). Antioxidant, anti-melanogenic and anti-wrinkle effects of *Phellinus vaninii*. Mycobiology, 47(4): 494–505.

INCI Decoder. (2020). https://incidecoder.com/ (Accessed on March 01, 2020).

Jedinak, A., Dudhgaonkar, S., Wu, Q.L., Simon, J. and Sliva, D. (2011). Anti-inflammatory activity of edible oyster mushroom is mediated through the inhibition of NF-κB and AP-1 signalling. Nutr. J., 10(52): 1–10.

Jenkins, G. (2002). Molecular mechanisms of skin ageing. Mech. Ageing Dev., 123: 801–810.

Kaewnarin, K., Suwannarach, N., Kumla, J. and Lumyong, S. (2016). Phenolic profile of various wild edible mushroom extracts from Thailand and their antioxidant properties, anti-tyrosinase and hyperglycaemic inhibitory activities. J. Func. Foods, 27: 352–364.

Kawagishi, H., Hamajima, K. and Inoue, Y. (2002). Novel hydroquinone as a matrix metallo-proteinase inhibitor from the mushroom, *Piptoporus betulinus*. Biosci. Biotechnol. Biochem., 66(12): 2748–2750.

Kawaguchi, Y., Tanaka, H., Okada, T., Konishi, H., Takahashi, M., Ito, M. and Asai, J. (1996). The effects of ultraviolet A and reactive oxygen species on the mRNA expression of 72-kDa type IV collagenase and its tissue inhibitor in cultured human dermal fibroblasts. Arch. Dermatol. Res., 288(1): 39–44.

Kieffer, M.E. and Efsen, J. (1998). Imedeen in the treatment of photoaged skin: An efficacy and safety trial over 12 months. J. Eur. Acad. Dermatol. Venereol., 11(2): 129–136.

Kim, D.U., Chung, H.C., Choi, J., Sakai, Y. and Lee, B.Y. (2018). Oral Intake of low-molecular-weight collagen peptide improves hydration, elasticity and wrinkling in human skin: a randomized, double-blind, placebo-controlled study. Nutrients, 10(7): 826. 10.3390/nu10070826.

Kim, B.C., Jeon, W.K., Hong, H.Y., Jeon, K.B., Hahn, J.H., Kim, Y.M., Numazawa, S., Yosida, T., Park, E.H. and Lim, C.J. (2007). The anti-inflammatory activity of *Phellinus linteus* (Berk. & M.A. Curt.) is mediated through the PKCδ/Nrf2/ARE signaling to up-regulation of heme oxygenase-1. J. Ethnopharmacol., 113(2): 240–247.

Kim, J.W., Kim, H.I., Kim, J.H., Kwon, O.C., Son, E.S., Lee, C.S. and Park, Y.J. (2016). Effects of ganodermanondiol, a new melanogenesis inhibitor from the medicinal mushroom *Ganoderma lucidum*. Int. J. Mol. Sci., 17(11): 1798. 10.3390/ijms17111798.

Kim, K.M., Kwon, Y.G., Chung, H.T., Yun, Y.G., Pae, H.O., Han, J.A., Ha, K.A., Kim, T.W. and Kim, Y.M. (2003). Methanol extract of *Cordyceps pruinosa* inhibits *in vitro* and *in vivo* inflammatory mediators by suppressing NF-κB activation. Toxicol. Appl. Pharmacol., 190(1): 1–8.

Kim, K.H., Park, S.J., Lee, J.E., Lee, Y.J., Song, C.H. Seong, Choi, H., Ku, S.K. and Kang, S.J. (2014). Anti-skin-aging benefits of exopolymers from *Aureobasidiumpullulans* SM2001. J. Cosmet. Sci., 65(5): 285–298.

Kozarski, M., Klaus, A., Vunduk, J., Zizak, Z., Niksic, M., Jakovljevic, D., Vrvic, M.M. and van Griensven, L.J.L.D. (2015). Nutraceutical properties of the methanolic extract of edible mushroom *Cantharelluscibarius* (Fries): Primary mechanisms. Food Funct., 6: 1875–1886.

Kumirska, J., Weinhold, M.X., Thöming, J. and Stepnowski, P. (2011). Biomedical activity of chitin/ chitosan based materials – Influence of physicochemical properties apart from molecular weight and degree of N-acetylation. Polymers, 3: 1875–1901.

Lassus, A., Jeskanen, L., Happonen, H.P. and Santalahti, J. (1991). Imedeen for the treatment of degenerated skin in females. J. Int. Med. Res., 19(2): 147–152.

Lee, C.M. (2016). Fifty years of research and development of cosmeceuticals: A contemporary review. J. Cosmet. Dermatol., 15(4): 527–539.

Lee, S.S., Tan, N.H., Fung, S.Y., Sim, S.M., Tan, C.S. and Ng, S.T. (2014). Anti-inflammatory effect of the sclerotium of *Lignosusrhinocerotis* (Cooke) Ryvarden, the tiger milk mushroom. BMC Compl. Alt. Med., 14: 359. 10.1186/1472-6882-14-359.

Liao, W.C., Hsueh, C.Y. and Chan, C.F. (2014). Antioxidative activity, moisture retention, film formation, and viscosity stability of *Auriculariafuscosuccinea*, white strain water extract. Biosci. Biotechnol. Biochem., 78: 1029–1036.

Liu, H. and He, L. (2012). Comparison of the moisture retention capacity of *Tremella* polysaccharides and hyaluronic acid. J. Anhui Agric. Sci., 40: 13093–13094.

Loden, M. (2005). The clinical benefit of moisturisers. J. Eur. Acad. Dermatol.Venereol., 19(6): 672–688.

Lourenço, A., Lobo, A.M., Rodríguez, B. and Jimeno, M.-L. (1996). Ceramides from the fungus *Phellinus pini*. Phytochemistry, 43: 617–620.

Lu, Y.Y., Ao, Z.H., Lu, Z.M., Xu, H.Y., Zhang, Z.M., Dou, W.F. and Xu, Z.H. (2008). Analgesic and anti-inflammatory effects of the dry matter of culture broth of *Termitomycesalbuminosus* and its extracts. J. Ethnopharmacol., 120(3): 432–436.

Maity, K., Kar, E., Maity, S., Gantait, S.K., Das, D., Maiti, S., Maiti, T.K., Sikdar, S.R. and Islam, S.S. (2011). Structural characterisation and study of immunoenhancing and antioxidant property of a novel polysaccharide isolated from the aqueous extract of a somatic hybrid mushroom of *Pleurotusflorida* and *Calocybeindica* variety APK2. Int. J. Biol. Macromol., 48: 304–310.

Meng, T.X., Zhang, C.F., Miyamoto, T., Ishikawa, H., Shimizu, K., Ohga, S. and Kondo, R. (2012). The melanin biosynthesis stimulating compounds isolated from the fruiting bodies of *Pleurotuscitrinopileatus*. J. Cosmet. Dermatol. Sci. Appl., 2: 151–157.

Michelot, D. and Melendez-Howell, L.M. (2003). *Amanita muscaria*: Chemistry, biology, toxicology, and ethnomycology. Mycol. Res., 107(2): 131–146.

Miyake, M., Yamamoto, S., Sano, O., Fujii, M., Kohno, K., Ushio, S., Iwaki, K. and Fukuda, S. (2010). Inhibitory effects of 2-amino-3H-phenoxazin-3one on the melanogenesis of murine B16 melanoma cell line. Biosci. Biotechnol. Biochem., 74: 753–758.

Mordor Intelligence. (2019). Asia-Pacific nutricosmetics market—Growth, trends, and forecast (2019–2024), ID: 4703930.

Murina, A.T., Kerisit, K.G. and Boh, E.E. (2012). Mechanisms of skin aging. Cosmet. Dermatol., 25: 399–402.

Murphy, G. and Nagase, H. (2008). Progress in matrix metalloproteinase research. Mol. Asp. Med., 29: 290–308.

Nagasaka, R., Ishikawa, Y., Inada, T. and Ohshima, T. (2015). Depigmenting effect of winter medicinal mushroom *Flammulnavelutipes* (higher Basidiomycetes) on melanoma cells. Int. J. Med. Mush., 17: 511–520.

Ospina, N.M., Alvarez, S.P.O., Sierra, D.M.E., Vahos, D.F.R., Ocampo, F.A.Z. and Orozco, C.P.O. (2015). Isolation of chitosan from *Ganoderma lucidum* mushroom for biomedical applications. J. Mater. Sci. Mater. Med., 26(3): 135.

Palma, M.L., Monteiro, C., Tavares, L., Julia, M. and Rodrigues, L.M. (2012). Relationship between the dietary intake of water and skin hydration. Biomed. Biopharm. Res., 9: 173–181.

Palma, L., Tavares, L. and Bujan, M.R.L. (2013). Diet related water seems to affect *in vivo* skin hydration and biomechanics. IUPS 2013 (Abstract B), Accessed June 17, 2015, p. 801.

Park, Y.M., Won, J.H., Kim, Y.H., Choi, J.W., Park, H.J. and Lee, K.T. (2005). *In vivo* and *in vitro* anti-inflammatory and anti-nociceptive effects of the methanol extract of *Inonotus obliquus*. J. Ethnopharmacol., 101(1-3): 120–128.

Patterson, M.L., Atkinson, S.J., Knäuper, V. and Murphy, G. (2001). Specific collagenolysis by gelatinase a, MMP-2, is determined by the hemopexin domain and not the fibronectin-like domain. FEBS Lett., 503: 158–162.

Puttaraju, N.G., Venkateshaiah, S.U., Dharmesh, S.M., Urs, S.M. and Somasundaram, R. (2006). Antioxidant activity of indigenous edible mushrooms. J. Agric. Food Chem., 54: 9764–9772.

Reed, R.E. (1964). The definition of cosmeceuticals. J. Cosmet. Sci., 13: 103–110.

Rijken, F., Kiekens, R.C. and Bruijnzeel, P.L. (2005). Skin-infiltrating neutrophils following exposure to solar-simulated radiation could play an important role in photoageing of human skin. Br. J. Dermatol., 152(2): 321–328.

Robinson, M.K., Binder, R.L. and Griffiths, C.E. (2009). Genomic-driven insights into changes in aging skin. J. Drugs Dermatol., 8(7 Suppl): s8–s11.

Rosado, C., Pinto, P. and Rodrigues, L.M. (2005). Modelling TEWL-desorption curves: A new practical approach for the quantitative in vivo assessment of skin barrier. Exp. Dermatol., 14(5): 386–390.

Sephora. (2020). https://www.sephora.com/(Accessed onMarch 01, 2020).

Shapiro, S.D., Kobayashi, D.K. and Ley, T.J. (1993). Cloning and characterization of a unique elastolytic metalloproteinase produced by human alveolar macrophages. J. Biol. Chem., 268: 23824–23829.

Sharma, V.K., Choi, J., Sharma, N., Choi, M. and Seo, S.Y. (2004). *In vitro* anti-tyrosinase activity of 5-(hydroxymethyl)-2-furfural isolated from *Dictyophoraindusiata*. Phytother. Res., 18(10): 841–844.

Shipley, J.M., Wesselschmidt, R.L., Kobayashi, D.K., Ley, T.J. and Shairo, S.D. (1996). Metalloelastase is required for macrophage-mediated proteolysis and matrix invasion in mice. Proc. Nat. Acad. Sci., 93: 3942–3946.

Statista. (2020). Growth rate of the global cosmetics market 2004–2019. Statista.com. March 18, 2020 (Accessed 1 April 2020).

Sun, J., He, H. and Xie, B.J. (2004). Novel antioxidant peptides from fermented mushroom *Ganoderma lucidum*. J. Agric. Food Chem., 52: 6646–6652.

Synytsya, A., Míčková, K., Synytsya, A., Jablonský, I., Spěváček, J., Erban, V., Kováříková, E. and Čopíková, J. (2009). Glucans from fruit bodies of cultivated mushrooms *Pleurotus ostreatus* and *Pleurotus eryngii*: Structure and potential prebiotic activity. Carbohydr. Polym., 76: 548–556.

Takahara, M., Kang, K., Liu, L., Yoshida, Y., McCormick, T.S. and Cooper, K.D. (2003). iC3b arrests monocytic cell differentiation into CD1c-expressing dendritic cell precursors: A mechanism for transiently decreased dendritic cells *in vivo* after human skin injury by ultraviolet B.J. Invest. Dermatol., 120(5): 802–809.

Takeuchi, H., Gomi, T., Shishido, M., Watanabe, H. and Suenobu, N. (2010). Neutrophil elastase contributes to extracellular matrix damage induced by chronic low-dose UV irradiation in a hairless mouse photoaging model. J. Dermatol. Sci., 60(3): 151–158.

Tamsyn, S.A.T., Pauline, H. and Declan, P.N. (2009). Anti-collagenase, anti-elastase and anti-oxidant activities of extracts from 21 plants. BMC Compl. Alt. Med., 9: 1–11.

Taofiq, O., Martins, A., Barreiro, M.F. and Ferreira, I.C.F.R. (2016). Anti-inflammatory potential of mushroom extracts and isolated metabolites. Tr. Food Sci. Technol., 50: 193–210.

Technavio. (2019). Nutricosmetics market by product, distribution channel, and geography—Forecast and analysis, 2020–2024. SKU: IRTNTR40134.

Uchida, R., Ishikawa, S. and Tomoda. H. (2014). Inhibition of tyrosinase activity and melanine pigmentation by 2-hydroxytyrosol. Acta Pharm. Sin. B, 4: 141–145.

Uchida, T., Kadhum, W.R., Kanai, S., Todo, H., Oshizaka, T. and Sugibayashi, K. (2015). Prediction of skin permeation by chemical compounds using the artificial membrane, Strat-M™. Eur. J. Pharm. Sci., 67: 113–118.

Ulta Beauty. (2020). https://www.ulta.com/ (Accessed 01 March 2020).

Urschitz, J., Iobst, S., Urban, Z., Granda, C., Souza, K.A., Lupp, C., Schilling, K., Scott, I, Csiszar, K. and Boyd, C.D. (2002). A serial analysis of gene expression in sun-damaged human skin. J. Invest. Dermatol., 119(1): 3–13.

US Food and Drug Administration. (2020). https://www.fda.gov/cosmetics (Accessed March 01, 2020).

Van, Q., Nayak, B.N., Reimer, M., Jones, P.J.H., Fulcher, R.G. and Rempel, C.B. (2009). Anti-inflammatory effect of *Inonotus obliquus, Polygala senega* L. and *Viburnum trilobum* in a cell screening assay. J. Ethnopharmacol., 125(3): 487–493.

Wolf, R., Wolf, D., Rudikoff, D. and Parish, L.C. (2010). Nutrition and water: Drinking eight glasses of water a day ensures proper skin hydration-myth or reality? Clin. Dermatol., 28(4): 380–383.

Won, S.Y. and Park, E.H. (2005). Anti-inflammatory and related pharmacological activities of cultured mycelia and fruiting bodies of *Cordyceps militaris*. J. Ethnopharmacol., 96(3): 555–561.

Yadav, T., Mishra, S., Das, S., Aggarwal, S. and Rani, V. (2015). Anticedants and natural prevention of environmental toxicants induced accelerated aging of skin. Environ. Toxicol. Pharmacol., 39: 384–391.

Yan, W., Zhang, L.L., Yan, L., Zhang, F., Yin, N.B., Lin, H.B., Huang, C.Y., Wang, L., Yu, J., Wang, D.M. and Zhao, Z.M. (2013). Transcriptome analysis of skin photoaging in chinese females reveals the involvement of skin homeostasis and metabolic changes. PLoS ONE, 8(4): e61946.

Yan, Z.F., Yang, Y., Tian, F.H., Mao, X.X., Li, Y. and Li, C.T. (2014). Inhibitory and acceleratory effects of *Inonotus obliquus* on tyrosinase activity and melanin formation in B16 melanoma cells. Evid. Based Comp. Alt. Med., 2014: 259836.

Yang, S.H., Liu, H.I. and Tsai, S.J. (2006). Edible *Tremella* polysaccharide for skin care. U.S. Patent US20060222608, October 05, 2006.

Yap, Y.H.Y., Abdul Aziz, A., Fung, S.Y., Ng, S.T., Tan, C.S. and Tan, N.H. (2014). Energy and nutritional composition of tiger milk mushroom (*Lignosustigris* C.S. Tan) sclerotia and the antioxidant activity of its extracts. Int. J. Med. Sci., 11(6): 602–607.

Yoon, K.N., Alam, N., Lee, J.S., Lee, K.R. and Lee, T.S. (2011). Detection of phenolic compounds concentration and evaluation of antioxidant and antityrosinase activity of various extract from *Lentinus edodes*. World Appl. Sci. J., 12: 1851–1859.

Yoshida, Y., Kang, K., Berger, M., Chen, G., Gilliam, A.C., Moser, A., Wu, L., Hammerberg, C. and Cooper, K.D. (1998). Monocyte induction of IL-10 and down-regulation of IL-12 by iC3b deposited in ultraviolet-exposed human skin. J. Immunol., 161(11): 5873–5879.

Zhang, Z.Y., Jin, B. and Kelly, J.M. (2007). Production of lactic acid from renewable materials by *Rhizopus* fungi. Biochem. Eng. J., 35: 251–263.

Zhang, K., Meng, X.Y., Sun, Y. and Guo, P.Y. (2013). Preparation of *Tremella, Speranskiaetuberculatae* and *Eriocaulonbuergerianum* extracts and their performance in cosmetics. Deterg. Cosmet., 36: 28–32.

Zillich, O.V., Schweiggert-Weisz, U., Eisner, P. and Kerscher, M. (2015). Polyphenols as active ingredients for cosmetic products. Int. J. Cosmet. Sci., 37(5): 455–464.

15

Mushrooms as Sources of Flavours and Scents

Ewa Moliszewska,[1,*] *Małgorzata Nabrdalik*[1] and *Julia Dickenson*[2]

1. INTRODUCTION

Mushrooms are consumed as foods, including nutraceuticals, functional foods or food-flavouring agents with a characteristic mushroom and umami taste shared among all cousins across the world. Many fungi are broadly used biotechnologically as sources of various substances considered as dietary and nutritional, being good sources of proteins and carbohydrates with low energy. They are also a source of a wide variety of applications, like cosmetic substances, plant protection and medicines (including antibiotics and other drugs). Fungi contain a massive variety of bioactive compounds, which are effective as antioxidants, anticancer and antimicrobial agents and as valuable sources of flavours and aromas (Tsai et al., 2009; Nöfer et al., 2018; Li et al., 2018b; Varghese et al., 2019; Sun et al., 2020). They are increasingly consumed as food all over the world and consumer interests are directed towards many new species for use in cuisine or for growing varieties that were previously found only in nature. This spreading interest in fungi and mushrooms means that demand for mushroom products has increased. They are versatile, used in a variety of fashions, sometimes fresh and at other times, processed in different ways. Mushroom processing can take many forms, such as drying, freezing, microwaving, marinating, cooking or frying. This aspect of versatility gives mushrooms an important status in the food industry, like breeding mycelia, selling mycelia, production of media, composts for mushroom growers, growing of mushrooms, storing, processing, transportation and trading.

[1] University of Opole, Faculty of Natural Sciences and Technology, Institute of Environmental Engineering and Biotechnology, B. Kominka St. 6A, 45-035 Opole, Poland.
[2] Polish-American Fulbright Commission, K.I. Gałczyńskiego St. 4, 00-362 Warsaw, Poland.
* Corresponding author: ewamoli@uni.opole.pl

Wild and cultivated mushroom species are a source of typical mushroom taste, like umami and kokumi. The umami taste belongs to a group of basic tastes, along with sweet, sour, salty and bitter. It usually enhances food's tastiness and is described as savoury and broth-like or meaty. The human tongue recognises this taste by employing the umami receptor, which shows low specificity and may bind to numerous diverse umami molecules. Currently, umami taste receptors have been identified as heteromeric receptors (T1R1/T1R3, mGluR1, mGluR4, taste-mGluR1, and taste-mGluR4). The T1R1/T1R3 receptors work primarily in the tip of the tongue, whereas the mGluRs work chiefly in the back of the tongue (Sun et al., 2020). Umami compounds have been identified as free L-amino acids, purine nucleotides, peptides, organic acids, amides and their derivatives, including derivatives of 5'-ribonucleotides (Zhang et al., 2013; Phat et al., 2016; Wang et al., 2020). A mixture of monosodium glutamate and 5'-nucleotides dissolved in water evokes a peculiar taste, which is rated as neutral or even unpleasant, but when added to food it becomes an efficient flavor-enhancer as well as a positive flavor-modulator (Sun et al., 2020; Zhang et al., 2020).

Mushrooms may differ in grades of umami taste due to the composition of compounds responsible for the taste. Path et al. (2016) found that *Pleurotus ostreatus* extract was assessed as the highest umami taste among the mushrooms tested in sensory examination, while *Flammulina velutipes* extract elicited the best results in electronic tongue measurements. The presence of umami compounds is influenced by many factors, like the species type, maturity stage, part of the mushroom (pileus or stipe), quality grade and storage duration. For instance, umami taste as a product of amino acid content in *Volvariella volvacea* may be twice that of *Agaricus bisporus* at its highest maturity stage (Zhang et al., 2013). It has been indicated that monosodium glutamate-like (MSG-like) amino acids, especially glutamic acid, are present in larger quantities than 5'-nucleotides in mushrooms; thus umami taste in mushrooms is built mostly by amino acidic components. This makes mushrooms a good source of amino acids, as they assist in avoiding synthetically obtained glutamic acid and sodium glutamate, which may be beneficial in human health (Zhang et al., 2013).

Recently the sixth taste, kokumi, has been understood as being implemented into all the taste groups. Translated from Japanese, 'kokumi' means 'rich taste' and in the flavour industry it is described as continuity, mouthfulness, richness and thickness. Kokumi substances are derivatives of glutamic acid and do not have a specific taste, like umami, but their presence intensifies the other five flavours. Hence, it would be more appropriate to consider it as a flavor-enhancer as opposed to a taste of its own. Peptides that contain the kokumi flavour include glutathione (γ-glutamylcysteinylglycin [γ-Glu-Cys-Gly]) and certain glutamyl peptides (e.g., γ-Glu-Val-Gly) (Ohsu et al., 2010; Zhang et al., 2013; Miyamura et al., 2015). In kokumi, the receptor responsible for binding these peptides is the calcium Ca-SR receptor and it is coupled with a G protein. Kokumi compounds directly activate Ca-SR, which can regulate satiety and modulate the appetite in a given product or dish, which is perceived as being rich in taste (Ohsu et al., 2010; Zhang et al., 2013; Amino et al., 2016). Studying fungal volatiles is already intriguing with the main goal being obtaining flavourful substances for the food industry. On the other hand, it may be rewarding to study fungal chemistry more intensively with added goals. For example, obtaining new fungal molecules with potentially interesting functions is necessary in the production of metabolites with medicinal properties. The profile of fungal volatiles is also fascinating owing to the presence of some compounds which point to the production of functionalised non-volatile metabolites with possible medicinal properties (Fraatz and Zorn, 2010; Lauterbach et al., 2019; Varghese et al., 2019). This chapter discusses the importance of mushrooms as flavours and scents in functional foods and medicinal products with their chemical composition.

2. Mushrooms may Smell like Flowers, Fruits and Food

Eating smooth, tangy yoghurt at breakfast, sharing bitter coffee or tea, cooking a tasty dinner – flavour is something that we experience every single day in our food. However, we do not often consider another important component of flavour: scent. Flavour comprises for the cumulative effects of a substance upon the human taste and smell sensors. The flavour and scent industry has been developing for over 160 years, based on experimentation and ancient knowledge passed down over generations. The main sources of aromas are whole plants, flowers, roots and fruits which yield essential oils and other volatile substances, but for many years, mushrooms were not considered a source of fragrance or aroma (Poucher, 1991; Moliszewska, 2014). But in the recent past mushrooms came to be recognised for their aromas and are sources of a broad range of smells, including flowery, fruity, vegetable-like, food-like and many others (Moliszewska, 2014).

Some mushroom odours can be pleasant and are considered atypical for fungi (Tables 1, 3). *Cortinarius suaveolens* has a strong orange blossom scent, which remains after drying; *Hygrophorus agathosmus* emits a cherry laurel scent and almond/marzipan odour (California Fungi, 2014; Moliszewska, 2014); *H. hyacenthinus* gives a jasmine scent, and *Clitocybe geotropa* releases a lavender aroma. Aniseed aroma is an odour identified in *Psatella arvensis, Agaricus sylvicola, Gloeophyllum odoratum* and *Trametes suaveolens* (Poucher, 1991; Snowarski, 2005, 2010; Moliszewska, 2014). This odour is created by p-anisaldehyde, although in some species it is combined with benzaldehyde and benzyl alcohol, which changes the profile of its odour (e.g., *Agaricus essettei* and *Gyrophragmium dunalii*) (Rapior et al., 2002; Fraatz and Zorn, 2010). In *Lentinellus cochleatus*, an anise-like aroma exists as well, but consists of more than one compound. The scent is created by a mixture of *p*-anisaldehyde, methyl *p*-anisate, methyl (Z)-*p*-methoxycinnamate and methyl (E)-*p*-methoxycinnamate (Rapior et al., 2002). A bitter almond-like scent may be found in *Clitocybe gibba* and *C. odora*, an almond-like scent with a note of flour in *Agaricus bitorquis*, and a fruit-like scent in *Lepista nuda, Inocybe erubescens, Russula emetica, Lactarius deliciosus, L. deterrimus* and *Fistulina hepatica*. A typical mushroom odour combined with an anise-like scent is characteristic for *Leucoagaricus leucothites* (Snowarski, 2005, 2010; Moliszewska, 2014). *Ceratocystis* species were found to possess fruit-like aromas, such as *C. variospora* which is a source of geraniol (Reineccius, 1994).

Another group of scents present in mushrooms is the aroma of food and vegetables (Tables 1, 3). For instance, a flour-like flavour is characteristic for *Tricholoma equestre, Calocybe gambosa, Catathelasma imperial* and *Entoloma sinuatum*. An

Table 1. Mushrooms as sources of different flavours and aromas

Mushroom	Flavour/Scent/Aroma	References
Agaricus bitorquis	almond-like and flour aroma	Poucher, 1991
Agaricus essettei	aniseed aroma	Rapior et al., 2002; Fraatz and Zorn, 2010
Agaricus sylvicola,	aniseed aroma	Snowarski, 2005, 2010
Amanita citrina	raw potato odour	Snowarski, 2005, 2010
Calocy begambosa	flour-like flavour	Poucher, 1991; Fraatz and Zorn, 2010
Catathelasma imperial	flour-like flavour	Poucher, 1991; Fraatz and Zorn, 2010
Catathelas maventricosum	cucumber, chicory and fenugreek aroma	Snowarski, 2005, 2010
Ceratocystis variospora	fruit-like aroma	Reineccius, 1994
Clitocy begeotropa	lavender aroma	Poucher, 1991; Snowarski, 2005, 2010
Clitocy begibba	bitter almond-like scent	Snowarski, 2005, 2010
Clitocy beodora	bitter almond-like scent	Snowarski, 2005, 2010
Clitopilus prunulus	cucumber, chicory and fenugreek aroma	Snowarski, 2005, 2010
Cortinarius suaveolens	orange blossom	California Fungi, 2014
Entoloma sinuatum	flour-like flavour	Poucher, 1991; Fraatz and Zorn, 2010
Gloeophyllum odoratum	aniseed aroma	Snowarski, 2005, 2010
Gyrophragmium dunalii	aniseed aroma	Rapior et al., 2002; Fraatz and Zorn, 2010
Hygrophorus agathosmus	cherry laurel and almond scent	California Fungi, 2014
Hygrophorus hyacenthinus	jasmine scent	Poucher, 1991
Hygrophorus russocoriaceus	cedar aroma	Fraatz and Zorn, 2010
Kalpuya brunnea	garlic-like and cheese-like aromas	Trappe et al., 2010
Lactarius camphoratus	melilot scent	Poucher, 1991
Lactarius glyciosmus	bergamotscent	Poucher, 1991
Lactarius helvus	chicory and fenugreek odour	Fraatz and Zorn, 2010
Lentinellus cochleatus	anise-like aroma	Rapior et al., 2002
Lyophyllum connatum	cucumber with a flour-like aroma	Snowarski, 2005, 2010
Macrocystidia cucumis	cucumber with a herring-like odour	Snowarski, 2005, 2010
Marasmius alliaceus	garlic odour	Guevara et al., 2013
Micromphale perforans	garlic odour	Guevara et al., 2013
Mycena galericulata	cucumber with a flour-like aroma	Snowarski, 2005, 2010
Mycena pura	cucumber with a herring-like odour	Snowarski, 2005, 2010
Pleurotus euosmus var. *euosmus*	tarragon aroma	Drawert et al., 1983
Pluteus cervinus	radish-like aroma	Snowarski, 2005, 2010
Polyporus umbellatus	dill-like aroma	Snowarski, 2005, 2010

Table 1 Contd. ...

...Table 1 Contd.

Mushroom	Flavour/Scent/Aroma	References
Psatella arvensis	aniseed aroma	Poucher, 1991; Snowarski, 2005, 2010
Trametes suaveolens	aniseed aroma	Snowarski, 2005, 2010
Tricholoma aurantium	cucumber, chicory and fenugreek aroma	Poucher, 1991
Tricholoma equestre	flour-like flavour	Poucher, 1991; Fraatz and Zorn, 2010
Tricholoma sulphureum	carbide-like smell	Snowarski, 2005, 2010
Tricholoma virgatum	cucumber, chicory and fenugreek aroma	Snowarski, 2005, 2010
Tuber beyerlei, T. castilloi, T. guevarai, T. lauryi, T. mexiusanum, T. miquihuanense, T. walker	garlic-like odour	Guevara et al., 2013
Tuber spp.	sulphurous odour	Kakumyan et al., 2020
Volvariella bombycina	cucumber with a herring-like odour	Snowarski, 2005, 2010
Volvariella speciosa	cucumber with a herring-like odour	Snowarski, 2005, 2010

intriguing aroma of cucumber, along with a characteristic chicory and fenugreek smell is the known fragrance of *Clitopilus prunulus* and *Catathelasma ventricosum* including *Tricholoma virgatum* and *T. aurantium* (Poucher, 1991; Fraatz and Zorn, 2010; Moliszewska, 2014). Similarly, cucumber combined with a flour-like aroma is typical for *Lyophyllum connatum* and *Mycena galericulata* and cucumber with a herring-like odour in *Macrocystis diyacucumis*. *Mycena pura, Volvariella speciosa, V. bombycine* and *Pluteus cervinus* present a radish-like aroma and the distinct odour of a raw potato is typical for *Amanita citrina*. For *Polyporus umbellatus*, a dill-like aroma is characteristic (Snowarski, 2005, 2010; Moliszewska, 2014). On the other hand, *Lactarius helvus*, known also as a maggi-pilz (maggi-mushroom), possesses a characteristic chicory and fenugreek smell and therefore is used as a spice (Fraatz and Zorn, 2010; Moliszewska, 2014). *Pleurotus euosmus* var. *euosmus* demonstrates a tarragon aroma (Drawert et al., 1983). A garlic-like odour seems to be quite popular among many mushrooms. It was found that some small truffles from Mexico and the USA (*Tuber beyerlei, T. castilloi, T. guevarai, T. lauryi, T. mexiusanum, T. miquihuanense* and *T. walker*) present a garlic odour (Guevara et al., 2013) and such odour was found in *Marasmius alliaceus* and *Micromphale perforans* too. Garlic-like and cheese-like aromas with distinct notes of mature Camembert were reported in *Kalpuya brunnea* (Trappe et al., 2010; Moliszewska, 2014).

Some mushrooms possess odours not typically characteristic for fungi (Tables 1, 3). For example, *Tricholoma sulphureum* presents a carbide-like smell (Snowarski, 2005, 2010; Moliszewska, 2014). Truffles give off a distinct mushroom aroma along with a characteristic sulphurous odour (Kakumyan et al., 2020). *Hygrophorus russocoriaceus* produces sesquiterpenes, which are responsible for a unique cedar aroma (Fraatz and Zorn, 2010; Moliszewska, 2014). On the other hand, *Melittis melissophyllum* L. subsp. *melissophyllum* (*Lamiaceae*), a flowering plant from central Italy, was found to be potential source of 1-octenol, a typical mushroom aroma compound and thus may be considered as a flavouring agent in the food industry (Maggi et al., 2009).

3. Biological Role of Smell in Mushroom Ecology

Mushrooms produce a broad array of aromas which vary depending on the specific species. They have characteristic odours and unique smells that are recognisable by mycologists and used to distinguish and identify different species (Kakumyan et al., 2020). For instance, Kakumyan et al. (2020) affirmed that recognition of *Pseudocolus fusiformis* according to its odour is possible (together with morphological and molecular identification) due to the unique absence of eight carbon volatiles. Mushroom taste and aroma are broadly used while picking mushrooms as a key test for determining edibility of the members of the *Agaricus* genus. Boletes' scent is recognised as enjoyable and appetising, but other (usually inedible) mushrooms or molds may produce malodorous smells considered as a stench.

Some characteristic odours may be used as a non-conventional identification tool by mycologists. For example, a garlic-like odour is characteristic to the members of genus *Mortierella* (Domsch et al., 1980) and a coconut-like fragrance to the members of genus *Trichoderma* (Rifai, 1969) and *Hypocrea caerulescens* (Jaklitsh et al., 2012). Fungal and mushroom odours result from the emission of volatile organic compounds, which are defined as low molecular weight compounds that evaporate at normal atmospheric temperatures and pressures (Kakumyan et al., 2020). Their biological roles remain mostly unclear, though some have been proven to play the role of eco-physiological attractants and signalling molecules in their natural environments. Attracting and signalling are employed by fungi to maintain interactions with other organisms or between members of the same species (Kakumyan et al., 2020). Profiles of chemical composition of volatile organic compounds change with aging and the maturation of mushrooms. As the mushroom body ages, the concentration of some crucial ingredients of their aroma (oct-1-en-3-ol and benzylic acid) also increases initially and then decreases. One of the main ecological roles of emitting these volatile compounds is to attract insects to obtain better dissemination of fungal spores. In some cases, like *Fomitopsis pinicola*, *Fomes fomentarius,* including the genera of *Aspergillus* and *Penicillium*, oct-1-en-3-ol plays a role as an aggregation hormone for some beetles. On the other hand, in shiitake (*Lentinula edodes*), it was found that the concentration of oct-1-en-3-ol decreases with increasing age of the fruiting bodies, while the amount of octan-3-one increases (Fraatz and Zorn, 2010).

Mushrooms also emit odours defined as unpleasant or faecal-like. The scent chemistry of a stinkhorn fungus (*Clathrus archeri*) mimics foetid odours and a scent of carrion and faeces in order to attract flies as effective agents of spore dispersal (Johnson and Jürgens, 2010). Volatile compounds responsible for this scent are highly odouriferous 3-chloroindole and indole. Kakumyan et al. (2020) considered the same reason for production of 3-methyl-butanol, 4-methyl-phenol, and volatile sulphur-containing compounds, including dimethyl disulphide, trisulphide and tetrasulphide by a mature fruiting body of *Pseudocolus fusiformis*. This group of scents contains compounds typical of carrion (oligosulphides) and of faeces (phenol, indole and *p*-cresol) (Johnson and Jürgens, 2010). Truffles also emit strong foul odours to attract flies and other animals to increase the likelihood of spore dispersal (Splivallo et al., 2011). The compound 3-chloroindole is known for its faecal-like odour when it is concentrated, but perplexingly becomes pleasant in highly diluted solutions (Poucher, 1991; Pildain et al., 2010). Foetid scent compounds were identified in species known for unpleasant odours (*Hygrophorus paupertinus, Tricholoma bufonium, T. inamoenum, T. lascivum, T. sulphureum, Lepiota bucknallii, Morchella conica, Coprinus picaceus, Boletus calopus* and *Gyrophragmium dunalii*) (Wood et al., 2003). Other mushroom species also produce different indole compounds before or after thermal processing (e.g., *Auricularia polytricha, Leccinum scabrum, Lentinula edodes, Macrolepiota procera* and *Suillus bovinus*); however, among them, 1-octen-3-ol, which is responsible for a pleasing mushroom aroma, is frequently present in minute amounts (Muszyńska et al., 2013).

A volatile 1-octen-3-ol shows a potential to inhibit germination of *Lecanicillium fungicola* spores and spores of other ascomycetes. *Lecanicillium fungicola* causes a bubble disease of *Agaricus bisporus*. It has been proven that 1-octen-3-ol is implicated in self-inhibition of fruiting body formation by button mushrooms, but on the other hand, temporary application of 1-octen-3-ol stimulates growth of bacterial populations, especially in *Pseudomonas* spp. *Pseudomonas* spp. and other bacteria were involved in mushroom formation of *A. bisporus*, as well as inhibition of *L. fungicola* spore germination. More fungi have been found to be sensitive to 1-octen-3-ol; on the potato dextrose agar, the radial growth of *Mycogone perniciosa* and *Trichoderma aggressivum,* and also some plant pathogens as *Rhizoctonia solani, Fusarium oxysporum* f. sp. *raphani* and *Botrytyis cinerea,* was inhibited. Thus, such an ecological role of 1-octen-3-ol may be used to control a broad range of fungal pathogens (Berendsen et al., 2013).

4. Chemistry of Mushrooms Smell and Taste

The mushroom flavour compounds are classified as alcohols, ketones, aldehydes and cyclic compounds (Chun et al., 2020; Sun et al., 2020) (Table 3). Based on the chemical composition of an edible mushroom profile, they may be divided into three groups: (1) a typical 'mushroom flavour' group with carbon-eight volatile derivatives (C8); (2) mushrooms containing volatile terpenoid derivatives; (3) mushrooms with sulphur-containing compounds (Fraatz and Zorn, 2010). In fact, compounds, such as 1-octen-3-ol, (E)-2-octen-1-ol and geranyl acetone, are mostly related to mushroom-like flavour (Chun et al., 2020).

4.1 The C8 Compounds

Typical flavour for edible mushrooms is produced by carbon-eight (C8) derivatives and among them, oct-1-en-3-ol was found to be the main compound responsible for the typical mushroom aroma. Other C8 derivatives, as 1-octanol, octan-3-ol, 1-octen-3-ol, octan-3-one, 1-octen-3-on and oct-2-en-1-ol, contribute significantly to the typical mushroom odour as well (Dijkstra and Wikén, 1976; Maga, 1981; Fraatz and Zorn, 2010; Sun et al., 2020). The precursor compound of the mushroom-like scent is the main mushroom fatty acid, linoleic acid. It is enzymatically converted to 1-octen-3-ol. Other C8 compounds, as 1-octen-3-one, 3-octanone and (E)-2-octenal are also possibly formed from linoleic acid (Schmidberger and Schieberle, 2020). Oct-1-en-3-ol was isolated for the first time by Freytag and Ney in 1968 (cf. Dijkstra and Wikén, 1976) from the *Aspergillus oryzae* (soy sauce and miso) and blue cheese fungi (*Penicillium* spp. or *P. roqueforti*) as the molecule responsible for the mushroom aroma. It was also identified as the first typical mushroom-like flavour compound with a unique sweetness and earthy taste. However, mushroom flavour is composed of an assembly of volatile compounds characteristic for numerous common edible mushrooms; for example, button mushrooms, meadow mushrooms, king boletes, chanterelles, truffles, honey mushrooms, parasol mushrooms and many others, including macromycetes which possess a characteristic assembly of volatile compounds. The composition of recognised compounds depends on the mushroom species and experimental methods and consists of several dozens to even 150 flavour compounds. Among them, a few C8 compounds were identified to be crucial for the overall flavour of edible mushrooms. The typical C8 alcohol is 1-octen-3-ol, which is also identified as a woody note and named 'mushroom alcohol' (Zhang et al., 2020). The aroma of compounds may vary according to the isomer type, e.g., (R) – (-)-oct-1-en-3-ol has a fruity mushroom-like scent, and presents more intensely than the (S) – (+)-isomer, of which a mouldy, grassy aroma is characteristic. The abundance of the R-isomer in edible Basidiomycetes is usually very high in *A. bisporus* (99 per cent) and in *Xerocomus badius* (82 per cent), which seems to be the lowest content of C8 derivatives in edible mushrooms (Zawirska-Wojtasiak, 2004; Fraatz and Zorn, 2010; Sun et al., 2020).

4.2 *Aromatic Compounds in Mushroom Aroma*

A separate group of mushroom aroma components consists of aromatic compounds, such as alcohols, phenols, aldehydes, ketones, pyrazines and some other chemicals; for instance sesquiterpenes are responsible for a cedar aroma (Dijkstra and Wikén, 1976; Fraatz and Zorn, 2010; Moliszewska, 2014). Pyrazines detected in boletes include methylpyrazine, 2,5-dimethylpyrazine, 2,6-dimethylpyrazine, 2-ethyl-6-methylpyrazine, trimethylpyrazine and 2,3-dimethylpyrazine (Zhang et al., 2020). Esters in dried mushrooms are present mainly as lactones and are found in multiple varieties of dried mushrooms, such as straw mushrooms (*Lentinus edodes* and *Tricholoma matsutake*) and their amount grows abundantly after drying. In some boletes, γ-valerolactone was found as an ester compound (Zhang et al., 2020).

The predominant identified aromatic aldehydes include 3-(methylthio) propionaldehyde, furfural, 2-thiophenecarboxaldehyde, hexanal and 1H-pyrrole-2-carboxaldehyde, benzaldehyde, *p*-anisaldehyde (4-methoxybenzaldehyde), benzyl acetate and 3-(methylthio) propionaldehyde; the last reported also in dried *Boletus* mushroom and pan-fried white mushrooms. Hexanal was also found in *Boletus edulis* as well as *Volvariella volvacea* mushrooms (Zhang et al., 2020). In *Clitopilus prunulus*, *Catathelasma ventricosum* and *Tricholoma virgatum*, (*E*)-non-2-enal is responsible for a cucumber-like odour with characteristic chicory and fenugreek notes (Dijkstra and Wikén, 1976; Fraatz and Zorn, 2010). The C8 derivatives may be considered mushroom ketones, which mainly include 3-octanone, 6-methyl-5-hepten-2-one and acetophenone. Additionally, the ketones, 2-pentanone, 2-heptanone and 2-nonanone are produced by *P. roqueforti* during ripening. They are probably responsible for the typical flavour of Roquefort, Cheddar and other cheeses (Reineccius, 1994).

A group of phenolic compounds detected in *Boletus* mushrooms include phenol, 2,3-dimethylphenol, 2,4-dimethylphenol, p-cresol, 3-ethylphenol, 2,6-dimethoxyphenol, 2,5-dimethylphenol, 2-methoxy-4-methylphenol and 4-ethyl-2-methoxyphenol, also known as benzyl alcohol. Phenolic compounds also include phenylethyl alcohol, which is a typical alcoholic compound with a floral note (Dijkstra and Wikén, 1976; Fraatz and Zorn, 2010; Zhang et al., 2020).

A unique aroma of certain mushrooms may be related to some characteristic compounds, e.g., 3-methylvaleric acid, pentanoic acid, tetradecanoic acid ethyl caproate, furfural, 2-thiophenecarboxaldehyde, 2-methylpropionic acid and 2,4-dimethylphenol (Zhang et al., 2020). Volatile compounds of *Agaricus bisporus* are some of the most well-studied and it was found that though C8 volatiles are present, the characteristic flavour is defined mainly by other important chemicals, such as benzaldehyde, limonene, N(2phenyethyl) acetamide, geranyl acetone and farnesyl acetone and (*E,E*)-farnesol (Fraatz and Zorn, 2010).

4.3 Nucleotides, Amino Acids and Other Compounds

The mushroom flavour and taste are also strongly influenced by monosodium glutamate-like (MSG-like) components, and 5'-nucleotides, which produce the most typical mushroom taste as well as the umami taste. Specifically, aspartic and glutamic acids (MSG-like components), 5'-guanosine monophosphate (5'-GMP), 5'-inosine monophosphate (5'-IMP), 5'-xanthosine monophosphate (5'-XMP) and 5'-adenosine monophosphate (5'-AMP) are responsible together with C8 components for the mushroom flavour. Differences in concentrations of these flavour components cause mushroom aromas to vary greatly, especially as high levels of 5'-guanosine monophosphate present a meaty flavour, which enhances the mushroom flavour much more than MSG-like components could (Chen, 1986; Tsai et al., 2009; Phat et al., 2016; Li et al., 2018a; Li et al., 2018b; Yin et al., 2019). A combination of amino acids like arginine, histidine, isoleucine, leucine, methionine, phenylalanine, tryptophan and valine endows a bitter taste; while lysine and tyrosine are tasteless amino acids; alanine, glycine, proline, serine and threonine provide a sweet note to the mushroom taste (Zhang et al., 2013; Li et al., 2018a; Li et al., 2018b).

Dijkstra and Wikén (1976) found that nucleotides, amino-acids and carbohydrates also contributed significantly to the mushroom flavour of *A. bisporus* (Dijkstra and Wikén, 1976). Free amino acids, especially aspartic acid and glutamic acid, are responsible for the umami flavour of edible fungi; thus they should also be considered important compounds of mushroom taste (Yang et al., 2020). Tsai et al. (2009) identified γ-aminobutyric acid (GABA) in three mushrooms (*Clitocybe maxima, Pleurotus ferulae* and *P. ostreatus*). It was also found in *A. blazei, A. cylindracea* and *B. edulis*, and as a biologically active compound would be beneficial not only to mushroom taste, but also for its dietary and therapeutic effects.

The first descriptive sensory flavour lexicon for fresh, dried, and powdered mushrooms of various commercially available species was developed by Chen et al. (2020). That lexicon describes, identifies, defines and references 27 flavour attributes for mushroom samples prepared as 'meat' and broth. Attributes are grouped into categories, such as musty (dusty/papery, earthy/humus, earthy/damp, earthy/potato, fermented, new leather, old leather, mold/cheesy, moldy/damp, mushroomy), and other attributes such as fishy, shellfish, woody, nutty, brown, green, cardboard, burnt/ashy, potato, umami, protein (vegetable), yeasty, bitter, salty, sweet aromatics, sour and astringent (Chen et al., 2020).

The aroma profile of mushrooms is variable, differing between species but also depending on conditions of production, time of growth and sources of nitrogen and carbon. Aroma changes occur due to the geographical origin, environmental differences, post-harvest processing, storing and cooking methods, age and part of fruiting body (caps, stalks or hyphae) (Dijkstra and Wikén, 1976; Fraatz and Zorn, 2010; Zhang et al., 2013; Moliszewska, 2014; Li et al., 2017; Sun et al., 2019). However, it was found that oyster mushrooms cultivated on four different woody substrates did not vary in aroma and flavour in sensory evaluation after frying and serving warm (Tisdale et al., 2006) (Table 2). On the other hand, oyster mushrooms obtained on cereal medium were more palatable than those cultivated on bagasse substrate, and overall single-carbon sources were better in essential amino acid production. However, on mixed-carbon sources, the 5'-nucleotides responsible for the umami taste were more abundantly produced (Li et al., 2017; Li et al., 2018b).

Table 2. The influence of substrate additives and treatment on the mushroom flavour features

Substrate treatment	Mushroom species	Flavour/Aroma/Taste	References
woody substrates	*Pleurotus ostreatus*	no differences	Tisdale et al., 2006
cereal medium vs. bagasse substrate	*Pleurotus ostreatus*	more palatable on cereals	Li et al., 2017; Li et al., 2018b
single-carbon sources	*Pleurotus ostreatus*	better to essential amino acid production	Li et al., 2017; Li et al., 2018b
mixed-carbon sources	*Pleurotus ostreatus*	more abundant production of 5'-nucleotides	Li et al., 2017; Li et al., 2018b
submerged cultures	*Agaricus bisporus* Boletes	less C8 compuunds	Dijkstra et al., 1972; Woźniak, 2007
liquid cultures with malt extract, phosphate and casein	*Agaricus bisporus*	moderate growth	Dijkstra et al., 1972
addition of vegetable oils and oleates to liquid cultures	*Agaricus bisporus*	mycelium showed the same flavour as fresh mushrooms although weaker	Dijkstra et al., 1972
linalool and coumarin added to the culture substrate	*Pleurotus euosmus* (the tarragon oyster mushroom)	intense sweet, flowery odour	Drawert et al., 1983
periodic illumination	*Nidula niveo-tomentosa*	raspberry compounds	Taupp et al., 2008

It was observed that mushrooms in storage have the highest respiration rate among all vegetables and fruits and this strongly influences the profile of flavour components. Enzymes, such as proteases and nucleases, are active and fruiting bodies change, developing spores as well as themselves in the process. As a result, an increase in the content of MSG-like amino acids and 5'-nucleotides is observed, which makes the umami taste more intense. For example, in button mushrooms collected at the button stage, the level of MSG-like amino acids increases twofold within 12 days of storage after harvest; however, 5'-nucleotide content grows only within the first three days. This observation demonstrates the importance of proper storage conditions for maintaining mushroom quality (Zhang et al., 2013).

Species differences in amino acids content responsible for umami taste show that higher amounts of these components are detected in *Craterellus cornucopioides, Pleurotus ostreatus, Boletus edulis, Morchella elata, Agaricus campestris, Macrolepiota procera, Flammulina velutipes, Cantharellus cibarius, Calocybe gambosa* and *Entoloma clypeatum* than in other edible mushrooms. On the other hand, the highest content of 5'-nucleotides is observed for *Craterellus cornucopioides, Tricholoma giganteum, Lentinus edodes, Dictyophora indusiata, Flammulina velutipesv* (Zhang et al., 2013). As mushrooms differ in flavour, it is important to know all the factors responsible for this variation. For instance, differences in a typical mushroom flavour may be caused by some specific molecules, as was observed in *Volvariella volvacea*. In this mushroom, one of the main volatile compounds is dihydro-β-ionone, which presents as an earthy, woody, mahogany smell. It also creates a characteristic mixture with camphene (woody, herbal, fir) and carvone (minty, licorice), as well as with typical mushroom scent compounds, like: 1-octen-3-one, 1-octen-3-ol, 3-octanol, 2-octanone, γ-undecalactone (fruity, peach, creamy), hexanal (fresh, green, fatty), 2-methylbutanal (musty, cocoa, coffee) and 2-nonanone (fresh, sweet, green) (Xua et al., 2019).

Usually, mushrooms are considered to be 'macromycetes', which includes many Basidiomycetes, and some Ascomycetes. However, Ascomycetes are less intensively examined as sources of scents and flavours. The Ascomycete genus *Daldinia* cf. *childiae*, collected in China, produces a few interesting and rare substances as well as typical mushroom odour molecules, like oct-1-en-3-ol (matsutake alcohol) and traces of cyclohexanol and cyclohexanone. Other contributing factors to its odour are the widespread volatile aromatic compounds, 2-phenylethanol and 1-phenylethanol acetophenone, but the main volatiles of *Daldinia* cf. *childiae* are 5-hydroxy-2-methyl-4-chromanone and 1,8-dimethoxynaphthalene. Compounds that evoke scents different to a typical mushroom scent include the biosynthetically related aliphatic compounds, such as 4-methylhexan-3-one, manicone and (4R,5R,6S)-5-hydroxy-4,6-dimethyloctan-3-one. Presence of some small heterocyclic compounds, represented by widespread 2-acetylfuran, 2-acetylthiazole, 2,5-dimethylpyrazine and the rare 2,4,6-trimethylpyridine, was previously only occasionally reported as a constituent of plant essential oils or from bacteria of the genus *Collimonas*, but never from fungi (Lauterbach et al., 2019).

Mushroom aldehydes and ketones are mushroom-odour components obtained through processes including the Strecker reaction of degradation of amino acids and lipids, the oxidative reaction of polyunsaturated fatty acids, the degradation of amino acids or the Maillard reaction observed during thermal-processing procedures or food storage. The Maillard reaction is a non-enzymatic browning process, which chemically occurs between reactive carbonyl groups of reducing sugars and amine groups of free amino acids, peptides or proteins (Chen et al., 2018). The reaction can also be performed at lower temperatures with pyridoxamine as a catalyst (Chen et al., 2018; Yin et al., 2019; Schmidberger and Schieberle, 2020). The Maillard reaction has also been reported as a way to form the 2-pentyl-furan and other furan compounds during thermal processing of mushrooms (Yin et al., 2019). Products of the Maillard reaction can confer kokumi taste, which enhances taste in general, and umami taste, which is caused by hydrolysed proteins (Chen et al., 2018). Interesting results were obtained by Chen et al. (2018) showing that 125°C was the optimal temperature for preparing the Maillard reaction products responsible for sensory properties of mushroom hydrolysate. They identified volatile compounds, such as 3-phenylfuran and 2-octylfuran, which are responsible for the caramel-like flavour, as well as 1-octen-3-ol, (E)-2-octen-1-ol and geranyl acetone, which were correlated to the mushroom aroma. Others, such as 2-thiophene-carboxaldehyde, 2,5-thiophene-dicarboxaldehyde and 3-methylbutanal, contribute a meat-like flavour. Additionally, the amount of free amino acids and 5'-GMP (the umami taste) were higher than in raw mushroom hydrolysate.

5. Edible Mushrooms and their Flavours

Edible mushrooms are cultivated worldwide with the major genera being *Agaricus*, *Pleurotus*, *Lentinula*, *Auricularia* and *Ganoderma*. The total world mushroom production reaches almost 40 million tons per year. China has emerged as the world's major mushroom producer, consumer and exporter, followed by the United States and the Netherlands, Poland, Spain, France, Italy, Ireland, Canada, the UK and India (Dudekula et al., 2020; Thakur, 2020). However, edible mushrooms are cultivated as well as collected from the natural environment. Those which are traditionally picked from the ground are boletes, chanterelles, morels and truffles. Hunting mushrooms is popular in eastern Europe, while in Sweden the activity has risen in popularity over the last 100 years, becoming widely accepted in the early 21st century, especially among the middle-class and the so-called hipster-generation (born in the 1990s). While in eastern and central Europe many people are familiar with edible and poisonous mushroom species, in Sweden, often the social media are used to identify those that are edible. Some people admit knowing only a few species, while some of them do not know any at all (Svanberg and Lindh, 2019; Kasper-Pakosz et al., 2016). European countries differ in their approach to mushroom hunting. For many, the main reason for picking mushrooms is just having fun or as a hobby while spending time in nature and secondarily, using the mushrooms as a food component. As food components, mushrooms serve as a valuable source of nutritional components [e.g., vitamins (B complex, D, tocopherols and ascorbic acid), polysaccharides (β-glucans), dietary fibres, terpenes, peptides and amino acids, glycoproteins, alcohols, mineral elements, unsaturated fatty acids and antioxidants]. Carbohydrate content of fruiting bodies of edible mushrooms comprises sugars (monosaccharides, their derivatives and oligosaccharides) and some amount of alcoholic sugars (mannitol and trehalose).

Antioxidant properties of mushrooms are mostly due to the content of phenolic compounds as well as tocopherol and ascorbic acid. The total lipid content in dried mushrooms is 20–30 g/kg, and they are rich in linoleic and oleic acids. It is worth noting that linoleic acid was reported as an anticarcinogenic compound in almost all stages of tumorigenesis on animal models. Minerals present in mushrooms are most often potassium, calcium, phosphorus and magnesium (Rathore et al., 2017; Li et al., 2018b). Although fungi as large-fruited mushrooms are edible for humans, they are not of vital importance for the diet but are still appreciated around the world for their unique flavours, aroma and taste. Boletus (king Bolete *Boletus edulis*) is appreciated for its specifically mild taste and pleasant mushroom scent and is one of the most popular mushrooms in Europe. The white button mushroom (*Agaricus bisporus*) is the most commonly cultivated and consumed species in the world (Fraatz and Zorn, 2010). Similarly, another important but not so famous species is the oyster mushroom (*Pleurotus ostreatus*).

5.1 *Boletes*

Boletes are valuable mushrooms picked in the wild and sold in markets afresh, dried or frozen (Aprea et al., 2015; Kasper-Pakosz et al., 2016). They are popular in European countries owing to their intensive aroma and are used to prepare sauces, soups, stuffing of dumplings and meat, and can also be added to a variety of dishes. Among boletes, the most popular and well-studied species is *Boletus edulis*. Common names for boletes may vary, depending on the country; they are called king boletes, boletes, porcini mushrooms, cepe and cep de Bordeaux. Boletes are rich in taste and nutritional compounds, such as proteins, minerals, fatty acids and vitamins (ascorbic acid and tocopherols). Fresh boletes present a semi-sweet and delicate umami taste but are generally consumed after various processing methods. In fresh boletes, the 1-octen-3-ol and 2,5-dimethylpyrazine are the most potent flavour compounds. Contribution of the *B. edulis* aroma includes 3-(methylthio) propionaldehyde and 2,6-dimethylpyrazine, while key aroma compounds in *B. aereus* include isovaleric acid, 2,6-dimethylpyrazine, benzene acetaldehyde and (E)-2-octenal. *Boletus auripes* seems to be rich in isovaleric acid, 3-ethylphenol and 2,6-dimethylpyrazine, while in *B. rubellus*, 3-methylvaleric acid, isovaleric acid and 2,3-dimethylpyrazine are the most aromatic compounds (Zhuang et al., 2020).

The dominant volatiles in dried boletes are the typical mushroom esters, aldehydes, acids, alcohols, pyrazines, ketones and phenols; however, each species differs in quantitative content of those compounds. The key aroma compounds identified in *B. edulis* are 3-(methylthio) propion aldehyde, 2,6-dimethylpyrazine and benzene acetaldehyde. Benzene acetaldehyde was identified in *B. edulis* by Zhuang et al. (2020) and seems to be a fundamental aroma-active aldehyde with a floral note. Notwithstanding, all boletes contain 1-octen-3-ol, 2,5-dimethylpyrazine and dozens of other potent aroma compounds in lower concentrations, contributing to their aroma profiles. Other boletes species contain some characteristic compounds which comprise their aromas. For example, key aroma compounds in *B. aereus* are isovaleric acid, 2,6-dimethylpyrazine, benzene acetaldehyde and (E)-2-octenal. These are the most aroma-active aldehydes, giving fresh and floral notes. On the other hand, isovaleric acid, 3-ethylphenol, 2,6-dimethylpyrazine and γ-valerolactone, with woody note, are the main flavouring compounds in *B. auripes,* but a smoky note is due to volatile phenols, dominated by 3-ethylphenol. *Boletus rubellus* presents a more sour aroma (isovaleric acid) and balsamic (benzoic acid), with contribution of 3-methylvaleric acid and 2,3-dimethylpyrazine. A roasted aroma is offered by 3-(methylthio) propion aldehyde, while 2,4-dimethylphenol is the most phenolic aroma-active compound (Zhueng et al., 2020).

5.2 Button Mushrooms

Agaricus bisporus is the most popular mushroom in the global production and market. It is available in two colours, white and brown, in a few sizes. Button mushrooms may be stored up to eight days in the refrigerator, but they are very vulnerable to microbial damage during production and storage. *Agaricus bisporus* contains nine essential amino acids, low fat content and a relatively high amount of carbohydrates and fibre in addition to vitamin D, which makes it the only vegan source of that vitamin. It presents a typical mushroom flavour composed on the basis of volatile and non-volatile substances, such as C8 compounds, soluble sugars, polyols, free amino acids, organic acids, 5'-nucleotides and monosodium glutamate (Zhang et al., 2018). As this mushroom is very popular as a food, it is processed in different ways. The whole C8 content in *A. bisporus* may vary between 44–98 per cent. The lesser part of its aroma is caused by low-boiling point volatiles, like benzaldehyde, benzyl alcohol, 1-octen-3-one, *n*-butyric acid and isovaleric acid (Dijkstra and Wikén, 1976; Fraatz and Zorn, 2010).

5.3 Oyster Mushrooms

Oyster mushroom is a common name for the *Pleurotus* species. They are the third largest group of commercially cultivated mushrooms. This group contains approximately 20 species within the genus *Pleurotus.* As Tisdale et al. (2006) assessed, the oyster mushroom may be cultivated on various woody substrates without a change in taste or aroma. They are popular because of their unique taste and flavour. The key odour compounds in *Pleurotus citrinopileatus*, *P. djamor*, *P. floridanus, P. ostreatus* and *P. sapidus* are C8 compounds, especially 1-octen-3-one, which provides a wet ground smell and 1-octen-3-ol, while in *P. cornucopiae,* 2-octenal was an additional key aroma compound (Yin et al., 2019). Volatile compounds in *Pleurotus* mushrooms include aldehydes, ketones, alcohols, ethers, acids, hydrocarbons, heterocyclic and aromatic compounds including free amino acids and 5'-nucleotides. Yin et al. (2019) identified 63 different volatile compounds and an important group in *Pleurotus* was aldehydes, such as hexanal, octanal, nonanal and benzene acetaldehyde. Benzaldehyde was found to be a major aldehyde in *P. cornucopiae* and *P. floridanus*, while hexanal was seen as the most abundant aldehyde in *P. citrinopileatus*, *P. djamor, P. ostreatus* and *P. sapidus.*

Pleurotus aroma is a combination of grassy and fatty odour of hexanal along with the meaty and grassy notes of nonanal and the nutty, fatty and oily octanal odour with the pleasant almond flavour of benzaldehyde. The same aroma notes are presented by alcohols, although they contribute less to the whole aroma profile if alcohols are saturated, while their influence is greater if they are unsaturated alcohols. Esters provide fruity, sweet, fresh or fatty flavours but two of them (2-propenoic acid-2-ethylhexyl ester and glycerol tricaprylate) are uncommon in mushrooms. The ketones, 3-octanone, 1-octen-3-one and 2-undecanone, were identified in *Pleurotus* mushrooms as key smell ingredients. The earthy and meaty odour is given by 2-pentyl-furan and naphthalene, which are present in *Pleurotus* spp. (Yin et al., 2019).

5.4 Chanterelles

The most popular species of chanterelles are *Cantharellus cibarius, C. cinnabarinus* and *C. tubaeformis*. Chanterelles occur worldwide and one of the most identifiable of this group is *C. cibarius*, which is found especially in eastern Europe. It is very popular, easy to identify and safely collected from natural sources during late summer and autumn. On the other hand, chanterelles from the Yunnan province of China are much more poorly known. Chanterelles present high nutritional quality and unique flavour, making them a popular ingredient in cooked and fried dishes prepared at private homes and restaurants (Li et al., 2018b). Apart from their tastes, chanterelles are also a source of nutritional and antioxidant compounds, including vitamin D, vitamin C and carotenes. *Cantharellus cibarius* contains β-carotene, lower amounts of lycopene and α-carotene and two other carotenoids, which are considered to be γ- and δ-isomers of carotene. As shown in Portugal, *C. cibarius* collected from natural sources may include up to 13.6 μg/g of β-carotene in dry mass (Muszyńska et al., 2016).

5.5 Truffles

Truffles are one of the most desirable mushrooms worldwide and regarded as a fine delicacy with unique flavour and high nutritional value. Their price is fairly high because they are mostly sourced from natural sites. For instance, *Tuber magnatum* reached a price of 8000 € per kilogram in 2000 and 2001 in Italy. Truffle-derived products have a short shelf-life and are only available during the truffle season. They are very exigent for growth conditions in the natural environment (e.g., sunlight, humidity, soil pH, surrounding flora and fauna) and even a slight change renders them unable to grow. Their uniqueness is not the only feature that is so in demand by the consumers; they also possess numerous physiological activities, such as antiviral, antibacterial, anti-inflammatory, antitumor, antioxidant and other protective properties (Feng et al., 2019; Lee et al., 2020).

Fruiting bodies of truffles (ascocarps) tend to be firm, dense and woody in comparison with other mushrooms, mostly exhibiting a soft and fragile nature. The mushroom market distinguishes white and black truffles. White truffles include *Tuber borchii, T. japonicum, T. latisporum, T. magnatum, T. maculatum, T. oregonense* and *Tirmania nivea,* while black truffles include *Tuber aestivum, T. brumale, T. himalayense, T. indicum, T. melanosporum, T. uncinatum* and *Terfezia claveryi* (Feng et al., 2019; Lee et al., 2020).

Truffles differ widely in aroma, depending heavily on their geographical origin and species type, and their aroma is one of the most important factors in assessing their quality. As of now, more than 200 different volatile substances have been reported in truffles, although only some of them contribute to their aroma and flavour profiles (Feng et al., 2019). Among the various flavour compounds in truffles, 2-methylbutanal, 3-methylbutanal, dimethyl disulphide, dimethyl sulphide and 2-methyl-1-propanol are the most common, although typical mushroom compound, 1-octen-3-ol is, also present (Lee et al., 2020). In white truffles (*T. magnatum*), bis-(methylthio)-methane is reported as the most critical aroma active compound, while in black truffle (*T. melanosporum*) it is either not detected or detected only in low amounts. Instead of that compound, black truffles contain high amounts of 1-octen-3-ol (Lee et al., 2020). The volatile compounds of three different freshly picked truffles from the Yunnan province of China were examined by Feng et al. (2019). They tested *Tuber sinensis*, *T. sinoalbidum* and *T. sinoexcavatum*, identifying 38 aroma-building compounds. In all the three species, dimethyl sulphide was the main aroma-producing compound. However the other compounds within the species differed. For *T. sinensis*, dimethyl disulphide, octanal and 1-octen-3-one were responsible for the aroma composition, while for *T. sinoalbidum* -3-octanone, bis(methylthio)methane, 1-octen-3-one, 3-octanol and 1-octen-3-ol, and for *T. sinoexcavatum*, 2-methyl-butanal, 2-methylbutanol, 3-(methylthio) propanal and benzene acetaldehyde (Feng et al., 2019).

Tuber borchii is known for its characteristic flavour in the mature stages. It was introduced in New Zealand for cultivation and it remains a model in *Tuber* research due to its short time of growth. Its flavour is mostly determined by thiophene volatiles, such as 3-methylthiophene and 3-methyl-4,5-dihydrothiophene (Splivallo et al., 2015). Zeppa et al. (2004) reported that the maximum concentration of thiophene volatiles occurs solely in fully matured fruiting bodies (71–100 per cent maturity) of these truffles. Interestingly, they also suggested that bacteria may be involved in aroma creation. Splivallo et al. (2015) proved that bacterial cells are present in both gleba and peridium between fungal cells but not inside them. These bacteria were recognised as α- and β-*Proteobacteria* and in *in vitro* tests, they were able to produce thiophene volatiles. These kinds of volatiles were neither produced by *T. borchii* mycelial cultures *in vitro* nor by bacteria grown together with mycelium, which suggests that production of thiophene volatiles is strongly correlated with maturity of fruiting bodies and occurs in storage at maturity. Research also showed that some bacteria are species-specific and do not produce such volatiles, even if precursors are present (Splivallo et al., 2015). The delicacy and highly prized aroma of tubers have influenced the food industry to develop synthetic tuber flavours for food aroma. Knowing the main components of an aroma, scent and taste assist in creating the perfect composition and combination of ingredients. Commercial farming companies developed research and cultivation techniques for this artificial flavour and their creation means that tubers' distinctive flavour is more popular and available to society (Lee et al., 2020).

5.6 Shiitake

Shiitake (*Lentinula edodes*) are commonly cultivated in Far East countries but they are also cultivated worldwide. Their fresh aroma differs from the aroma of the dried form. Drying of shiitake is the most frequently used process for their preservation and allows the mushroom's characteristics and unique shiitake aroma to develop, due to the presence of 5'-guanosine monophosphate. Dried ones also contain higher amounts of vitamin D than fresh shiitake (Dermiki et al., 2013). The aroma of fresh samples of *L. edodes* is characterised by a mixture of alcohols, ketones and sulphur compounds. The composition of 3-octanol (sweet, oily and nutty odour) with 3-octanone (sweet, fruity, earthy, cheesy and mushroom odour) and 1-propanethiol create the distinctive mushroom and earthy smell of fresh shiitake (Qin et al., 2020). Similar observations were made by Schmidberger and Schieberle (2020), corroborating that 1-octen-3-ol is the main ingredient of this mushroom smell composition, followed by phenyl acetaldehyde, phenyl acetic acid, and 1-octen-3-one. In this composition, the sulphur compounds (1,2,3,5,6-pentathiepane, 1,2,4,5-tetrathiane, and 1,2,4,6-tetrathiepane) are present in low amounts. Aldehydes, products of amino acids degradation [e.g., phenyl acetaldehyde, 3-(methylthio) propanal and 3-methylbutanal] were also detected, although they were not previously reported as constituents of raw shiitake (Schmidberger and Schieberle, 2020). Tests performed by Dermiki et al. (2013) proved that aqueous extracts of dried *L. edodes* may be used as taste and flavour enhancers for meat formulations. They demonstrated that extracts prepared at 70°C lost the major shiitake volatile compounds through volatilisation and after that, they contain smaller concentrations of lanthionine - a cyclic sulphur compound, but also less of 1-octen-3-ol, 1,3-dithiethane and dimethyl disulphide. Nevertheless, incorporating that extract into ground meat caused significantly higher levels of savoury taste due to the presence of 5'-ribonucleotides in cooked meat (Dermiki et al., 2013).

6. Does Aroma Differ in Different Parts of a Mushroom?

The umami ingredients present in the pileus, stripes, fruiting body and mycelia of mushrooms vary extensively, depending on the mushroom species. In mushroom processing, stalks are often treated as waste although as Li et al. (2018a) suggested they may be used as a valuable byproduct. It was proven that in the case of shiitake, stalks may contain fibre, calcium and carbohydrates, which are significantly higher in content than in the caps, although content of amino acids responsible for such tastes (umami, sweet and bitter) and flavour-creating 5'-nucleotides (5'-GMP+5'-IMP+5'-XMP+5'-AMP) is lower. The main volatile compounds for *L. edodes* showed no significant difference in caps and stalks, although the composition of aroma differed [e.g., stalks do not contain (Z)-2-octen-1-ol, eucalyptol, undecanal, 2,4-bis(1,1-dimethylethyl)-phenol, (E)-2-decenal and 2-phenylpropenal, while in caps they are present]. The main C8 components are present in caps as well as stalks, although with some compositional differences (e.g., the amount of octen-3-ol is about 40 per cent higher in stalks than in caps) (Li et al., 2018a). Pelusio et al. (1995) and Fraatz and Zorn (2011) also noted a pronounced difference between the concentrations of octan-3-ol, octan-3-one, and oct-2-en-1-ol in caps and the stalks of *A. bisporus*. The predominant C8 volatile in the caps was octan-3-one, while oct-1-en-3-ol dominated in the stalks. In contrast, more oct-1-en-3-ol was produced in the cap and gills than in the stalk when the mushroom samples were incubated at room temperature for 5 min after homogenisation (Fraatz and Zorn, 2010).

7. Mushroom Processing and Preservation vs. Flavours

Food preservation is a mean of preventing spoilage in long-term storage, often used also for mushrooms. Mushrooms are preserved by drying, osmotic dehydration, canning, freezing, irradiation; however, these methods result in colour change, loss of flavour and aroma including texture and nutritive changes (Pal et al., 2017; Sun et al., 2019). Zhang et al. (2019) found, while working with *Pleurotus geesteranus*, that the best temperature for short-term storage of mushrooms is 5°C as at this temperature the umami taste is better preserved than at other temperatures, like 0 or 10 or 20°C. The composition of mushroom flavour compounds changes even with such simple processing procedures, as chopping or grinding owing to the release of enzymes. In chopped mushrooms, higher levels of 3-octanone are identified, while in homogenised samples, 1-octen-3-ol is more frequent. Thermal processing usually leads to reduction in the concentration of C8 compounds along with formation of new compounds, mainly due to the Maillard reaction (Chun et al., 2020). Heating methods, like microwaving, boiling and autoclaving, usually result in loss of ingredients which are responsible for the umami taste in mushrooms, while irradiation has only a mild effect on umami taste constituents (Zhang et al., 2013). All processing methods influence the intensity of mushrooms' flavour, mostly due to the loss of volatiles and some decomposition of taste compounds. This should be considered in the commercial production of mushroom concentrates. In some cases, C8 volatiles are not present or are present only in trace amounts, although the flavour of processed mushrooms is described as 'similar to mushroom and yeast'.

7.1 Drying

Drying (natural and oven-drying) is the foremost worldwide method used in mushroom preservation. Drying tends to change mushrooms' flavour characteristics. For example, black trumpet mushrooms' (*Pleurotus eryngii*) aroma is changed to a dusty/papery, earthy/humus, earthy/damp and old leather character when dried and used to make broth. However, in contrast, portobello mushrooms (*Agaricus bisporus*) showed much smaller increase in earthy/damp aroma, combined with a dusty/papery and moldy/damp character after drying (Chun et al., 2020). The change of mushrooms' flavour is due to the activation of enzymes during drying, leading to the production of sulphur-containing compounds that change the whole aroma composition (Qin et al., 2020). There are also other methods, such as freeze drying, which is an emerging technology for drying mushrooms. This method results in better quality than dehydration by other methods because mushrooms almost do not lose aroma and flavour. The method proposed by Pal et al. (2017) consists of several stages of freeze drying (fresh harvested mushrooms are consequently processed in steps including washing, blanching in hot water or steam, steeping, slicing, freezing for –32°C for four to six hours under 4.7 mm Hg pressure to obtain a final product of 3 per cent moisture content) (Pal et al., 2017).

Changes in aroma during the drying process of shiitake (*Lentinus edodes*) were described by Qin et al. (2020). The early stage of drying shiitake revealed a strong garlic and rotten-egg odour, which is mainly due to the emission of sulphur compounds, such as dimethyl trisulfide, thioanisole, 2,3,5-trithiahexane and also cyclic sulphur compounds, including lenthionine, a unique compound found in this mushroom. According to the research by Qin et al. (2020), in the early stages of drying (0.5–1.5 hour), the concentration of sulphur compounds increases and the combination of these with the presence of hydrogen sulphide and sulphur dioxide causes a negative sulphur perception of the mushrooms. However, linear sulphur compounds are responsible for the unique aroma of fresh and dried *L. edodes*. The composition of sulphur compounds mainly consists of dimethyl trisulphide (fresh onion odour) together with thioanisole and 2,3,5-trithiahexane, which together are responsible for a characteristic garlic flavour. Lenthionine (1,2,3,5,6-pentathiepane; also known as 1,2,3,5,6-pentathiacycloheptane) was detected only in the early stages of drying. Dermiki et al. (2013) confirmed the loss of lenthionine and other volatile compounds in high-temperature processing of shiitake.

The middle-phase of drying shiitake (2–3.5 hours) reveals a sautéed aroma, with no rotten-egg odour. This phase is characterised by a high ester content, including methyl butyrate, ethyl propanoate, ethyl isobutyrate, ethyl 2-methylbutyrate and ethyl isovalerate. In that period, the concentration of sulphur compounds [e.g., 2-(methylthio)ethanol, 2,4-dithiapentane and methyl-allyl- disulfide] decreases (Qin et al., 2020). In the last stage of drying (over 3.5 hr), the sautéed aroma transitions to burn, which is mainly due caramelisation and the Maillard reaction of esters and aromatic compounds. Within this drying stage, the concentration of sulphur compounds also decreases [e.g., 2-(methylthio) ethanol, 2,4-dithiapentane and methyl-allyl-disulphide] (Qin et al., 2020). The whole drying process changes the flavour profile of shiitakes and this is a result of the strong loss of alcohols, such as 3-octanol (responsible for sweet, oily and nutty odour) and ketones, like 3-octanone (responsible for sweet, fruity, earthy, cheesy and mushroom odour) including 1-propanethiol. This is most presumably due to evaporation within the drying process (Qin et al., 2020). Zhao et al. (2019) examined five types of drying methods commercially used in China (freeze drying, far-infrared radiation drying, heat pump drying, hot air drying and hot air combined with instant controlled pressure drop drying) and they found that far-infrared radiation drying was most suitable to produce dried shiitakes, which yields a product with rich flavour and nutrients, including appearance and texture. However, they did not test the profiles of the volatile compounds present in the dried mushrooms in detail. For *Suillus granulatus*, a wild edible mushroom with a strong umami taste, the best drying method for preservation of taste was vacuum drying (Zhao et al., 2020).

In many cases, it was shown that processing changes the aroma profile of mushrooms, like that of boletes. Thomas (1973) showed that after drying or dehydration, it is possible to produce new volatile compounds, such as pyrazines, pyrroles, 3-(methylthio) propionaldehyde and others. Zhang et al. (2018) also indicated that the drying of boletes significantly changes the odour profile. Sensory description of dried boletes flavour identified five componential aromas, which are indicated as roasted, buttery, floral, smokey and woody (Zhang et al., 2020). The roasted flavour is related to pyrazines (2,5-dimethylpyrazine 2,6-dimethylpyrazine) and 2-thiophenecarboxaldehyde, a sulphur-containing aldehyde that gives a roasted or roasted meat note. The smoky note shows a positive correlation with phenols (Zhang et al., 2020). Zhuang et al. (2020) found that among the four dried bolete varieties, *Boletus aereus* contains the richest composition of volatile compounds. Boletes convective dried in hot-air at 70°C for 200 min, reaching a moisture content of about 6.5 per cent, vary in flavour parameters depending on the species. *Boletus edulis* achieved roasted (2-ethyl-3,5-dimethylpyrazine) and buttery (γ-hexalactone) flavour, while *B. aereus* shows woody note (cedrol) and *B. auripes* presents an evident floral (β-ionone) and smoky (4-ethyl-2-methoxy-phenol) aroma. In contrast, *B. rubellus* was seen to possess weak sensory attributes (Zhuang et al., 2020).

7.2 Cooking

Consumers cook mushrooms through a wide variety of methods, depending on tradition and cuisine, but mushroom cooking is practiced all over the world. The taste and flavour compounds and their precursors occur naturally in raw mushrooms, but utilising increasing temperature in cooking leads to their release and initiates changes of each compound, causing changes in flavour (bringing out the specific taste of cooked mushroom dishes). Thermal treatments enable an increase in the release of certain compounds and their further chemical or enzymatic reactions. As Sun et al. (2019) demonstrated, cooking improves the nutritional value of mushrooms due to the increase in the amounts of available macro- and micro-nutrients, including polysaccharides, polyphenols and amino acids. However, the effect of cooking on taste and flavour compounds varies, depending on the mushroom species. In this group of chemical compounds, free amino acids and 5'-nucleotides seem to be the most important. According to model tests made for *A. bisporus* by Rotola-Pukkila et al. (2019), 70°C is the recommended temperature for the most suitable taste and flavour components in cooked mushrooms. The same observations were made by Dermiki et al. (2013) and confirmed that mushroom extracts prepared in higher temperatures, such as 70°C, provide a better umami taste and higher levels of savoury taste due to 5'-ribonucleotides. During cooking, flavour compounds are divided between the mushroom body and the extract (sauce/extract) and at 70°C, the samples of mushroom may reach the highest concentration of total 5'-nucleotides and flavour 5'-nucleotides. Increasing temperature during cooking causes the total amount of free amino acids to decrease due to further chemical reactions, which also exhibits a strong effect on the taste. The same effect of cooking on the taste compounds is observed in the case of wild mushrooms, such as *Cantharellus cibarius, C. tubaeformis, Lactarius trivialis* and *Suillus variegates*, confirming the previous observations made for button mushrooms. Free amino acid content decreases and the umami-enhancing components change on increase in the cooking temperature. Cooking also reduces the amount of free amino acids that are responsible for the bitter taste in wild mushrooms. The umami-enhancing nucleotide 5'-GMP is not present in raw wild mushrooms, but occurs in raw cultivated mushroom, *Agaricus bisporus*. For wild mushroom species, 5'-GMP appears only in cooked samples and its concentration is mainly dependent on enzymatic activity during heating. Comparison of the wild mushroom species tested by Rotola-Pukkila et al. (2019) reveals that the highest concentration of 5'-GMP occurs in cooked *L. trivialis* and is even higher than in cooked *A. bisporus*. A separate analysis of the cooked mushrooms and cooking extract revealed that the distribution of the taste compounds is even in

Table 3. Chemical compounds responsible for the flavour and aroma and their descriptions. Source of the descriptions – The Good Scents Company Information System (http://www.thegoodscentscompany.com/index.html)

Component (in order of appearance in the text)	Flavours and odours
C8 derivatives	General mushroom flavour and scents
benzaldehyde	Odour: strong sharp sweet bitter almond cherry Flavour: sweet, oily, almond, cherry, nutty and woody
(R) – (-)-oct-1-en-3-ol	Flavour: fruity mushroom-like
(S) – (+)-oct-1-en-3-ol	Flavour: mouldy, grassy
indole	Odour: animal floral mothball fecal naphthalene Flavour: animal, fecal, naphthyl, with earthy, perfumey, phenolic and chemical nuances
dimethyl disulfide	Odour: sulfurous vegetable cabbage onion Flavour: sulfurous cabbage, malt, cream
p-cresol	Odour: phenolic narcissus animal mimosa Flavour: phenolic
1-octen-3-ol	Odour: mushroom earthy green oily fungal raw chicken Flavour: mushroom, earthy, fungal, green, oily, vegetative, umami sensation and savory brothy
(E)-2-octen-1-ol	Odour: green vegetable Flavour: fatty oily sweet fruity
(Z)-2-octen-1-ol	Odour: sweet floral
geranyl acetone	Odour: fresh rose leaf floral green magnolia aldehydic fruity Flavour: floral fruity tropical green pear apple banana citrus
1-octen-3-one	Odour: herbal mushroom earthy musty dirty Flavour: intense creamy earthy mushroom with fishy and vegetative nuances
3-octanone	Odour: fresh herbal lavender sweet mushroom Flavour: mushroom, ketonic, cheesy and moldy with a fruity nuance
(E)-2-octenal	Odour: fresh cucumber fatty green herbal banana waxy green leaf Flavour: sweet green citrus peel spicy cucumber oily fatty brothy
sesquiterpenes	cedar aroma (Dijkstra and Wikén, 1976; Fraatz and Zorn, 2010) Odour: orange peel
6-methyl-5-hepten-2-one	Flavour: citrus green musty lemongrass apple
methylpyrazine	Odour: nutty cocoa roasted chocolate peanut green Flavour: nutty brown roasted musty astringent
2,5-dimethylpyrazine	Odour: cocoa roasted nutty beefy roasted beefy woody grassy medicinal Flavour: musty potato cocoa nutty fatty oily
2,6-dimethylpyrazine	Odour: cocoa roasted nutty meaty roasted meaty coffee Flavour: nutty coffee cocoa musty bready meaty
2-ethyl-6-methylpyrazine	Not determined
trimethylpyrazine	Odour: nutty nut skin earthy powdery cocoa potato baked potato peanut roasted peanut hazelnut musty Flavour: raw nut skin vegetable cocoa toasted earthy chocolate coffee

Table 3 Contd. ...

...Table 3 Contd.

Component	Flavours and odours
2,3-dimethylpyrazine	Odour: nutty nut skin cocoa peanut butter coffee walnut caramellic roasted Flavour: nutty nut skin peanut cocoa coffee roasted coffee walnut corn chip bready
γ-valerolactone	Odour: herbal sweet warm tobacco cocoa woody Flavour: tonka coumarinic tobacco cocoa chocolate dark chocolate coconut
3-(methylthio) propionaldehyde	Odour: musty potato tomato earthy vegetable creamy Flavour: musty tomato potato vegetable moldy cheesy onion beefy brothy egg nog seafood
furfural	Odour: sweet woody almond bread baked Flavour: brown sweet woody bready nutty caramellic burnt astringent
2-thiophenecarboxaldehyde	Odour: sulfurous Flavour: almond bitter almond cherry
1H-pyrrole-2-carboxaldehyde	Odour: musty beefy coffee
p-anisaldehyde (4-methoxybenzaldehyde),	sweet powdery mimosa floral hawthorn balsamic Flavour: creamy powdery vanilla spicy marshmallow
benzyl acetate	Odour: sweet floral fruity jasmin fresh Flavour: fruity, sweet, with balsamic and jasmin floral undernotes
hexanal	fresh, green, fatty (Xua et al., 2019) Odour: fresh green fatty aldehydic grass leafy fruity sweaty Flavour: green, woody, vegetative, apple, grassy, citrus and orange with a fresh, lingering aftertaste
(E)-non-2-enal	Odour: cucumber odour with characteristic chicory and fenugreek notes fatty green cucumber aldehydic citrus Flavour: green soapy cucumber melon aldehydic fatty
acetophenone	Odour: sweet pungent hawthorn mimosa almond acacia chemical Flavour: powdery, bitter almond cherry pit-like with coumarinic and fruity nuances
2-pentanone	Odour: sweet fruity ethereal wine banana woody Flavour: sweet, fruity and banana-like with a fermented nuance
2-heptanone	Odour: sweet fruity ethereal wine banana woody Flavour: sweet, fruity and banana-like with a fermented nuance
2-nonanone	fresh, sweet, green (Xua et al. 2019) Odour: fresh sweet green weedy earthy herbal Flavour: cheesy, green, fruity, dairy, dirty, buttery
2,3-dimethylphenol	Odour: phenolic chemical musty Flavour: phenolic
2,4-dimethylphenol	Odour: smoky roasted Flavour: burnt roasted
3-ethylphenol	Odour: musty Flavour: phenolic burnt truffle

Table 3 Contd. ...

...Table 3 Contd.

Component	Flavours and odours
2,6-dimethoxyphenol	Odour: smoky phenolic balsamic bacon powdery woody Flavour: sweet medicinal creamy meaty vanilla spicy
2,5-dimethylphenol	Odour: sweet naphthyl phenolic smoky bacon Flavour: musty chemical stringent phenolic
2-methoxy-4-methylphenol	Odour: spicy clove vanilla phenolic medicinal leathery Flavour: vanilla spicy clove woody leathery chemical
4-ethyl-2-methoxyphenol	Odour: spicy smoky bacon phenolic clove Flavour: woody smoky spicy sweet vanilla
3-methylvaleric acid	Odour: animal sharp acidic cheesy green fruity sweaty Flavour: sour cheesy fresh fruity
pentanoic acid	Odour: acidic sweaty rancid Flavour: acidic dairy milky cheesy
tetradecanoic acid (myristic acid)	Odour: waxy fatty soapy coconut Flavour: waxy fatty soapy creamy cheesy
ethyl caproate (ethyl hexanoate)	Odour: sweet fruity pineapple waxy green banana Flavour: sweet pineapple fruity waxy banana green estery
2-methylpropionic acid	Odour: acidic sour cheesy dairy buttery rancid Flavour: acidic sour cheesy cheesy limburger cheese dairy creamy
limonene	Odour: citrus herbal terpenic camphoreous
N(2-phenyethyl) acetamide	Not determined
farnesyl acetone	Flavour: fruity wine floral creamy
(E,E)-farnesol	Odour: mild muguet floral sweet lily waxy
dihydro-β-ionone	earthy, woody, mahogany (Xua et al. 2019) Odour: earthy woody mahogany orris dry amber Flavour: woody, seedy, berry raspberry, with leafy, spicy nuances
3-octanol	Odour: earthy mushroom herbal melon citrus woody spicy minty Flavour: musty, mushroom, earthy, creamy dairy
γ-undecalactone	fruity, peach, creamy (Xua et al. 2019) Flavour: fruity peach creamy fatty lactonic apricot ketonic coconut
2-methylbutanal	musty, cocoa, coffee (Xua et al. 2019) Odour: musty cocoa coffee nutty Flavour: musty rummy nutty cereal caramellic fruity
2-phenylethanol	Odour: floral rose rose dried rose Flavour: floral sweet rose bready
1-phenylethanol	Odour: fresh sweet almond gardenia hyacinth Flavour: chemical medicinal balsamic vanilla woody
5-hydroxy-2-methyl-4-chromanone	Not determined
1,8-dimethoxynaphthalene	Not determined
4-methylhexan-3-one	Not determined, serves as pheromone of ants (Dhotare et al. 2000)
manicone ((E)-4,6-dimethyl-4-octen-3-one)	Pheromone of *Manica* spp. (Hymenoptera) (https://pherobase.com/database/compound/compounds-detail-manicone.php)

Table 3 Contd. ...

...Table 3 Contd.

Component	Flavours and odours
(4R,5R,6S)-5-hydroxy-4,6-dimethyloctan-3-one	Not determined
2-acetylfuran	Odour: sweet balsamic almond cocoa caramellic coffee Flavour: sweet nutty roasted sweet baked
2-acetylthiazole	Odour: nutty popcorn peanut roasted peanut hazelnut Flavour: corn chip musty
2,4,6-trimethylpyridine	Not determined
2-pentylfuran	Odour: fruity green earthy beany vegetable metallic Flavour: green waxy musty cooked caramellic
3-phenylfuran	Flavour: caramel-like (Chen et al. 2018)
2-octylfuran	Flavour: caramel-like (Chen et al. 2018)
2,5-thiophene-dicarboxaldehyde	Not determined
3-methylbutanal	Odour: ethereal aldehydic chocolate peach fatty Flavour: fruity dry green chocolate nutty leafy cocoa
benzeneacetaldehyde	Odour: green sweet floral hyacinth clover honey cocoa Flavour: honey sweet floral chocolate cocoa spicy
isovaleric acid	Odour: sour stinky feet sweaty cheese tropical Flavour: cheesey, dairy, creamy, fermented, sweet, waxy and berry
octanal	Odour: aldehydic waxy citrus orange peel green fatty Flavour: aldehyde, green with a peely citrus orange note
nonanal	Odour: waxy aldehydic rose fresh orris orange peel fatty peely Flavour: effervescent, aldehydic citrus, cucumber and melon rindy, with raw potato and oily nutty and coconut like nuances
2-propenoic acid-2-ethylhexyl ester	Not determined
glycerol tricaprylate	Not determined
2-undecanone	Odour: waxy fruity creamy fatty orris floral Flavour: waxy, fruity with creamy cheese like notes
naphthalene	Odour: pungent dry tarry
dimethyl disulphide,	sulfurous vegetable cabbage onion Flavour: sulfurous cabbage malty creamy
dimethyl sulphide	Odour: sulfurous onion sweet corn vegetable cabbage tomato green radish Flavour: sulfurous vegetable tomato corn asparagus dairy creamy minty
2-methyl-1-propanol (isobutanol)	Odour: ethereal winey Flavour: fusel whiskey
bis-(methylthio)-methane (truffle sulfide)	Odour: garlic sulfurous green spicy mushroom Flavour: alliaceous sulfurous fresh onion garlic vegetable cabbage spicy mustard horseradish
2-methyl-butanal	Odour: musty cocoa coffee nutty Flavour: musty rummy nutty cereal caramellic fruity

Table 3 Contd. ...

...Table 3 Contd.

Component	Flavours and odours
2-methylbutanol	roasted winey onion fruity fusel alcoholic whiskey
3-(methylthio) propanal	Odour: musty potato tomato earthy vegetable creamy Flavour: musty tomato potato vegetable moldy cheesy onion beefy brothy egg nog seafood
3-methyl-4,5-dihydrothiophene	Not determined
1-propanethiol	Odour: cabbage gassy sweet onion Flavour: alliaceous fresh meaty onion sulfurous
phenylacetaldehyde	Odour: green sweet floral hyacinth clover honey cocoa Flavour: honey sweet floral chocolate cocoa spicy
1,2,3,5,6-pentathiepane	Not determined
1,2,4,5-tetrathiane	Not determined
1,2,4,6-tetrathiepane	Not determined
1,3-dithiethane	Not determined
eucalyptol	Odour: eucalyptus herbal camphoreous medicinal Flavour: minty camphoreous cooling eucalyptus medicinal
undecanal	Odour: waxy soapy floral aldehydic citrus green fatty fresh laundry Flavour: waxy, aldehydic, soapy with a citrus note and slight laundry detergent nuance
2,4-bis(1,1-dimethylethyl)-phenol	Not determined
(E)-2-decenal	Odour: waxy fatty earthy coriander green mushroom aldehydic Flavour: waxy fatty earthy coriander mushroom green pork fatty
2-phenylpropenal	Odour: sweet spicy aldehydic aromatic balsamic cinnamyl resinoushoney powdery Flavour: spicy sweet aromatic aldehydic honey cinnamyl resinous
methyl butyrate,	Odour: fruity apple sweet banana pineapple Flavour: impacting, fusel, fruity and estry with a cultured dairy, acidic depth
ethyl propanoate,	Odour: sweet fruity rummy juicy fruity grape pineapple Flavour: ethereal fruity sweet winey bubble gum apple grape
ethyl isobutyrate,	Odour: sweet ethereal fruity alcoholic fusel rummy Flavour: pungent, etherial and fruity with a rum-and egg nog-like nuance
ethyl 2-methylbutyrate	Odour: sharp sweet green apple fruity Flavour: fruity fresh berry grape pineapple mango cherry
ethyl isovalerate	Odour: fruity sweet apple pineapple tutti frutti Flavour: sweet, fruity, spice, metallic and green with a pineapple and apple iift
2-(methylthio)ethanol	Odour: sulfurous meaty
2,4-dithiapentane	Odour: garlic sulfurous green spicy mushroom Flavour: alliaceous sulfurous fresh onion garlic vegetable cabbage spicy mustard horseradish

Table 3 Contd. ...

...Table 3 Contd.

Component	Flavours and odours
methyl-allyl-disulfide	Odour: alliaceous garlic onion green onion
2-ethyl-3,5-dimethylpyrazine	Odour: roasted burnt almond roasted nutty coffee Flavour: sweet nutty caramellic coffee corn cocoa potato
γ-hexalactone	buttery (Zhuang et al. 2020) Odour: herbal coconut sweet coumarin tobacco Flavour: sweet, creamy, vanilla-iike with green lactonic powdery nuances
cedrol	woody note (Zhuang et al. 2020) Odour: cedarwood woody dry sweet soft Flavour: woody amber floral cedar ambrette musk powdery
β-ionone	Floral (Zhuang et al. 2020) Odour: floral woody sweet fruity berry tropical beeswax Flavour: woody, berry, floral, green and fruity
4-ethyl-2-methoxy-phenol	Odour: smoky (Zhuang et al. 2020)
2,5-thiophenedicarboxaldehyde	Flavour: meat-like (Chen et al. 2018)
(Z,E)-α-farnesene	Not determined for that isomere
α-bisabolol	Odour: floral peppery balsamic clean
(+)-valencene	Odour: sweet fresh citrus grapefruit woody orange dry green oily Flavour: orange citrus fruity juicy citrus peel peely woody
methyl cinnamate	Odour: sweet balsam strawberry cherry cinnamon Flavour: cinnamyl balsamic spicy fruity mango papaya cherry
raspberry ketone (4-(4-hydroxyphenyl)butan-2-one	Odour: sweet berry jam raspberry ripe floral Flavour: fruity, jammy, berry, raspberry and blueberry with seedy, cotton candy nuances

both at a cooking temperature of 90°C. However, cultivated species vary in content of taste components in raw as well as in cooked samples. So also the wild mushrooms showed strong differences in composition of taste components; *L. trivialis* and *S. variegatus* showed higher concentrations of taste-enhancing compounds compared to *Cantharellus cibarius* and *C. tubaeformis* (Rotola-Pukkila et al., 2019).

7.3 Freezing

Low temperatures are used to freeze food products or freeze dry them, which allows for long-term preservation in a nearly fresh form. However, as Campo et al. (2017) suggest, any freezing steps involve the presence of off-odour methional. Freezing is used as a method of choice for mushroom preservation as well to keep them fresh and ready to use easily, in contrast to freeze-dried mushrooms, which need to be sliced and dehydrated. Nonetheless, freeze-drying makes it possible to preserve the aroma profile of the mushroom so that it smells nearly the same as its fresh counterparts. For example, in freeze-dried truffles (*Tuber melanosporum*) dimethyl-sulphide and dimethyl-disulphide are identified, just as in the aroma of fresh truffles (Campo et al., 2017). However, as Campo et al. (2017) proved, freeze-dried truffles still vary from those that are frozen or canned. Freezing leads to the degradation of nutritional components, especially vitamins and minerals, including changes in sensory characteristics, although some special pre-treatments, particularly blanching, are a crucial requirement in mushroom preparation for storing them frozen (Jaworska and Bernas, 2009; Bernas and Jaworska, 2016).

8. Are Mushroom Flavours Useful?

Mushrooms have been suggested as producers of many industrially and medically valuable substances, such as proteins and carbohydrates. They are also considered as a possible source of aromas and flavours (Table 2). Thus, several patents have been reported for fungi, including mushrooms, for methods and processes of spice production and food preparation. Typically, the mushroom compounds, oct-1-en-3-ol and octan-3-one, may be produced via lipid degradation in submerged cultures by many mushrooms. Woźniak (2007) described a procedure for cultivating the mycelia of three varieties of King Bolete. She recommended production of mycelium first, followed by drying the mycelial biomass for use as a food supplement, instead of using dried wild mushrooms. This recommendation seems to be advantageous although the concentration of C8 compounds in cultured mycelium is significantly lower than in fresh fruiting bodies. The same effect was observed by Dijkstra et al. (1972) for *Agaricus bisporus* grown in liquid cultures. They designed a medium for submerged cultures using malt extract, phosphate and casein and achieved a moderate level of growth on media containing glucose, asparagine, phenylalanine, vitamins and minerals. Mycelial growth was stimulated by vegetable oils and oleates and obtained samples of mycelium showed the same flavour as fresh mushrooms, although weaker. On the other hand, liquid cultures of *Pleurotus ostreatus* mycelia exhibit a broad composition of flavour substances, depending on the age and culture conditions, although oct-1-en-3-ol and octan-3-one are the predominant compounds. Cultures of *Pleurotus florida* (=*P. ostreatus* cf. *florida*) produce fruiting bodies with a sweet anise and almond-like odour owing to the synthesis of *p*-anisaldehyde and benzaldehyde, and the natural compound of *P. florida,* oct-1-en-3-ol, is present only in minute quantities (Taupp et al., 2008; Fraatz and Zorn, 2010).

Mushroom mycelia demonstrate a significant availability of umami ingredients, which makes them a good source of these compounds in artificial cultures and as mushroom substitutes in food production (Zhang et al., 2013). Free amino acids released from mushrooms may be used as enhancers of umami taste, as shown by Poojary et al. (2017). They successfully used a specific enzymatic extraction procedure from six different mushrooms, including shiitake (*L. edodes*), oyster (*P. ostreatus*), tea tree mushroom (*Agrocybe aegerita*) and white, brown and portobello champignons (*A. bisporus*) to obtain free amino acids, including those responsible for the umami taste. *Agaricus bisporus* (both white and brown varieties) is the best source of umami taste, which is generally the alternative option to using monosodium glutamate in the food industry (Poojary et al., 2017). The tarragon oyster mushroom (*P. euosmus*) produces an intense sweet, flowery odour when linalool and coumarin are added as the main ingredients in culture (Drawert et al., 1983). Chen et al. (2018) demonstrated that mushroom hydrolysates could be used as sources of flavour and taste compounds obtained in the Maillard reaction. Changes in chemical composition seen in relatively high temperature yielded typical mushroom compounds, like 1-octen-3-ol, 2-octen-1-ol and nonanal. Heating of mushroom hydrolysates also allowed researchers to identify nitrogen- and sulphur-containing compounds, such as 3-phenylfuran and 2-octylfuran (caramel-like flavour), as well as 2-thiophene-carboxaldehyde and 2,5-thiophenedicarboxaldehyde (meat-like flavours). Their experiment showed that mushroom flavour and umami taste composition may be enriched by heating and through initiation of the Maillard reaction on mushroom hydrolysate (Chen et al., 2018).

Basidiomycetes may also be considered a good source of sesquiterpenes, although not many mushrooms have been explored as such. *Schizophyllum commune* was recognised for its ability to produce several sesquiterpenes, such as (Z,E)-α-farnesene, ar-curcumene and α-bisabolol (Ziegenbein et al., 2006; Scholtmeijer et al., 2014). The *S. commune* possesses the mevalonate pathway required for the synthesis of the farnesyl diphosphate, a substrate for sesquiterpene production. The *thn* mutant produces less schizophyllan and proteins that hamper isolation and purification of homologous and heterologous products. The *thn* mutation also causes several morphological changes, such as the production of a pungent smell, a higher radial growth rate and a lower biomass. Introduction of a valencene synthase gene in the *thn* mutant strains resulted in a four-fold increase in production of the sesquiterpene (+)-valencene, both in mycelium and in fruiting bodies, with no schizophyllan production compared to wild-type transformants. In addition, due to the absence of schizophyllan, the n-dodecane isolation from the culture medium increased six-fold to seven-fold in the *thn* mutant strains (Scholtmeijer et al., 2014). Results of the research by Scholtmeijer et al. (2014) gave quite a new biotechnological source of aromatic compounds and sesquiterpene drugs, displaying a new avenue for future investigation in Basidiomycetes.

Mushroom substances produced biotechnologically have dual applications in food as well as pharmaceutical industries. Currently, there is a huge demand for melanin inhibitors in chemotherapy, food and cosmetic products preservation owing to intense melanin formation which is often a problem in chemotherapies or preservation. As Satooka et al. (2017) demonstrated, methyl cinnamate inhibits enzymatic and cellular melanin formation in murine B16-F10 melanoma cells without affecting the cell growth. Methyl cinnamate, 1-octen-3-ol and indole were recognised in the European *Tricholoma caligatum* and methyl cinnamate is also one of the constituents of volatile compounds in mycelium and sporocarp of American matsutake (*Tricholoma magnivelare*) (Wood and Lefevre, 2007; Satooka et al., 2017). Another example of a flavour compound which can be obtained from Basidiomycete culture is a raspberry ketone (4-(4-hydroxyphenyl) butan-2-one), which is produced from L-phenylalanine. The source for this compound is *Nidula niveo-tomentosa* (a bird's nest fungus) grown in submerged cultures. Biotechnological growth of mycelia also yields betuligenol. These compounds are used as food flavours as well as diet ingredients (Taupp et al., 2008; Moliszewska, 2014). Growers of fungi in artificial media should proceed carefully with properly composed media and growth conditions of culture for retention of flavour or aroma (chemical and physical conditions). For example, the production of a culture of raspberry compounds by *N. niveo-tomentosa* requires periodic illumination as researched by Taupp et al. (2008) (Table 2).

9. Conclusion and Future Directions

Mushrooms serve as potential alternate sources of flavour, fragrance, aroma, odour and scent to plant-derived products. These products constitute raw materials in food and in the pharmaceutical and cosmeceutical industries. Explicit study of mushroom volatile compounds and their precise chemistry responsible for properties, such as flavour and taste, provides the basics for their broad use in several applications. Some of the chemical constituents in mushrooms contributing to their flavour include C8 volatiles, amino acids and nucleotides. Mushrooms also possess pleasant flavours, like those of garlic, coconut and fruit owing to their precise chemical components (Fig. 1). Nevertheless, mushrooms are also known to produce unpalatable flavour and odour to attract insects and other animals, which perpetuate and disseminate their spores in order to broaden their territory. These properties differ from mushroom to mushroom, with variables such as geographic conditions, wild versus cultivated mushrooms, techniques of cultivation, different parts of the fruiting body, cultured mycelia, methods of processing and methods of preservation. Studies on flavours and odours of mushrooms need basic knowledge of chemistry of specific compounds and their application in different industries. Intriguingly, flavour compounds of mushrooms may also be useful to attract specific insects or animals (like pheromones), which have potential applications in biodiversity studies as well as to understand the natural chemical cues between mushrooms and animals. Greater knowledge about mushroom aromatic compounds and their chemical components has far-reaching importance and such knowledge can be applied to biotechnological innovations in several industries.

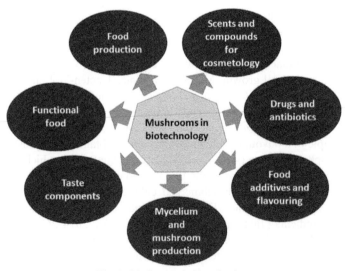

Fig. 1. Mushrooms in biotechnology

References

Amino, Y., Nakazawa, M., Kaneko, M., Miyaki, T., Miyamura, N. et al. (2016). Structure-CaSR-activity relation of kokumi γ-glutamyl peptides. Chem. Pharmaceut. Bull., 64(8): 1181–1189.

Aprea, E., Romano, A., Betta, E., Biasioli, F., Cappellin, L. and Fanti, M. (2015). Volatile compound changes during shelf life of dried *Boletus edulis*: Comparison between SPME-GC-MS and PTR-ToF-MS analysis. J. Mass Spectrom., 50(1): 56–64.

Berendsen, R.L., Kalkhove, S.I.C., Lugones, L.G., Baars, J.J.P., Wösten, H.A.B. and Bakker, P.A.H.M. (2013). Effects of the mushroom-volatile 1-octen-3-ol on dry bubble disease. Appl. Microbiol. Biotechnol., 97: 5535–5543. 10.1007/s00253-013-4793-1.

Bernas, E. and Jaworska, G. (2016). Vitamins profile as an indicator of the quality of frozen *Agaricus bisporus* mushroom. J. Food Comp. Anal., 49: 1–8.

California Fungi – *Hygrophorus agathosmus*. http://www.mykoweb.com/CAF/species/Hygrophorus_agathosmus.html (Accessed, June 2014).

Campo, E., Marco, P., Oria, R., Blanco, D. and Venturini, M.E. (2017). What is the best method for preserving the genuine black truffle (*Tuber melanosporum*) aroma? An olfactometric and sensory approach. LWT, 80: 84–91. 10.1016/j.lwt.2017.02.009.

Chen, X., Yu, J., Cui, H., Xia, S., Zhang, X. and Yang, B. (2018). Effect of temperature on flavor compounds and sensory characteristics of Maillard reaction products derived from mushroom hydrolysate. Molecules, 23: 247. doi:10.3390/molecules23020247.

Chun, S.S., Chambers, E. and Han, I. (2020). Development of a sensory flavour lexicon for mushrooms and subsequent characterization of fresh and dried mushrooms. Foods, 9: 980. 10.3390/foods9080980.

Dermiki, M., Phanphensophon, N., Mottram, D.S. and Methven, L. (2013). Contributions of non-volatile and volatile compounds to the umami taste and overall flavour of shiitake mushroom extracts and their application as flavour enhancers in cooked minced meat. Food Chem., 141: 77–83. 10.1016/j.foodchem.2013.03.018.

Dhotare, B., Hassarajani, S.A. and Chattopadhyay, A. (2000). Convenient synthesis of (3R, 4S)-4-Methyl-3-hexanol and(S)-4-Methyl-3-hexanone, the pheromones of ants. Molecules, 5: 1051–1054.

Dijkstra, F.I.J., Scheffers, W.A. and Wikien T.O. (1972). Submerged growth of the cultivated mushroom, *Agaricus bisporus*. Ant. Leeuwenh., 38: 329–340.

Dijkstra, F.Y. and Wikén, T.O. (1976). Studies on mushroom flavours. ZeitschriftfürLebensmittel-Untersuchung und Forschung, 160: 255–262.

Domsch, K.H., Gams, W. and Anderson, T.-H. (1980). Compendium of soil fungi. vol. 1, Academic Press, London.

Drawert, F., Berger, R.G. and Neuhäuser, K. (1983). Biosynthesis of flavor compounds by microorganisms 4. Characterization of the major principles of the odor of *Pleurotus euosmus*. Eur. J. Appl. Microbiol. Biotechnol., 18: 124–127.

Dudekula, U.T., Doriya, K. and Devarai, S.K. (2020). A critical review on submerged production of mushrooms and their bioactive metabolites. 3 Biotech., 10: 337. 10.1007/s13205-020-02333-y.

Feng, T., Shui, M., Song, S., Zhuang, H., Sun, M. and Yao, L. (2019). Characterisation of the key aroma compounds in three truffle varieties from China by flavouromics approach. Molecules, 24: 3305. 10.3390/molecules24183305.

Fraatz, M.A. and Zorn, H. (2010). Fungal flavours. pp. 249–268. *In*: Hofrichter, M. (ed.). The Mycota X, Industrial Applications. 2nd ed., Springer-Verlag, Berlin, Heidelberg.

Guevara, G., Bonito, G., Trappe, J.M., Czares, E., Williams, G. et al. (2013). New North American truffles (*Tuber* spp.) and their ectomycorrhizal associations. Mycologia, 105(1): 194–209. 10.3852/12-087.

Jaklitsh, W.M., Stadler, M. and Voglmayr, H. (2012). Blue pigment in *Hypocreacaerulescens* sp. nov. and two additional new species in sect. *Trichoderma*. Mycologia, 104(4): 925–941. 10.3852/11-327.

Jaworska, G. and Bernas, E. (2009). The effect of preliminary processing and period of storage on the quality of frozen *Boletus edulis* (Bull: Fr.) mushrooms. Food Chem., 113: 936–943.

Johnson, S.D. and Jürgens, A. (2010). Convergent evolution of carrion and faecal scent mimicry in fly-pollinated angiosperm flowers and a stinkhorn fungus. South Afr. J. Bot., 76: 796–807.

Johnson, S.D. and Jürgens, A. (2010). Evolution of carrion and faecal scent mimicry in fly-pollinated angiosperm flowers and a stinkhorn fungus. South Afr. J. Bot., 76: 796–807.

Kakumyan, P., Suwannarach, N., Kumla, J., Saichana, N., Lumyong, S. and Matsui, K. (2020). Determination of volatile organic compounds in the stinkhorn fungus *Pseudocolus fusiformis* in different stages of fruiting body formation. Mycoscience, 61: 65–70.

Kasper-Pakosz, R., Pietras, M. and Łuczaj, Ł. (2016). Wild and native plants and mushrooms sold in the open-air markets of southeastern Poland. J. Ethnobiol. Ethnomed., 12: 45. 10.1186/s13002-016-0117-8.

Lauterbach, L., Wang, T., Stadler, M. and Dickschat, J.S. (2019). Volatiles from the ascomycete *Daldinia* cf. *childiae* (Hypoxylaceae), originating from China. Med. Chem. Commun., 10: 726–734.

Lee, H., Nam, K., Zahra, Z. and Farooqi, M.Q.U. (2020). Potentials of truffles in nutritional and medicinal applications: A review. Fungal Biol. Biotechnol., 7: 9. 10.1186/s40694-020-00097-x.

Li, S., Wang, A., Liu, L., Tian, G., Wei, S. and Xu, F. (2018a). Evaluation of nutritional values of shiitake mushroom (*Lentinus edodes*) stipes. J. Food Measure. Character., 12: 2012–2019. 10.1007/s11694-018-9816-2.

Li, W., Chen, W., Yang, Y., Zhang, J., Feng, J. et al. (2017). Effects of culture substrates on taste component content and taste quality of *Lentinula edodes*. Int. J. Food Sci. Technol., 52: 981–991.

Li, X., Guo, Y., Zhuang, Y., Qin, Y. and Sun, L. (2018b). Nonvolatile taste components, nutritional values, bioactive compounds and antioxidant activities of three wild Chanterelle mushrooms. Int. J. Food Sci. Technol., 53: 1855–1864.

Maga, J. (1981). Mushroom flavor. J. Agric. Food Chem., 29(1): 4–7.

Maggi, F., Bílek, T., Lucarini, D., Papa, F., Sagratini, G. and Vittori, S. (2009). *Melittismelissophyllum* L. subsp. *melissophyllum* (Lamiaceae) from central Italy: A new source of a mushroom-like flavor. Food Chem., 113: 216–221.

Miyamura, N., Kuroda, M., Mizukoshi, T., Kato, Y., Yamazaki, J. et al. (2015). Distribution of a kokumi peptide, γ-Glu-Val-Gly, in various fermented foods and the possibility of its contribution to the sensory quality of fermented foods. Ferm. Technol., 4(2): 1000121. 10.4172/2167-7972.1000121.

Moliszewska, E. (2014). Mushroom flavour. Acta Universitatis Lodziensis, Folia Biol. Oecol., 10: 80–88.

Muszyńska, B., Sułkowska-Ziaja, K. and Wójcik, A. (2013). Levels of physiological active indole derivatives in the fruiting bodies of some edible mushrooms (Basidiomycota) before and after thermal processing. Mycoscience, 54: 321–326.

Muszyńska, B., Mastej, M. and Sułkowska-Ziaja, K. (2016). Biological function of carotenoids and their occurrence in the fruiting bodies of mushrooms. Med. Int. Rev., 27(107): 113–122.

Nöfer, J., Lech, K., Figiel, A., Szumny, A. and Carbonell-Barrachina, Á.A. (2018). The influence of drying method on volatile composition and sensory profile of *Boletus edulis*. Journal of Food Quality, https://doi.org/10.1155/2018/2158482.

Ohsu, T., Amino, Y., Nagasaki, H., Yamanaka, T., Takeshita, S. et al. (2010). Involvement of the calcium-sensing receptor in human taste perception. J. Biol. Chem., 285(2): 1016–1022.

Pal, P., Singh, A.K., Kumari, D., Rahul, R., Pandey, J.P. and Sen, G. (2017). Study of biochemical changes on freeze-dried and conventionally dried white button mushroom as a sustainable method of food preservation. pp. 149–155. *In*: Mukhopadhyay, K., Sachan, A. and Kumar, M. (eds.). Applications of Biotechnology for Sustainable Development. Springer Nature, Singapore. 10.1007/978-981-10-5538-6-18.

Path, C., Moon, BoK. and Lee, C. (2016). Evaluation of umami taste in mushroom extracts by chemical analysis, sensory evaluation, and an electronic tongue system. Food Chem., 192: 1068–1077. 10.1016/j.foodchem.2015.07.113.

Pelusio, F., Nilsson, T., Montanarella, L., Tilio, R., Larson, B. et al. (1995). Headspace solid-phase microextraction analysis of volatile organic sulfur compounds in black and white truffle aroma. J. Agric. Food Chem., 43: 2138–2143.

Pildain, M.B., Coetzee, M.P.A., Wingfield, B.D., Wingfield, M.J. and Rajchenberg, M. (2010). Taxonomy of *Armillaria* in the Patagonian forests of Argentina. Mycologia, 102(2): 392–403. 10.3852/09-105.

Poojary, M.M., Orlien, V., Passamonti, P. and Olsen, K. (2017). Enzyme-assisted extraction enhancing the umami taste amino acids recovery from several cultivated mushrooms. Food Chem., 234: 236–244.

Poucher, W.A. (1991). Poucher's perfumes, cosmetics and soaps, vol. 1. The Raw Materials of Perfumery, 9th ed., and revised by A.J. Jouhar, Springer Science + Business Media, Dordrecht.

Qin, L., Gao, J.-X., Xue, J., Chen, D., Lin, S.-Y. et al. (2020). Changes in aroma profile of shiitake mushroom (*Lentinus edodes*) during different stages of hot air drying. Foods, 9: 444. 10.3390/foods9040444.

Rapior, S., Talou, T., Pélissier, Y. and Bessiére, J.-M. (2002). The anise-like odour of *Clitocybe odora, Lentinellus cochleatus* and *Agaricu essettei*. Mycologia, 94(3): 373–376.

Rathore, H., Prasad, S. and Sharma, S. (2017). Mushroom nutraceuticals for improved nutrition and better human health: A review. Pharma Nutrition, 5: 35–46.

Reineccius, G. (1994). Source Book of Flavours. 2nd ed., Springer Science + Business Media, Dordrecht.

Rifai, Y. (1969). A revision of the genus *Trichoderma*. Mycol. Pap., 116: 1–56.

Rotola-Pukkila, M., Yang, B. and Hopia, A. (2019). The effect of cooking on umami compounds in wild and cultivated mushrooms. Food Chem., 278: 56–66.

Satooka, H., Cerda, P., Kim, H.J., Wood, W.F. and Kubo, I. (2017). Effects of matsutake mushroom scent compounds on tyrosinase and murine B16-F10 melanoma cells. Biochem. Biophys. Res. Com., 487: 840–846.

Schmidberger, P.C. and Schieberle, P. (2020). Changes in the key aroma compounds of raw shiitake mushrooms (*Lentinula edodes*) induced by pan-frying as well as by rehydration of dry mushrooms. J. Agric. Food Chem., 68: 4493–4506.

Scholtmeijer, K., Cankar, K., Beekwilder, J., Wösten, H.A.B., Lugones, L.G. and Bosch, D. (2014). Production of (+)-valencene in the mushroom-forming fungus *S. commune*. Appl. Microbiol. Biotechnol., 98: 5059–5068. 10.1007/s00253-014-5581-2.

Snowarski, M. (2005). Atlas grzybów (in Polish), Wyd. Pascal, Bielsko-Biała.

Snowarski, M. (2010). Spotkania z przyrodą. Grzyby (in Polish). MULTICO Oficyna Wydawnicza, Warszawa.

Splivallo, R., Deveau, A., Valdez, N., Kirchhoff, N., Frey-Klett, P. and Karlovsky, P. (2015). Bacteria associated with truffle-fruiting bodies contribute to truffle aroma. Environmental Microbiology, 17(8): 2647–2660. 10.1111/1462-2920.12521.

Sun, L., Zhang, Z., Xin, G., Sun, B., Bao, X. et al. (2020). Advances in umami taste and aroma of edible mushrooms. Tr. Food Sci. Technol., 96: 176–187.

Sun, Y., Lv, F., Tian, J., Ye, X. Qian, Chen, J. and Sun, P. (2019). Domestic cooking methods affect nutrient, phytochemicals, and flavour content in mushroom soup. Food Sci. Nutr., 7: 1969–1975.

Sun, Y., Lv, F., Tian, J., Ye, X., Chen, J. and Sun, P. (2019). Domestic cooking methods affect nutrient, phytochemicals, and flavor content in mushroom soup. Food. Sci. Nutr., 7: 1969–1975. 10.1002/fsn3.996.

Svanberg, I. and Lindh, H. (2019). Mushroom hunting and consumption in twenty-first century post-industrial Sweden. J. Ethnobiol. Ethnomed., 15: 42. 10.1186/s13002-019-0318-z.

Taupp, D.E., Nimtz, M., Berger, R.G. and Zorn, H. (2008). Stress response of Nidulaniveo-tomentosa to UV-A light. Mycologia, 100(4): 529–538. 10.3852/07-179R.

Thakur, M.P. (2020). Advances in mushroom production: Key to food, nutritional and employment security: A review. Ind. Phytopathol., 73: 377–395. 10.1007/s42360-020-00244-9.

Thomas, A.F. (1973). An analysis of the flavour of the dried mushroom, *Boletus edulis*. J. Agric. Food Chem., 21: 955–958. 10.1021/jf60190a032.

Tisdale, T.E., Miyasaka, S.C. and Hemmes, D.E. (2006). Cultivation of the oyster mushroom (*Pleurotus ostreatus*) on wood substrates in Hawaii. World J. Microbiol. Biotechnol., 22: 201–206. 10.1007/s11274-005-9020-5.

Trappe, M.J., Trappe, J.M. and Bonito, G.M. (2010). *Kalapuya brunnea* gen & sp. nov. and its relationship to the other sequestrate genera in *Morchellaceae*. Mycologia, 102(5): 1058–1065. 10.3852/09-323.

Tsai, S.-Y., Huang, S.-J., Lo, S.-H., Wu, T.-P., Lian, P.-Y. and Mau, J.-L. (2009). Flavour components and antioxidant properties of several cultivated mushrooms. Food Chem., 113: 578–584.

Varghese, R., Dalvi, Y.B., Lamrood, P.Y., Shinde, B.P. and Nair C.K.K. (2019). Historical and current perspectives on therapeutic potential of higher Basidiomycetes: An overview. 3 Biotech., 9: 362. 10.1007/s13205-019-1886-2.

Wang, W., Zhou, X. and Liu, Y. (2020). Characterisation and evaluation of umami taste: A review. Tr. Anal. Chem., 127. 115876. 10.1016/j.trac.2020.115876.

Wood, W.F., Smith, J., Wayman, K. and Largent, D.L. (2003). Indole and 3-chloroindole: The source of the disagreeable odour of *Hygrophorus paupertinus*. Mycologia, 95(5): 807–808.

Wood, W.F. and Lefevre, C.K. (2007). Changing volatile compounds from mycelium and sporocarp of American matsutake mushroom, *Tricholoma magnivelare*. Biochem. Syst. Ecol., 35(9): 634–636.

Woźniak, W. (2007). A characteristic of mycelium biomass of edible boletus. Acta Mycol., 42(1): 129–140.

Xua, X., Xua, R., Jiaa, Q., Fenga, T., Huang, Q. et al. (2019). Identification of dihydro-β-ionone as a key aroma compound in addition to C8 ketones and alcohols in *Volvariella volvacea* mushroom. Food Chemistry, 293: 333–339. 10.1016/j.foodchem.2019.05.004.

Yang, R., Li, Q. and Hu, Q. (2020). Physicochemical properties, microstructures, nutritional components, and free amino acids of *Pleurotus eryngii* as affected by different drying methods. Sci. Rep., 10: 121. 10.1038/s41598-019-56901-1.

Yin, Ch., Fan, X., Fan, Z., Shi, D., Yao, F. and Gao, H. (2019). Comparison of non-volatile and volatile flavor compounds in six *Pleurotus* mushrooms. J. Sci. Food Agric., 99: 1691–1699. 10.1002/jsfa.9358.

Zawirska-Wojtasiak, R. (2004). Optical purity of (R)-(–)-1-octen-3-ol in the aroma of various species of edible mushrooms. Food Chem., 86(1): 223–118.

Zeppa, S., Gioacchini, A.M., Guidi, C., Guescini, M., Pierleoni, R. et al. (2004). Determination of specific volatile organic compounds synthesised during *Tuber borchii* fruit body development by solid-phase microextraction and gas chromatography/mass spectrometry. Rapid Com. Mass Spectrom., 18: 199–205.

Zhang, H., Huanga, D., Pu, D., Zhanga, Y., Chen, H. et al. (2020). Multivariate relationships among sensory attributes and volatile components in commercial dry porcini mushrooms (*Boletus edulis*). Food Res. Int., 133: 109112. 10.1016/j.foodres.2020.109112.

Zhang, K., Pu, Y.-Y. and Sun, D.-W. (2018). Recent advances in quality preservation of postharvest mushrooms (*Agaricus bisporus*): A review. Tr. Food Sci. Technol., 78: 72–82. 10.1016/j.tifs.2018.05.012.

Zhang, Y., Venkitasamy, Ch., Pan, Z. and Wang, W. (2013). Recent developments on umami ingredients of edible mushrooms—A review. Tr. Food Sc. Technol., 33: 78–92. 10.1016/j.tifs.2013.08.002.

Zhao, X., Wei, Y., Gong, X., Xu, H. and Xin, G. (2020). Evaluation of umami taste components of mushroom (*Suillus granulatus*) of different grades prepared by different drying methods. Food Sci. Hum. Wellness, 9: 192–198. 10.1016/j.fshw.2020.03.003.

Zhao, Y., Bi, J., Yi, J., Jin, X., Wu, X. and Zhou, M. (2019). Evaluation of sensory, textural, and nutritional attributes of shiitake mushrooms (*Lentinula edodes*) as prepared by five types of drying methods. J. Food Proc. Eng., 42: e13029. 10.1111/jfpe.13029.

Zhuang, J., Xiao, Q., Fenga, T., Huang, Q., Ho, C.-T. and Song, S. (2020a). Comparative flavor profile analysis of four different varieties of *Boletus* mushrooms by instrumental and sensory techniques. Food Res. Int., 136: 109485. 10.1016/j.foodres.2020.109485.

Ziegenbein, F.C., Hanssen, H.-P. and König, W.A. (2006). Chemical constituents of the essential oils of three wood-rotting fungi. Flavour. Frag. J., 21: 813–816.

Immunoceuticals

16

Macrofungal Polysaccharides as Immunoceuticals in Cancer Therapy

Sujata Chaudhuri[1],* and *Hemanta Kumar Datta*[2]

1. INTRODUCTION

Fungi have traditionally been used as medicine as early as 3000 BC. They were regarded as a valuable resource of natural antibiotics and other biologically active compounds, especially after the discovery of penicillin in 1929. In several Asian countries, mushrooms are valued as edible and medicinal resources. Macrofungi, such as *Ganoderma lucidum*, *Lentinus edodes* and *Tremella fuciformis* have been used for hundreds of years and extensively cultivated. Most of our knowledge about mushrooms as food and medicinal agents comes from these species. Mushrooms are currently popular as food worldwide, with increasing awareness of them constituting a healthy diet. They are gluten-free with low fat content and contain polysaccharides, glycoproteins, proteoglycans, terpenoids, fatty acids, niacin, potassium, riboflavin, selenium, vitamin D and dietary fibres besides possessing low calories.

The cell wall of mushrooms contains different types of polysaccharides with various biological properties, such as antitumor, antimicrobial, antidiabetic, antioxidant and immune-modulating activities. The biologically active constituent molecules of the macro fungal organelles and their secondary metabolites are an area of intense interest and research. Several major substances with immune-modulating and antitumor activities have been isolated from them. Anti-tumor activity of the

[1] Department of Botany, University of Kalyani, B-16/212, Kalyani 741235, West Bengal, India.
[2] School of Chemical Sciences, Indian Association for the Cultivation of Science, 2A &2B Raja S.C. Mullick Road, Kolkata 700032, India.
* Corresponding author: sujatachaudhuri@gmail.com

fruit bodies of mushrooms and mycelial extracts have been evaluated, using different cancer cell lines. The polysaccharide extracts of many mushrooms have shown potent antitumor activity against sarcoma 180, mammary adenocarcinoma 755, leukemia L-1210, etc. The antitumor essence was later discovered to be a type of β-D-glucan, a polysaccharide yielding D-glucose only by acid hydrolysis (Mizuno, 1999).

Medicinal fungal research is focused on the discovery of compounds that can modulate the biologic response of immune cells. Such compounds that appear to stimulate the human immune response are being sought for the treatment of cancer, immunodeficiency disease or for generalised immunosuppression following drug treatment. Mushroom polysaccharides are regarded as biological response modifiers (BRM) (Wasser, 2002). This means that they cause no harm and place no additional stress on the body, but help the body to adapt to various environmental and biological stresses.

Many studies strongly suggest the future prospects of fungal polysaccharides as immunoceuticals for the prevention and treatment of various types of cancers. These polysaccharides also have the ability to heal oxidative damage and suppress neuro-inflammation. Hence they also have a great potential to be used in the development of remedies for neurodegenerative conditions (Trovato et al., 2017; Ferrão et al., 2017). Further research for development of polysaccharide(s) formulations, which can stimulate multiple immune routes to treat cancer and other diseases, is essential for discovery of unique immunoceuticals in future.

2. Fungal Polysaccharides

The fungal cell wall contains polysaccharides as a structural component. The two major types include the rigid fibrillar chitin (or cellulose) and the matrix consisting of β-glucan, α-glucan and glycoproteins (Ruiz-Herrera, 1956). Mushroom polysaccharides are mainly glucans, such as (1 3), (1 6)-β-glucans and (1 3)-α-glucans. Most are heteropolysaccharides or heteroglycans with two or more different monosaccharide units. About half the mass of the fungal cell wall consists of ß-glucans (Seviour et al., 1992; Klis et al., 2001; McIntosh et al., 2005). Non-cellulosic ß-glucans are either linear or branched and consist of a backbone of glucose residues usually joined by ß-(1-3) linkages, to which glucose side chain residues are often attached. Hetero-β-D-glucans are linear polymers of glucose with other D-monosaccharides (Wasser, 2002). The side chains of heteroglucans may contain glucuronic acid, xylose, galactose, mannose and arabinose or ribose as a main component or in different combinations. Some bind to protein residues and form polysaccharide-protein (PSP) and polysaccropeptide krestin (PSK) complexes.

Glycans are polysaccharides with units other than glucose in their backbone. They are known by their sugar component in the backbone as either galactans, fucans, xylans ormannans. Heteroglycans contain glucose as a main component and arabinose, mannose, galactose, fucose and xylose or glucuronic acid as side chains in different combinations. Details of some bioactive polysaccharides of macrofungi are given in Table 1.

Table 1. Bioactive macrofungal polysaccharides (#, Commercially developed polysaccharide products).

Fungi	Source	Type	Main bioactivity	References
Agaricus blazei	Fruit body and mycelium	Glucan, heteroglycan, glucan protein and glucomannan-protein	Antitumor	Mizuno, 1992; Mizuno, 1998
Armillariella tabescens	Mycelium	Heteroglycan	Antitumor	Kiho et al., 1992
Auricularia auricula	Fruit body	Glucan	Hyperglycemia, immunomodulatin and antitumor	Ukai et al., 1982; Ukai et al., 1983
Clitopilus caespitosus	Fruit body	Glucan	Antitumor	Liang et al., 1996
Cordyceps sp.	Fruit body, mycelium and culture broth	Glucan and heteroglycan	Antitumor, immunomodulating and hyperglycemia	Hsu et al., 2002
Dictyophora indusiata	Fruit body	Heteroglycan, mannan and glucan	Antitumor and hyperlipidemia	Hara et al., 1991
Flammulina velutipes	Fruit body and mycelium	Glucaneprotein complex and glycoprotein	Antitumor, antiflammatory, antiviral and immunomodulating	Zeng, 1990
Ganoderma applanatum	Fruit body	Glucan	Antitumor	Nakashima et al., 1979
Ganoderma lucidum	Fruit body and culture broth	Heteroglycan, mannoglucan and glycopeptide	Hyperglycemia, immunomodulating, antitumor and antioxidative	Miyazaki and Nishijima, 1981; Mizuno, 1997
Grifola frondosa	Fruit body	Proteoglycan, glucan, heteroglycan andgrifolan#	Immunomodulating, antitumor, antiviral and hepatoprotective	Zhuang et al., 1993; Cun et al., 1994; Zhuang et al., 1994
Hericum erinaceus	Fruit body and mycelium	Heteroglycan and heteroglycanpeptide	Hyperglycemia, immunomodulating and antitumor	Kawagishi et al., 1990; Mizuno, 1992; Mizuno, 1998
Inonotus obliquus	Fruit body mycelium	Glucan	Antitumor and immunomodulating	Kim et al., 2005
Lentinus edodes	Culture broth and fruit body	Mannoglucan, polysaccharide-protein, glucan andlentinan#	Immunomodulating, antitumor and antiviral	Chihara, 1969; Chihara et al., 1970; Hobbs, 2000
Morchella esculenta	Fruit body	Heteroglycan	Hyperglycemia and antitumor	Duncan et al., 2002
Marasmiellus palmivorus	Fruit body	Heteropolysaccharide	Immunomodulating and antitumor	Datta et al., 2019
Omphalia lapidescens	Fruit body	Glucan	Antiflammatory and immunomodulating	Saito et al., 1992

Table 1 Contd. ...

...Table 1 Contd.

Fungi	Source	Type	Main bioactivity	References
Peziza vericulosa	Fruit body	Proteoglycan and glucan	Immunomodulating and antitumor	Mimura et al., 1985
Phellinus linteus	Fruit body	Glucan	Antitumor	Kim et al., 2004
Pleurotus citrinopileatus	Fruit body	Galactomannan	Antitumor	Wang et al., 2005
Pleurotus tuber-regium	Sclerotium and mycelium	ß-D-glucan	Hepato-protective and anti-breast cancer	Zhang et al., 2001, 2003, 2006
Polypours umbellatus	Mycelium	Glucan	Antitumor and immunomodulating	Yang et al., 2004
Polystictus versicolar	culture broth, mycelium and fruit body	Heteroglycan, glycopeptide, krestin (PSK)#	Immunomodulating, antitumor, hyperglycemia and antiflammatory	Cui and Chisti, 2003
Schizophyllum commune	Mycelium	Glucan and schizophyllan#	Antitumor	Yamamoto, 1981
Sclerotinia sclerotiorum	Sclerotium	Glucan, scleroglucan (SSG)#	Antitumor	Palleschi et al., 2005
Trametes robiniophila	Mycelium	Proteoglycan	Immunomodulating, hepatoprotective and anticancer	Zhang, 1995
Tremella fuciformis	Fruit body, mycelium and culture broth	Heteroglycan	Hyperglycemia, immunomodulating and antitumor	Huang, 1982
Tremella aurantialba	Fruit body and mycelium	Heteroglycan	Immunomodulating and hyperglycemia	Liu et al., 2003
Tricholoma mongolium	Fruit body	Glucan	Antitumor	Wang et al., 1996

3. Factors Influencing Bioactivity of Polysaccharides

Fungal polysaccharides are known for their antitumor activity, which fluctuates with their chemical structure and physical properties (Zong et al., 2012). The structural variability depends on the monosaccharide residues, their sequence and the types of glycosidic linkages present and is directly correlated with the bioactivity of polysaccharides. The differences in bioactivity are also influenced by the ability of the polysaccharide to solubilise in water, its size, branching rate and form.

Majority of the biologically active polysaccharides are β-D-glucans and hetero-β-D-glucans, though some α-D-glucans have also been ascribed with similar activities (Mizuno, 1999; Gao et al., 1996). Structural features as β-(1-3) linkages in the backbone (main chain) of the glucan and additional β-(1-6)-branch points are needed for antitumor activity (Wasser, 2002). The ß-glucans with only (1-6) glycosidic linkages have little or no activity. In general, higher molecular weight glucans have been reported to be more effective than those with low molecular weight against tumors (Mizuno et al., 1996; Mizuno, 1999). The $(1\rightarrow3)$- and $(1\rightarrow6)$-β-D-glucans are the major polysaccharides from *Ganoderma lucidum* with antitumor activity (Bohn et al., 1995). An alkaline-soluble polysaccharide from *Flammulina velutipes* was found to be effective against sarcoma S-180 *in vivo* but not *in vitro* (Leung et al., 1997). Several hetero-β-D-glycans, such as glucurono-β-D glucan, arabinoxylo-β-D-glucan, xylo-β-D-glucan, manno-β-D-glucan and xylomanno-β-D-glucan, also showed strong antitumor activity (Wasser et al., 1999). These polysaccharides showed enhanced antitumor, antibacterial, antiviral, anticoagulatory and wound-healing properties. A purified endo-polysaccharide, an alpha linked fucoglucomannan isolated from *Inonotus obliquus* mycelia was reported to cause activation of B-cells and macrophages (Kim et al., 2006). A bioactive heteropolysaccharide (MFPS1) from the wild non-edible mushroom, *Marasmiellus palmivorus* with arabinose as the dominating monosaccharide and repeating units of three arabinose, two glucose, two galactose and one mannose was reported (Datta et al., 2019).

The β-glucan possesses different structural features and differs in the degree of branching (DB), molecular weight (MW) and conformation, which significantly influences its bioactivities (Yadomae et al., 2000; Han et al., 2020). In addition, there are several other biologically active mushroom polysaccharides, which are attached with protein or peptide molecules, such as glycopeptides, proteoglucans or glycoproteins. They have been reported to induce phagocytosis and stimulate the release of cytokines, such as TNF-α and various types of interleukins (Chen et al., 2007).

The β-glucans exhibit a range of conformations, like helices (single, double, or triple), random coils, worm-like and rod-like or aggregates. Many studies have suggested that triple-stranded and helical-chain polysaccharides, with a few exceptions, exhibit greater anticancer activity than those in random coils or rods (Meng et al., 2016; Zhang et al., 2018; Guo et al., 2019). Liu et al. (2011) based on their study of structural elucidation of a water-soluble polysaccharide from *Agaricus blazei* (ABP-W1), reported it to be a triple helix when in aqueous solution and stated that this type of conformation improves the solubility and anticancer activity because of better interaction between the polysaccharide and cellular receptors. The helical-structured polysaccharides from *Antrodia camphorate* exhibited a strong antitumor activity (Chen et al., 2001). Similarly, a random coil conformation polysaccharide from *Cordyceps militaris* also showed considerable antitumor activity (Yu et al., 2007; Lee et al., 2010a).

The HEB-AP Fr I, a β-mannan with a laminarin-like triple helix conformation isolated from *Hericium erinaceus* exhibited immune-stimulant property via macrophage activation (Lee et al., 2009a). The FII-1 (from *H. erinaceus*) also possessed antitumor activity (Mizuno et al., 1992). Intracellular Dox accumulation was triggered by HE from *H. erinaceus* (Lee et al., 2009a). Lentinan, a β-glucan from *Lentinula edodes* is a bioactive immunomodulator and its conformation was important for immune-stimulating activity (Bohn et al., 1995; Mizuno, 1999). The helical conformation is the main criterion for the immune-potentiating activity. The highly water soluble ß-glucan, referred as SCG from *Sparassis crispa*, is a triple helix polysaccharide and possibly exists as an irregular single helical conformation. This conformation is responsible for cytokine production and the synthesis of inducible nitric oxide synthase (iNOS), nitric oxide (NO) and type I collagen (Tada et al., 2007;

Kwon et al., 2009). PSK isolated from *Trametes versicolor* is a β-glucan with peptide having antitumor activity and stimulates cytokines production (Enshasy et al., 2013).

Studies also suggest that the distribution of the branch units along the backbone chain also plays a crucial role in its activity (Bohn et al., 1995). The DB determines the tertiary structure of polysaccharides. Active polysaccharides exhibit a large range of DB from 0.02 to 0.75. According to Bohn and BeMiller (1995), when the DB ranges from 0.2 to 0.33, β-(1 → 3)-d-glucans display a significant antitumor activity. Maximum immune-modulating and antitumor activities were achieved with a DB of 0.32 (Bae et al., 2013). Schizophyllan, from *Schizophyllum commune*, an extensively studied polysaccharide with significant immunopotentiating activity, has a DB of 33 per cent (Chihara et al., 1970). The polysaccharide from *Tremella fuciformis* stimulated human monocytes to release interleukins, such as IL-1 and IL-6 and TNF *in vitro* (Gao et al., 1998). The main chain and additional β-(1→6) branch points of β-(1→3) linkages of glucans are the prime factors in antitumor action. However, polysaccharides with (1→6) linkages β-glucans invariably have lower activity. The spectrum of β-(1→3) and (1→6)-glucosidic linkages have specific characteristics and show immune-stimulating activity, including macrophage activation and expression of IL-1β and TNF-α and NO production (Mizuno et al., 1992). A linear water-soluble (1→3)-β-D-glucan from *Auricularia auricula* showed extremely strong antitumor activity (Zhang et al., 1995). The α-mannan or α-glucan have also been reported to possess anticancer activity (Hara et al., 1991; Zhang and Cheung, 2002).

Molecular weight is another factor, which influences the antitumor property of a polysaccharide (Zhong et al., 2013). The MW may range from low to high, depending on the source. For example, soluble bioactive β-glucans from two edible mushrooms *Grifola frondosa* and *Lentinus edodes* were estimated to have MW of about 23 and 400 kDa, respectively (Masuda et al., 2009; Sasaki and Takasuka, 1976). In general, high molecular weight mushroom polysaccharides have better anticancer activity (Mizuno, 1999; Wasser, 2002; El Enshasy et al., 2013; Paterson and Lima, 2014) though there are some exceptions. According to reports, some low MW polysaccharides may also have considerable anticancer property.

To initiate antitumor mechanism, the polysaccharides must interact with its receptors. High MW polysaccharides may make better connection with the receptors than those with low MW. However, some exceptions have also been recorded, such as the bioactivity of (1→3)-α-glucuronoxylomannans from *Tremella fuciformis* is not dependent on MW for its activity. The hydrolysed product of glucuronoxylomannans with MW ranging from 53–1000 kDa has intact fractions with antitumor activity (Gao et al., 1996). The activity of (1→3)-β-glucans extracted from *G. frondosa* changed in relation to their MW. The highest MW fraction (800 kDa) exhibited the highest antitumor and immunomodulatory potential (Ren et al., 2012). A water-soluble glucan fraction from *H. erinaceus* with MW exceeding 100 kDa exhibited anti-artificial pulmonary metastatic tumor and immunoenhancing effects (Wang et al., 2001).

A 200 kDa polysaccharide from *Flammulina velutipes* exhibited potent antitumor activity against sarcoma S-180 *in vivo* (Leung et al., 1997). Another immune-mediated antitumor polysaccharide SCG (*S. crispa*) with MW exceeding 2000 kDa stimulated the macrophages and dendritic cells directly to produce TNF-α and IL-12 and indirectly triggered CD4+ Thymus lymphocytes (T cells) via the production of IFN-γ and granulocyte macrophage - colony stimulating factor (GM-CSF) (Harada et al., 2006; Tada et al., 2007).

Münzberg et al. (1995) reported that schizophyllan with MW range of 1–5 kDa showed the highest and most effective biological activity. A low MW fraction (20 kDa) isolated from the fruit body of *A. blazei* was found to possess tumor-specific cytocidal and immune-potentiating activity (Fujimiya et al., 1999). Polysaccharides from *Tremella fuciformis* with MW ranging from 53 to 1000 kDa induced human monocytes to express IL6 as capably as the nonhydrolysed fraction (Leung et al., 1997). MFPS1 from *Marasmiellus palmivorus* with a high molar weight of ~ 1,45,000 g/mol was reported to trigger both innate and adaptive immune system. It induced pro-inflammatory response and activated the anticancer immune surveillance system (Datta et al., 2019). More research is required to unravel the correlation between MW and the anticancer mechanism of mushroom polysaccharides.

Promising bioactive polysaccharides have also been reported from fungi other than mushrooms and macrofungi. *Pseudoepicoccum cocos*, an ascomycetous pathogen of coconut, yielded relatively high MW polysaccharides, which at low dosage exhibited good antitumor activities *in vivo*. A 26–268 kDa polysaccharide with extended chain conformation could increase the antitumor activity by raising the chance of interaction with the immune system (Chen et al., 2009). Similarly, a low MW branched β-glucan isolated from *Sclerotium rolfsii,* aubiquitous pathogen, increased the secretion of TNF-α and stimulated the proliferation of lymphocytes (Bimczok et al., 2009).

3.1 Mechanisms of Antitumor Activity

The mechanisms by which fungal polysaccharides may exert antitumor effect include: (1) cancer-preventing activity; (2) Immuno-enhancing activity; and (3) direct tumor-inhibition activity by inducing apoptosis of tumor cells (Fig. 1). The actual mechanism(s) is however, composed of an intricate series of reactions, including the activation of innate and adaptive immune system (Chan et al., 2009).

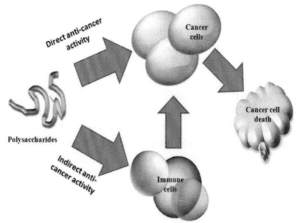

Fig. 1. Mechanisms of antitumor activity of mushroom polysaccharides.

Cancer cells multiply rapidly and spread to the other organs and lymphatic system of the human body, initiating metastasis in the later stages of cancer. This stage is vital and needs to be targeted to avoid proliferation. Though the different components, such as cytokines, interleukins and others of the innate immune system initially act as anticancer immune surveillance circuit (Shaked, 2019), in the later stages of spread, it fails to recognise the cancer cells and becomes less active and is not capable of reversing the situation (Almand et al., 2000; Dunn et al., 2004; Gabrilovich, 2004; Grivennikov et al., 2010; Harjes, 2019). Various types of fungal polysaccharides with different chemical structures have been reported to modify the immune responses against cancer cells (*in vitro* and *in vivo*) in different cell line models. This is performed mainly by modifying the natural defense pathways by the activation of effector cells, such as T lymphocytes, B lymphocytes, cytotoxic T lymphocytes, natural killer (NK) cells and macrophages. These cells release cytokines, such as tumor inducing factor a (TNF-a), interferon c and interleukin 1b (IL-1b) immediately and initiate active immune response by way of antiproliferative properties leading to apoptosis and differentiation in tumor cells by production of NO, reactive oxygen intermediates and interleukins (Barbosa et al., 2019; Barbosa and de Carvalho, 2020). Therefore, to reinstall the unbiased immune system with anticancer property, an external stimulus becomes essential. Macrofungal polysaccharides have been reported to work as an external stimulator under these conditions (Wasser, 2002; Schepetkin and Quinn, 2006; Moradali et al., 2007) and once again reinstall the anticancer immune surveillance system while eliminating the neoplastic cells from the host.

Mushroom polysaccharides possess certain physicochemical characteristics, which help them to interact and bind with the receptors present on the surface of the tumor cells to induce apoptosis of the cancer cells (Zhang et al., 2014). The β-glucans from mushrooms exhibit substantial immune-stimulatory activities, ranging from high nutraceutical to pharmaceutical levels (Vetvicka and Vetvickova, 2012; Giavasis, 2014; Rathore et al., 2017). Some mushroom-derived β-glucans are being used currently for cancer immune-chemotherapy (Friedman, 2016; Sugiyama, 2016), which increases the survival rate of cancer patients and improves their life quality.

Various studies have reported that polysaccharides execute their antitumor activity indirectly through the host's immune system, instead of a direct cytotoxic effect (Bohn et al., 1995). Polysaccharides guide the immune system to fight against biological stresses and develop anticancer immunity. Thus, they work as BRM (Zhang et al., 2007). Studies have shown that these polysaccharides can turn on the innate immune system to block the cancer progression. The polysaccharide-activated effector cells, such as macrophages, T-lymphocytes, B-lymphocytes, cytotoxic T-lymphocytes (CTL) and natural killer cells produce cytokines, like TNF-α, IFN-c and IL-1β. Cytokines have anti-proliferative effect, trigger apoptosis, differentiate tumor cells and stimulate the production of reactive nitrogen, oxygen intermediates and interleukins (Collins et al., 1991; Paterson, 2006). Lucas et al. (1957) first reported that mushroom polysaccharides possess antitumor property. Since then, different types of polysaccharides have been isolated from a variety of mushrooms to ascertain their antitumor property. For example, lentinan can stimulate the human immune system (Hobbs, 2000).

4. Immunomodulating Activities

4.1 *Adaptive Immunity*

The adaptive immune system is also called the acquired immune system or specific immune system. Mushroom polysaccharides can accelerate the cytotoxic potential of NK cells by raising the release of TNF-α and IFN-γ from macrophages and lymphocytes, respectively (Kuo et al., 2006) (Fig. 2). Thymus, macrophage phagocytosis, humoral antibody production and delayed-type sensitivity response are the major evidences that support the immunopotentiation of polysaccharides (Schepetkin and Quinn, 2006). Some studies have also reported that polysaccharides from mushroom fruit bodies could stimulate cytokine release through toll-like receptor-4-modulated protein kinase signalling pathway (Guo et al., 2009). Oral or intraperitoneal administration of polysaccharide of *Antrodia camphorata* (AC-PS) in mice exhibited a higher spontaneous proliferation of the splenocytes. This proliferation was further improved by phyto-haemagglutinin treatment. In AC-PS-treated mice, IL-12 was markedly increased while the expression of cytokines, like IL-6, TNF-α and IFN-γ, was also moderately improved (Liu et al., 2004).

Administration of CS-PCS3-II, a polysaccharide obtained from the plant pathogenic microfungi *P. cocos,* in mice promoted immune reaction to antigen and it was evident from the higher phagocytic index that there was a dose-dependent rise in humoral antibody production. CS-PCS3-II could also trigger T cell-mediated secretion of various cytokines and kills cancer cells specifically (Chen et al., 2010). In *P. cocos* polysaccharide (PC-II) treated endothelial cells, IFN-c suppression induced IP-10 protein release in a dose-dependent manner, suggesting the regulation of inflammation by PC-II (Lu et al., 2010). Both CS-PCS3-II and PC-II thus exhibited a significant anticancer property.

Polysaccharides of *Ganoderma lucidum* can increase the expression of cytokines generated from human monocytes, namely macrophages and T-lymphocytes. In *G. lucidum* polysaccharide-treated mice, elevated expression of cytokines (IFN-γ, IL-4 and IL-6) from spleen lymphocytes were noted (Wang et al., 1997; Guggenheim et al., 2014). Some reports indicated that a proteoglycan fraction from *G. lucidum* worked as an immunostimulatory drug, activating the B-cells to improve the immune response of tumor patients (Ye et al., 2011).

Another important antitumor polysaccharide from *Grifola frondosa* (common name, Maitake), whose one fraction was found to improve the activation of helper T-cells by maintaining the balance between Th1 and Th2 cells, thus worked as an immunotherapeutic agent for cancer patients by activating their cellular immune system (Kodama et al., 2003). Intramuscular administration of schizophyllan to mice increased the differentiation of cytotoxic T lymphocytes. The IL-1, 2, and 3 and IFN-γ production enhanced the activity of NK, spleen, lymphoid and bone marrow cells to execute antitumor activity. Thus, the main anticancer activity of schizophyllan is due to stimulation of cytokines (Suzuki et al., 1982; Tsuchiya et al., 1989). Lentinan and schizophyllan are almost similar in their biological activities and their mode of action (Jong et al., 1991). However, lentinan has the capability to recover the immune stimulation to the normal level and can block malignant transformation (Chihara, 1992). Lull et al. (2005) reported that lentinan is an oriented adjuvant dealing with T-cells and could control the balance between Th1/2 and Th1 via IL-12 production. Furthermore, in stomach-cancer patients, lentinan-induced slow down of T-lymphocyte differentiation and reduced regulatory T cells (Treg cell) activity by inhibition of prostaglandin (Aoki, 1984). In the spleen, lentinan increased the production of activated cytotoxic T-lymphocytes (Ye et al., 2011). In peripheral blood mononuclear cells, lentinan induced the production of IL-1α, IL-1β and TNF-α (Chihara, 1992). The host defence mechanism is activated by lentinan through colony stimulating factor (CSF), TNF-α, interferon (IFN), IL-1 and IL-3, released from T-cells and macrophages, which result in maturation, differentiation and proliferation of immunocompetent cells (Sakagami et al., 1988). In addition, lentinan regulated the permeation of activated NK cells and cytotoxic T-lymphocytes to kill the tumors (Suzuki et al., 1994).

The PSK from *Trametes versicolor* also developed both cellular and humoral immunity and induced the expression of cytokines, such as TNF-α, IL-1, IL-6 and IL-8 (Tzianabos, 2000; Hsieh and Wu, 2001). These cytokines could be responsible for the rise of T-cell cytotoxicity against tumor cells, stimulation of antibody production from B-cells or promotion of expression of receptors for IL-2 on T-cell surface (Kato et al., 1995). The SCG from *Sparassis crispa* boosted the response of hematopoiesis (Harada et al., 2003). Thus, it suggests that mushroom polysaccharides influence the adaptive immunity by activating the B- and T-cell functions and inhibit promotion and progression of tumors by hindering tumor angiogenesis. An increase in the IFN-γ and IL-12 was recorded when peripheral blood mononuclear cell (PBMC) and mice macrophage (RAW 264.7) were treated with the heteropolysaccharide (MFPS1) from *Marsmiellus palmivorus*. The IFN-γ was secreted from Th1 cell and maintained M1 phenotype of the macrophage. The IL-12 was released by M1 macrophage to sustain the Th1 phenotype, while IL-12 over-expression was statistically significant in RAW 264.7 cells, but not so in PBMC. The over-expression of these two cytokines indicated that a pro-inflammatory antitumor immune-surveillance was initiated by MFPS1 through M1Φ. The upregulated cytokines stimulated the innate and adaptive immune system against cancer cells (Datta et al., 2019).

4.2 *Innate Immunity*

Generally, the first line of defence is innate immunity or non-specific immunity. The prime members of this system that work as caretakers are the macrophages, neutrophils, NK cells and dendritic cells (Fig. 2). Innate immunity is always regulated by chemical messengers and cytokines and stimulated by activating the inflammatory and acute phase responses (Chihara, 1992; Trinchieri, 2003). The β-glucans isolated from different fungi have been reported to stimulate the immune functions by acting as pathogen-associated molecular patterns (PAMP) on cell membrane receptors (Brown and Gordon, 2005). The innate immune responses in macrophages activated via a type II transmembrane protein receptor Dectin-1 can interact with a variety of β-(1→3) and β-(1→6)-glucans. This follows a series of biochemical reactions,

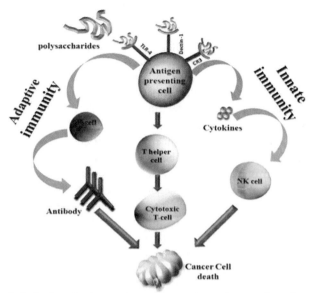

Fig. 2. Activation of anti-cancer immune system by mushroom polysaccharides.

including activation of phagocytosis, production of reactive oxygen species and induction of inflammatory cytokines (Willment et al., 2005). Another important β-glucan receptor, complement receptors [CR3 also called macrophage 1 antigen (Mac-1) or CD11b/CD18], are commonly found on the surface of immune effector cells, such as macrophages, neutrophils and NK cells (Ross et al., 1999).

4.3 Anticancer Mechanism of Macrophages

Macrophages control the antitumor immune responses and may be the first line of defence against tumors. Antitumor response turns on a series of events, starting with inflammatory response by macrophages and activating the cytotoxic lymphoid system via NKs and dendritic cells (DC) (De Palma and Lewis, 2013). Oral administration of β-glucans leads to their moving to the proximal small intestine, where they are recognised by the macrophages *in vivo* or *in vitro* (Chan et al., 2009). The β-glucans break down into micromolecular fragments and are transported to the bone marrow and endothelial cells (Hong et al., 2004). The macrophages then discharge the micromolecular fragments, which are recognised by the circulating granulocytes, monocytes and DC. As a result, the innate immune system is activated (De Silva et al., 2012). The activated macrophages interact with intracellular pathogens and activate the NK cells and T-lymphocytes via cytokine production to implement the antitumor effect by different mechanisms (Munz et al., 2005). The NK cells destroy the cell membranes of tumor cells by secreting specific chemical substances (De Silva et al., 2012).

Polysaccharides from *Ganoderma lucidum* were found to activate the bone marrow-derived macrophages in a dose-dependent manner. These activated macrophages markedly increased the phagocytotic activity and induced the release of IL1β and NO (Zhang et al., 2011). Additionally, polysaccharides (*G. lucidum*) treated human umbilical cord blood mononuclear cells differentiated themselves considerably into macrophages and NK cells (Chien et al., 2004). Treatment of human acute monocytic leukemia cell lines with these polysaccharides activated the caspases and tumor protein gene (p53) and resulted in improved macrophage differentiation. Researchers have used the changes in cell adherence, cell cycle arrest, increased expression of differentiation markers and down-regulation of myeloperoxidase (MPO) to establish the differentiation. The MPO is an enzyme that is exclusively synthesised in myeloid and monocytic cells and the down-regulation of MPO activity is an indication of macrophage differentiation (Lin and Austin, 2002; Hsu et al., 2011). Macrophage phagocytosis is also registered in lentinan and PSK treatments. Schizophyllan also enhances the production of macrophages and lymphocytes. Besides, activation of phagocytes significantly increased the release of reactive oxygen species (ROS) and pro-inflammatory cytokines, such as IL-6, IL-8 and TNF-α could be recognised by the high expression of CD11b and CD69L markers on the leukocyte surface (Suzuki et al., 1982; Kubala et al., 2003). The polysaccharide from *G. frondosa*, Grifolan, activated the macrophage and enhanced gene expression of IL-6, IL-1 and TNF-α *in vitro* (Jong et al., 1991). A polysaccharide from *Lentinus lepideus* was reported to effectively restore a radiation-damaged bone marrow system by increased expression of IL-1, IL-6 and GM-CSF (Jin et al., 2003). Galactomannan from *Morchella esculenta* increased the macrophage activity through enhanced nuclear factor kappa-light-chain-enhancer of activated B cells (NF-κB)-expression in THP-1 human monocytic cells. The activation of the transcription factor, NF-κB, regulated the cell survival and suppressed the apoptotic potential of chemotherapeutic drugs (Moradali et al., 2007). Fucogalactan derived from *Sarcodon aspratus* augmented the expression of TNF-α and NO in macrophages of mice *in vitro* (Moradali et al., 2007). The induction of NO and TNF-α by activated macrophages was found to have a cytotoxic effect on the malignant cells. *Hericium erinaceus* polysaccharide, the HEB-AP Fr I, also activated the macrophages efficiently by up-regulating TNF-α, IL-1β and NO release (Lee et al., 2009a). The WEHE from *H. erinaceum* could increase the expression of iNOS. Increased iNOS activated NF-κB by creating micromolar concentration of NO and activated macrophages (Son et al., 2006). Activated NF-κB-induced enzymes, like cyclooxygenase-2, lead to further DNA damage by increased reactive oxygen species (Li and Verma, 2002). In addition, ATOM from *Agaricus blazei* increased the number of peritoneal macrophages in tumor-bearing mice (Moradali et al., 2007).

4.4 *The Anticancer Mechanism of Natural Killer Cells*

The NK cells effectively inhibit tumor development, growth and metastasis (Hayakawa and Smyth, 2006). A polysaccharide from *H. erinaceum* was reported to be responsible for the lysis of mouse T cell lymphoma (Yac1 cells) by activated NK cells of splenocytes (Piontek et al., 1985; Yim et al., 2007). It showed indirect activation of NK cells by promoting other immunomediators or cellular components (Yim et al., 2007). Oral administration in mice with Hep3B hepatoma cell line, with powder from *Phellinus linteus* showed immune-modulatory and antitumor effects through increased release of IL-12, IFN-γ and TNF-α, which activated the NK cells (Jeong et al., 2011). Intramuscular administration of schizophyllan in mice notably improved the NK cell activity in the spleen cells (Suzuki et al., 1982). The MFPS1 also activated the NK cells to directly interact with the cancer cells to execute cytotoxicity in the latter (Datta et al., 2019).

4.5 Anticancer Mechanism of DCs

Lentinan interacts with DC to develop immunomodulation and antitumor activity. Further, the DC combine with NK cells to execute a key function of the removal of tumor cells (Chihara, 1992). *Antrodia camphorata* polysaccharide (GF2) was reported to stimulate the maturation of DC cells, which in turn exhibited an up-regulated expression of the MHC class II and CD86 molecules and enhanced IL-10 and IL-12 production (Liu et al., 2010). The Sparan (SCG) derived from *S. crispa* on the other hand, induced the MHC-II expression and NO production in DC and macrophages. In addition, it also up-regulated the expression of pro-inflammatory cytokines, such as IL-12, IL-1β, TNF-α and IFN-α/β by the DC. Among these cytokines, IL-12 is the chief factor in the initiation of Th1 immune response (Kim et al., 2010). Administration of PSK showed direct interaction with tumor cells and eliminated the transformed cells via the inflammatory response (Mizutani and Yoshida, 1991). The PSK was responsible for the rise in DC and T-cell capacity. Moreover, the PSK influenced the phenotypic and functional maturation of dendritic cells from human CD14+ cells (Nio et al., 1991).

4.6 Direct Tumor Inhibition Activity

4.6.1 Induced Apoptosis of Cancer Cell

Apoptosis is regulated by the Bcl-2 class proteins. Based on the number of Bcl-2-homology domains, they are divided into three sub-families: (1) containing Bcl-2, Bcl-xL and Bcl-w having antiapoptotic activity and sharing sequence homology; (2) consisting of Bax, Bad and Bak with proapoptotic activity and subfamily; (3) containing Bik and Bid, also with proapoptotic activity.

Bax has a key role in forming oligodimers in the apoptotic process (Kinnally and Antonsson, 2007). The polysaccharide from *Cordyceps militaris* (WECM) was reported to trigger apoptosis by increasing the level of Bax and decreasing Bcl-2 in a dose-dependent manner in human lung carcinoma (Park et al., 2009). Tumourigenesis is an imbalance between proliferation and apoptosis. A study indicated that the direct addition of *G. lucidum* polysaccharide B (GL-B) into a tumor cell-culture medium neither controlled the proliferation nor triggered apoptosis in murine Sarcoma cancer S-180 and human leukemia HL-60 cell lines *in vitro*, but could do so when the serum of GL-B was injected into mice bearing S-180 and HL-60 tumors. Splenocyte and peritoneal macrophage-conditioned medium with GL-B blocked, however, the proliferation of HL-60 and triggered apoptosis *in vitro* (Cao and Lin, 2006). *Ganoderma lucidum* polysaccharide (GLP) reduced integrin expression to inhibit cancer cell adhesion and tumor cell proliferation (Lu et al., 2001). Another study revealed that *Amauroderma rude* also induced the apoptosis and suppressed the development of colony formation (Jiao et al., 2013). Lentinan reduced tumor growth by cytokine-mediated immune-modulation and down-regulated caspase-3 in mice with liver cancer (Lull et al., 2005). On the other hand, grifolin up-regulated death-associated protein kinase 1 (DAPK1) and triggered apoptosis via the p53–DAPK1 pathway (Luo et al., 2011). Maitake (*G. frondosa*) has been reported to induce apoptosis in MCF-7 human breast cancer cells (Martin and Brophy, 2010). Previous studies have shown that polysaccharides from *Inonotus obliquus* (Chaga mushroom) triggered the apoptosis by up-regulating caspase-3 in HT-29 colon cancer cells and in melanoma cells, both under *in vitro* and *in vivo* conditions (Lee et al., 2009b; Youn et al., 2009). Schizophyllan induced apoptosis by increasing the levels of caspase-3 protein *in vivo* (Mansour et al., 2012). A water-soluble β-glucan isolated from *Auricularia auricula-judae* (AG) treated-mice showed S-180 tumor cell apoptosis by significantly decreasing Bcl-2 and increasing Bax levels (Ma et al., 2010). Polysaccharides from *A. camphorate* (AC-PS) inhibited the proliferation of human leukemic U937 cells significantly and showed antitumor activity against S-180 tumor in the ICR mice model. The *in vivo* cytolytic activity of splenocytes was enhanced and mice serum interleukin-12 levels significantly increased by AC-PS treatment (Liu et al., 2004). The direct tumor inhibition activity, i.e., tumor apoptosis of mushroom polysaccharides clearly works in a dose-dependent way. The Bcl-2 family, DAPK1 and caspase-3 are the key proteins involved in direct tumor inhibition. The indirect antitumor activity of the heteropolysaccharide MFPS1 from *M. palmivorus* triggering apoptosis in cancer cells under *in vitro* as well as *in vivo* conditions was reported by Datta et al. (2019). Apoptosis was activated by depletion of intracellular ROS. The up-regulation of ROS stimulated the monocytes to secrete ILs, which in turn triggers apoptosis in cancer cells.

4.6.2 Anti-Angiogenesis

Angiogenesis is a rate-limiting multistep process in tumor progression and occurs early in tumor growth (Folkman, 2004). *Ganoderma lucidum* polysaccharides inhibit angiogensis by suppressing NO, which is a key factor in angiogensis. The Gl-PP, a polysaccharide protein complex from *G. lucidum*, inhibited angiogenesis by reducing the release of pro-angiogenic factors and diminishing the development of vascular endothelial cells (Cao and Lin, 2006). These polysaccharides could also inhibit prostate cancer angiogenesis through mitogen-activated protein kinase (MAPK) and protein kinase B signalling. They could also alter the phosphorylation of extracellular signal-regulated kinases 1/2 and Akt kinases. Another report revealed that *G. lucidum* polysaccharides block the morphogenesis of the capillary by preventing the release of angiogenic factors, namely vascular endothelial growth factor (VEGF) and transforming growth factor b1 (TGF-b1) performing a key step in angiogenesis in cancer development (Stanley et al., 2005). A water-soluble polysaccharide of *G. lucidum* spores was reported to inhibit Lewis lung cancer cells in mice by modification of the MAPK pathway and spleen tyrosine kinase Syk-dependent TNF-α and IL-6 release in murine resident peritoneal macrophages (Guo et al., 2009). *Antrodia cinnamomea* polysaccharides showed anti-angiogenic effects in the endothelial cells by markedly suppressing VEGF interaction with VEGF receptor, blocking VEGFR2 phosphorylation and cyclin D1 expression (Cheng et al., 2005). Another study showed that a polysaccharide from *Phellinus linteus* proved to be a novel inhibitor of angiogenesis and worked as an immune potentiator in inhibiting cancer cell adhesion (Lee et al., 2010b).

4.6.3 Cell Cycle Arrest

The cell cycle progression is controlled at several irreversible transition points or check-points. Cyclins and cyclin-dependent kinases (CDK) are the main regulators in the eukaryotic cell cycle check-points (Sherr, 1996). Studies have shown that grifolin arrests cells in the G1-phase because of the constraint of the ERK1/2 or the ERK5 pathway (Ye et al., 2011). However, schizophyllan-induced cell arrest at the G2/M phase. It significantly induced tyrosine15 (Y15) phosphorylation of CDK1 without hampering the levels of CDK1 protein and stimulated the p53 expression (Aleem, 2011). During mitosis, CDK1 serves as an important regulator that is bound to cyclin B (Desai et al., 1992). Therefore, schizophyllan increased Y15 phosphorylation by deactivating CDK1, which may have been responsible for G2/M cell cycle arrest (Taylor and Stark, 2001). Polysaccharides derived from *Agaricus blazei* and *Coriolus versicolor* induced cell arrest at the G2/M phase and G0/G1 phase, respectively in tumor cells (Hsieh et al., 2004; Jin et al., 2006). Akihisa et al. (2004) reported that polysaccharides from *Poria cocos* showed inhibitory effect on calf DNA polymerase α and rat DNA polymerase β. The antitumor activity was mainly due to the increased lymphocyte population levels coupled with increase in CD4+ cell. The mature T-cells, which were the Th cells with CD5+ activated cytotoxicity.

Proliferation of CD4+ cells and activation and maturation of T-cell (CD34-/CD5+) was stimulated in MFPS1-treated PBMC cells. The CD14+ up-regulation specified the activation of macrophages by the polysaccharide. Increased B-cell population and interaction of B-cell and T-cell was also observed with MFPS1 treatment. This interaction promoted further activation of the immune system to kill the cancer cells (Datta et al., 2019). They further proposed the immune-assisted anticancer mechanism of MFPS1 (Fig. 3).

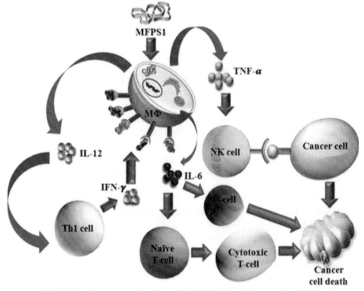

Fig. 3. Anticancer mechanism of MFPS1 (from Datta et al., 2019).

Conclusion

Notwithstanding the remarkable progress made in recent years in our understanding of the activation immune response of the host by fungal polysaccharides and the possible mechanism(s) involved, yet many challenges remain. Since there is a general consensus that the polysaccharide length and structure have a major role in the host's recognition and response, purification and characterisation of these polysaccharides is also a great challenge. The degrees of purification and the method of extraction influence the physiological activity of the polysaccharides. Thus, ideally, the structural characteristics, solubility, size, molecular weight, frequency of side-chain branching and conformation should be specified while reporting the outcome of different experiments on the immunomodulatory activity of such molecules. The properties of polysaccharides would help researchers in antitumor drugs development and vaccine production. It would also greatly help in synthesising such polymers. Synthetic glucans could provide a unique opportunity to investigate their immunomodulating activities. The use of polysaccharides as adjuvants to chemotherapy is very promising since it is known to reduce oxidative stress and the side effects of chemotherapy. Natural macrofungal polysaccharides thus are a viable option as immunoceticals or anticancer immunopotentiators and potent adjuvants and demand intense research efforts to continue.

References

Akihisa, T., Mizushina, Y., Ukiya, M., Oshikubo, M., Kondo, S., Kimura, Y., Suzuki, T. and Tai, T. (2004). Dehydrotrametenonic acid and dehydroeburiconic acid from *Poria cocos* and their inhibitory effects on eukaryotic DNA polymerase α and β. Biosci., Biotechnol., Biochem., 68(2): 448–450.

Aleem, E. (2011). The mushroom extract schizophyllan reduces cellular proliferation and induces G2/M arrest in MCF-7 human breast cancer cells. Life Sci. J., 8(4): 777–784.

Almand, B., Resser, J.R., Lindman, B., Nadaf, S., Clark, J.I., Kwon, E.D., Carbone, D.P. and Gabrilovich, D.I. (2000). Clinical significance of defective dendritic cell differentiation in cancer. Clin. Canc. Res., 6: 1755–1766.

Aoki, T. (1984). Lentinan. Immune modulation agents and their mechanisms. Immunology Studies, 25: 62–77.

Bae, I.Y., Kim, H.W., Yoo, H.J., Kim, E.S., Lee, S., Park, D.Y. and Lee, H.G. (2013). Correlation of branching structure of mushroom β-glucan with its physiological activities. Food Res. Int., 51(1): 195–200.

Barbosa, J.R., dos Santos Freitas, M.M., da Silva Martins, L.H. and de Carvalho Junior, R.N. (2019). Polysaccharides of mushroom *Pleurotus* spp.: New extraction techniques, biological activities and development of new technologies. Carbohyd. Polym., 229: 115550.

Barbosa, J.R. and de Carvalho Junior, R.N. (2020). Occurrence and possible roles of polysaccharides in fungi and their influence on the development of new technologies—Review. Carbohyd. Polym., 246: 116613.

Bimczok, D., Wrenger, J., Schirrmann, T., Rothkötter, H.J., Wray, V. and Rau, U. (2009). Short chain region-selectively hydrolysed scleroglucans induce maturation of porcine dendritic cells. Appl. Microbiol. Biotechnol., 82(2): 321–331.

Bohn, J.A. and BeMiller, J.N. (1995). $(1 \rightarrow 3)$-β-D-Glucans as biological response modifiers: A review of structure-functional activity relationships. Carbohyd. Polym., 28(1): 3–14.

Brown, G.D. and Gordon, S. (2005). Immune recognition of fungal β-glucans. Cell., Microbiol., 7(4): 471–479.

Cao, Q.Z. and Lin, Z.B. (2006). *Ganoderma lucidum* polysaccharides peptide inhibits the growth of vascular endothelial cell and the induction of VEGF in human lung cancer cell. Life Sci., 78(13): 1457–1463.

Chan, G.C.F., Chan, W.K. and Sze, D.M.Y. (2009). The effects of β-glucan on human immune and cancer cells. J. Hematol. Oncol., 2(1): 25.

Chen, J.C., Lin, W.H., Chen, C.N., Sheu, S.J., Huang, S.J. and Chen, Y.L. (2001). Development of *Antrodia camphorata* mycelium with submerge culture. Fungal Sciences, 16: 7–22.

Chen, J. and Seviour, R. (2007). Medicinal importance of fungal β-(1→ 3),(1→ 6)-glucans. Mycol. Res., 111(6): 635–652.

Chen, X., Xu, X., Zhang, L. and Zeng, F. (2009). Chain conformation and anti-tumor activities of phosphorylated (1→ 3)-β-d-glucan from *Poria cocos*. Carbohyd. Polym., 78(3): 581–587.

Chen, X., Zhang, L. and Cheung, P.C.K. (2010). Immunopotentiation and anti-tumor activity of carboxymethylated-sulphated β-(1→3)-D-glucan from *Poria cocos*. Int. Immunopharmacol., 10(4): 398–405.

Cheng, J.J., Huang, N.K., Chang, T.T., Wang, D.L. and Lu, M.K. (2005). Study for anti-angiogenic activities of polysaccharides isolated from *Antrodia cinnamomea* in endothelial cells. Life Sci., 76(26): 3029–3042.

Chien, C.M., Cheng, J.L., Chang, W.T., Tien, M.H., Tsao, C.M., Chang, Y.H., Chang, H.Y., Hsieh, J.F., Wong, C.H. and Chen, S.T. (2004). Polysaccharides of *Ganoderma lucidum* alter cell immunophenotypic expression and enhance CD56+ NK-cell cytotoxicity in cord blood. Bioorg. Med, Chem., 12(21): 5603–5609.

Chihara, G. (1969). The antitumor polysaccharide lentinan: An overview. pp. 687–694. *In*: Aoki, T. et al. (eds.). Manipulation of Host Defense Mechanism. Nature, 222.

Chihara, G., Hamuro, J., Maeda, Y.Y., Arai, Y. and Fukuoka, F. (1970). Fractionation and purification of the polysaccharides with marked antitumor activity, especially lentinan from *Lentinus edodes* (Berk.) Sing. (an edible mushroom). Canc. Res., 30(11): 2776–2782.

Chihara, G. (1992). Immunopharmacology of lentinan, a polysaccharide isolated from *Lentinus edodes*: Its application as a host defence potentiator. Int. J. Ori. Med., 17(5): 57–77.

Collins, M.D., Rodrigues, U., Ash, C., Aguirre, M., Farrow, J.A.E., Martinez-Murcia, A., Phillips, B.A., Williams, A.M. and Wallbanks, S. (1991). Phylogenetic analysis of the genus Lactobacillus and related lactic acid bacteria as determined by reverse transcriptase sequencing of 16S rRNA. FEMS Microbiol. Lett., 77(1): 5–12.

Cui, J. and Chisti, Y. (2003). Polysaccharopeptides of *Coriolus versicolor*: Physiological activity, uses and production. Biotechnol. Adv., 21(2): 109–122.

Cun, Z., Mizuno, T., Ito, H., Shimura, K., Sumiya, T. and Kawade, M. (1994). Antitumor activity and immunological property of polysaccharides from the mycelium of liquid-cultured *Grifola frondosa*. J. Jap. Soc. Food Sci. Technol., 41: 724–733.

Datta, H.K., Das, D., Koschella, A., Das, T., Heinze, T., Biswas, S. and Chaudhuri, S. (2019). Structural elucidation of a heteropolysaccharide from the wild mushroom *Marasmiellus palmivorus* and its immune-assisted anticancer activity. Carbohyd. Polym., 211: 272–280.

De Palma, M. and Lewis, C.E. (2013). Macrophage regulation of tumor responses to anticancer therapies. Canc. Cell, 23(3): 277–286.

De Silva, D.D., Rapior, S., Fons, F., Bahkali, A.H. and Hyde, K.D. (2012). Medicinal mushrooms in supportive cancer therapies: An approach to anti-cancer effects and putative mechanisms of action. Fungal Diversity, 55(1): 1–35.

Desai, D., Gu, Y. and Morgan, D.O. (1992). Activation of human cyclin-dependent kinases *in vitro*. Mol. Biol. Cell, 3(5): 571–582.

Duncan, C., Pugh, J., Pasco, G., David, N., Ross, S. and Samir, A. (2002). Isolation of a galactomannan that enhances macrophage activation from the edible fungus *Morchella esculenta*. J. Agric. Food Chem., 50: 5683–5685.

Dunn, G.P., Old, L.J. and Schreiber, R.D. (2004). The immunobiology of cancer immunosurveillance and immunoediting. Immunity, 21(2): 137–48.

El Enshasy, H.A. and Hatti-Kaul, R. (2013). Mushroom immunomodulators: Unique molecules with unlimited applications. Tr. Biotechnol., 31(12): 668–677.

Ferrão, J., Bell, V., Calabrese, V., Pimentel, L. and Pintado, M. (2017). Impact of mushroom nutrition on microbiota and potential for preventative health. J. Food Nutr. Res., 5: 226–233.

Folkman, J. (2004). Endogenous angiogenesis inhibitors. Acta Pathol. Microbiol. Immunol. Scand., 112(7-8): 496–507.

Friedman, M. (2016). Mushroom polysaccharides: Chemistry and antiobesity, antidiabetes, anticancer and antibiotic properties in cells, rodents and humans. Foods, 5: 80.

Fujimiya, Y., Suzuki, Y., Katakura, R. and Ebina, T. (1999). Tumor-specific cytocidal and immunopotentiating effects of relatively low molecular weight products derived from the basidiomycete, *Agaricus blazei* Murill. Anticanc. Res., 19(1A): 113–118.

Gabrilovich, D. (2004). Mechanisms and functional significance of tumour-induced dendritic-cell defects. Nat. Rev. Immunol., 4(12): 941–952. 10.1038/nri1498.

Gao, Q., Berntzen, G., Jiang, R., Killie, M.K. and Seljelid, R. (1998). Conjugates of *Tremella* polysaccharides with microbeads and their TNF-stimulating activity. Planta Medica, 64(06): 551–554.

Gao, Q.P., Jiang, R.Z., Chen, H.Q., Jensen, E. and Seljelid, R. (1996). Characterisation and cytokine stimulating activities of heteroglycans from *Tremella fuciformis*. Planta Medica, 62(04): 297–302.

Giavasis, I. (2014). Bioactive fungal polysaccharides as potential functional ingredients in food and nutraceuticals. Curr. Opin. Biotechnol., 26: 62–173.

Grivennikov, S.I., Greten, F.R. and Karin, M. (2010). Immunity, inflammation and cancer. Cell, 140: 883–899.

Guggenheim, A.G., Wright, K.M. and Zwickey, H.L. (2014). Immune modulation from five major mushrooms: Application to integrative oncology. Integrat. Med. Clin. J., 13(1): 32.

Guo, L., Xie, J., Ruan, Y., Zhou, L., Zhu, H., Yun, X., Jiang, Y., Lü, L., Chen, K., Min, Z. and Wen, Y. (2009). Characterisation and immunostimulatory activity of a polysaccharide from the spores of *Ganoderma lucidum*. Int. Immunopharmacol., 9(10): 1175–1182.

Guo, M.Z., Meng, M., Duan, S.Q., Feng, C.C. and Wang, C.L. (2019). Structure characterisation, physicochemical property and immunomodulatory activity on RAW264.7 cells of a novel triple-helix polysaccharide from *Craterellus cornucopioides*. Int. J. Biol. Macromol., 126: 796–804.

Han, B., Baruah, K., Cox, E., Vanrompay, D. and Bossier, P. (2020). Structure-functional activity relationship of β-Glucans from the perspective of immunomodulation: A Mini-review. Front. Immunol., 11: Article # 658.

Hara, C., Kumazawa, Y., Inagaki, K., Kaneko, M., Kiho, T. and Ukai, S. (1991). Mitogenic and colony-stimulating factor-inducing activities of polysaccharide fractions from the fruit bodies of *Dictyophora indusiata* Fisch. Chem. Pharmaceut. Bull., 39: 1615–1616.

Harada, T., Miura, N., Adachi, Y., Nakajima, M., Yadomae, T. and Ohno, N. (2002). Effect of SCG, 1, 3-β-D-glucan from *Sparassis crispa* on the hematopoietic response in cyclophosphamide-induced leukopenic mice. Biol. Pharmaceut. Bull., 25(7): 931–939.

Harada, T., Kawaminami, H., Miura, N.N., Adachi, Y., Nakajima, M., Yadomae, T. and Ohno, N. (2006). Cell to cell contact through ICAM-1-LFA-1 and TNF-α synergistically contributes to GM-CSF and subsequent cytokine synthesis in DBA/2 mice induced by 1, 3-β-D-glucan SCG. J. Inter. Cytok. Res., 26(4): 235–247.

Harjes, U. (2019). Helping tumour antigens to the surface. Nat. Rev. Canc., 19: 608.

Hayakawa, Y. and Smyth, M.J. (2006). Innate immune recognition and suppression of tumors. Adv. Canc. Res., 95: 293–322.

Hobbs, C. (2000). Medicinal value of *Lentinus edodes* (Berk.) Sing. (Agaricomycetideae). A literature review. Int. J. Med. Mush., 2(4): 287–302.

Hong, F., Yan, J., Baran, J.T., Allendorf, D.J., Hansen, R.D., Ostroff, G.R., Xing, P.X., Cheun, N.K. and Ross, G.D. (2004). Mechanism by which orally administered β-(1→3)-glucans enhance the tumoricidal activity of antitumor monoclonal antibodies in murine tumor models. J. Immunol., 173: 797–806.

Hsieh, T.C. and Wu, J.M. (2001). Cell growth and gene modulatory activities of Yunzhi (Windsor Wunxi) from mushroom *Trametes versicolor* in androgen-dependent and androgen-insensitive human prostate cancer cells. Int. J. Oncol., 18(1): 81–89.

Hsieh, T.C., Kunicki, J., Darzynkiewicz, Z. and Wu, J.M. (2004). Effects of extracts of *Coriolus versicolor* on cell-cycle progression and expression of Interleukins-1β,-6, and -8 in promyelocytic HL-60 leukemic cells and mitogenically stimulated and non-stimulated human lymphocytes. J. Alt. Compl. Med., 8(5): 591–602.

Hsu, J.W., Yasmin-Karim, S., King, M.R., Wojciechowski, J.C., Mickelsen, D., Blair, M.L., Ting, H.J., Ma, W.L. and Lee, Y.F. (2011). Suppression of prostate cancer cell rolling and adhesion to endothelium by 1α,25-dihydroxyvitamin D3. Am. J. Pathol., 178(2): 872–80.

Hsu, T., Shiao, L., Hsieh, C. and Chang, D. (2002). A comparison of the chemical composition and bioactive ingredients of the Chinese medicinal mushroom Dong Chong Xia Cao, its counterfeit and mimic, and fermented mycelium of *Cordyceps sinensis*. Food Chem., 78: 463–469.

Huang, N. (1982). Cultivation of *Tremella fuciformis* in Fujian, China. Mush. Newslett. Trop., 2: 2–5.

Jeong, J.W., Jin, C.Y., Park, C., Hong, S.H., Kim, G.Y., Jeong, Y.K., Lee, J.D., Yoo, Y.H. and Choi, Y.H. (2011). Induction of apoptosis by cordycepin via reactive oxygen species generation in human leukemia cells. Toxicol. *In Vitro*, 25(4): 817–824.

Jiao, C., Xie, Y.Z., Yang, X., Li, H., Li, X.M., Pan, H.H., Cai, M.H., Zhong, H.M. and Yang, B.B. (2013) Anticancer activity of Amauroderma rude. PLoS ONEe, 8(6): e66504.

Jin, C.Y., Choi, Y.H., Moon, D.O., Park, C., Park, Y.M., Jeong, S.C., Heo, M.S., Lee, T.H., Lee, J.D. and Kim, G.Y. (2006). Induction of G2/M arrest and apoptosis in human gastric epithelial AGS cells by aqueous extract of *Agaricus blazei*. Oncol. Rep., 16(6): 1349–1355.

Jin, M., Jeon, H., Jung, H.J., Kim, B., Shin, S.S., Choi, J.J., Lee, J.K., Kang, C.Y. and Kim, S. (2003). Enhancement of repopulation and hematopoiesis of bone marrow cells in irradiated mice by oral administration of PG101, a water-soluble extract from *Lentinus lepideus*. Exp. Biol. Med., 228(6): 759–766.

Jong, S.C. (1991). Immunomodulatory substances of fungal origin. J. Immunol. Immunopharmacol., 11: 115–122.

Kato, M., Hirose, K., Hakozaki, M., Ohno, M., Saito, Y., Izutani, R., Noguchi, J., Hori, Y., Okumoto, S., Kuroda, D. and Nomura, H. (1995). Induction of gene expression for immunomodulating cytokines in peripheral blood mononuclear cells in response to orally administered PSK, an immunomodulating protein-bound polysaccharide. Canc. Immunol. Immunother., 40(3): 152–156.

Kawagishi, H., Kanao, T., Inagaki, R., Mizuno, T., Shimura, K. and Ito, H. (1990). Formulation of a potent antitumor (1-6)-ß-D glucan-protein complex from *Agaricus blazei* fruiting bodies and antitumor activity of the resulting products. Carbohyd. Polym., 12: 393–404.

Kiho, T., Shiose, Y., Nagai, K. and Ukai, S. (1992). Polysaccharides in fungi. XXX. Antitumor and immunomodulating activities of two polysaccharides from the fruiting bodies of *Armillariella tabescens*. Chem. Pharmaceut. Bull., 40: 2110–2114.

Kim, G., Choi, G., Lee, S. and Park, Y. (2004). Acidic polysaccharide isolated from *Phellinus linteus* enhances through the up-regulation of nitric oxide and tumor necrosis factor-alpha from peritoneal macrophages. J. Ethnopharmacol., 95: 69–76.

Kim, H.S., Kim, J.Y., Ryu, H.S., Park, H.G., Kim, Y.O., Kang, J.S., Kim, H.M., Hong, J.T., Kim, Y. and Han, S.B. (2010). Induction of dendritic cell maturation by β-glucan isolated from *Sparassis crispa*. Int. Immunopharmacol., 10(10): 1284–1294.

Kim, Y., Han, L., Lee, H., Ahn, H., Yoon, Y. and Jung, J. (2005). Immuno-stimulating effect of the endo-polysaccharide produced by submerged culture of *Inonotus obliquus*. Life Sci., 77: 2438–2456.

Kim, Y.O., Park, H.W., Kim, J.H., Lee, J.Y., Moon, S.H. and Shin, C.S. (2006). Anti-cancer effect and structural characterisation of endo-polysaccharide from cultivated mycelia of *Inonotus obliquus*. Life Sci., 30, 79(1): 72–80.

Kinnally, K.W. and Antonsson, B. (2007). A tale of two mitochondrial channels, MAC and PTP, in apoptosis. Apoptosis, 12(5): 857–868.

Klis, F.M., Groot, P.D. and Hellingwerf, K. (2001). Molecular organisation of the cell wall of *Candida albicans*. Med. Mycol., 39(1): 1–8.

Kodama, N., Komuta, K. and Nanba, H. (2003). Effect of maitake (*Grifola frondosa*) D-fraction on the activation of NK cells in cancer patients. Journal of Medicinal Food, 6(4): 371–377.

Kubala, L., Ruzickova, J., Nickova, K., Sandula, J., Ciz, M. and Lojek, A. (2003). The effect of (1→3)-β-D-glucans, carboxymethylglucan and schizophyllan on human leukocytes *in vitro*. Carbohyd. Res., 338(24): 2835–2840.

Kuo, M.C., Weng, C.Y., Ha, C.L. and Wu, M.J. (2006). *Ganoderma lucidum* mycelia enhance innate immunity by activating NF-κB. J. Ethnopharmacol., 103(2): 217–222.

Kwon, A.H., Qiu, Z., Hashimoto, M., Yamamoto, K. and Kimura, T. (2009). Effects of medicinal mushroom (*Sparassis crispa*) on wound healing in streptozotocin-induced diabetic rats. Am. J. Sur., 197(4): 503–509.

Lee, J.S., Cho, J.Y. and Hong, E.K. (2009a). Study on macrophage activation and structural characteristics of purified polysaccharides from the liquid culture broth of *Hericium erinaceus*. Carbohyd. Polym., 78(1): 162–168.

Lee, S.H., Hwang, H.S. and Yun, J.W. (2009b). Antitumor activity of water extract of a mushroom, *Inonotus obliquus*, against HT-29 human colon cancer cells. Phytother. Res., 23(12): 1784–1789.

Lee, J.S., Kwon, J.S., Yun, J.S., Pahk, J.W., Shin, W.C., Lee, S.Y. and Hong, E.K. (2010a). Structural characterisation of immunostimulating polysaccharide from cultured mycelia of *Cordyceps militaris*. Carbohyd. Polym., 80(4): 1011–1017.

Lee, Y.S., Kim, Y.H., Shin, E.K., Kim, D.H., Lim, S.S., Lee, J.Y. and Kim, J.K. (2010b). Anti-angiogenic activity of methanol extract of *Phellinus linteus* and its fractions. J. Ethnopharmacol., 131(1): 56–62.

Leung, M.Y.K., Fung, K.P. and Choy, Y.M. (1997). The isolation and characterisation of an immunomodulatory and anti-tumor polysaccharide preparation from *Flammulina velutipes*. Immunopharmacol., 35(3): 255–263.

Li, Q. and Verma, I.M. (2002). NF-κB regulation in the immune system. Nat. Rev. Immunol., 2(10): 725–734.

Liang, Z., Miao, C. and Zhang, Y. (1996). Influence of chemical modified structures on antitumor activity of polysaccharides from *Clitopilus caespitosus*. Zhong Guo Yao Xue Za Zhi, 31: 613–615.

Lin, K.M. and Austin, G.E. (2002). Functional activity of three distinct myeloperoxidase (MPO) promoters in human myeloid cells. Leukemia, 16(6): 1143.

Liu, C., Xie, H., Su, B., Han, J. and Liu, Y. (2003). Anti-thrombus effect on the fermented products of mycelium from *Tremella aurantialba*. Nat. Prod. Res. Develop., 3: 35–37.

Liu, J., Zhang, C., Wang, Y., Yu, H., Liu, H., Wang, L., Yang, X., Liu, Z., Wen, X., Sun, Y. and Yu, C. (2011). Structural elucidation of a heteroglycan from the fruiting bodies of *Agaricus blazei* Murill. Int. J. Biol. Macromol., 49(4): 716–720.

Liu, J.J., Huang, T.S., Hsu, M.L., Chen, C.C., Lin, W.S., Lu, F.J. and Chang, W.H. (2004). Antitumor effects of the partially purified polysaccharides from *Antrodia camphorata* and the mechanism of its action. Toxicol. Appl. Pharmacol., 201(2): 186–193.

Liu, K.J., Leu, S.J., Su, C.H., Chiang, B.L., Chen, Y.L. and Lee, Y.L. (2010). Administration of polysaccharides from *Antrodia camphorata* modulates dendritic cell function and alleviates allergen-induced T helper type 2 responses in a mouse model of asthma. Immunology, 129(3): 351–362.

Lu, H., Uesaka, T., Katoh, O., Kyo, E. and Watanabe, H. (2001). Prevention of the development of preneoplastic lesions, aberrant crypt foci, by a water-soluble extract from cultured medium of *Ganoderma lucidum* (Reishi) mycelia in male F344 rats. Oncol. Rep., 8(6): 1341–1345.

Lu, M.K., Cheng, J.J., Lin, C.Y. and Chang, C.C. (2010). Purification, structural elucidation, and anti-inflammatory effect of a water-soluble 1, 6-branched 1, 3-α-d-galactan from cultured mycelia of *Poria cocos*. Food Chem., 118(2): 349–356.

Lucas, E.H., Ringler, R.L., Byerrum, R.U., Stevens, J.A., Clarke, D.A. and Stock, C.C. (1957). Tumor inhibitors in *Boletus edulis* and other Holobasidiomycetes. Antib. Chemother., 7(1): 1–4.

Lull, C., Wichers, H.J. and Savelkoul, H.F. (2005). Anti-inflammatory and immunomodulating properties of fungal metabolites. Mediat. Inflamm., 2005(2): 63–80.

Luo, X.J., Li, L.L., Deng, Q.P., Yu, X.F., Yang, L.F., Luo, F.J., Xiao, L.B., Chen, X.Y., Ye, M., Liu, J.K. and Cao, Y. (2011). Grifolin, a potent antitumour natural product upregulates death-associated protein kinase 1 DAPK1 via p53 in nasopharyngeal carcinoma cells. Eur. J. Canc., 47(2): 316–325.

Ma, Z., Wang, J., Zhang, L., Zhang, Y. and Ding, K. (2010). Evaluation of water soluble β-D-glucan from *Auricularia auricular-judae* as potential anti-tumor agent. Carbohyd. Polym., 80(3): 977–983.

Mansour, A., Daba, A., Baddour, N., El-Saadani, M. and Aleem, E. (2012). Schizophyllan inhibits the development of mammary and hepatic carcinomas induced by 7, 12 dimethylbenz (α) anthracene and decreases cell proliferation: Comparison with tamoxifen. J. Canc. Res. Clin. Oncol., 138(9): 1579–1596.

Martin, K.R. and Brophy, S.K. (2010). Commonly consumed and specialty dietary mushrooms reduce cellular proliferation in MCF-7 human breast cancer cells. Exp. Biol. Med., 235(11): 1306–1314.

Masuda, Y., Matsumoto, A., Toida, T., Oikawa, T., Ito, K. and Nanba, H. (2009). Characterisation and antitumor effect of a novel polysaccharide from *Grifola frondosa*. J. Agric. Food Chem., 57(21): 10143–10149.

McIntosh, M., Stone, B.A. and Stanisich, V.A. (2005). Curdlan and other bacterial $(1\rightarrow 3)$-β-D-glucans. Appl. Microbiol. Biotechnol., 68(2): 163–173.

Meng, X., Liang, H. and Luo, L. (2016). Antitumor polysaccharides from mushrooms: Review on the structural characteristics, antitumor mechanisms and immunomodulating activities. Carbohyd. Res., 424: 30–41.

Mimura, H., Ohno, N., Suzuki, I. and Yadomae, T. (1985). Purification, antitumor activity, and structural characterisation of ß-1,3-glucan from *Peziza vesiculosa*. Chem. Pharmaceut. Bull., 33: 5096–5099.

Miyazaki, T. and Nishijima, M. (1981). Studies on fungal polysaccharides. XXVII. Structural examination of a water-soluble, antitumor polysaccharide of *Ganoderma lucidum*. Chem. Pharmaceut. Bull., 29: 3611–3625.

Mizuno, T. (1992). Antitumor active polysaccharides isolated from the fruiting body of *Hericium erinaceum*, and edible, and medicinal mushroom called Yamabushitake or Houtou. Biosci. Biotechnol. Biochem., 56: 349–357.

Mizuno, T., Wasa, T., Ito, H., Suzuki, C. and Ukai, N. (1992). Antitumor-active polysaccharides isolated from the fruiting body of *Hericium erinaceum*, an edible and medicinal mushroom called Yamabushitake or Houtou. Biosci. Biotechnol. Biochem., 56(2): 347–348.

Mizuno, T. (1997). Anti tumor mushrooms *Ganoderma lucidum, Grifora frondosa, Lentinus edodes*, and *Agaricus blazei*. Gendaishorin, Tokyo, 188–193.

Mizuno, T. (1998). Bioactive substances in Yamabushitake, the *Hericium erinaceum* fungus, and its medicinal utilisation. Food Food Ingr. Jap. J., 167: 69–81.

Mizuno, T. (1999). The extraction and development of antitumor-active polysaccharides from medicinal mushrooms in Japan. Int. J. Med. Mush., 1(1): 9–29.

Mizutani, Y. and Yoshida, O. (1991). Activation by the protein-bound polysaccharide PSK (Krestin) of cytotoxic lymphocytes that act on fresh autologous tumor cells and T24 human urinary bladder transitional carcinoma cell line in patients with urinary bladder cancer. J. Urol., 145(5): 1082–1087.

Moradali, M.F., Mostafavi, H., Ghods, S. and Hedjaroude, G.A. (2007). Immunomodulating and anticancer agents in the realm of macromycetes fungi (macrofungi). Int. Immunopharmacol., 7(6): 701–724.

Münzberg, J., Rau, U. and Wagner, F. (1995). Investigations on the regioselective hydrolysis of a branched β-1, 3-glucan. Carbohyd. Polym., 27(4): 271–276.

Nakashima, S., Umeda, Y. and Kanada, T. (1979). Effects of polysaccharides from *Ganoderma applanatum* on immune responses. I. Enhancing effect on the induction of delayed hypersensitivity in mice. Microbiol. Immunol., 23: 501e513.

Nio, Y., Shiraishi, T., Tsubono, M., Morimoto, H., Tseng, C.C., Imai, S. and Tobe, T. (1991). *In vitro* immunomodulating effect of protein-bound polysaccharide, PSK on peripheral blood, regional nodes, and spleen lymphocytes in patients with gastric cancer. Canc. Immunol. Immunother., 32(6): 335–341.

Palleschi, A., Bocchinfuso, G., Coviello, T. and Alhaique, F. (2005). Molecular dynamics investigations of the polysaccharide scleroglucan: First study on the triple helix structure. Carbohyd. Res., 340: 2154–2162.

Park, S.E., Yoo, H.S., Jin, C.Y., Hong, S.H., Lee, Y.W., Kim, B.W., Lee, S.H., Kim, W.J., Cho, C.K. and Choi, Y.H. (2009). Induction of apoptosis and inhibition of telomerase activity in human lung carcinoma cells by the water extract of *Cordyceps militaris*. Food Chem. Toxicol., 47(7): 1667–1675.

Paterson, R.R.M. (2006). *Ganoderma*—A therapeutic fungal biofactory. Phytochem., 67(18): 1985–2001.

Paterson, R.R.M. and Lima, N. (2014). Biomedical effects of mushrooms with emphasis on pure compounds. Biomed. J., 37(6): 357–368.

Piontek, G.E., Taniguchi, K., Ljunggren, H.G., Grönberg, A., Kiessling, R., Klein, G. and Kärre, K. (1985). YAC-1 MHC class I variants reveal an association between decreased NK sensitivity and increased H-2 expression after interferon treatment or *in vivo* passage. J. Immunol., 135(6): 4281–4288.

Rathore, H., Prasad, S. and Sharma, S. (2017). Mushroom nutraceuticals for improved nutrition and better human health: A review. Phar. Nutr., 5: 35–46.

Ren, L., Perera, C. and Hemar, Y. (2012). Antitumor activity of mushroom polysaccharides: A review. Food Func., 3(11): 1118–1130.

Ross, G.D., Větvička, V., Yan, J., Xia, Y. and Větvičková, J. (1999). Therapeutic intervention with complement and β-glucan in cancer. Immunopharmacol., 42(1): 61–74.

Ruiz-Herrera, J. (1991). Fungal Cell Wall: Structure, Synthesis and Assembly. CRC Press.

Saito, K., Nishijima, M., Ohno, N., Yadomae, T. and Miyazaki, T. (1992). Structure and antitumor activity of less-branched derivatives of analkali-soluble glucan isolated from *Ompharlia lapdescence.* Chem. Pharmaceut. Bull., 40: 261–263.

Sakagami, Y., Mizoguchi, Y., Shin, T., Seki, S., Kobayashi, K., Morisawa, S. and Yamamoto, S. (1988). Effects of an anti-tumor polysaccharide, schizophyllan, on interferon-γ and interleukin 2 production by peripheral blood mononuclear cells. Biochem. Biophys. Res. Comm., 155(2): 650–655.

Sasaki, T. and Takasuka, N. (1976). Further study of the structure of lentinan, an anti-tumor polysaccharide from *Lentinus edodes.* Carbohyd. Res., 47: 99–104.

Schepetkin, I.A. and Quinn, M.T. (2006). Botanical polysaccharides: Macrophage immunomodulation and therapeutic potential. Int. Immunopharmacol., 6(3): 317–333.

Seviour, R.J., Stasinopoulos, S.J., Auer, D.P.F. and Gibbs, P.A. (1992). Production of pullulan and other exopolysaccharides by filamentous fungi. Crit. Rev. Biotechnol., 12(3): 279–298.

Shaked, Y. (2019). The pro-tumorigenic host response to cancer therapies. Nat. Rev. Canc., 19: 667–685.

Sherr, C.J. (1996). Cancer cell cycles. Science, 274(5293): 1672–1677.

Stanley, G., Harvey, K., Slivova, V., Jiang, J. and Sliva, D. (2005). *Ganoderma lucidum* suppresses angiogenesis through the inhibition of secretion of VEGF and TGF-β1 from prostate cancer cells. Biochem. Biophys. Res. Comm., 330(1): 46–52.

Sugiyama, Y. (2016). Polysaccharides. pp. 37–50. *In*: Yamaguchi, Y. (ed.). Immunotherapy of Cancer. Springer, Berlin.

Suzuki, M., Arika, T., Amemiya, K. and Fujiwara, M. (1982). Cooperative role of T lymphocytes and macrophages in anti-tumor activity of mice pretreated with schizophyllan (SPG). Jap. J. Exp. Med., 52(2): 59–65.

Suzuki, M., Iwashiro, M., Takatsuki, F., Kuribayashi, K. and Hamuro, J. (1994). Reconstitution of anti-tumor effects of lentinan in nude mice: Roles of delayed-type hypersensitivity reaction triggered by CD4-positive T cell clone in the infiltration of effector cells into tumor. Canc. Sci., 85(4): 409–417.

Tada, R., Harada, T., Nagi-Miura, N., Adachi, Y., Nakajima, M., Yadomae, T. and Ohno, N. (2007). NMR characterisation of the structure of a β-(1→ 3)-D-glucan isolate from cultured fruit bodies of *Sparassis crispa.* Carbohyd. Res., 342(17): 2611–2618.

Taylor, W.R. and Stark, G.R. (2001). Regulation of the G2/M transition by p53. Oncogene, 20: 1803–1815.

Trinchieri, G. (2003). Interleukin-12 and the regulation of innate resistance and adaptive immunity. Nat. Rev. Immunol., 3(2): 133–146.

Trovato, A., Pennisi, M., Crupi, R., Di Paola, R. and Alario, A. (2017). Neuroinflammation and mitochondrial dysfunction in the pathogenesis of Alzheimer's disease: Modulation by *Coriolus versicolor* (Yun-Zhi) nutritional mushroom. J. Neurol. Neuromed., 2: 19–28.

Tsuchiya, Y., Igarashi, M., Inoue, M. and Kumagai, K. (1989). Cytokine-related immunomodulating activities of an anti-tumor glucan, sizofiran (SPG). J. Pharm. Bio-Dynam., 12(10): 616–625.

Tzianabos, A.O. (2000). Polysaccharide immunomodulators as therapeutic agents: Structural aspects and biologic function. Clin. Microbiol. Rev., 13(4): 523–533.

Ukai, S., Morisaki, S., Goto, M., Kiho, T., Hara, C. and Hirose, K. (1982). Polysaccharides in fungi. VII. Acidic heteroglycans from the fruiting bodies of *Auricularia auricula-judae.* Chem. Pharmarceut. Bull., 30: 635–649.

Ukai, S., Kiho, T., Hara, C., Morita, M., Goto, A. and Imaizumi, N. (1983). Polysaccharides in fungi: XIII. Antitumor activity of various polysaccharides isolated from *Dictyophora indusiata, Ganoderma japonicum, Cordyceps cicadae, Auricularia auricula-judae* and *Auricularia* sp. Chem. Pharmarceut. Bull., 31: 741–749.

Vetvicka, V. and Vetvickova, J. (2012). Combination of glucan, resveratrol and vitamin C demonstrates strong anti-tumour potential. Anticanc. Res., 32: 81–87.

Wang, H., Ooi, V., Ng, T., Chiu, K. and Chang, S. (1996). Hypotensive and vasorelaxing activities of a lectin from the edible mushroom *Tricholoma mongolicum.* Pharmacol. Toxicol., 79: 318–323.

Wang, J., Hu, S., Liang, Z. and Yeh, C. (2005). Optimisation for the production of water-soluble polysaccharide from *Pleurotus citrinopileatus* in submerged culture and its antitumor effect. Appl. Microbiol. Biotechnol., 67: 759–766.

Wang, J.T., Zhang, D.W. and Shen, J.Y. (2001). Modern thermodynamics in CVD of hard materials. Int. J. Refract. Met. Hard Mat., 19(4): 461–466.

Wang, Q., Wang, F., Xu, Z. and Ding, Z. (2017). Bioactive mushroom polysaccharides: a review on monosaccharide composition, biosynthesis and regulation. Molecules, 22(6): 955.10.3390/molecules22060955.

Wang, S.Y., Hsu, M.L, Hsu, H.C., Tzeng, C.H., Lee, S.S., Shiao, M.S. and Ho, C.K. (1997). The anti-tumor effect of *Ganoderma lucidum* is mediated by cytokines released from activated macrophages and T lymphocytes. Int. J. Canc., 70(6): 699–705.

Wang, S.Y., Hsu, M.L., Hsu, H.C., Lee, S.S., Shiao, M.S. and Ho, C.K. (1997). The anti-tumor effect of *Ganoderma lucidum* is mediated by cytokines released from activated macrophages and T lymphocytes. Int. J. Canc., 70(6): 699–705.

Wasser, S.P. and Weis, A.L. (1997). Medicinal mushrooms. p. 39. *In*: Nevo, E. (ed.). *Ganoderma lucidum* (Curtis: Fr.), P. Karst. Peledfus Publishing House, Haifa, Israel.

Wasser, S.P. and Weis, A.L. (1999). Medicinal properties of substances occurring in higher basidiomycetes mushrooms: current perspectives. Int. J. Med. Mush., 1(1): 47–50.

Wasser, S.P. (2002). Medicinal mushrooms as a source of antitumor and immunomodulating polysaccharides. Appl. Microbiol. Biotechnol., 60(3): 258–274.

Willment, J.A., Marshall, A.S., Reid, D.M., Williams, D.L., Wong, S.Y., Gordon, S. and Brown, G.D. (2005). The human β-glucan receptor is widely expressed and functionally equivalent to murine Dectin-1 on primary cells. Eur. J. Immunol., 35(5): 1539–1547.

Yadomae, T. (2000). Structure and biological activities of fungal beta-1, 3-glucans. J. Pharmaceut. Soc. Jap., 120(5): 413–431.

Yamamoto, T. (1981). Inhibition of pulmonary metastasis of lewis lung carcinoma by a glucan, schizophyllan. Invas. Metast., 32: 71.

Yang, L., Wang, R., Liu, J., Tong, H., Deng, Y. and Li, Q. (2004). The effect of *Polyporus umbellatus* polysaccharide on the immunosuppression property of culture supernatant of S180 cells. Chin. J. Cell. Mol. Immunol., 20: 234–237.

Ye, L., Zheng, X., Zhang, J., Tang, Q., Yang, Y., Wang, X., Li, J., Liu, Y. and Pan, Y. (2011). Biochemical characterisation of a proteoglycan complex from an edible mushroom *Ganoderma lucidum* fruiting bodies and its immunoregulatory activity. Food Res. Int., 44(1): 367–372.

Yim, M.H., Shin, J.W., Son, J.Y., Oh, S.M., Han, S.H., Cho, J.H., Cho, C.K., Yoo, H.S., Lee, Y.W. and Son, C.G. (2007). Soluble components of *Hericium erinaceum* induce NK cell activation via production of interleukin-12 in mice splenocytes. Acta Pharmacol. Sin., 28(6): 901–907.

Youn, M.J., Kim, J.K., Park, S.Y., Kim, Y., Park, C., Kim, E.S., Park, K.I., So, H.S. and Park, R. (2009). Potential anticancer properties of the water extract of *Inonotus obliquus* by induction of apoptosis in melanoma B16-F10 cells. J. Ethnopharmacol., 121(2): 221–228.

Yu, R., Yang, W., Song, L., Yan, C., Zhang, Z. and Zhao, Y. (2007). Structural characterisation and antioxidant activity of a polysaccharide from the fruiting bodies of cultured *Cordyceps militaris*. Carbohydrate Polymers, 70(4): 430–436.

Zeng, Q. (1990). The antitumor activity of *Flammulina velutipes* polysaccharide (FVP), edible fungi of China. Szechuan Inst. Mat. Med., 10: 2–19.

Zhang, H., Nie, S., Guo, Q., Wang, Q., Cui, S.W. and Xie, M. (2018). Conformational properties of a bioactive polysaccharide from *Ganoderma atrum* by light scattering and molecular modelling. Food Hydrocoll., 84: 16–25.

Zhang, L. (1995). The effect of *Trametes robiniophila* Murr. (TRM) substantial composition on immune function of mice. Acta Edulis Fungi, 2: 35–40.

Zhang, L. and Yang, L. (1995). Properties of *Auricularia auricula-judae* β-D-glucan in dilute solution. Biopolymers, 36(6): 695–700.

Zhang, M., Cheung, P.C.K. and Zhang, L. (2001). Evaluation of mushroom dietary fibre (non-starch polysaccharides) from sclerotia of *Pleurotus tuber-regium* (Fries) Singer as a potential antitumor agent. J. Agric. Food Chem., 49: 5059–5062.

Zhang, M., Zhang, L. and Cheung, P.C.K. (2003). Molecular weight and anti-tumor activity of the water-soluble mushroom polysaccharides isolated by hot water and ultrasonic treatment from the sclerotia and mycelia of *Pleurotus tuber-regium*. Carbohyd. Res., 56: 123–128.

Zhang, M., Chiu, L.C.M., Cheung, P.C.K. and Ooi, V.E.C. (2006). Growth-inhibitory effects of a ß-glucan from the mycelium of *Poria cocos* on human breast carcinoma MCF-7 cells: cell cycle arrest and apoptosis induction. Oncol. Rep., 15: 637–643.

Zhang, M., Cui, S.W., Cheung, P.C.K. and Wang, Q. (2007). Antitumor polysaccharides from mushrooms: a review on their isolation process, structural characteristics and antitumor activity. Tr. Food Sci. Technol., 18(1): 4–19.

Zhang, P. and Cheung, P.C. (2002). Evaluation of sulfated *Lentinus edodes* α-(1→ 3)-D-glucan as a potential antitumor agent. Biosci. Biotechnol. Biochem., 66(5): 1052–1056.

Zhang, S., Nie, S., Huang, D., Feng, Y. and Xie, M. (2014). A novel polysaccharide from *Ganoderma atrum* exerts antitumor activity by activating mitochondria-mediated apoptotic pathway and boosting the immune system. J. Agric. Food Chem., 62: 1581–1589.

Zhang, Y., Li, S., Wang, X., Zhang, L. and Cheung, P.C. (2011). Advances in lentinan: Isolation, structure, chain conformation and bioactivities. Food Hydrocoll., 25(2): 196–206.

Zhong, W., Liu, N., Xie, Y., Zhao, Y., Song, X. and Zhong, W. (2013). Antioxidant and anti-aging activities of mycelial polysaccharides from *Lepista sordida*. Int. J. Biol. Macromol., 60(6): 355–359.

Zhuang, C., Mizuno, T., Shimada, A., Ito, H., Suzuki, C. and Mayuzumi, Y. (1993). Antitumor protein-containing polysaccharides from a Chinese mushroom Fengweigu or Houbitake, *Pleurotus sajor-caju* (Fr.) Sing. Biosci. Biotechnol. Biochem., 57: 901–906.

Zhuang, C., Mizuno, T., Ito, H., Shimura, K., Sumiya, T. and Kawade, M. (1994). Antitumor activity and immunological property of polysaccharides from the mycelium of liquid-cultured *Grifola frondosa*. Nippon Shokuhin Kogyo Gakkaishi, 41: 724–735.

Zong, A., Cao, H. and Wang, F. (2012). Anticancer polysaccharides from natural resources: A review of recent research. Carbohyd. Polym., 90: 1395–1410.

Index

Printed and bound by CPI Group (UK) Ltd, Croydon, CR0 4YY

17/10/2024

01775709-0008